Nano and Micro-Scale Energetic Materials

Nano and Micro-Scale Energetic Materials

Propellants and Explosives

Volume 1

Edited by Weiqiang Pang and Luigi T. DeLuca

WILEY VCH

Editors

Prof. Weiqiang Pang
Xi'an Modern Chemistry Research Institute
Science and Technology on Combustion and Explosion Laboratory
No. 168 Zhangba East Road
710065 Xi'an
China

Prof. Luigi T. DeLuca
Politecnico di Milano
Space Propulsion Laboratory (RET)
via G. La Masa 34
21056 Milan
Italy

Cover Image: © djero.adlibeshe yahoo.com/Shutterstock

Library of Congress Card No.: applied for

British Library Cataloguing-in-Publication Data
A catalogue record for this book is available from the British Library.

Bibliographic information published by the Deutsche Nationalbibliothek
The Deutsche Nationalbibliothek lists this publication in the Deutsche Nationalbibliografie; detailed bibliographic data are available on the Internet at <http://dnb.d-nb.de>.

Print ISBN: 978-3-527-35206-7
ePDF ISBN: 978-3-527-83532-4
ePub ISBN: 978-3-527-83533-1
oBook ISBN: 978-3-527-83534-8

Typesetting Straive, Chennai, India
Printing and Binding CPI Group (UK) Ltd, Croydon, CR0 4YY

C117195_030723

Contents

Preface

An accidental explosion somewhere in China, during the Chin (221 BCE–207 BCE) or Han dynasty (206 BCE–220 CE), probably in the far 220 BCE, marked the beginning of the solid energetic materials (EMs) development on our planet. Chinese alchemists had discovered black powder (or something very close, according to historians[1]): a mixture of potassium nitrate (KNO_3, also known as saltpeter), sulfur, and charcoal. For centuries, the development of fireworks or firecrackers to scare evil spirits was the fortuitous driving force for pyrotechnics, solid propellants, and solid explosives as well. Primordial solid rocket motors (SRMs) were built, particularly in India in the late 1700s (cast-iron tubes), with gravimetric specific impulse (I_s) reaching at most ≈80 seconds under standard operating conditions.

Only after the discovery of liquid nitroglycerin (NG) by Sobrero (Italy) much later, in 1846, the development of EMs took different and targeted directions, although somewhat intermingled, for pyrotechnics, propellants, and explosives. Progress was promoted in all solid energetic areas. As far as rocket and gun propulsion are concerned, Nobel (Sweden) prepared in 1888 the first double-base smokeless propellant, called ballistite, using a gel of NG and nitrocellulose (NC). Brouns (Russia) developed in 1925 the first widely used smokeless solid propellant for SRMs based on pyroxylin (a highly flammable mixture of NC) and conceived in 1933 an energetically more efficient double-base propellant based on NG (known as N-powder). In 1942, Parson (USA) manufactured the first castable composite propellant based on potassium perchlorate. In 1955, Atlantic Research Corporation (ARC, USA) successfully burned the first metalized propellant using 20 μm particle size aluminum (Al) and ammonium perchlorate (AP) bound by polyvinylchloride (AP/PVC/Al = 59/20/21 in mass%). The last two accomplishments resorted to microscale energetic materials (μEMs) and paved the way for SRM to reach a delivered $I_s = 235$ seconds. In 1959, the implementation by Gen and Miller (Russia) of nano-scale energetic materials (nEMs) at the Institute of Chemical Physics (ICP) was markedly fruitful for nano-thermites and other energetic applications. However, for propulsion nEMs are still limited to laboratory research except for special ingredients such as catalysts and ballistic modifiers. Following these milestones, during last about five decades since 1970 (when the first hydroxyl-terminated

1 PengFei Li, IASPEP Conference, Qingdao, China, 17 Sep. 2015.

polybutadiene [HTPB] test flights were carried out and the first combustion tests of AP/bitumen loaded with 90 nm nAl were reported), the progress of solid EMs for rocket propulsion has been modest at industrial level for both solid motors and hybrid engines. The binder of composite solid propellants is the only ingredient that went through several variants even for large-scale applications: the introduction of energetic plasticizer made it possible for SRMs to go past a delivered I_s = 250 seconds. Latest evolution is the use of ammonium dinitramide (ADN) as replacement for AP to develop the so-called "green propellants"; this phase is still in progress, but rocket systems testing is expected in the near future.

Regarding commercial/industrial explosives, black powder started being replaced by NG-based formulations (dynamites) only after 1870. Later, in the twentieth century, both dynamite and black powder were replaced by explosives based on mixes of high explosives (for example, trinitrotoluene [TNT]) or combustible powders (for example, Al) with ammonium nitrate (AN). After several adjustments, in 1950 manufacturers started to develop safer water-filled AN explosive suspensions that were waterproof (slurry) and cheap, as well as safe, and non-water-resistant mixtures of AN with liquid hydrocarbons, such as the common ammonium nitrate fuel oil (ANFO). At the end of the twentieth century effective, safe, environment-friendly aqueous solutions of AN emulsified with a fuel - industrial explosives – were established in many countries.

Regarding military explosives, trinitrotoluene [(TNT) ($C_7H_5N_3O_6$)], prepared in 1863, was the standard explosive during World War I. Afterward, new explosives were developed, such as pentaerythritol tetranitrate [(PETN) ($C_5H_8N_4O_{12}$)] prepared in 1894, hexogen [(RDX) ($C_3H_6N_6O_6$)] recognized as an explosive only in 1920, and octogen [(HMX) ($C_4H_8N_8O_8$)] prepared in 1943. A recent development is the production of reduced sensitivity RDX achieved by careful recrystallization of the material to get nearly defect-free crystals. In this way, the RDX sensitivity could be significantly reduced and may even be better than that of standard HMX (for example, see the very insensitive VI-RDX by Eurenco).

Polymer-bonded explosives (PBXs) were developed to reduce the sensitivity of the explosive crystals by embedding them in a rubber-like polymeric matrix. In PBX formulation, particles of an explosive powder are bound together in a matrix using small quantities (5–14% mass%) of a synthetic polymer. After an initial attempt in 1947, the first PBX composition was fabricated at the Los Alamos Scientific Laboratories in 1952 by embedding RDX crystals in polystyrene plasticized with dioctyl phthalate. Afterward, a range of binder systems was developed, but polyurethanes remain the most used. HTPB cured with polyisocyanates made available polyurethane rubber with high strain capability and low glass-to-rubber transition (GRT) temperatures. The HTPB binder is able to bear up to 90 mass% of explosive particle filling, another advantage for high-performance warheads. Disadvantages are the very negative oxygen balance and the fact that HTPB acts as an inert binder. Binders were also tested with energetic groups attached along the polymer chain, mainly azide groups as with glycidyl azide polymer (GAP). However, no product was able to replace HTPB, because of limited strain capability and too high GRT temperatures. All curable binder-based PBXs are of the "cast-cure" type, meaning that the liquid slurry is cast in the molds or warhead forms and cured there. Another type of PBX, cheaper and suitable for mass production, is the pressed one, whereby a thermoplastic binder bears the high explosive particles; the slurry is pressed into the molds or warhead forms. PBX techniques are normally used for explosive materials that are not easily melted for casting. Young's modulus of the explosive crystal is typically one order of magnitude higher than that of the binder. Thus, the deformation characteristic of the composite is primarily influenced by the binder. Heating from the plastic deformation of the binder generates local hot spots and serves as an ignition mechanism.

Novel μEMs containing special functional groups and nEMs today appear suitable for application in a wide range of energetic systems. They have the potential to radically affect the characteristics of EM formulations currently in use. Specifically, they promise much higher energy densities and explosive yields in comparison to conventional materials without necessarily degrading sensitivity, stability, energy release, and mechanical properties. A new application is the manufacture of μEM inside binders. The novel concepts just mentioned are quickly described in the book outline that follows. In general, μEMs and nEMs exhibit excellent properties and good application results due to their small size effect, crystal quality, high surface energy, and high surface activity. Adding μEMs and nEMs to composite explosives or solid propellants can significantly improve their performance and tailoring of the properties for specific applications. However, enough results to precisely define limitations and advantages are presently not available. Most advances in nanotechnology are still in an exploratory phase. There is a huge difference between lab-scale demonstration and large-scale application, in terms of cost and viability. And it is not yet clear how much of the nano-sized materials at the large scale will ever be practical in view of such factors as cost, long-term stability, homogeneity, the need for a strong, sustained industrial base, and handling/safety issues.

This edited book *Nano and Micro-scale Energetic Materials: Propellants and Explosives* explores innovative nEMs and μEMs ingredients as well as formulations. It provides a comprehensive survey of advanced technology on preparation, characterization, combustion, and performance in potential applications for propellants and explosives. The 22 chapters of this volume summarize the most recent achievements of leading research groups active in twelve countries (Algeria, Canada, China, Czechia, Germany, India, Israel, Italy, Japan, Russia, South Africa, and USA). The book can be divided into seven parts, somewhat arbitrarily due to the multiple and overlapping technical aspects covered in many of the contributions.

Part 1 includes two chapters (1 and 2) intended to provide readers with fundamentals propaedeutic to the volume. An inspiring discussion, stressing similarities and differences between composite propellants and explosives, is presented in Chapter 1. A stimulating review of high-pressure burning rate experiments, as predictive capabilities from theory and modeling are still not sufficient, is conducted in Chapter 2.

Part 2 includes four chapters (3–6) focused on new high-energy materials. Current perspectives of novel organic nEM composites (based on nanosized HMX, RDX, and CL-20) as powerful ingredients of new generation, together with the advantages of nano co-crystallization in developing innovative explosives and propellants, even in large-scale industrial production, are discussed in Chapter 3. In Chapter 4, new CL-20 clathrates are examined (in terms of synthesis, thermal stability, and combustion behavior) to improve their oxygen balance; the resulting compounds, being energetically superior to ε-CL-20, are of interest as powerful explosives and energetic fillers for rocket propellants. New formulations for rocket propulsion are discussed in Chapter 5. A comprehensive presentation of TKX-50, pure or in many variants/dual versions, as an ingredient suitable for both propellants and explosives is reported in Chapter 6.

Part 3 includes three chapters (7–9) focused on high-energy metal-based pyrotechnic nanocomposites. Metastable intermixed composites (MICs) are treated in Chapter 7, focusing on fabrication into three typical structures: the randomly distributed structure, the multilayered structure, and the core–shell structure. Nanothermites in general are examined in Chapter 8, presenting a summary of recent findings for newly introduced nanothermite mixtures and the challenges that persist. Al/CuO nanothermites are discussed in Chapter 9, investigating the agglomeration of nanoparticles under combustion with microscopic imaging with a high spatial (1 μm) and temporal (55 μs per frame) resolution and suggesting two approaches to alleviate agglomeration by replacing spherical CuO particles with high length-to-diameter CuO wires and adding 2.5 mass% unreactive carbon fiber wires.

Part 4 includes four chapters (10–13) dedicated to rocket propulsion. A useful survey of GAP features, as a ubiquitous ingredient affirmed for solid propulsion and promising for hybrid propulsion, is offered in Chapter 10. For the current solid propellant formulations meant for space launchers, by far the most important goal is to overcome the limitations of AP due to chlorine (Chapter 11) and of Al due to agglomeration (Chapter 12). For both issues, several options are being considered, but none is today operative at large scale. Chapter 13 presents the results of an experimental investigation on laser ignition of various energetic materials, containing nanopowders of Al/B, Al/Fe, Al/Ti, and Al/Cu, in air at ambient conditions. Changing the composition of the nano-bimetal additives makes it possible to significantly modify the ignition delay of aluminized propellants.

Part 5 includes two chapters (14 and 15) dedicated to solid propellants and solid fuels for in-space propulsion and power. Use of metals for space propulsion and power systems is discussed in Chapter 14, and an introduction to micropropulsion including innovative solid propellants (thermites and primer explosives) for space microthrusters is offered in Chapter 15.

Part 6 includes three chapters (16–18) treating primary and secondary explosives. An extensive review of new high explosives and melt-casts explosives is offered in Chapter 16. Pyrotechnic alternatives to primary explosive-based initiators with particular attention to nanocomposites combining the energy density of the thermites with the reaction rate of the explosive are reviewed in Chapter 17. Laser initiation of light-sensitive primary explosives by means of environment-friendly photosensitive EMs with low initiation thresholds is treated in Chapter 18.

Part 7 includes four chapters (19–22) dealing with sensitivity and mechanical properties of explosives. A detailed review of the initiation micromechanism of mainly organic energetic materials is offered in Chapter 19. Ignition of explosives under low-velocity impact is discussed in Chapter 20, focusing on the relation between the stochastic microcrack characteristics (length and density) and ignition; the ignition probability is calculated based on Monte Carlo method and the simulations match the test results well. Ignition of HMX and RDX single crystals under different impact conditions is treated in Chapter 21, showing that the RDX molten phase plays an important role in the viscous hotspots formation; modeling the coupled mechanical-thermal-chemical responses is proposed for both impacted HMX and RDX crystals. In Chapter 22, four high explosive formulations (HMX-based PBXs with energetically augmented HTPB binder) containing higher amounts of bimodal HMX filler have been investigated. Dynamic mechanical analysis in torsion mode was carried out at several deformation frequencies, since the particle-enriched binder causes significant differences in loss factor shape. Special characterization techniques were applied for loss factor (exponentially modified Gauss function) and glass–rubber temperature shift (modified Arrhenius equation). The purpose of the particle-enriched HTPB is to get a type of energetic binder and to increase its density; this is important for creating a more uniform shock front in the detonation.

It turns out that most propellants and explosives today in use employ a very small number of ingredients: RDX, HMX, AP, Al, NC, and a handful of other ingredients used in combination with them. The prominent importance of metals, either micron-sized or nano-sized, is a common trait to all areas: nanoscale energetic composites (Chapters 7–9), solid propellants (Chapters 12–15), and explosives (Chapters 20–21). Propellants benefitted from the use as co-oxidizers of RDX and HMX, initially developed as explosives, while explosives benefitted from the use of HTPB and cast-cure technology for PBXs compositions to improve the safety properties of melt-cast explosives (chapter 1). For both propellants and explosives production, relevant performance parameters, cost, safety, health, environment, and aging overall need to be mastered and harmonically optimized in just a single product. However, these multiple constraints are inherently in conflict with each other. For example, often - but not always – higher performance implies higher sensitivity while green formulations imply less performance.

While the search for more energetic molecules is restricted by the basic chemical constraints of the CHON systems, new and additional requests emerged. High-energy density and green features before and after burning (i.e. nonpolluting + nontoxic + no submicrometric particulate emissions) are important requirements. Multilevel optimization requires painstaking attention and should also be adapted to the specific mission objectives. For example, space pollution should be accounted for space access launchers (Chapter 11); micropropulsion peculiarities should be considered for secondary in-space propulsion (Chapter 15); and optical propellant properties should be examined for laser-assisted or driven combustion (Chapter 15) and explosives initiation (Chapter 18). Sometimes the complexity of the matter points out unusual solutions: for example, high ingredient density implies less performance in terms of I_s, but the Tsiolkovsky equation may equally be satisfied in terms of the vehicle velocity increment (deltaV) needed for a given propulsive mission.

To improve the quality of this book, many experts were involved as external reviewers of the chapters in addition to the editors. It is our great pleasure to thank the following international reviewers for their substantial help in raising the quality of the book. Without their efforts, the publication of this volume would have been impossible. The list of experts includes Jai P. Agrawal, Alice Atwood, Valery A. Babuk, Sara Cerri, Marc Comet, Stanislaw Cudzilo, Mauricio Ferrapontoff Lemos, Edward L. Dreizin, Walter W. Focke, Alon Gany, Nick G. Glumac, Keiichi Hori, Mikhail A. Ilyushin, Stefan Kelzenberg, Thomas M. Klapötke, Sebastian Knapp, David Lempert, Nikita V. Muravyev, Carole Rossi, Valery P. Sinditskii, Denis Spitzer, Niklas Wingborg, Richard Yetter, Gregory Young, Vladimir E. Zarko, Svatopluk Zeman; ChongWei An, PengWan Chen, SongQi Hu, ChengDong Kong, HongQi Nie, Hui Ren, Jie Sun, Ning Wang, Wen Ao, YuShi Wen, JinSheng Xu, KangZen Xu, QiLong Yan, RongJie Yang, YanJing Yang, KaiLi Zhang, Long Zhang, JianGuo Zhang, WenChao Zhang, XianXu Zheng.

Finally, we especially thank Dr. ShaoYu Qian, Program Manager, and Ms. Katherine Wong, Senior Managing Editor, from John Wiley & Sons for their patience and efficient support. Their assistance made the publication of this book much easier than expected. The editors sincerely hope that the research efforts described in this book will continue to exhibit fast growth and lead to maturation of advanced technologies conducive of higher performance of propellants and explosives for better space exploration and defense. We also sincerely wish that this joint international effort will help all readers to gain a better understanding of the puzzling intricacies and appealing secrets of nanoscale and microscale EMs as well as of the perplexing difficulties but also fascinating horizons of space propulsion.

WeiQiang Pang, Xi'an, China
Luigi T. DeLuca, Milan, Italy
04 May 2022

Le cose sono unite da legami invisibili. Non puoi cogliere un fiore senza turbare una stella. **Things are united by invisible bonds. You can't pick a flower without upsetting a star.**

(Galileo Galilei, 1564–1642)

About the Editors

Prof. Dr. Weiqiang Pang is the group leader of design, simulation, and application of solid propellants at Xi'an Modern Chemistry Research Institute in Xi'an, China. He is a visiting Scholar of Politecnico di Milano and spent most of his career working on innovative energetic materials exploration, and their combustion and application in solid propellants. Professor Weiqiang Pang is an editorial board member of *Chinese Journal of Explosives & Propellant*, *Journal of Solid Rocket Technology*, and *Equipment Environmental Engineering*. He has authored 11 books and over 150 scientific publications, and has 60 patents.

Prof. Luigi T. DeLuca received his Ph.D. degree from Princeton University, USA, in 1976. He founded the Space Propulsion Laboratory (SPLab) and worked until his retirement as a Professor of Aerospace Propulsion at Politecnico di Milano, Milan, Italy. Prof. Luigi T. DeLuca has authored over 15 books and 300 scientific publications. He is Honorary Professor at Omsk State Technical University, Omsk, Russia; Honorary Fellow of the High Energy Materials Society of India (HEMSI), India; a Fellow of the Korean Government Brain Pool Program and Visiting Professor at Konkuk University, Seoul, Korea; and currently a Visiting Professor at Nanjing University of Science & Technology, China.

Part I

Fundamentals

1

Composite Heterogeneous Energetic Materials: Propellants and Explosives, Similar but Different?

Levi Gottlieb

RAFAEL, Dep. M4 POB 2250, Haifa 1021, Israel

1.1 Introduction

The discipline of energetic materials (EMs), is truly an interdisciplinary subject, spanning from basic sciences such as thermodynamics, combustion and detonation physics, organic, physical, and polymer chemistry to more applied disciplines such as material, chemical, mechanical, and aeronautical engineering. In this review, composite propellants and explosives will be compared with an emphasis on similarities and differences in material aspects and characteristics of operation. This chapter will be concerned with "composite" system where an energetic material is embedded in a polymeric matrix. In a strict sense melt-cast explosive and heterogenous gelled propellants (composite modified double base [CMDB] propellants [1]) may also classified as composite energetics but will not be discussed (Figure 1.1).

A schematic representation of the EM's mode of reaction [2], which shows the main difference between explosives and propellants, was formulated by the late Dr. Manfred Held [3, p. 379]. The scheme (Figure 1.2) shows the ultimate purpose of explosives as blast, shattering, and accelerating fragmented slivers (brisance and heave) and propellants as the means of accelerating objects as shown in Figure 1.2.

EM's are metastable compounds or mixture of substances that contain both fuel and oxidizer and react readily in a redox reaction with release of energy and hot gases as long as the rate of heat generation is higher than the heat dissipation to the surrounding. EMs compositions are designed to provide energy on demand, by converting stored chemical energy to mechanical work. The reactions of EM's are classified according to the rate of advancement of the process in the material as combustion, deflagration,[1] and detonation.[2] A combustion of a liquid hydrocarbon fuel releases ~6 times more energy than an explosive but the energy release rate of an explosives is $\sim 10^6$ faster. Therefore, the main differences between combustion, deflagration, and detonation are the reaction timescales. Although the total energy available for any configuration is fixed by the type and mass of energetics, the

1 Latin: de + flagrare = from + to burn away.
2 Latin: de + tonare = from + to thunder.

Nano and Micro-Scale Energetic Materials: Propellants and Explosives, First Edition.
Edited by Weiqiang Pang and Luigi T. DeLuca.

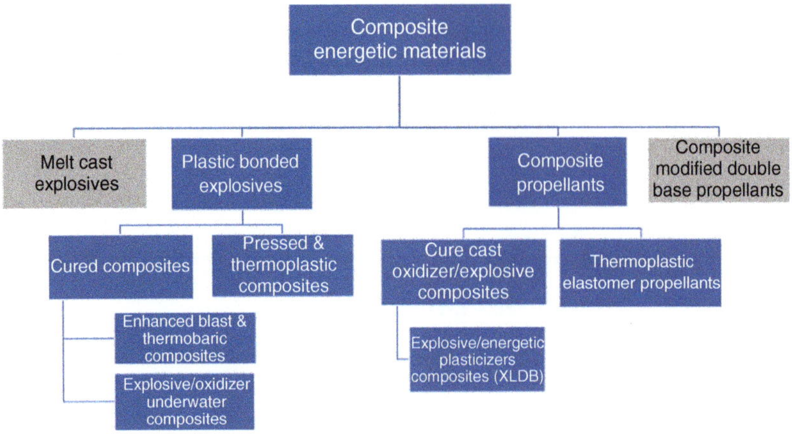

Figure 1.1 **Classification of composite energetics systems** (Gray shaded composites are not included in this review).

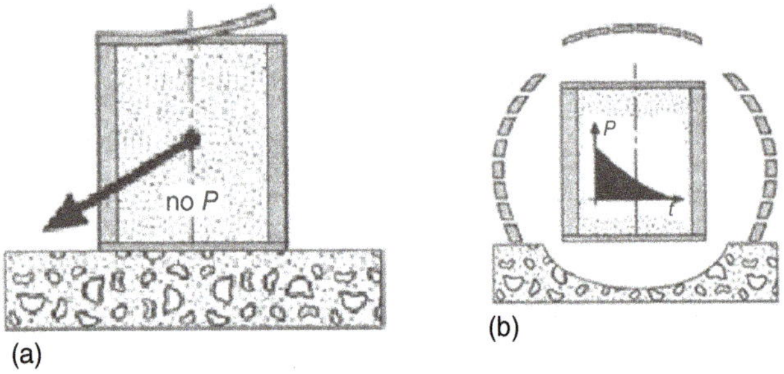

Figure 1.2 Response of a propellant (a) and an explosive (b). Source: Lee [2]. Reproduced with permission of Springer Nature.

difference in power output between burning and detonation is so large as to completely change the nature of the event [4]. A schematic representation of the differences in pressure history and mechanical effect of these three modes of reaction on the surroundings is shown in Figure 1.3 [2].

Propellants react by deflagration involving a complex ensemble of processes involving solid and gas-phase reactions. From solid phase melt, thermal decomposition of the binder and oxidizer, to premixed and diffusion flames in the gas phase as described by the Beckstead-Derr-Price (BDP)[3] combustion model shown in Figure 1.4 [5, 6].

Propellant deflagration is essentially a diffusion wave with a velocity proportional to diffusivity and reaction rate. Explosives are described to react by detonation, which is a leading shock front followed by a chemical reaction zone described by the Zel'dovich-von Neumann Doring (ZND) detonation[4] model [7]. They will initiate directly under the influence of a shockwave where the material is compressed and heated above the decomposition temperature through adiabatic compression. The shock is sustained by the thrust generated by the expansion of the gaseous

3 Beckstead–Derr–Price Model.
4 Zel'dovich–von Neumann DÖring Model.

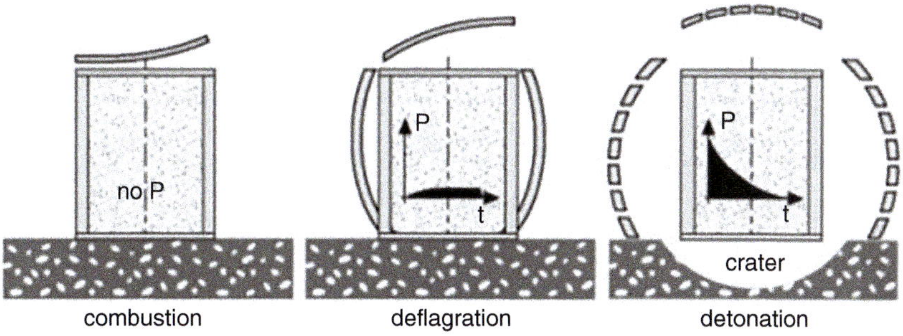

Figure 1.3 Pressure history and mechanical effect of the basic modes of reaction of EM according to rate; combustion, deflagration, and detonation. Source: Lee [2]. Reproduced with permission of Springer Nature.

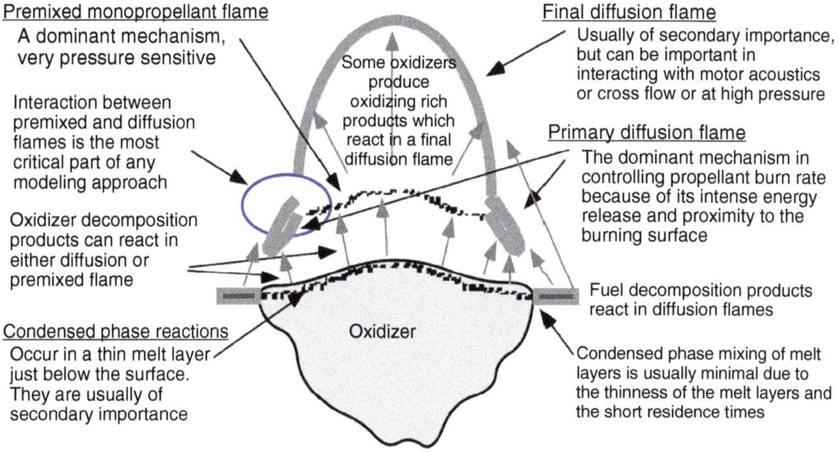

Figure 1.4 Multiple flame BDP propellant combustion model. Source: Beckstead et al. [5]. Reproduced with permission of Elsevier Technology Journals.

products in the reaction zone. This "semi-closed-loop" process is accelerating the compression wave to faster than the speed of sound in the material. Eventually, these complex chains of events will reach the point where released energy from the decomposition reaction equals the dissipated energy to the surroundings plus the energy needed to compress the material and keep the shockwave moving. A steady detonation wave is obtained in this stationary state.

1.2 Structure and Composition

Composite energetics are essentially heterogeneous materials with a solid oxidizer or a solid explosive incorporated in a polymer matrix. The polymer matrix provides the fuel component in the case of propellants and the necessary form and rigidity to the grain in both. Micrographs showing the heterogenous structure of a composite propellant [8] and a plastic bonded explosive (PBX) [9] (Figure 1.5) show the similarity between them.

Typical examples of composite energetics appear in Table 1.1.

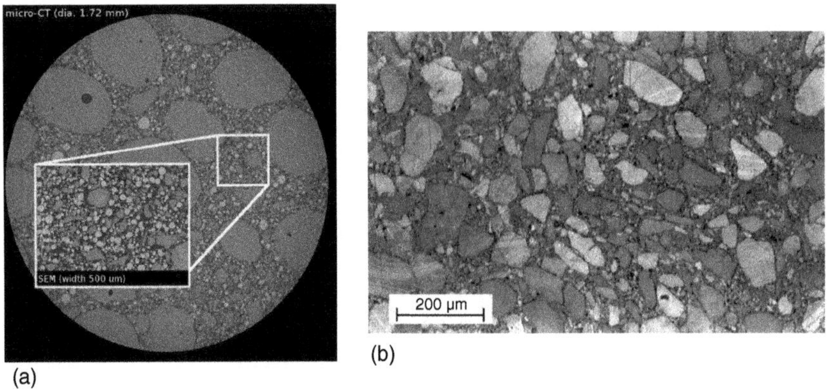

Figure 1.5 Micrographs of a composite energetics. (a) – propellant. Source: Ksoiba et al. [8]. Reproduced with permission of John Wiley and Sons. (b) – PBX. Source: Rae et al. [9]. Reproduced with permission of the Royal Society of London.

Table 1.1 Typical compositions of composite propellants and explosives.

Composite	Application	Oxidizer/ explosive	Energetic additive	Metallic fuel	Binder	Method of manufacturing
Propellants	Solid rocket motor (SRM)	AP, AN, KP, ADN	RDX, HMX, NQ	Al	HTPB, CTPB, PBAN, GAP	Blending, slurry casting and cure
	Gun propellant	RDX, NC	TAGzT, GUDN		Cellulose Acetate, NC, BDNPF/A	Solvent/ plasticizer assisted extrusion/ kneading
	Gas generators	KN, AN, KP, AP	5-AT, TAGN		HTPB, DB	Blending, slurry casting, and cure
	Energetic thermoplastic elastomers	AP, AN	RDX, NQ, CL-20		Poly-BAMO/ AMMO PGN, GAP	Cast-cure/ extrusion mixing
Explosives	Cast Cure PBX	RDX, HMX, NTO, CL-20	AP	Al	HTPB, GAP, PolyNiMMO	Blending and slurry casting and cure
	Pressed PBX	RDX, HMX, HNS, TATB	—	—	Viton, Teflon Estane, PPG	Solvent-assisted elastomer coating and Pressing
	Thermoplastic Elastomers Based	RDX, HMX			Polyisobutylene PDMS	Extrusion mixing

1.2.1 Energetic Fillers

The area of energetic materials has gone an "explosive" expansion becoming a legitimate chemical pursuit. Some of the major areas of interest in looking for new materials are (i) Nitrogen-based heterocyclic neutrals and ionic or ionic liquids; (ii) Green, chlorine, and heavy metal-free materials; (iii) Energetic co-crystals; (iv) High-temperature materials; (v) All nitrogen materials. It is therefore not surprising that the search for materials with better performance, lower sensitivity, no toxicity, and minimal effect on the environment is an ongoing effort. Two relevant directions can be identified, nitrogen-heterocyclic-based explosive substances replacing the sensitive conventional nitramines with performance at least as good as RDX and oxidizer free of ammonium perchlorate (AP) burdens. Motivations for AP replacement include tactical shortcomings (HCl condensation trail in humid climate), serious ecological influences (HCl contribution to acid rain and ozone depletion), and toxicological effects (perchlorate accumulation in the thyroid gland) inherent with perchlorates and their combustion products [10]. AP oxidizer replacement by CHNO oxidizers is not new in propellant materials development (i.e. cross linked double base [XLDB]-nitramines-propellants). However, high density, high heat of formation, high oxygen balance, and good thermal stability AP replacements is still a substantial challenge. Another obstacle that hampered the introduction of inorganic AP replacements as hydrazinium nitroformate (HNF) and ammonium dinitramide (ADN) is the incompatibility with conventional isocyanates curing agents [10]. It seems that the search for new and improved fillers for both propellants and explosives may converge to heterocyclic-based EMs (Figure 1.6). A semi-chronological chart exemplifying trend in oxidizer and explosive substances development toward a possible unification of fillers is shown in Figure 1.6. The sought new heterocyclic materials [11–14], some with record high number of nitro

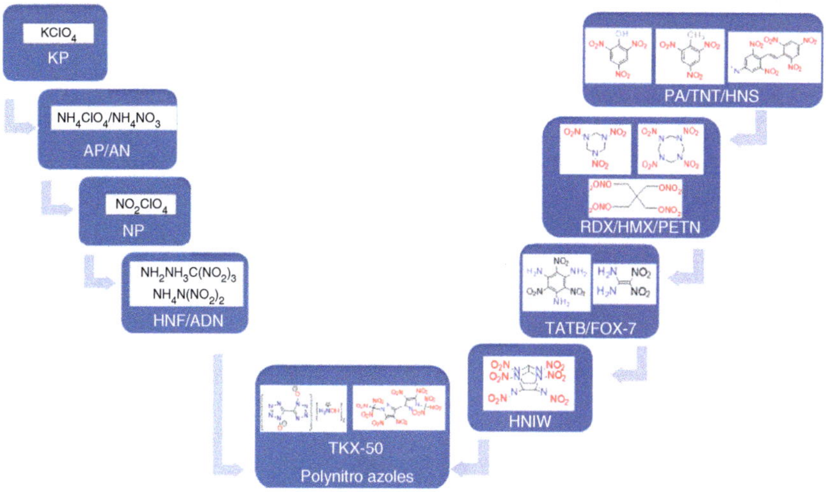

Figure 1.6 Semi-chronological development of oxidizers and explosive fillers.

groups although have some promise, but, as can be judged from the synthetic difficulty, and the relatively low melting points, are not going to be readily introduced as commodities.

1.2.2 Binder Systems

Binders have a crucial role in propellants and PBXs. In both systems, they are used to convey desirable mechanical properties to the composition and have a major influence on performance as energetic material and a critical impact on sensitivity. The constant strive for better mechanical safety and performance has generated a rather large body of work aimed at improving these traits. The three main components in binder systems (polymer, surface-active materials, and plasticizer) will be discussed and compared in brief in this section.

PBX's sensitivity [15] and detonation characteristics [16] are influenced by the polymeric binder system. In propellants, they are also an indispensable part as the fuel component in the combustion reaction. Solid propellant performance history [17, 18] also shows the tight linkage between binder identify and delivered specific impulse. Figure 1.7 shows some of the polymers developed for propellants and PBXs formulations (c. 1940–2000). The materials are categorized by the implied mode of application. Cured refers to casting and chemical crosslinking, pressed refers to precoating and mold pressing and thermoplastic elastomeric materials usually refers to kneading and extruding. Continuing development was driven for bettering production, mechanical properties performance, aging, and cost.

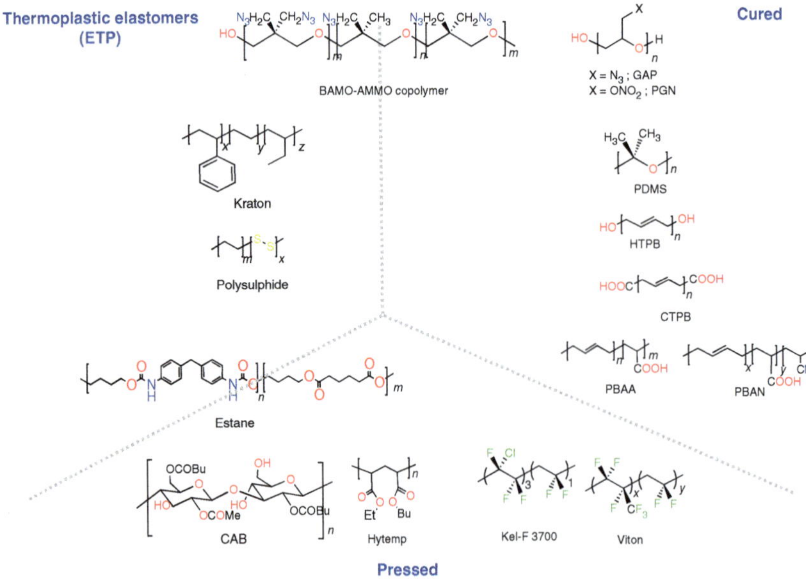

Figure 1.7 Polymeric binders developed for propellants and explosives compositions. Materials on the division lines may be used in more than one manner.

1.2.2.1 Binder Systems for Cast Cure Propellants and Explosives

The first castable composite propellant, developed by J.W. Parsons at the Guggenheim Aeronautical Laboratory at the California Institute of Technology (GALCIT),[5] was based on potassium perchlorate embedded in asphalt pitch [19]. The asphalt was replaced by polysulfides that can be oxidatively cured forming disulfide bonds. The material had good elastomeric properties performing at wider temperature range compared to non-cured asphalt. However, polysulfides are incompatible with aluminum reacting slowly while storage. Thus, improving performance by addition of aluminum fuel necessitated replacing polysulfides with a polymer binder compatible with aluminum. Soon a series of polybutadiene-based curable elastomers (polybutadiene acrylic acid [PBAA], polybutadiene acrylic acid acrylonitrile [PBAN], carboxy terminated polybutadiene [CTPB]) have been developed [20]. These materials suffered from rather poor reproducibility of mechanical properties (PBAA and PBAN),[6] elevated cure temperatures, and use of toxic curing agents (epoxides and aziridines). The search continued until hydroxy terminated polybutadiene (HTPB)-based polyurethanes, the still running state of the art in propellant rocket motors appeared [21, 22]. The use of HTPB and cast-cure technology was also implemented for PBXs in an effort to improve the safety properties of melt-cast explosives in agreement with the insensitive munitions (IM) guidelines. Traditional melt-cast explosives were sought to be replaced by PBXs using the logic and guidelines of composite propellants [23]. In fact, some propellant formulations can be actually considered as PBXs and the intended use is more a matter of initiation, detonator for explosives, and ignitor for propellants. Three basic types of propellants have been considered: (i) HTPB/AP/Al, (ii) HTPB/AP/Al/nitramines, and (iii) polyester/nitrate esters (NE)/nitramines (minimum smoke XLDB). Type (i) compositions are common and easy to formulate up to high-solid loadings. The addition of aluminum contributes to air blast and the materials are not prone to sympathetic detonation. However, the low detonation velocity necessitates an addition of a high explosive to enhance performance. Type (ii) propellants enable quite high-solid loadings but not as high as type (i) and there is a need for specialized bonding agents. These compositions also show high detonation velocity but the susceptibility to sympathetic detonation needed some reformulation using an insensitive high explosive. Type (iii) materials were not considered since they can hold only a limited amount of solid materials compared to the other types and they are also prone to sympathetic detonation although the detonation velocity is high. On the basis of such considerations, PBXs replacements of traditional melt-cast explosive were developed using propellants technology and shown to be an insensitive high-performance material (Table 1.2).

The next stage of binder development is energetic polymers with explosophores on the backbone (Figure 1.8). Substituting inert binders with energetic elastomers enables the same level of performance with less filler. Although the binders are

---

5 GALCIT became later to the Jet Propulsion Laboratory (JPL).

6 Copolymers with acrylic acid produce random distribution of carboxyl units on the backbone. Crosslinking through these randomly distributed moieties limits the control and reproducibility of mechanical properties of the cured matrix.

Table 1.2 PBX's developed as melt-cast explosive replacements[a)] using propellants guidelines.

Sample	Composition (%)	ρ (g/cm^3)	V_D (m/s)	P_D[b)] (GPa)
Tritonal	TNT/Al/ 80/20	1.79	6400	17.5
AFX-930	HTPB/RDX/NQ/Al (16/32/37/15)	1.61	6700	22.0
AFX-931	HTPB/RDX/AP/Al (16/32/37/15)	1.61	6200	19.3
PBXN-111[c)] [24]	HTPB/RDX/Ap/Al (12/20/43/25)	1.80	5640/7064[d)]	21.6
Comp. B	TNT/Wax/RDX (36/1/63)	1.65	7800	28.0
AFX-960	HTPB/HMX/NQ (14/56/30)	1.63	7700	23.3
H-6	TNT/Wax/RDX/Al (29.5/5/44/21)	1.76	7650	26.0
PBXN-109	HTPB/RDX/Al (16/64/20)		7500	
Octol	TNT/HMX (75/25)	1.81	8643	31.4
PBXN-110	HTPB/HMX (12/88)	1.672	8330	

a) Shaded rows are melt-cast explosives that were replaced by a PBX.
b) Calculated.
c) Made in RAFAEL.
d) Calculated Explo-5.

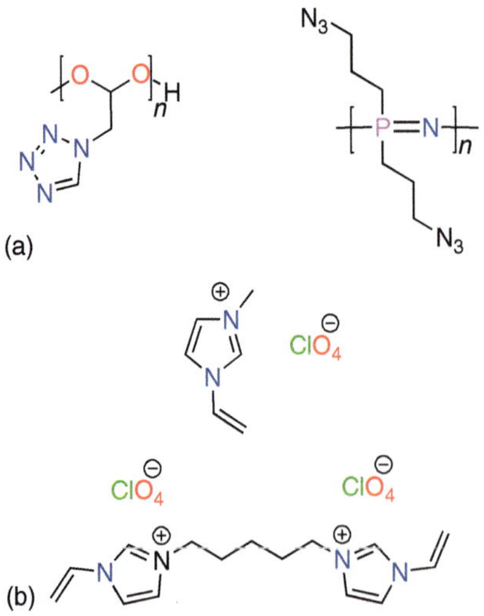

Figure 1.8 Nitrogen-rich energetic polymers. (a): polytetrazole and polyphosphazene, (b): UV-3D printable energetic monomers.

energetic, they still offer the composition substantial capabilities of absorbing and dissipation of energy, thus yielding composites with lower sensitivity. Large variety of polymers has been prepared mainly with pendent azido and nitrato groups [25, 26]. Glycidyl azide polymer (GAP)[7] is still the most explored energetic binder as it is commercially available. The increasing number of nitrogen-rich species explored in energetic materials is also reflected by attempts to introduce nitrogen-rich polymers as polytetrazoles and polyphosphazenes as energetic binders (Figure 1.8) [27]. Yet another trend in energetic materials applications is the emerging 3D printing technology. Additive manufacturing techniques bring the promise of on-demand manufacturing of energetic devices from their ingredients with macroscopic, mesoscopic, and microscopic control over structure, morphology, composition, and thus performance.

1-Vinylimidazolium-based mono and bifunctional energetic monomers have been prepared (Figure 1.8b) that are UV polymerizable and inducive to 3D printing. The new monomers were 3D printed alone and in the presence of AP producing high-resolution structures with very low postproduction shrinkage [28].

1.2.2.2 Pressed PBXs

PBXs were first developed to create materials that could be pressed to high bulk density approaching 95–96% theoretical maximum density (TMD). With that, still withstand temperature fluctuations, mechanical deformations, or any other lifecycle events and perform on-demand, concomitant with acceptable safety properties. Favorable binders have high enough bonding energy between the filler and binder fulfilling the above requirements.

A great deal of effort has been invested in understanding the surface and adhesion properties of explosives and binders. One approach was to explore the acid–base compatibility of the binder–filler duo by measuring polymer adsorption on explosive filler surface from an acidic, basic, or neutral solvent. Judging from the degree of polymer deposition according to the acid–base nature of the solvent–polymer–surface triad, matching explosive-binder system could be selected to enhance adhesion and improve the mechanical properties. A more rigorous approach relays on actually calculating the work of adhesion[8] between the filler and polymer and using the quantitative data for selection. This type of study was conducted using TATB [29], an explosive with both acidic (NO_2) and basic (NH_2) groups. TATB surface matching is complicated, although the material contains hydrogen donors and acceptors, however, they are involved in intra and intermolecular bonds and are less conducive to adhesion to other surfaces resulting in a crystal structure showing marked difference according to the crystallographic planes. Molecular dynamic (MD) simulations [30] with fluoro binders have shown a marked difference between planes. The [001] plane offers only limited surface area for bonding mainly through π–π interaction. The [100] plane on the other hand is a rough surface enabling penetration of the polymer melt to the crystal filler

7 See Chapter 4 in this book.
8 Work of adhesion can be calculated form the surface tension of the materials and is given by $W_{ad} = \gamma_{poly} + \gamma_{filler} - \gamma_{Interface}$.

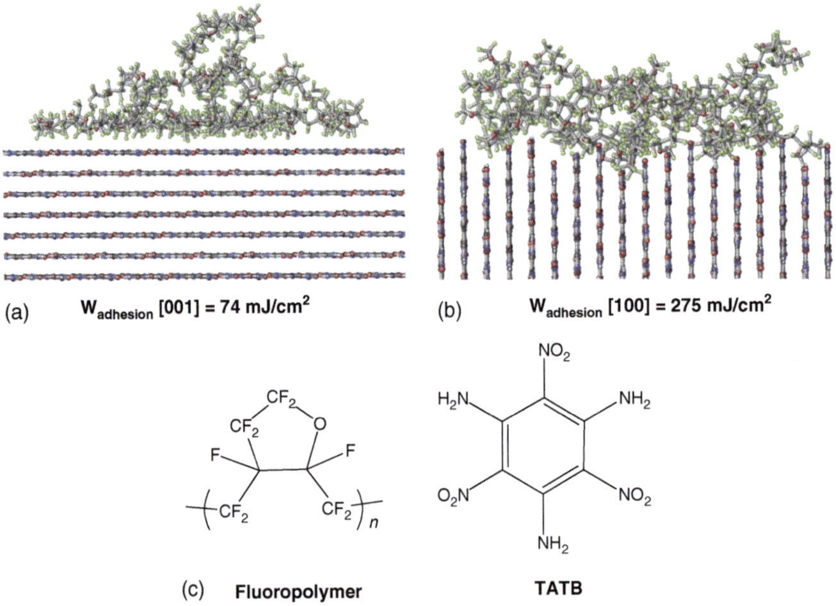

Figure 1.9 MD simulations of fluoropolymer on TATB [001, 100] crystallographic planes. Snapshot time: ~300 ps at $T > T_g$ (polymer) Fluoropolymer chain structure on TATB surfaces. (a) - [001] surface; (b) - [100] surface; [c] - Fluoropolymer and TATB structure. Source: Geee et al. [30]. Reproduced with permission of American Chemical Society.

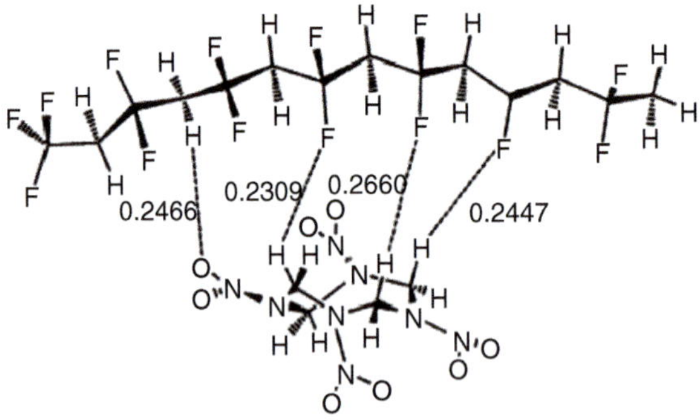

Figure 1.10 Interaction between a fluoropolymer and HMX (distances given in nm). Source: Xiao et al. [31]. Reproduced with permission of Springer Nature.

with stabilizing interactions with the basic and acidic groups. Figure 1.9 shows the systems after 300 ps equilibration. The improved interaction is well reflected in the work of adhesion on [100] plane.

As may be expected the main interactions between a fluoropolymer (FP) and a nitramines explosive are hydrogen bonds between O—H atoms and F—H atoms as depicted in Figure 1.10 [31].

These simulations not only enable probing the effect of polymer structure (type and monomer ratio) [32] but also the interaction of different crystallographic planes of the solid explosive with the polymer [33]. A natural outcome of MD simulations stems from the atomistic level resolution enabling an accurate allocation of the atoms that participate in the surface bonding for different polymer aggregation

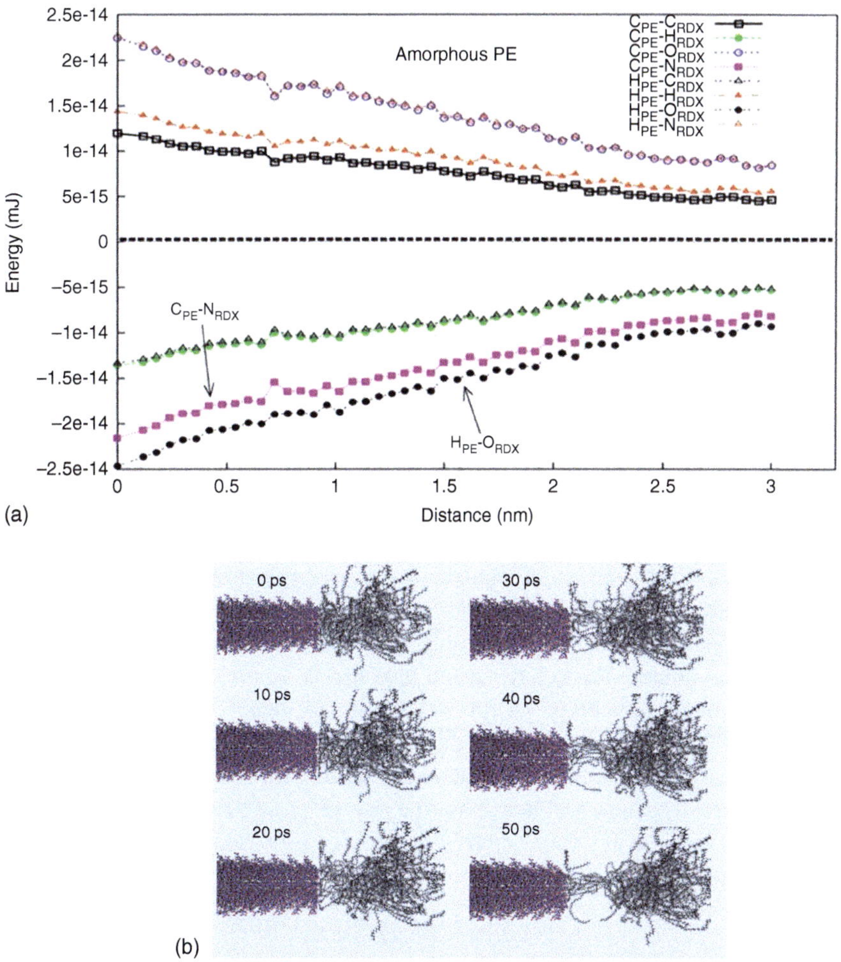

Figure 1.11 MD simulations of amorphous RDX–PE-pressed PBX. (a) Atom-atom pair contribution to the total interaction energy as function of distance for amorphous PE (b) MD simulation of snapshots of RDX-PE configuration taken every 10ps during tensile loading. Source: Taylor et al. [34]. Reproduced with permission of American Chemical Society.

state, i.e. crystalline or amorphous [34]. A nice example showing the merits of the simulation is the RDX–PE interaction study in Figure 1.11, showing the atom–atom pair contribution to the total surface interaction as function of the distance between RDX and amorphous PE. Snapshots of the configuration taken at 10 ns intervals exemplifies the separation of RDX and PE surfaces under tensile loading of amorphous (Figure 1.11) and crystalline (not shown) PE. The atom–atom curve shows that most of the stabilizing energy comes from oppositely charged surface atoms H_{PE} and O_{RDX} atoms C_{PE} and N_{RDX} atoms.

The same atoms are responsible for the surface interaction with crystalline PE, but the elongation is much lower with zero interaction after 0.6 nm versus >300 nm for the amorphous binder.

1.2.2.3 Plasticizers

Plasticizers are typically one of the minor ingredients in propellant and explosive compositions but with a major effect on production and properties of these materials. Plasticizers bring down the viscosity of a liquid polymer and/or lowering the melt viscosity of thermoplastic elastomers, making the compositions easier to mix [35]. Thus, it improves workability and enables higher solid loadings of the polymer with fillers explosives or oxidizers, which is crucial for the maximum performance. Plasticizers also improve the mechanical properties of composite materials by bringing down the modulus and tensile strength concomitantly increases elongation. Often lowering of the glass transition temperature is observed that facilitates the materials behavior at low temperatures. The need to avoid the cold temperature brittleness is sometimes met by using mixture of plasticizers as K10[9] or bis(2,2-dinitro propyl)formal/acetal (BDNPF/A) forming eutectic mixtures. The most common plasticizers in use in explosives and propellants are aliphatic mono or di esters. Typical examples are dioctyl adipate (DOA)[10] and isodecyl pelargonate (IDP) [36]. Nitrate esters (NEs), nitric acid esters of polyols are more frequently used in high-energy smokeless propellant compositions [1] where aluminum is omitted inspite of its contribution to energy release since it is a source of secondary smoke keeping comparable performance to aluminized propellants. In these cases, a polar polymer binder as polycaprolactone (PCL) is needed to achieve a homogenous binder. These have found also use in several PBXs compositions [37]. Low vulnerability PBXs[11] with 50% additional blast energy compared to tritonal,[12] a PBXs containing AP for underwater applications,[13] or an enhanced blast thermobaric composition[14] [37] all obtained using trimethylolethane trinitrate (TMETN)/PCL. A major problem with the use of plasticizers is the migration of plasticizer, influencing the composite properties [38]. Some typical examples of actual energetic composites are shown in Table 1.3 according to the type of composite.

1.2.3 Surface Active Materials (SAMs)

Composite propellants and PBXs contain large amounts of solid materials with high-surface area engulfed by a polymer often with different polarity and surface characteristics. Thus, the importance of surface properties on the structural and safety traits of these composites. Poor surface compatibility also puts difficulties on production as high viscosities and poor flowability. Moreover, if the high solid filler is not properly covered by the organic binder, defects and cavities may occur.

Filler–binder interactions are influenced by a variety of factors such as particle size, volumetric loading, and predominantly bond strength. Mechanical properties

9 K10 or Rowanite 8001 is a 65/35 of 2,4-dinitroethylbenzene/2,4,6-trinitroethylbenzene.
10 DOA is the acronym for bis(2-ethylhexyl) adipate.
11 i.e. RX-35-EK: 39.5%HMX/28%Al/24.8%TMETN/7.7%PCP.
12 Tritonal is a melt-cast explosive (80/20 TNT/Al).
13 PBXW-123: see composition in Table 1.4.
14 PBXIH-135EB: 42%HMX/33%Al/25%PCP+TMETN.

Table 1.3 Composition of propellants and PBXs.

Sample	Name	RDX (%)	HMX (%)	AP (%)	Al (%)	Binder	(%)	Plasticize	(%)	References
Pressed PBX	PBXN-11		96			Hytemp	1.0	DOA	3.0	
	LX-09		93			PNDPA	4.6	FEFO	2.4	[39]
	PBX-9205	92				Polystyrene	6.0	DOP	2.0	[39]
	PBXN-9501		95			Estane	2.5	BDNPF/A	2.5	
Cured Cast PBX	PBXW-123			44.8	30.2	PCL	6.2	TMETN	18.8	[37]
	PBXC-126		79			GAP	8.4	TMETN/TEGDN	8.0/2.5	[3, p. 712]
	PBXW-128		77			HTPB	11.0	IDP	11.0	[3, p. 713]
	GHX-76	42		25	15	GAP		BDNPF/A	[Poly+Pla]=18	[40]
Propellants	CSP-1			65	18	HTPB	9.5	DES	2.0	[41]
	E1	15		49	20	HTPB	11.0	DOA	5.0	[42]
	E2	15		49	20	HTPB-*b*-CPL	11.0	NG	5.0	[42]
Thermoplastic Bonded Propellants	E3			64	24	Polyisoprene		HC oil ($C_{19}H_{35}$)	[Isoprene+oil]=12	[43]
	E4	18.5		40	18	BAMO/AMMO	15	BuNENA	5	[44]

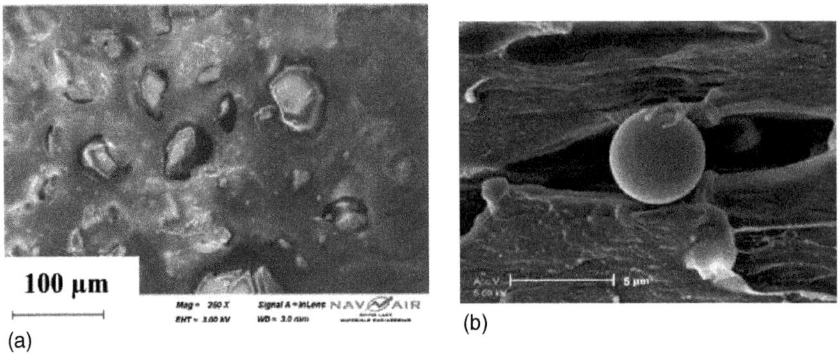

Figure 1.12 Dewetting of composite matrices by electron microscopy. (a) – HTPB/CL-20. Source: Kalman et al. [45]. Reproduced with permission of John Wiley and Sons.; (b) – Glass beads/polypropylene. Source: Fu et al. [46]. Reproduced with permission of Elsevier.

deteriorate with poor filler–binder adhesion decreasing tensile and compressive strengths and reducing the elongation the material can withstand without permanent damage. In cases where the filler–binder adhesion is poor and the binder is sufficiently elastic and can elongate without tearing, it will extend sufficiently eventually breaking free from the solid filler forming voids known as dewetting. Dewetting was nicely visualized in HTPB/CL-20 [45] and glass beads/polypropylene [46] composites as shown Figure 1.12.

The formation of voids may have a detrimental effect on safety and performance, especially in PBXs. It was argued that exposed explosive substances in a PBX may reach the nitramines filler sensitivity [47]. Moreover, voids can act as initiation sites when adiabatically compressed, thus increasing vulnerability to shock and low-impact assaults. Structural integrity is also highly important for propellants where the ability to survive long-term temperature changes during storage, rapid heating, and pressure changes during ignition and combustion are crucial for avoiding cracks and increased burn surfaces which may be of catastrophic consequences. Dewetting is characterized by a large extension of the binder prior to rupture resulting in high extensibilities and low tensile strength canceling filler reinforcement. The effect of bonding agents (BA) can be demonstrated by plotting the ratio of the strain at maximum strength (ε_m) to the strain at failure (ε_f) as function of the surface active material (SAM) content as depicted in Figure 1.13. When no SAM is present $\varepsilon_m/\varepsilon_f \ll 1$, indicating early separation of the filler from the binder and continued elongation of the polymeric binder. As the content of SAM is increased, the elongation ratio approaches 1, indicating the filler and binder approaching a unison. This reaches a plateau since all is needed is a monomolecular layer coverage of the filler. Data for collagen pre-coated 1,3,5,7-tetranitro-1,3,5,7-tetrazacyclooctane (octogen) (HMX) (13 μm) and 1,3,5-trinitro-1,3,5-triazacyclohexane (hexogen) (RDX) (14 μm) [48] enables demonstration of the effect of SAM content on the $\varepsilon_m/\varepsilon_f$ ratio.

Although void formation was concluded in early studies to be mainly the consequence of cohesive failure in the binder [49] and therefore a hard layer around

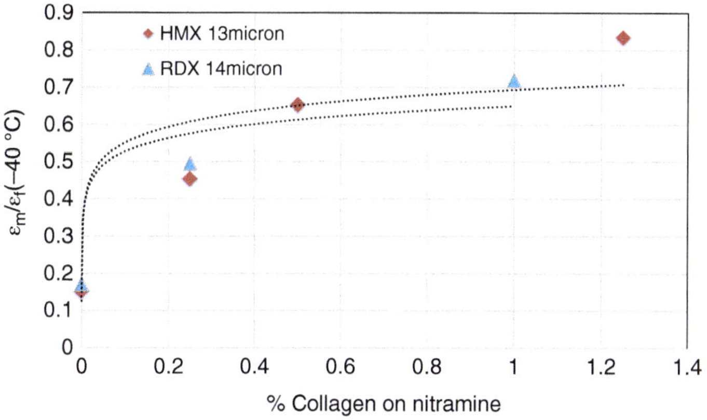

Figure 1.13 $\varepsilon_m/\varepsilon_f$ (−40 °C) for collagen-coated HMX and RDX.

the filler is needed for preventing binder peel from the solid particle surface, tuning filler and binder properties so they can interact (i.e. acid–base) improves mechanical properties [50].

SAMs, form an intermediate layer connecting the filler with the binder, can be classified according to the interaction with the filler and binder system as Wetting Agents (WA) and Bonding Agents (BA) as depicted in Figure 1.14.

WA's are materials that adhere to the filler surface through secondary physical interactions such as hydrogen bonds, dipole–dipole, and donor–acceptor interactions. Such materials usually contain a polar head group and an oily hydrophobic tail that enables lubrication of the filler by the binder as the hydrophobic tail is soluble in the binder system stabilized by dispersion forces. BA's enable not only

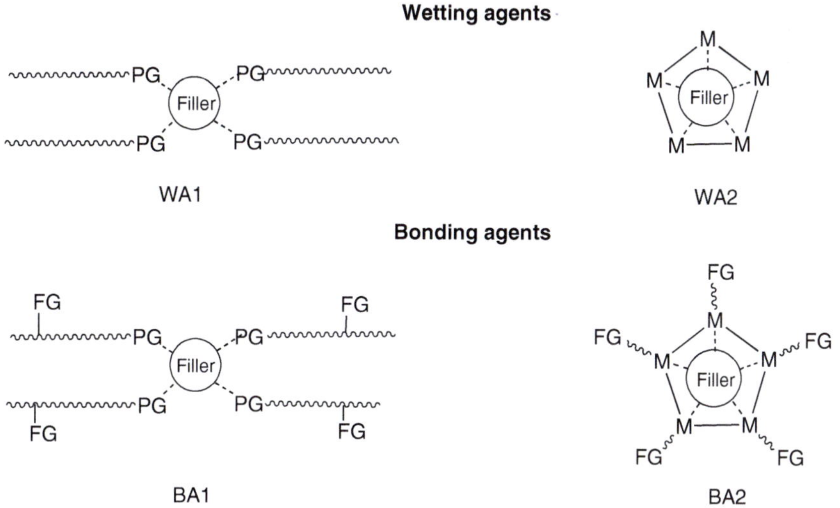

Figure 1.14 Modes of interactions of SAM with filler in composite EM. PG – polar group, FG – functional group, M – monomer.

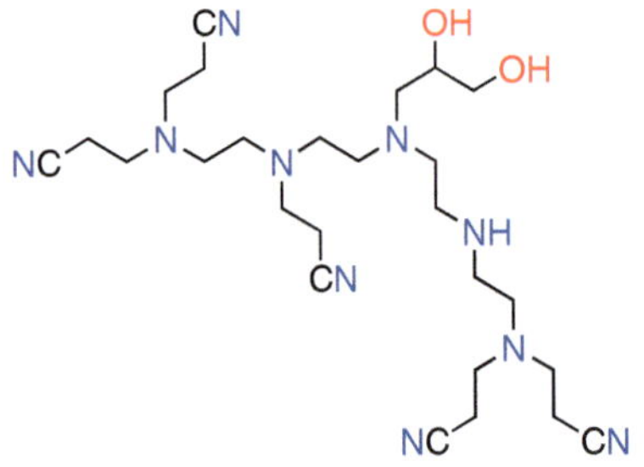

Figure 1.15 TEPANOL, BA1 type SAM for AP propellants.

interaction with the filler surface but also have other functionalities that allow the BA to react with curing agents thus reinforcing the filler–binder interaction by covalent bonds. SAM can be divided according to the mode they form. Premade SAM materials (WA1/BA1) or in situ formed SAM (WA2/BA2).

Lecithin [51] is a typical example of a premade WA1 type SAM with the polar choline phosphate head and a lipophilic fatty acid tail used in PBXs compositions. TEPANOL, is an example of a BA1 type SAM for AP propellants (Figure 1.15)[15] [52] anchors to the ammonium salt particles surface through electrostatic bonds that are formed by an acid–base reaction of a stronger tertiary amine base with the acidic ammonium ion:

$$R_3N + NH_4{}^+X^- \rightarrow R_3NH^+X^- + NH_3; X = ClO_4{}^-, NO_3{}^-, SO_4{}^{-2}$$

TEPANOL may also act as wetting agents in PBXs through interaction with the nitramines moiety, but is prone to incompatibility by reaction with the nitramines group probably by exchanging an NO_2 group between amines, especially with RDX [47, 53]. WA2 and BA2 (in situ formed) type materials are formed in situ in the presence of the filler and as a response to the filler surface chemical properties. Typical examples of such materials are the aziridine-based compound as HX-752 and tris-(2-methyl-1-aziridinyl) phosphine oxide (MAPO) and RX (Figure 1.16).

These undergo acid-induced homo-polymerization of the acyl/phosphine aziridine moiety on ammonium salts surface [54, 55] covering the inorganic polar particle with an organic mesophase that is physically compatible with the nonpolar binder [56]. The unreacted acyl aziridine groups in HX-752 can react with OH

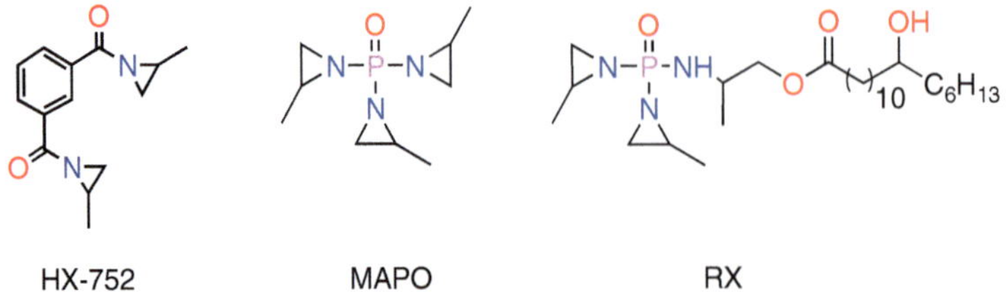

Figure 1.16 Acid-activated aziridine wetting (WA2) and bonding agents (BA2).

15 Idealized structure, the material is actually a mixture of compounds with the shown functionality.

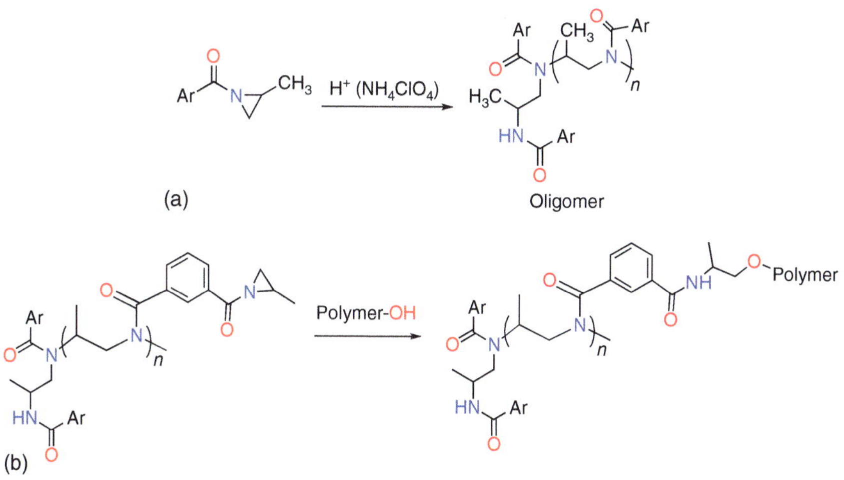

Figure 1.17 Homo-polymerization of acyl aziridine (a) and crosslinking reaction of the mesophase with OH functuionality (b).

nucleophiles on HTPB binder affording crosslinking of the covered particles to the matrix [57] defining HX-752 as a bonding agent of type BA2 (Figure 1.17).

In OH terminated prepolymers, MAPO with the same active aziridine moiety, acts as a wetting agent formed in situ (WA2) since it does not react with the OH groups [58]. In COOH terminated prepolymers (i.e. CTPB), MAPO on the other hand acts as a bonding agent capable of crosslinking by phosphine aziridine ring opening with carboxyl groups (forming a crosslinking reaction product, $(Az)_2$–PO–NH–CH(CH_3)–CH_2–O–CO–Poly) [59]. RX is formed by reaction of MAPO with a limited amount of 12-hydroxystearic acid [60], keeping part of the aziridines for bonding to AP and forming an encoring hydroxyl group for attachment to the binder by reaction with crosslinking isocyanate.

Heterocyclic neutral SAMs (Figure 1.18) were designed to match nitramines that lack an active functional group (beyond explosophores) and can rely only on secondary interactions.

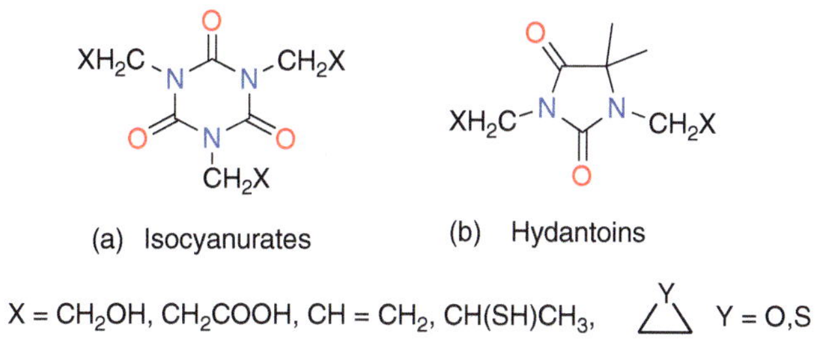

Figure 1.18 Isocyanurates and hydantoin wetting (WA1) and bonding agents (BA1) for neutral energetic filler (nitramines). Isocyanurate (a) and hydantoin (b) wetting (WA1) and bonding agents (BA1) for neutral energetic fillers (nitramines).

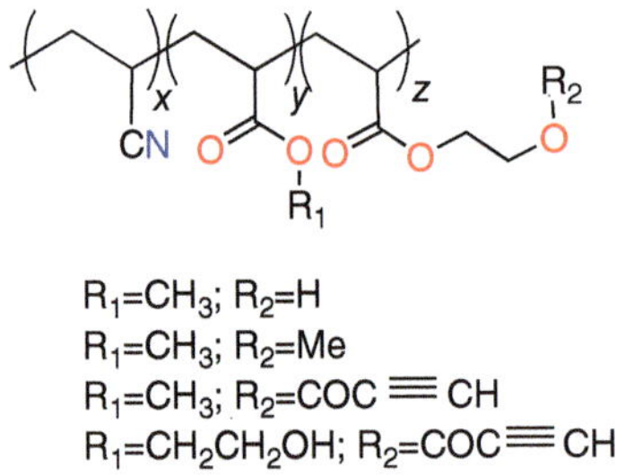

Figure 1.19 Neutral polymeric bonding agents (NPBA's) for nitramine fillers in polar binder systems.

Properly substituted hydantoin and isocyanurates are capable of reacting with the curative isocyanate [61], thus these heterocyclic materials fit WA1 and BA1 according to the mode of substitution.

Smokeless propellant composites that do not contain AP, contain energetic ingredients such as nitrate esters plasticizers (NE) or energetic binders which are needed to boost performance. In the inevitably polar binder system, the substantial solubility of the SAM in the binder prevents the strong adherence to the solid filler yielding composites with poor mechanical properties. Polymeric SAM were developed for coating of the explosives enabling adhesion to the explosive filler and bonding to the binder system. In attempt to eliminate a pre-coating step, neutral polymeric bonding agents (NPBA's; Figure 1.19), [62–64] were developed as SAM that will allow adherence to nitramine fillers as part of the mixing process without a precoating step. NPBA's show an enhancement of the tensile strength and elastic modulus of low signature propellants, and have similar properties as composite with NC precoated nitramines

Pressed PBX's that contain nitramines (RDX, HMX, CL-20) or aromatic explosives with alternating amino and nitro substituents (TATB, LLM-105) usually do not contain bonding agents [65], and the binder itself can be regarded as the bonding agent gluing the enclosed particles to each other. Nevertheless, more experimental and sometimes intuitive [66] approaches to surface bonding enhancement for pressed PBX's have been reported, mainly using polymer pre-coating the energetic fillers. A prevalent choice is polydopamine (PDA), formed by an in situ oxidative polymerization on the explosive filler (Figure 1.20). The effect of PDA coating on TATB [68], HMX [67], and LLM-105 [69] was thoroughly studied and some substantial improvements in tensile and compressive strengths and creep strain of the corresponding fluoropolymer bound PBXs were reported. Recently, even double coating of explosives with PDA as one of the coating layers with hyper-branched polymers (HBP), or wax have been explored for surface adhesion and mechanical properties enhancement. (TATB@PDA)@HBP [70], (HMX@Wax)@PDA [71], (CL-20@PDA)@HBP [72], show modest improvements in mechanical properties but good improvements in safety. The effect of PDA coating on the explosive filler surface was investigated using a variety of methods beyond the actual measurements

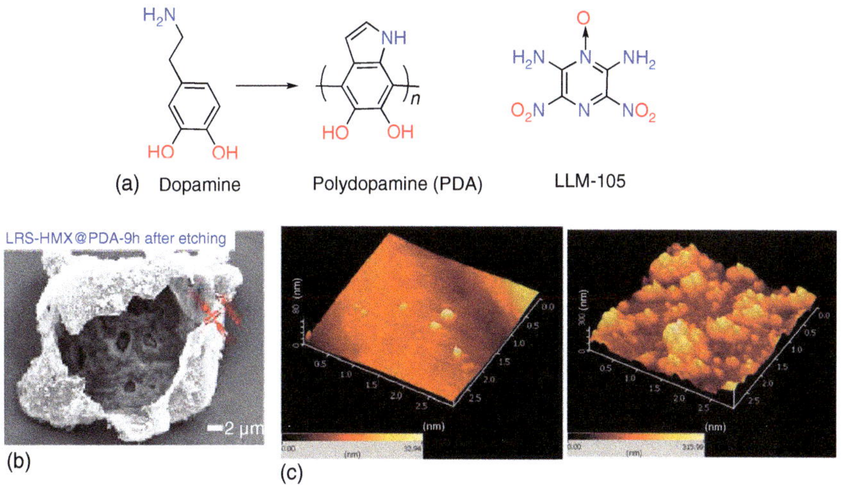

Figure 1.20 Dopamine polymerization and coating of explosives. (a): in situ oxidation of dopamine to PDA in the presence of LLM-105. (b): PDA shell left after selective dissolution of HMX from HMX@PDA. Source: Lin et al. [67]. Reproduced with permission of MDPI Open Accesses. (c): AFM surface topography of LLM-105@PDA. Source: He et al. [68]. Reproduced with permission of The Royal Society of Chemistry.

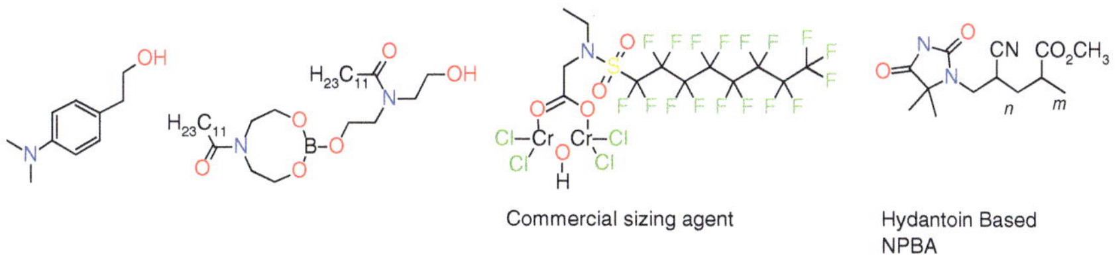

Figure 1.21 Bonding agents tested for TATB.

of mechanical and creep behaviors. Atomic force microscopy (AFM) Profile of coated materials showed increased roughness of the surface (Figure 1.20) [68]. Selective dissolution of the HMX from HMX@PDA [67], resulted in a hollow shell showing an imprint of the crystal surface (Figure 1.20). From these studies, it was concluded that surface bonding is the culmination of two mechanisms, chemical bonding, mainly hydrogen bonding, some acid–base interaction and π–π stacking when relevant (LLN-105 and TATB) but also physical interlocking due to surface roughening.

TATB which is somewhat reluctant to surface bonding (see Section 1.2.2.2) has attracted some efforts improving surface adhesion using SAM. These were chosen in a completely intuitive fashion [66] in contrary to predesigned SAM [73, 74] with limited success (Figure 1.21).

An attempt to generalize a judicious selection of a SAM according to the composite energetic is given in Table 1.4.

Table 1.4 Selection of SAM according to energetic system.

Composite	Oxidizer	Nitramine	Binder	SAM Type	Typical Example
Propellants	AP	—	HTPB	BA1/BA2/WA2	TEPANOL/ HX-752/MAPO
	AP	RDX/HMX	HTPB	BA1/BA2	HX-752/MAPO
	—	RDX/HMX	Polyether/ polyester/ nitrate Ester	BA1	NPBA/PDA
PBX's	—	RDX/HMX	HTPB	WA1/BA1	Lecithin/ hydantoins
	AP	RDX/HMX	HTPB	BA1/BA2	Isocyanurate/ HX-752
	—	RDX/HMX	Fluoropolymer/ estane	Usually not needed	—

1.3 Performance

In a sense, detonation chemistry can be considered as a decomposition reaction followed by a disproportionate redox, with a simple Arrhenius type correlation of the initial rate with an inverse temperature dependence [75]. These occur in the high-pressure small volume zone, making the detonation chemistry practically a unimolecular process. In contrary, propellants reacts by diffusion-controlled[16] higher-order reactions [6].

These differences explain the different trends in reaction performance as function of the content of explosive and oxidizer (expressed as oxygen balance [OB]). The scaled performance of a series of "synthetic composites" containing AP/HTPB as propellant mimic and HMX/HTPB as PBX mimic were calculated using EXPLO05-V6.05 and Chemical equilibrium with applications (CAE)[17] software. The ratio of the velocity of detonation (VoD) of the composite relative to HMX, VoD(PBX)/VoD(HMX), and ratio of the specific impulse (I_{sp}) of the composite relative to AP, I_{sp}(propellant)/I_{sp}(AP) are shown in Figure 1.22.

The scaled VoD shows a monotonic rise up to the OB of the pure explosive that is the maximum output PBXs can achieve. Propellants on the contrary that rely on diffusion-controlled reactions between fuel and oxidizer reach a maximum at the perfect stoichiometric ratio (OB = 0%), above that the region of monopropellant combustion begins with lower efficiency up to OB = +34%, corresponding to AP itself.

16 Unimolecular reactions do occur in propellants (oxidizer monopropellant flame and binder thermolysis) but the output comes mainly from diffusion-controlled fuel and oxidizer mixed flame.
17 Nasa Lewis Research Center's computer program.

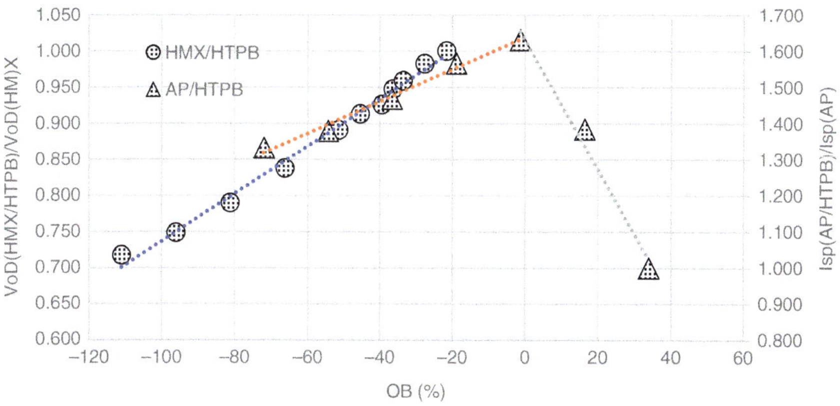

Figure 1.22 Scaled performance as function of oxygen balance for a propellant (AP/HTPB) and a PBX (HMX/HTPB).

1.4 Sensitivity

Propellants and explosives store chemical energy that is transformed into mechanical energy on demand. Inadvertent initiation of these materials can be a result of a variety of stimuli such as mechanical impact, friction, heat, flame, or electrostatic discharge. One of the main concerns during development and manufacturing and even during operation of energetics is the sensitivity to these stimuli resulting in accidental initiation. A variety of procedures for sensitiveness[18] testing were developed [76], however direct comparison is difficult since the response is highly dependent on particle size and composition, test regime, and machine design that practically eliminates the ability to compare different apparatus. Among the sensitivity, tests drop weight impact (or drop hammer) attracted the most attention since it seems to be more amiable to modeling and prediction. It was shown that the usual criteria for impact sensitivity the 50% Go/No-Go energy[19] ($E_{50\%} = m_w gh$ [J]) is at best an indirect and inaccurate measure of the responsive value. When the kinetic energy of the falling weight is separated from the plastic energy [77], which is translated to deforming, shear heating, and hot spots formation inducing initiation it turned to be in the order of ~50% of the energy for 50% probability for initiation measured by the Go/No-Go criterium. In a series of elegant works Filed et al. [78, 79] published the results of impact initiation of various energetic compositions while optically interrogating the chain of events during the test using a transparent

18 Sensitiveness is probably a better term than sensitivity describing the likelihood of accidental initiation of an energetic. Sensitivity is more closely associated with reliability of initiation defining the initiation response map of a device [2]. However, since the literature uses mainly sensitivity this term will still be used also.

19 $E_{50\%} = m_w gh$ (J); m_w – weight mass (kg), $g = 9.81$ (m^2/s) gravitational acceleration, h – height (m).

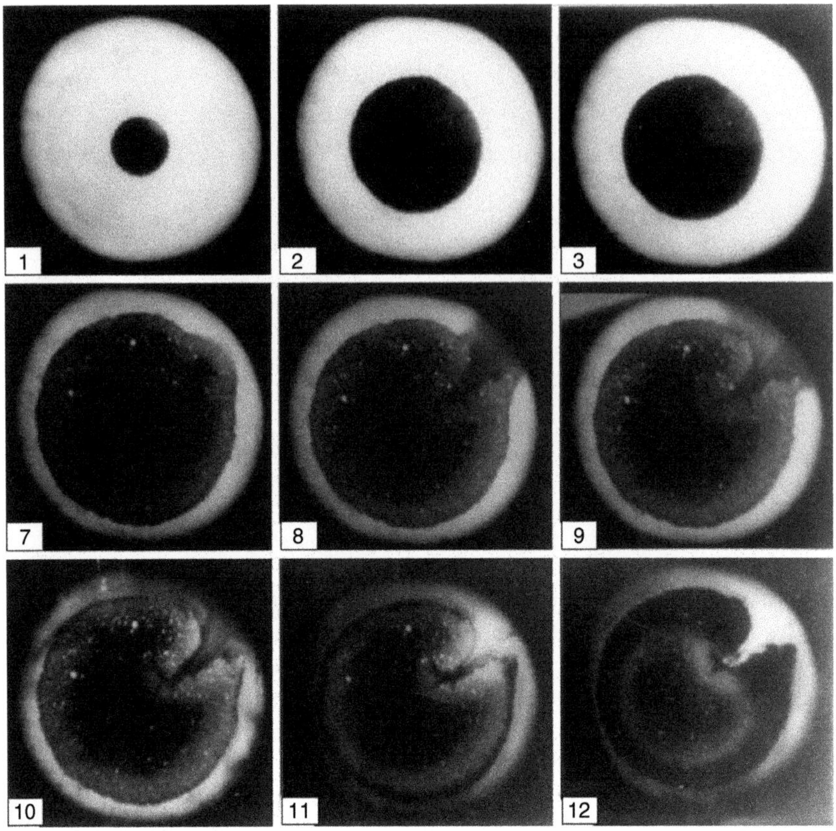

Figure 1.23 Sequence of events in an impact sensitivity test of a propellant sample. Composition: 33%HTPB/66%AP (30 mm). Frames: 0, 350, 420, 581, 595, 630,770, 812 μs. Source: Balzer et al. [80]. Reproduced with permission of The Royal Society of Chemistry.

glass anvil. Figures 1.23 and 1.24 are snapshots of an impact event of a composite propellant and a PBX.

In both cases, the initial impact causes spreading of the samples but the PBX shows visible tearing (on the edges) after 273 μs (frame 3) in contrary to the propellant sample that spreads more smoothly. Gasification and burst events after ignition are evident at 581 μs (frame 7) in the propellant and after 476 μs (frame 8) in the PBX. Another major difference is the fact that ignition in the propellant occurred in just a few sites venting off when the weight lifted form the periphery inwards. In the PBX however, it seems to be a complete surface event causing consumption of the materials throughout the sample.

1.4.1 Sensitivity Correlations

In contrary to the large body of quantitative structure–property relationships (QSPR) correlations of impact sensitivity of explosives substances with their chemical structure [81, 82], semiempirical-based molecular properties [83–85], or solid structure [86–88], much less successful correlations exist for sensitivity with

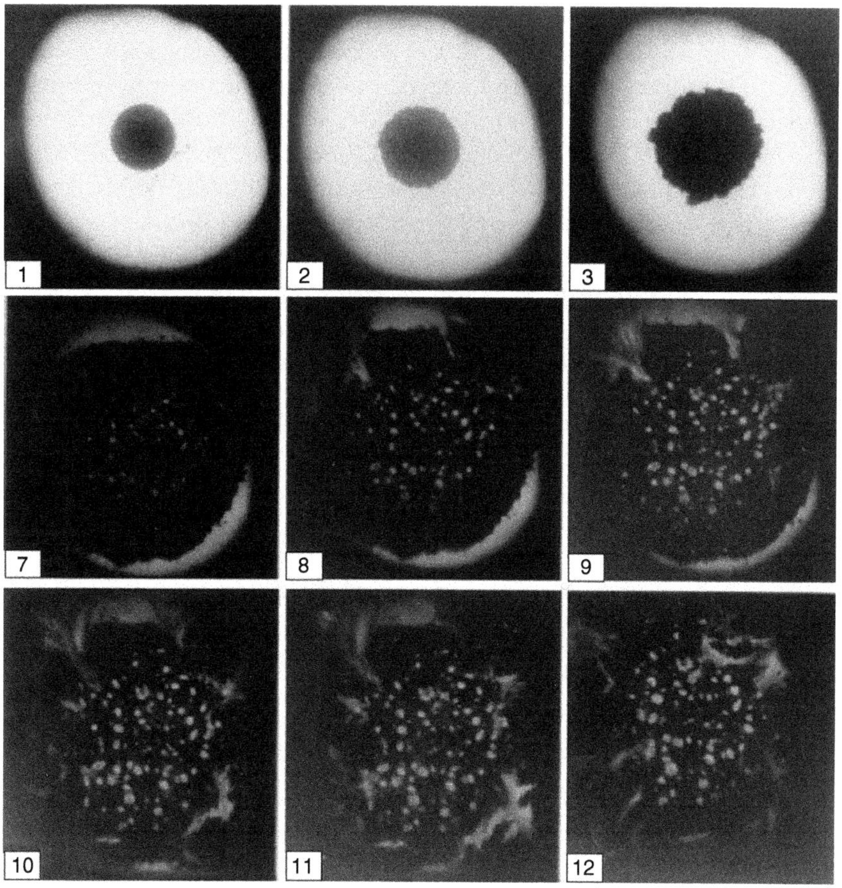

Figure 1.24 Sequence of events in an impact sensitivity test of a PBX sample. Composition: 40.5% AP(400 μm)/23.8%Al/15.0%RDX/14.1%TNT/7.1%CTBN[20] Frames: 0, 105, 273, 469, 476, 490, 504, 511, 560 μs respectively. Source: Balzer et al. [80]. Reproduced with permission of The Royal Society of Chemistry.

actual compositions. Although much effort was invested in sorting correlations between physical characteristics of composite energetics, especially PBXs with sensitivity data, $[(\text{VoD})^2\rho \sim \ln(\text{impact});\ Q_{\text{det.}} \times \rho \sim \text{Friction};\ \Delta H_\text{f} \sim \text{Impact}]$ no general type correlations are to be found [89]. Usually, these are not general and need to subdivided according to material composition i.e. type of binder and filler. Recently the significance of binder stiffness as expressed by elastic modulus on 50% drop energy of HMX-based PBXs was shown [90]. The report also correlated the 50% drop energy with work of adhesion of the binders and concluded the elastic modulus is the significant factor. This is not surprising, since stiff binders are less adopted for energy absorption. However, the correlation shows a rather steep decay to a constant value limiting the usefulness of the correlation. Another MD simulation of the modes of energy absorption of common binders for PBXs showed

20 CTBN – Carboxy terminated polybutadiene acrylonitrile copolymer.

a positive correlation between the impact sensitivity data of the composites and the energy absorption by the binder [91]. Earlier attempts of factor analysis of multiple regressions of a variety of sensitivity tests (impact, friction, ignition temperature, detonability, etc.) coupled with principal component analysis were reported but cannot be examined in detail from the report [92]. Yet in a more recent report [93] performance data (detonation velocity, gurney energy, and plate dent test) were combined with density yielding a unified scale of performance. In a similar way sensitivity data (impact, thermal decomposition, and gap test) were combined separately with density forming a unified sensitivity scale. The performance score versus sensitivity score were plotted in an attempt to reproduce the basic holding paradigm in EM's (which current research efforts trying to depart from) the better the performance the higher the sensitivity, however, the correlation of the unified performance versus sensitivity does not seem to hold.

Impact and friction sensitivities of in-house composites, RAFAEL propellants (RP), and RAFAEL explosives (RE) were measured using BAM[21] impact and friction machines. Bruceton's up and down statistical analysis for determining the 50% probability of impact initiation and 6 no fire (6NF) instead of the usual 1 in 6 criteria for friction were used for measuring sensitivity. The sensitivity data were expressed as Figure of Insensitivity (FoI) for impact and friction against RDX standard [94, p. 63] are given in Table 1.5. In a naive attempt to suggest a more general type correlation between sensitivity and materials properties the product of FoI(impact) with FoI(friction) was charted as function of the product of the absolute value of oxygen balance (OB) and TMD on a logarithmic scale as depicted in Figure 1.25. The absolute value of the product OB×ρ was used in order allow a logarithmic scale and depicted in reverse order to emphasize a lower absolute value is actually higher OB (for OB < 0). The somewhat surprising result is that a rather good correlation ($r^2 = 0.87$) was obtained. The rationale behind this phenomenological correlation was to gather separately "chemistry" (oxygen balance) and "composition" (density) data versus sensitivity data. This may be refined with binder mechanical properties or other traits but it seems to hold without regard to type of composite (propellant or explosive). This correlation also reemphasizes that better performance is linked to lower insensitivity.

The direct comparison between propellants and explosives is almost impossible. Therefore, the standard score[22] for FoI for impact and friction were calculated and plotted (Figure 1.26) to show that, compared to the average value, the explosives formulations are more impact sensitive (6/4) than propellant formulations (2/3) and vice versa for friction on average. However, it cannot be stated that this is a completely general trend since not enough data is available in literature.

---

21 Bundesanstalt fur Materialfprufung und Forshung.

22 Standard score = $\frac{[\text{Value}]\ [\text{Mean}]}{[\text{Standard deviation } (\sigma)]}$

Table 1.5 FoI and standard score for Rafael explosives (RE) and Rafael Propellants (RP) Measured on BAM Machines.

	Explosives									XLDB propellant	Propellants				
Composite	**RE-1**	**RE-2**	**RE-3**	**RE-4**	**RE-5**	**RE-6**	**RE-7**	**RE-8**	**RE-9**	**RP-1 XLDB**	**RP-2**	**RP-3**	**RP-4**	**RP-5**	**RE-6**
Energetic filler	HMX	HMX	RDX	RDX	RDX	HMX	HMX	RDX	RDX	NE/HMX	AP/RDX	AP/RDX	AP	AP/RDX	AP
Binder	HTPB	HTPB	HTPB	PDMS	HTPB	HTPB	HTPB	HTPB	HTPB	PPG	HTPB	HTPB	HTPB	HTPB	HTPB
Figure of insensitivity (impact)	126.3	101.5	85.8	130.7	108.0	213.9	190.3	122.4	156.8	137.78	21.8	77.3	131.5	129.8	108.9
Figure of insensitivity (friction)	8.8	14.7	10.4	8.3	13.3	13.5	13.5	13.5	3.0	10.62	1.2	2.0	1.5	3.5	1.7
FoI (impact) standard score	−0.28	−0.90	−1.30	−0.17	−0.74	1.92	1.33	−0.38	0.49	0.01	−1.57	−0.36	0.82	0.78	0.33
FoI (friction) standard score	−0.60	1.05	−0.16	−0.74	0.66	0.71	0.71	0.71	−2.24	−0.10	−0.91	0.02	−0.54	1.68	−0.26

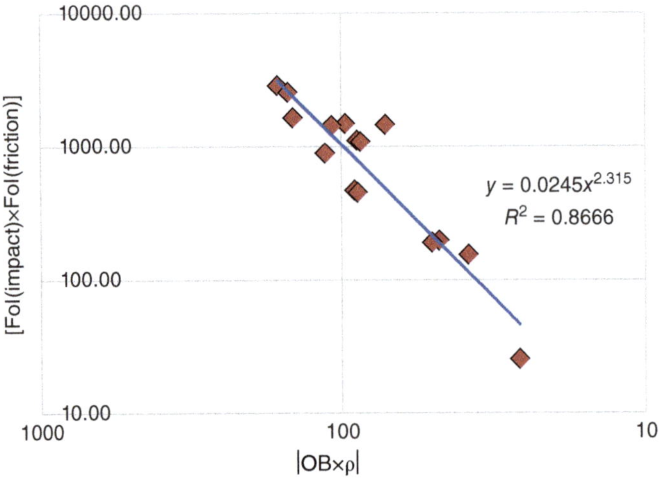

Figure 1.25 Sensitivity correlation for composites: ln[Fol(Impact)×Fol(Friction)] versus ln(|ρ×OB|).

1.4.2 Transfer to Detonation in Propellants and PBXs

1.4.2.1 Factors Determining Transfer to Detonation

Solid propellant detonability has been a major concern over the years attested by violent, although rare accidents [95]. These concerns were amplified especially as one of the major methods for increasing propellant performance is by adding a high explosive as RDX or HMX to the material formulation. Adding increasing amounts of RDX to a PBAN/AP[23] propellant diminishes the minimum diameter at which a steady-state detonation can be sustained (critical diameter – d_c) [96]. Explosive disposal of HTPB/AP/Al class 1.3 propellants in large-caliber motors also resulted occasionally in full-fledged detonations [97] although routine safety test for determining explosivity as large-scale gap test (LSGT) do not respond at "0" cards and therefore these propellants were generally considered "safe".

Much effort was devoted to the experimental, theoretical and quantitative formulation, of the processes underlying deflagration to detonation transfer (DDT) [98, 99]. Most of the data and insights were gained from experimenting and modeling the process in explosives [100, 101]. The DDT process is a complicated phenomenon, from the mode of ignition through modes of combustion and system response that are naturally heavily dependent on material factors[24] [4]. However, there seems to be a consensus on the major role porous and granulated medium has on the probability that a DDT event may indeed occur. A DDT process in a solid rocket motor is usually rationalized by cracking and formation of regions of granular or porous material.

23 69%AP/15%Al/16%PBAN.

24 Among these are: gas permeability of the solid bed, which is a function of material composition, porosity, particle size and distribution, bed compaction, heat transfer rate, velocity and viscosity of the product gases, physical dimensions, and defects and weakness in the material.

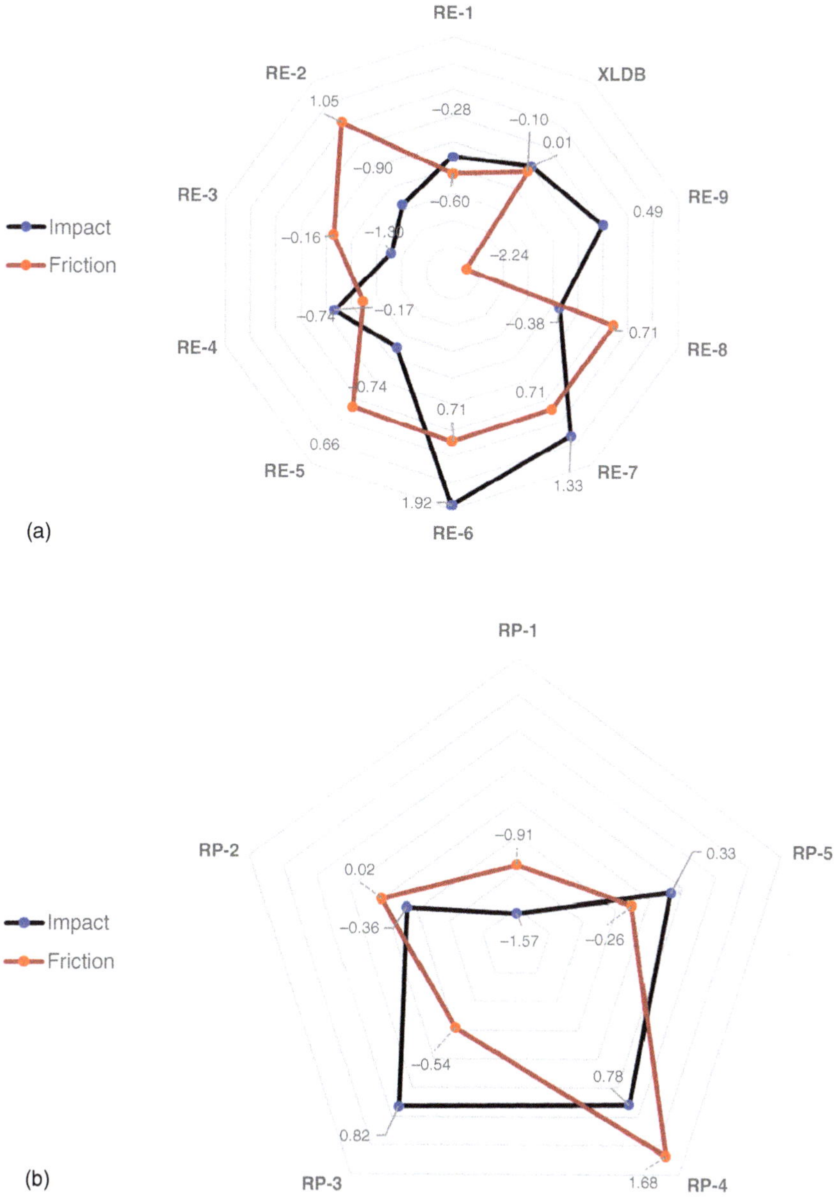

Figure 1.26 Standard score distribution of impact and friction figure of insensitivity (FoI) data for RAFAEL Explosives (RE)-(a) and RAFAEL Propellants (RP)-(b).

The key requirements for DDT to occur are a large surface-to-volume (S/V) ratio and porosity of the material. The S/V ratio combines the particle's size and shape into a single variable and both influence the material permeability and compaction that determine detonability [102, 103]. The run to detonation (L_{DDT}), the minimum length needed for a stable detonation to evolve as function of S/V

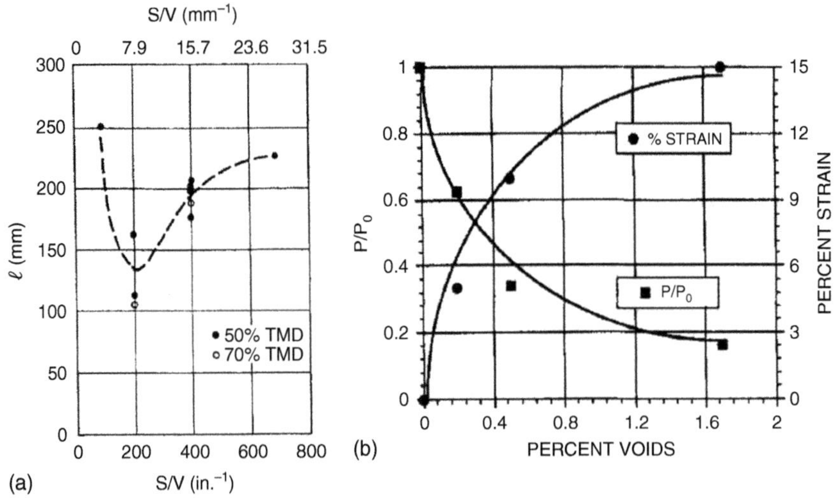

Figure 1.27 Effect of *S/V* and void content on propellant response. (a): L_{DDT} as function of *S/V* ratio. Source: Bernecker et al. [103]. Reproduced with permission of American Institute of Aeronautics and Astronautics (AIAA). (b): Increased ease of response as function of void content. Source: Boggs et al. [95]. Reproduced with permission of American Institute of Aeronautics and Astronautics (AIAA).

is shown in Figure 1.27a. L_{DDT} shows a minimum (or a maximum probability) for an intermediate *S/V* ratio. The decreased probability for detonation on both sides of the curve can be rationalized by not compact enough material leading to dissipation and to compact material not allowing pressure wave to evolve and transfer the decomposition wave (see Figure 1.29). An increased ease of response as the void content increases is demonstrated using a 1.3 classified (no mass explosion hazard) propellant with 88% solids. A 15% change in strain causes 1.7% increase in void content. This in turn resulted in ~5 time decrease in pressure needed to induce response (ignition) of the propellant, compared to unstressed material (Figure 1.27b) [95].

Composite propellants have comparable or even higher energy density than conventional explosives, however, the detonability characteristics are different. Composite propellants are heterogeneous and a prerequisite for energy release is partial pyrolysis and evaporation of the binder fuel and diffusive mixing with gaseous products of the oxidizer before the potential energy of the system can be exploited. Pores and gas pockets function as hot spots supporting detonation. However composite propellants contain only little porosity, also the decomposition kinetics during explosion of a monopropellant oxidizer are considerably slower than high explosives. This results in a low efficiency towards initiation by shock heating in the time frame necessary for propagating a detonation. Another important factor that has a major influence on propellant detonability is the geometrical shape of the propellant grain referring to the bore shape. Critical diameter is usually defined for a filled cylindrical charge, however, often propellant charges have more complicated geometries i.e. circular or a star-shaped hollow cores. For these cases, the concept of "critical geometry" instead of critical diameter was adopted. Critical geometry (δ_c) is simply the diameter of the propellant area transformed from any shape to a circular shape[25] [104]. The critical geometry from a series of go/no-go detonation experiments using PBAN/AP/8%RDX propellants with different core shapes and sizes was determined and the ratio of critical geometry to the critical diameter was determined to be 0.92 ± 0.03 thus enabling estimating the critical dimensions of a composite regardless of shape.[26] The concept of critical geometry is probably a valid method for predicting the detonability of an energetic material when the critical diameter of the material is known. Another consideration of the effects internal geometry has on the detonability of propellant grains during motor operation is the speed of material consumption reaching a subcritical dimension.

Another phenomenon that was thoroughly examined was the delayed detonation through shock, termed XDT that occurs when a projectile or metal fragment impacts a rocket motor with a perforated propellant. XDT which also refers to unknown mechanism to detonations does not only take longer to evolve (≥45 μs) compared to shock to detonation (<20 μs) but also seems to occur in materials with lower stimulus to undergo shock to detonation (XDT >160 cards) compared to SDT for the same materials in non perforated configuration (<160 cards) in the card gap test [105].

Model studies and hydrodynamic simulations have managed to give an insight into the processes and events that control XDT. The projectile or fragment penetrating the propellant at a high velocity creates a bubble of material at the rear face. Impact of the expanding bubble on the second layer of the propellant results in reaction that for sufficiently energetic material will transition to detonation, which propagates back through the dispersed bubble material. The reactions are controlled by the velocity of the impacting object-generating damage and heat to cause ignition of the propellant web and the probability of the debris reacting in the perforation before breaking up and losing most of the energy. Thus XDT is bound by the perforation dimensions and velocity of the impacting object [106, 107].

1.4.2.2 DDT Description

When a flame from a surface-deflagrating propellant reaches a high void or granulated region, it will accelerate into this medium igniting more propellant by convective heat transfer[27] and intrusive combustion from the burning front. The rising pressure generates a pressure wave that travels away from the burning front at a speed of sound in the compacted material. Early on [108] the distinction

---

25 Any surface (S) with circumference (P) translated to an equivalent critical diameter δ_c.
$\delta_c = (4S/P)_c$; For circular central perforation it can be shown that the critical web for detonation is given by $w_c = \delta_c/2$.

26 $\delta_c = 0.92 d_c$; This may of course change for different materials.

27 Convective and conductive combustions are two modes of burning of the propellant. Convective combustion is a fast rate reactive flow process where hot gases infiltrate into a porous area in the solid material. The velocity of convective burn is a function of permeability of the porous material, driving pressure gradient and viscosity and temperature of the flow. Conductive combustion is a slow rate, laminar, pressure dependent, surface combustion sustained by heat conduction form the flame to the condensed fuel.

between connected and non-connected pores influence on detonation probability was realized. Since pressure rate depends on the exposed propellant area, in composites with connected pores, gases can diffuse and increase the total burning surface and therefore increase sensitivity to shock and detonation. In grains with non-connected pores, flame spreading is much more difficult until the high pressure may cause rupture of the interpore area. This is suggestive that a temperature dependence is expected for DDT development. Lowering temperature increases brittleness of the material and new interpore area may be formed more easily, increasing sensitivity to detonation after ignition. Rubbery materials are less affected by low temperature-induced brittleness and the sensitivity is expected to be lower compared to tough materials as DB's.

For modeling DDT, three basic cases were considered by Krier et al. [109–112] according to the degree of granularity of the solid bed: (i) High granularity-substantial granularity of the grain. As the pore content is increased charge diameter required for propagating a steady-state detonation decreases. In the limit of sufficient porosity, the critical diameter is very small. This means that the local regions containing this high porosity may undergo detonation; (ii) Confined granularity–granularity is confined to a small area. It is assumed that the undamaged region is impermeable to the flow of hot gases from the burning zone of fragmented material. This implies that only stress waves can be transmitted across the interface. The impermeable material can still detonate with pressures needed for initiation of detonation smaller than the shock pressure needed for a homogenous material. Voids and discontinuities act as hot spots that initiate the non-connected porous material. However, it was postulated such localized regions of detonation will not propagate further than the porous region. Rather it is more probable that it will blow the propellant grain apart and thus provide a large portion of surface area for subsequent rapid deflagration and a thermal explosion [113]. Energetic binders that contain reactive groups (i.e. ONO_2, N_3), less gas diffusion will be required to release its energy and the time to detonation, or run to detonation lengths will decrease. DDT may occur in much smaller charge diameters than expected. This case seems to hold to most of DDT occurrences. (iii) Non-Granular materials- end burning grain that is impermeable to the flow of hot gases. Pressure rise in burning propellants is not rapid enough to cause a shock initiation but may do so in case of explosives.

DDT process can be graphically described in time–distance domain (Figure 1.28) [4]. It starts with the ignition of the composite by either a thermal or mechanical impact causing a compaction wave in the material. After intrusion of high-velocity gases produced from the initial combustion, formation of the high-density zone (~90% TMD termed "piston") is observed. Permeability of the solid bed and the rate of pressurization that is a function of pressure, accessible surface area and the burn rate exponent, control the piston acceleration, which affects the rapidity of formation and strength of the eventual shock. The superposition of compression waves causes an increase in density to ~100%TMD in a region above the combustion front and below the initial compression wave. This ~100%TMD zone (termed "plug") effectively stops the combustion wave from further advancing but the high compression and temperature result in a shock. With sufficient time, distance, and

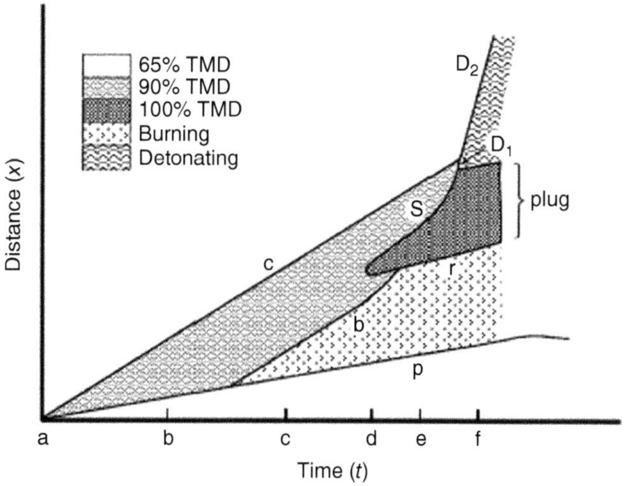

Figure 1.28 Time–distance description of a DDT process. p-piston compression (ignition pressure wave); c-compaction wave; b- burning wave; S-shock wave; r-rearward shock (retonation); D-detonation. Source: McAfee [4]. Reproduced with permission of Springer Nature.

above some critical diameter favoring further concentration of the released energy for material decomposition rather than dissipation to the surrounding a shock will sustain yielding a "full-blown" detonation.

A recent study describes in detail the DDT process in pristine and thermally damaged (185 °C → 25 °C) HMX-based PBX-9501[28] [114]. As noticed already earlier, undamaged highly consolidated low porosity[29] ($\phi_{initial}$ = 0.02) PBX 9501, did not undergo DDT. However, the thermally damaged material with increased porosity attributed to volume change due to HMX polymorphic transition ($\beta \rightarrow \delta$) and some binder degradation (Figure 1.29; $\phi_{damaged}$ = 0.05) underwent DDT. The probability for a DDT event as expressed in DDT run length (L_{DDT}) as function of porosity are schematically shown in Figure 1.29. It was suggested that the dependence of L_{DDT} in porosity is not a simple continuum but is better described by a jump in the governing mechanism controlling the phenomena. The comprehensive discussion rationalizes the different behaviors as function of porosity in terms of the possibilities for a convective and compressive-intrusive burning mode, initiating reaction by hot spot ignition and thermal explosion eventually overtaking the flame front before transitioning to detonation. In case of materials with low porosity ($\phi_{damaged}$ = 0–0.04), the bed is compacted so convective burning into the material bed cannot occur. It was proposed in this cases that compressive and deconsolidative[30] burn modes will

28 PBX 9501: 95% HMX, 2.5% Estane, 2.5% BDNPF/A.

29 Porosity = Volume fraction of voids; $\phi = \frac{\rho_{TMD} - \rho_{Geometrical}}{\rho_{TMD}}$.

30 Deconsolidative combustion is observed when a flame is advancing in a high-density low porosity compacted energetic material that causes particles or chunks of the composite propellant or explosive to detach and flow in the opposite direction to the advancing flame front. The

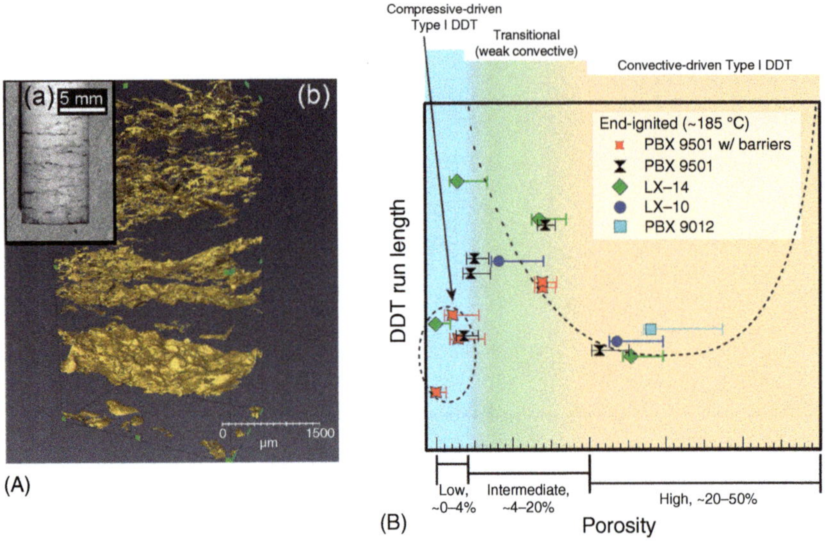

Figure 1.29 (A): Damaged PBX 9501 in a tube. (a) Large view, (b) Micro-CT **(B): L_{DDT} as function of porosity.** (Source: Parker et al. [114]. Reproduced with permission of Elsevier Science.

satisfy these conditions, amplified by hot spots formation due to debonding and filler cracking. Intermediate stressed porosity materials ($\phi_{damaged} = 0.04$–0.20) seem to have longer run to detonation distances compared to materials with low and high porosities suggesting that there is enough exploitable porosity here to permit slow, non-vigorous convective burn propagation for a considerably longer duration before the onset of thermal explosion behind the flame infiltration front. The delayed thermal runaway causing a delayed compressive burning and compaction that overtakes the flame infiltration front before transitioning to detonation. Therefore, it was concluded that the reason for longer DDT run lengths in this range of porosities is due to the extended time and distance required to produce the thermal explosion. For the high porosity ($\phi_{damaged} = 0.2$–0.5) cases, gases can infiltrate the bed efficiently and high-surface-area convective burn mode can form an impulsive thermal and physical explosion, causing effective compaction further along the grain. This porosity range is the domain of classic convective-driven type DDT. High porosity however, needs much higher pressure and time to consolidate before an efficient shock wave can form.

We have tested DDT development conditions induced by a shaped charge plasma jet for different types of propellants. A high-energy XLDB-type propellant (HMX/NE/Polyester) was compared to a conventional HTPB/AP propellant using the experimental configuration depicted in Figure 1.30a [115]. The tested propellant

---

detached material is consumed in the exhaust plume. This is a mode of increasing surface area and mass burn.

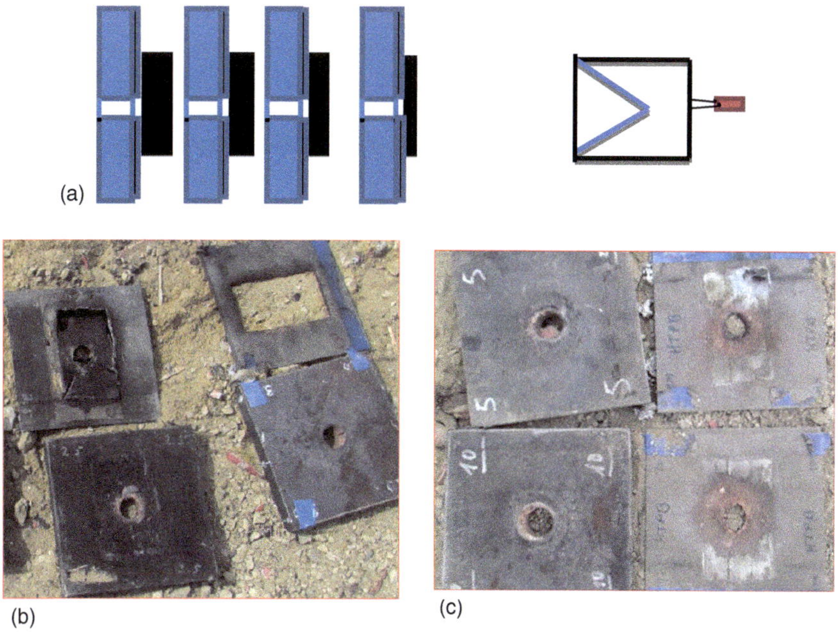

Figure 1.30 Shape Charge Plasma Jet-Induced DDT Experimental Configuration and Results. (a): Experimental Configuration. (b): Left Full Detonation in XLDB Propellant; (c): No detonation in HTPB/AP Propellants.

and a thin perforated steel plate were fastened to a thick perforated steel plate with a Styrofoam spacer (to decay reflected shockwaves). The perforation assured that the jet would only interact with the tested material. The propellant width changed gradually to determine the critical width for initiation. Firing of the shaped charge caused initiation of XLDB propellant above a certain width and not at all in the case of HTPB/AP propellant. XLDB propellant that underwent above a certain width, a full detonation, causing a complete cutting of the thin steel plate. With the lower energy propellant only a hammering effect around the perforation is observed as a result of the shock wave transfer not undergoing detonation.

1.5 Summary

Propellants and explosives are quite similar by appearance but not by response or mechanism of reaction, therefore the differences are more profound than the similarities. Table 1.6 summarizes the main differences between the two composites.

Table 1.6 Similarities and differences between propellants and explosives.

Criteria	Propellant	PBX
Appearance	Composite heterogenic mixture	Composite Heterogenic Mixture
Chemical reaction	Combustion (deflagration); intermolecular redox reaction	Combustion (detonation); intramolecular (disproportionative) redox reaction
Kinetics	Diffusion controlled	Unimolecualr controlled. Enhanced blast and thermobaric explosives contain a diffusive intermolecular metal combustion stage
Velocity @ reaction time	Subsonic (2–50 mm s^{-1}); “Long” (10^{-2}–10^{-3} s)	Supersonic (4–10 km s^{-1}); “Short” (10^{-6} s)
Maximum pressure	0.7–100 MPa	7000–70 000 MPa
Output power density	Low power: 10^6 W cm^{-3}	High power: 10^{10} W cm^{-3}
Initiation	Flame, spark, friction, high temperature	Shock wave; thermal (deflagration to detonation transition); mechanical impact
Propagation	Thermal reactions: conductive, convective, compression burning, and flame intrusion	Shock wave propagation supported by exothermic decomposition
Pressure dependence and confinement	Dependent on pressure and affected by confinement	Not dependent on external pressure heavily dependent on confinement
Transitions to detonation	Can undergo transition to detonation and explode	Can undergo transition to detonation when deflagrated in confinement. Does not undergo transition to deflagration from detonation, remains unreacted
Size	Not size dependent	Size dependent; size ≥ failure (critical) diameter
Physical effects	Object propulsions	Blast and shattering
Applications	Rocket motors, gun shells, gas generators	Demolition and destruction

List of Abbreviations

5AT	5-Aminotetrazole
6NF	6-No fire
ADN	Ammonium dinitramide
AFM	Atomic force microscopy
AMMO	Poly(3-azidomethyl-3-methyloxetane)
AN	Ammonium nitrate

AP	Ammonium perchlorate
BA	Bonding agent
BAM	Bundesanstalt fur Materialfprufung und Forshung
BAMO	Bis(azidomethyl)oxetane
BDNPF/A	Bis(2,2-dinitro propyl)formal/acetal
BDP	Beckstead-Derr-Price
CAE	Chemical equilibrium with applications
CMDB	Composite modified double base
CPL	Polycaprolactone
CTBN	Carboxy terminated polybutadiene acrylonitrile copolymer
CTPB	Carboxy terminated polybutadiene
DB	Double base
DDT	Deflagration to detonation transfer
DOA	Dioctyl adipate
DOP	Dioctyl phtalate
EBX	Enhanced blast explosive
EMs	Energetic materials
FEFO	Bis(2-fluoro-2,2-dinitroethyl)formal
FG	Functional group
GAP	Glycidyl azide polymer
GUDN	Gunylurea dinitramide
HBP	Hyper branched polymers
HC	Hydrocarbon
HMX	1,3,5,7-tetranitro-1,3,5,7-tetrazacyclooctane (octogen)
HNF	Hydrazinium nitroformate
HNS	Hexanitro stilbene
HTPB	Hydroxy terminated polybutadiene
IDP	Isodecyl pelargonate
IM	Insensitive munitions
KN	Potassium nitrate
KP	Potassium perchlorate
LSGT	Large scale gap test
MAPO	Tris-(2-methyl-1-aziridinyl) phosphine oxide
MD	Molecular dynamics
NC	Nitro cellulose
NE	Nitrate esters
NENA	2-Nitroxy ethyl nitramine
NPBA	Neutral polymeric bonding agents
NQ	Nitro quanidine
NTO	3-Nitro-1,2,4-triazol-5-One
OB	Oxygen balance
PBAA	Polybutadiene acrylic acid
PBAN	Polybutadiene acrylic acid acrylonitrile
PBXs	Plastic bonded explosives
PCL	Polycapro lactone

PDA	Polydopamine
PDMS	Polydimethyl siloxane
PE	Poly ethylene
PEG	Poly ethylene glycol
PGN	Poly glycidyl nitrate
PPG	Poly propylene glycol
QSPR	Quantitative structure-property relationships
RDX	1,3,5-Trinitro-1,3,5-triazacyclohexane (hexogen)
SAMs	Surface active materials
SDT	Shock to detonation transition
SRM	Solid rocket motor
TAGN	Tri amino guanidine nitrate
TATB	1,3,5-Triamino-2,4,6-trinitrobenzene
TBX	Thermobaric explosives
TMD	Theoretical maximum density
TMETN	Trimethylolethane trinitrate
TNT	1,3,5-Trinitrotoluene
WA	Wetting agent
XDT	Delayed detonation through shock
XLDB	Cross linked double base
ZND	Zel'dovich-von Neumann Doring detonation model

References

1 Davenas, A. (1993). The main families and use of solid propellants. In: *Solid Rocket Propulsion Technology* (ed. A. Davenas), 329–367. Oxford: Pergamon Press.

2 Lee, P.R. (1998). Hazard assessment of explosives and propellants. In: *Explosive Effects and Applications* (ed. J.A. Zukas and W.P. Walters), 259–339. New-York: Springer-Verlag.

3 Koch, E.-C. (2021). Held, Manfred (1933-2011). In: *High Explosives Propellants Pyrotechnics*, 712. Berlin: De Gruyter.

4 McAfee, J.M. (2010). The deflagration-to-detonation transition. In: *Shock Wave Science and Technology Reference Library*, vol. 5 (ed. B.W. Asay), 483–535. Berlin, Springer-Verlag: *Non Shock Initiation of Explosives.*

5 Beckstead, M. 2000. An overview of combustion mechanisms and flame structures for advanced soild propellants. In: *36th AIAA/AMSE/SAE/ASEE Joint Propulsion Conference and Exhibit*, Huntsville.

6 Beckstead, M.W. (1993). Solid propellants combustion mechanisms and flame structure. *Pure and Applied Chemistry* 65: 297–307.

7 Lee, J.H. (2008). *The Detonation Phenomenon.* Cambridge: Cambridge University Press.

8 Ksoiba, G.D., Wixom, R.R., and Oehlschlaeger, M.A. (2017). High-fidelity microstructural characterization and performance modeling of aluminized composite propellant. *Propellants, Explosives, Pyrotechnics* 42: 1387–1395.

9 Rae, P.J., Goldrin, H.T., Palmer, S.J. et al. (2002). Quasi-static studies of the deformation and failure of beta-HMX based polymer bonded explosives. *Proceedings of the Royal Society of London A* 458: 743–762.

10 Trache, D., Klapotke, T.M., Maiz, L. et al. (2017). Recent advances in new oxidizers for solid rocket propulsion. *Green Chemistry* 19: 4711–4736.

11 Zhao, G., Yin, P., Kumar, D. et al. (2019). Bis(3-nitro-1-(trinitromethyl)-1H-1,2,4-triazol-5-yl)methanone: an applicable and very dense oxidizer. *Journal of the American Chemical Society* 141: 19581–19584.

12 Dalinger, I.L., Suponitsky, K.Y., Shkineva, T.K. et al. (2018). Bipyrazole bearing ten nitro groups-a-novel highly dense oxidizer for forward-looking rocket propulsions. *Journal of Materials Chemistry A* 6: 14780–14786.

13 Dalinger, I.L., Shkineva, T.K., Vatsadze, I.A. et al. (2021). Novel energetic CNO oxidizer: pernitro-substituted pyrazole-furazan framework. *FirePhysChem* 1: 83–89.

14 Klapotke, T.M. and Suceska, M. (2021). Theoretical evaluation of TKX-50 as an ingredient in rocket propellants. *Zeitschrift für Anorganische und Allgemeine Chemie* 647: 572–574.

15 Bao, P., Li, J., Han, Z. et al. (2021). Comparing the impact safety between two HMX-based PBX with different binders. *FirePhysChem* 1: 139–145.

16 Li, X., Chen, S., Wang, X. et al. (2017). Effect of polymer binders on safety and detonation properties of epsilon-CL-20 bases pressed-polymer-bonded explosives. *Materials Express* 7: 209–215.

17 Umholtz, P. D. (1999). The history of solid rocket propulsion and aerojet. In: *35th AIAA/ASME/SAE/ASEE Joint Propulsion Conference and Exhibit*, Los Angels.

18 DeLuca, L.T. (2017). Highlights of solid rocket propulsion history. In: *Chemical Rocket Propulsion* (ed. L. DeLuca, T. Shimada, V. Sinditskii and M. Calabro), 1015–1032. Springer.

19 Last edited 22 June 2022. Jack Parsons (Rocket Engineer). en.wikipedia.org./wiki/Jack_Parsons_(rocket engineer) (Accessed 2021).

20 Ang, H.G. and Pisharath, S. (2012). Polymers as binders and plasticizers-historical perspective. In: *Energetic Polymers and plasticizers for Enhancing Performance*, 1–17. Weinheim: Wiley-VCH.

21 Mastrolia, E.J. and Klager, K. (1969). Solid propellants based on polybutadiene binders. In: *Propellant Manufacture, Hazard and Testing* (ed. C. Boyers and K. Klager), 122–164. Washington: American Chemical Society.

22 Ajaz, A.G. (1995). Hydroxyl-terminated polybtadiene telechelic polymer (HTPB): binder for solid rocket propellants. *Rubber Chemistry and Technology* 68: 481–506.

23 Lynch, R. (1990). Development of insensitive high explosives using propellant technology. In: *AIAA/SAE/ASME/ASEE 26 Joint Propulsion Conference*. FL: Orlando.

24 T. Yarom, A. Kalkstein, M. Rav-Hon and S. Mandelbaum, "Design and Testing of Underwater IM Explosive Charge". www.Ark/Mega_Folder/Explosive/MISC/NDIA.org_Insensitive_Munitions_and_Energetic_Materials_Symposium_2000/1550/designandtestingpdf.pdf.

25 Paraskos, A.J. (2017). Energetic polymers: synthesis and applications. In: *Energetic Materials. Challenges and Advances in Computational Chemistry and Physics* (ed. M. Shukla, V. Boddu, J. Steevens, et al.), 91–134. Springer.

26 Badgujar, D.M., Talawar, M.B., Zarko, V.E., and Mahulikar, P.P. (2017). New directions in the area of modern energetic polymers: an overview. *Combustion, Explosion and Shock Waves* 53: 371–387.

27 Cheng, T. (2019). Review of novel energetic polymers and binders – high energy propellant for the new space race. *Designed Monomers and Polymers* 22: 54–65.

28 Sevilia, S., Young, M., Grinstein, D. et al. (2019). Novel, printable energetic polymers. *Macromolecular Materials and Engineering* 304: 1900018.

29 Bower, J.K., Kolb, J.R., and Pruneda, C.O. (1980). Polymeric coatings effect on surface activity and mechanical behavior of high explosives. *Industrial and Engineering Chemistry Product Research and Development* 19: 326–329.

30 Gee, R.H., Maiti, A., Bastea, S., and Fried, L.L. (2007). Molecular dynamics investigation of adhesion between TATB surface and amorphous fluoropolymers. *Macromolecules* 40: 3422–3428.

31 Xiao, G., Fang, G., Ji, G., and Xiao, H. (2005). Simulation investigations in the binding energy and mechanical properties of HMX-based polymer-bonded explosives. *Chinese Science Bulletin* 1: 21–26.

32 Tao, J. and Wang, T. (2017). Molecular dynamic simulation for fluoropolymers applied in epsilon-CL-20- based explosive. *Journal of Adhesion Science and Technology* 31: 250–260.

33 Gee, R.H., Maiti, A., Bastea, S., and Fried, L.E. (2007). Molecular dynamic investigation of adhesion between TATB surface and amorphous fluoropolymers. *Macromolecules* 40: 3422–3428.

34 Taylor, D.E., Strawhecker, K.E., Shanholtz, E.R. et al. (2014). Investigations of the intermolecular forces between RDX and polyethlene by force-distance spectroscopy and molecular dynamics simulations. *The Journal of Physical Chemistry. A* 118: 5083–5097.

35 Sikdar, A.K. and Reddy, S. (2013). Review on energetic thermoplastic elastomers (ETPEs) for military science. *Propellants, Explosives, Pyrotechnics* 38: 14–28.

36 Agrawal, J.P. (2010). *High Energy Materials*. Weinheim: Wiley- VCH.

37 Vadhe, P.P., Pawar, R.B., Sinha, R.K. et al. (2008). Cast aluminized explosives (review). *Combustion, Explosion, Shock Waves* 44: 461–477.

38 Gottlieb, L. and Bar, S. (2003). Migration of plasticizer between bonded propellant interfaces. *Propellants, Explosives, Pyrotechnics* 28: 12–17.

39 Dobratz, B.M. and Crawford, P.C. (1985). *LLNL Explosives Handbook: Properties f Chemical Explosives and Explosive Stimulants*. Livermore CA: Lawrence Livermore National laboratory.

40 Keicher, T., Happ, A., Kretschmer, A. et al. (1999). Influence of aluminium/ammonium perchlorate on the performance of underwater explosives. *Propellants, Explosives, Pyrotechnics* 24: 140–143.

41 Pang, W.Q., DeLuca, L.T., Wang, K. et al. (2021). Effect of hydroborate iron additives (BH-Fe) on the properties of composite solid rocket propellants. *Journal of Physics: Conference Series* 1721: 012006.

42 Lysien, K., Stolarczyk, A., and Jarosz, T. (2021). Solid propellant formulations: a review of recent progress and utilized components. *Materials* 14: 6657.

43 Babuk, V.A., Vassiliev, V.A., and Sviridov, V.V. (2001). Propellant formulation factors and metal agglomeration in combustion of aluminized solid rocket propellant. *Combustion Science and Technology* 163: 261–289.

44 Zhang, C., Luo, Y.-J., Jiao, Q.-J. et al. (2014). Application of the BAMO-AMMO alternative block energetic thermoplastic elastomer in composite propellants. *Propellants, Explosives, Pyrotechnics* 39: 689–693.

45 Kalman, K. and Essel, J. (2017). Influence of particle size on the combustion of CL-20/HTPB propellants. *Propellants, Explosives, Pyrotechnics* 42: 1–8.

46 Fu, S.Y., Feng, X.Q., Lauke, B., and Mai, Y.W. (2008). Effect of particle size, particle/matrix interface adhesion and particle loading on mechanical properties of particulate-polymer composites. *Composites: Part B* 39: 933–961.

47 Bellerby, J.M. and Kiriratnikom, C. (1989). Explosive-binder adhesion and dewetting in nitramine-filled energetic materials. *Propellants, Explosives, Pyrotechnics* 14: 82–85.

48 Allen, H.C. (1983) Bonding agents for nitramines in rocket propellants. US Patent 438263, 21 June 1983.

49 Oberth, A.E. and Bruenner, R.S. (1965). Tear phenomena around solid inclusions in castable elastomers. *Transactions of the Society of Rheology* 9: 165–185.

50 Stockelhuber, K.W., Svistkov, A.S., Pelevin, A.G., and Heinrich, G. (2011). Impact of filler surface modification on large scale mechanics of styrene butadiene/silica rubber composites. *Macromolecules* 44: 4366–4381.

51 van-der Hijden, A., ter Horst, J., Kenddrick, J. et al. (2005). Crystallization. In: *Energetic Materials* (ed. U. Teipel), 53–157. Weinheim: Wiley-VCH.

52 Oberth, E. A. and Bruenner, R. S. (1976). Bonding agents for polyurethanes. US Patent 4000023, 28 December 1976.

53 Singh, A., Singla, P., Sahoo, S.C., and Soni, P.K. (2020). Compatibility and thermal decomposition behavior of an epoxy resin with some energetic compounds. *Journal of Energetic Materials* 38: 432–444.

54 Quacchia, R.H., Johnson, D.E., and Di Milo, A.J. (1967). Rearrangement of 1-benzoyl-2-ethylaziridine on solid inorganic potassium and ammonium salts. *I&EC Product Research and Development* 6: 268–273.

55 Wei, H., Shenhui, L., Shenliang, X., Gen, T., and Hongxu, L. (2008). Investigation on the Interaction Between Bonding Agent MAPO and Oxidizer AP. In: *39th International Annual Conference of ICT- Frauhnhofer Institut fur Chemische Technologie*, Karlsruhe.

56 Tomalia, D.A. and Sheetz, D.P. (1966). Homopolymerization of 2-alkyl- and 2-aryl-2-oxazoline. *Journal of Polymer Science Part A-1* 4: 2253–2265.

57 Broline, B. M. (1985). Acylaziridine Reactivity in Liner and Propellant Environments. 11 October 1985. apps.dtic.mil/sti/citations/ADP004985 (accessed 2020).

58 Hasegawa, K., Takizuka, M., and Fukuda, T. (1983). Bonding agents for AP and nitramine/HTPB composite propellants. In: *AIAA/SAE/ASME 19th Joint Propulsion Conference*. Washington: Seattle.

59 Mastrolia, E.J. and Klager, K. (1969). Solid propellants based on polybutadiene binders. In: *Propellants Manufacture Hazards and Testing*, vol. 88 (ed. C. Boyars and K. Klager), 122–164. Washington DC: American Chemical Society.

60 Seiamareli, J., Holanda, J. A., Dutra, R. C., Lourenco, V. L. and Iha, K. (2003). RX bonding agent – study of its natural ageing. In: *34th Int. Ann. Conf. ICT*, Karlsruha, 2003.

61 Yadav, A., Pant, C.H., and Das, S. (2020). Research Advances in Bonding Agents for Composite Propellants. *Propellants, Explosives, Pyrotechnics* 45: 1–11.

62 Kim, C.S., Youn, C.H., Noble, P.N., and Gao, A. (1992). Development of neutral polymeric bonding agents for propellants with polar composites filled with organic nitramine crystals. *Propellants, Explosives, Pyrotechnics* 17: 38–42.

63 Kim, C.S., Noble, P.N., Youn, C.H. et al. (1992). The mechanism of filler reinforcement from addition of neutral polymeric bonding agents to energetic polar propellants. *Propellants, Explosives, Pyrotechnics* 17: 51–58.

64 Landsem, E., Jensen, T.L., Hansen, F.K. et al. (2012). Neutral Polymeric Bonding Agents (NPBA) and their use in smokeless composite rocket propellants based on HMX-GAP-BuNENA. *Propellants, Explosives, Pyrotechnics* 37: 581–591.

65 Yan, Q.L., Zeman, S., and Elbeih, A. (2012). Recent advances in thermal analysis and stability evaluation of insensitive plastic bonded explosives (PBXs). *Thermochica Acta* 537: 1–12.

66 Bailey, A., Bellerby, J.M., Kinloch, S.A. et al. (1992). The identification of bonding agents for TATB/HTPB polymer bonded explosives. *Philosophical Transactions of the Royal Society of London. Series A: Physical and Engineering Sciences* 339: 321–333.

67 Lin, C., Gong, F., Yang, Z. et al. (2019). Core-shell structured HMX@Polydopamine energetic microspheres: synergistically enhanced mechanical, thermal and safety performance. *Polymers* 11: 568.

68 He, G., Yang, Z., Pan, L. et al. (2017). Bioinspired interfacial reinforcement of polymer-based energetic composites with high loading of solid explosive crystals. *Journal of Materials Chemistry A* 5: 13499–13510.

69 Lin, C., Wen, Y., Huang, X. et al. (2020). Tuning the mechanical performance efficiently of various LLM-105 based polydopamine modification. *Composite Part B* 186: 107824.

70 Zeng, C., Lin, C., Zhang, J. et al. (2019). Grafting hyperbranched polyester on the energetic crystals: enhanced mechanical properties in highly-loaded polymer based composites. *Composites Science and Technology* 184: 107842.

71 Lin, C., Zeng, C., Wen, Y. et al. (2020). Litchi-like core-shell HMX@HPW@PDA microparticles for polymer-bonded energetic composites with low sensitivity and high mechanical properties. *Applied Materials & Interfaces* 12: 4002–4013.

72 Yang, X., Gong, F., Zhang, K. et al. (2021). Enhanced creep resistance and mechanical properties for CL-20 and FOX-7 based PBXs by crystal surface modification. *Propellants, Explosives, Pyrotechnics* 46: 572–578.

73 Lin, C., Liu, J., He, G. et al. (2015). Non-linear viscoelastic properties of TATB-based polymer bonded explosives modified by a neutral polymeric bonding agent. *RSC Advances* 5: 35811–35820.

74 Li, F., Ye, L., Nie, F., and Liu, Y. (2007). Synthesis of boron-containing coupling agents and its effect on the interfacial bonding of fluoropolymer/TATB composite. *Journal of Applied Polymer Science* 105: 777–782.

75 Hamilton, B.W., Steele, B.A., Sakano, M.N. et al. (2021). Predicted reaction mechanisms, products speciation, kinetics and detonation properties of the insensitive explosive 2,6-diamino-3,5,-dinitropyrazine-1-oxidde (LLM-105). *The Journal of Physical Chemistry. A* 125: 1766–1777.

76 Suceska, M. (1995). *Test Methods for Explosives*. New York: Springer Verlag.

77 Coffey, C.S. and DeVost, V.F. (1995). Impact testing of explosives and propellants. *Propellants, Explosives, Pyrotechnics* 20: 105–115.

78 Field, J.E. (1992). Hot spot mechanisms for explosives. *Accounts of Chemical Research* 25: 489–496.

79 Wally, S.M., Field, J.E., Biers, R.A. et al. (2015). The use of glass anvils in drop-weight studies of energetic materials. *Propellants, Explosives, Pyrotechnics* 40: 361–365.

80 Balzer, J.E., Siviour, C.R., Walley, S.M. et al. (2004). Behaviour of ammonium perchlorate based propellants and a polymer-bonded explosive under impact loading. *Proceedings of the Royal Society of London A* 460: 781–806.

81 Mathieu, D. (2016). Physics-based modeling of chemical hazard in regulatory framework: comparison with quantitative structure-property relationship (QSPR) methods for impact sensitivities. *Industrial and Engineering Chemistry Research* 55: 7569–7577.

82 Mathieu, D. and Alaime, T. (2014). Predicting impact sensitivities of nitro compounds on the basis of a semi-empirical rate constant. *The Journal of Physical Chemistry. A* 118: 9720–9726.

83 Bondarchuk, S.V. (2018). Quantification of impact sensitivity based on solid-state derived criteria. *The Journal of Physical Chemistry. A* 122: 5455–5463.

84 Jensen, T.L., Moxnes, J.F., Unneberg, E., and Christensen, D. (2020). Models for predicting impact sensitivity of energetic materials based on the triger linkage hypothesis and arrhenius kinetics. *Journal of Molecular Modeling* 26: 65.

85 Joy, J., Danovich, D., and Shaik, S. (2021). Nature of the trigger linkage in explosive materials is a charge-shift bond. *The Journal of Organic Chemistry*.

86 Dlott, D.D. and Fayer, M.D. (1990). Shocked molecular solids: vibrational up pumping, defect hot spot formation and the onset of chemistry. *The Journal of Physical Chemistry* 92: 3798–3812.

87 Bernstein, J. (2018). Ab initio study of energy transfer rates and impact sensitivities of crystalline explosives. *The Journal of Chemical Physics* 148: 084502.

88 Tian, B., Xiong, Y., Chen, L., and Zhang, C. (2018). Relationship between the crystal packing and the impact sensitivity of energetic materials. *CrystEngComm* 20: 837–848.

89 Yan, Q.L. and Zeman, S. (2012). Theoretical evaluation of sensitivity and thermal stability for high explosives based on quantum chemistry methods: a brief review. *International Journal of Quantum Chemistry* 1–13.

90 Wickham, J.A., Beaudoin, S.P., and Son, S.F. (2020). The Role of Adhesion and Binder Stiffness in the Impact Sensitivity of Cast Composite Energetic Materials. *Journal of Applied Physics* 128: 214902-1–214902-7.

91 Brochu, D., Abou-Rachid, H., Soldera, A., and Brisson, J. (2017). Sensitivity of polymer-bonded explosives from molecular modeling data. *International Journal of Energetic Materials and Chemical Propulsion* 16: 367–382.

92 Ek, S. (1976). Sensitivity of explosive substances a multivariate approach. In: *Proceedings of the 6th Symposium (International) on Detonation*, 272–280. Coronado: California, Office of Naval Research-Department of Navy.

93 Licht, H. (2000). Performance and sensitivity of explosives. *Propellants, Explosives, Pyrotechnics* 25: 126–132.

94 Akhavan, J. (1998). *The Chemistry of Explosives.* The Royal Society of Chemistry.

95 Boggs, T.L., Atwood, A.I., Lindfors, A.J. et al. (2000). Hazard associated with solid propellants. In: *Solid Propellant Chemistry, Combustion and Motor Interior Ballistics*, Reston, Virginia (ed. V. Yang, T.B. Brill and W.-Z. Ren), 221–264. American Institute of Aeronautics and Astronautics.

96 Mellor, A., Boggs, T.L., Covino, J. et al. (1988). Hazard initiation in solid rocket and gun propellants and explosives. *Progress in Energy and Combustion Science* 14: 213–244.

97 Merrill, C. (1992). Large Class 1.3 Rocket Motor Detonation Character. apps.dtic.mil/sti/citations/ADA529554 (Accessed 2022).

98 Bdzil, J.B., Menikof, R., Son, S.F. et al. (1999). Two-phase modeling of deflagration-to-detonation transition in granular materials: a critical examination of modeling issues. *Physics of Fluids* 11: 378–402.

99 Baer, M.R. and Nunziato, J.W. (1986). A two phase mixture theory for the deflagration-to-detonation transition (DDT) in reactive granular materials. *International Journal of Multiphase Flow* 12: 861–889.

100 Gifford, M.J., Luebcke, P.E., and Field, J.E. (1999). A new mechanism for deflagration-to-detonation in porous granular explosives. *Journal of Applied Physics* 86: 1749–1753.

101 Veerbek, R. and van der Steen, A. (1990). Linking experimental and theoretical results on the deflagration to detonation transition. *Propellants, Explosives, Pyrotechnics* 15: 35–37.

102 A. G. Butcher, B. D. Hopkins and N. J. Robinson (1979). Fundamental experiments on DDT. In: *15th JANNAF Combustion Meeting*, Newport, Rhode Island.

103 Bernecker, R.R. (1984). The deflagration-to-detonation transition process for high-energy propellant-a review. *AIAA Journal* 24: 82–91.

104 Elwell, R.B., Irwin, O.R., Salzman, P.K., and Valor, N.H. (1968). Recent progress in evaluating the detonation characteristics of solid composite propellant. *AIAA Journal* 6: 489–796.
105 Brunet, J. (1993). Safety characteristics of solid propellants and hazards of solid rocket motors. In: *Solid Rocket Propulsion Technology* (ed. A. Davenas), 303–327. Elsevier.
106 Finnegan, S.A., Pringle, J.K., Schulz, J.C. et al. (1993). Impact-induced delayed detonation in an energetic material debris bubble formed at an air gap. *International Journal of Impact Engineering* 14: 241–254.
107 Finnegan, S.A., Pringle, J.K., Atwood, A.I. et al. (1995). Characterization of impact-induced violent reaction behavior in solid rocket motors using a planar motor test model. *International Journal of Impact Engineering* 17: 311–322.
108 Amster, A.B., Noonan, E.C., and Bryan, G.J. (1960). Soild propellant detonability. *American Rocket Society Journal (ARS Journal)* 30: 960–964.
109 Butler, P.B., Lembeck, M.F., and Krier, H. (1982). Modeling of shock development and transition to detonation initiated by burning in porous propellant beds. *Combustion and Flame* 46: 75–93.
110 Krier, H., Butler, P.B., and Coyne, D.W. (1982). *Modeling of Deflagration to Shock to Detonation Transition (DSDT) in Porous High Energy Soild Propellants and Explosives*. Urbana: Air Force Office of Scientific Research (AFSC).
111 Krier, H., Butler, B.P., and Cudak, C. (1984). *Deflagration to Shock to Detonation Transition of Energetic Propellants*. Urbana, IL: Air Force Office of Scientific Research.
112 Krier, H. and Stewart, J.R. (1985). *Prediction of Detonation Transition in Porous Explosives from Rapid Compression Loadings*. Urbana, IL: Air Force Office of Scientific Research.
113 Salzman, P.K., Irwin, O.R., and Andersen, W.H. (1965). Theoretical detonation characteristics of solid composite propellants. *AIAAJ* 3: 2230–2238.
114 Parker, G.R., Heatwole, E.M., Holmes, M. d. et al. (2020). Deflagration-to-detonation transition in hot HMX and HMX-based polymer-bonded explosives. *Combustion and Flame* 215: 295–308.
115 RAFAEL (2015). Internal Report. Shaped Charge Jet Induced Detonation of XLDB Propellant. Unpublished Results, Haifa.

2

High-Pressure Combustion Studies of Energetic Materials

Richard A. Yetter and Eric Boyer

The Pennsylvania State University, Department of Mechanical Engineering, University Park, PA, USA

2.1 Introduction

The dependence of burning rate on pressure is of fundamental importance to ballistic properties. The linear burning rate of a propellant or explosive is often described over a specific pressure range by the empirical equation $r_b = aP^n$. The parameter a is often considered a function of initial temperature, whereas the exponent n is independent of temperature and related to the overall reaction order. Burning rate data of propellants and explosives have been measured for years using various experimental techniques. For example, the photocinemicrographic and closed bomb combustion methods are two common approaches for measuring the burning rate. In the photocinemicrographic method, the energetic material sample, usually in the form of a strand, is pressurized in the combustion vessel to a desired pressure, and a video of the sample regression is recorded for analysis of the burning rate. A small sample is used such that the partial volume of gas produced from combustion does not contribute significantly to the overall volume of the chamber, and thus, the sample is assumed to burn at nearly constant pressure. The upper limit for these windowed chamber experiments has generally been about 50 MPa, although most of the limited data available are for 7 MPa and lower. For pressures higher than 50 MPa, the closed bomb technique is used in which the sample, in the form of a powder or strand, burns in a relatively small volume without observation. Pressure time data are used to deduce the burning rate from a model developed to describe the experiment that must account for variable thermochemistry, heat losses, ignition, and flame spreading within the sample. The methods for measuring burning rates have been reviewed in several papers, e.g. [1–3]. Considerable effort is devoted to reducing the uncertainties in these measurements, as well as to extending their usage for example from steady state to transient conditions. Most of the measurement techniques have been described for burning of solid propellants but are generally equally applicable to liquid propellant columns in tubes.

In the present chapter, the goal is not to review the various measurement techniques but to emphasize first the importance of studying the burning process under

Nano and Micro-Scale Energetic Materials: Propellants and Explosives, First Edition.
Edited by Weiqiang Pang and Luigi T. DeLuca.

extreme conditions, that is, under conditions beyond the propellant's intended usage to further develop fundamental understanding of combustion under all conditions, and secondly, the importance of visually observing the combustion process to this understanding and obtaining accurate burning rate measurements. The chapter begins by presenting burning rate measurements of various liquid and solid propellants, followed by visual observations of their combustion behavior, and finally a discussion of their pressure-dependent characteristics.

2.2 Burning Rates as a Function of Pressure

Examples of the burning rate dependence on pressure of several monopropellants are shown in Figures 2.1–2.3. Figure 2.1 shows the burning rates of octahydro-1,3,5,7-tetranitro-1,3,5,7-tetrazocine (HMX) and 1,3,5-trinitroperhydro-1,3,5-triazine (RDX) as a function of pressure, where it is observed that the empirical burning rate equation can nearly correlate the data with a single pressure exponent. However, near 12 MPa, some variation in the value of n exists in these measurements and a new value of "a" is attained. Three different experiments were required to achieve the entire pressure range for HMX [4].

Figure 2.2 shows regression rate data for ammonium perchlorate (AP) at ambient temperature where the low-pressure deflagration limit is shown to be approximately 2 MPa. The negative slope of the curve between ~12 and 25 MPa has been explained by unstable combustion and changes in the chemical mechanism. Above 25 MPa, the burning rate again increases with pressure, having a greater pressure exponent than the pressure dependence below 12 MPa.

The pressure dependence below the self-deflagration limit can be achieved from counterflow burning rate experiments as illustrated in Figure 2.3, in which hydrocarbon gases are counter flowed against the decomposition gases from the AP monopropellant flame to produce a stable diffusion flame that supports the

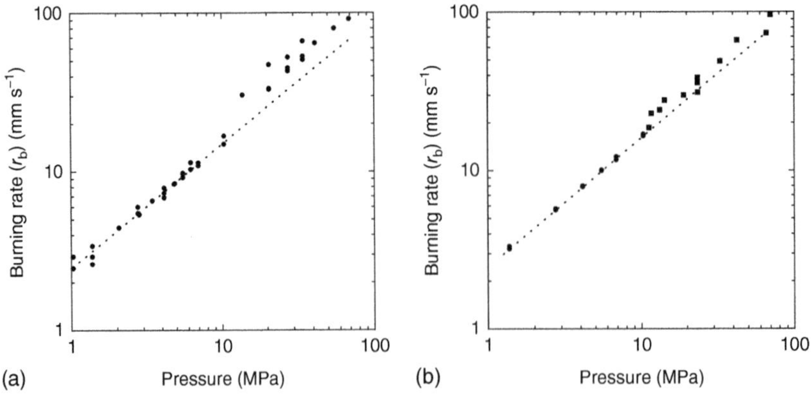

Figure 2.1 (a) HMX burning rate at ambient temperature; (b) RDX burning rate at ambient temperature. Source: Data from Atwood et al. [4]. The dashed lines are extrapolations from the low-pressure data.

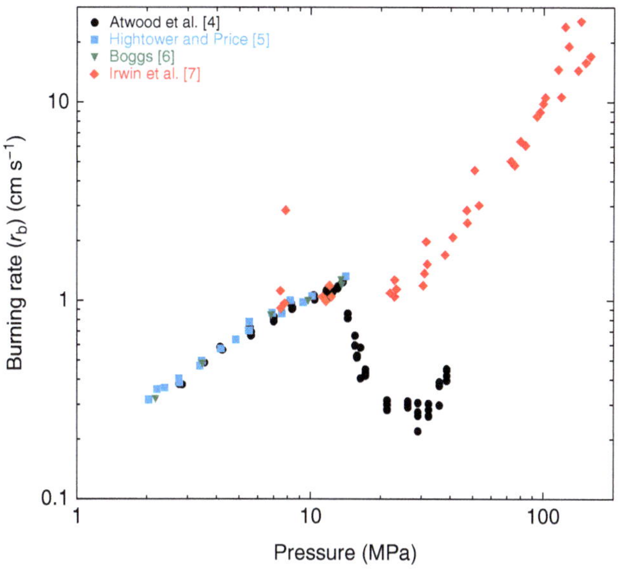

Figure 2.2 AP burning rate at ambient temperature. Source: Data from Atwood et al. [4], Hightower and Price [5], Boggs [6], and Irwin et al. [7].

monopropellant flame. As seen in the figure, between 0.1 and 1 MPa, the pressure exponent is nearly the same as above the self-deflagration limit from 2 to 12 MPa, whereas the pre-exponential factor a is about a factor of 5 lower. The range of burning rate values at a specific pressure, such as 0.1, 0.45, or 0.85 MPa, is a result of varying the strain rate by approximately a factor of 4 [8] illustrating the relative importance of strain rate versus pressure in the figure. In between 1 and 2 MPa, Figure 2.3 shows a high-pressure exponent. It is interesting to note that the AP burning rate in the counterflow experiment essentially represents the burning rate of an infinite diameter particle in an AP-based composite propellant. However, the experiment illustrates how modifications to the design can be used to obtain more information at extreme conditions, in this case, low pressures, on burning behavior to aid in developing models.

Two examples of burning rates for composite propellants with AP or HMX as oxidizers and hydroxyl-terminated polybutadiene (HTPB) as the fuel binder are shown in Figure 2.4. A slope break is observed in the vicinity of 20 MPa for both composite propellants, which is near the pressures at which the monopropellant oxidizers themselves exhibit a slope break. These two composite propellants were formulated with large oxidizer particles having nominal diameters of 90 μm for AP and 195 μm for HMX. Large oxidizer particles can magnify the break, whereas smaller particles appear to wash out the break. Composite propellants with small oxidizer particles generally yield higher burning rates at lower pressures, and a decrease in the pressure exponent at the transition is often observed rather than an increase as shown in Figure 2.4. For composite propellants with a bimodal or greater particle distribution, the large particles, which are necessary for processibility, have been reported to be the cause of the slope break.

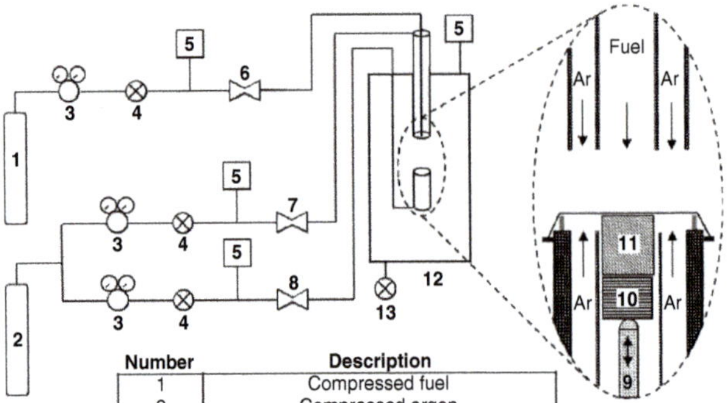

Number	Description
1	Compressed fuel
2	Compressed argon
3	Two-stage pressure regulator
4	Electro-pneumatic solenoid valve
5	Setra pressure transducer
6	Fuel orifice
7	Fuel co-flow orifice
8	Oxidizer co-flow orifice
9	Linear variable displacement transducer
10	Teflon spacer
11	Oxidizer pellet
12	Pressure vessel
13	Exhaust valve

(a)

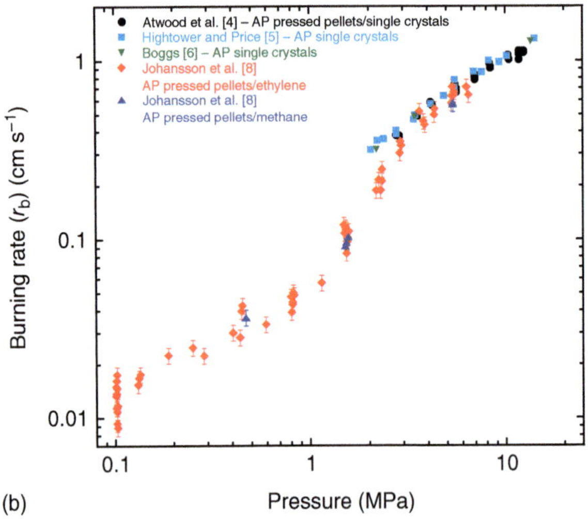

(b)

Figure 2.3 (a) Experimental facility for pressurized counterflow strand burning measurements. (b) AP burning rate below the low-pressure self-deflagration limit supported by counterflow diffusion flame. Note at pressures above the self-deflagration limit, the AP burning rate reestablishes the values of Figure 2.2 [8]. Source: Data from Atwood et al. [4], Hightower and Price [5], Boggs [6], and Johansson et al. [8].

The burning rates of JA2 propellant, which is composed of nitrocellulose, nitroglycerin, and diethylene glycol dinitrate, are shown in Figure 2.5 for pressures between 10 and 600 MPa. These measurements were taken in closed bomb experiments with break wires (or pressurization) and from visual observations of the regressing burning surface (ultra-high-pressure optical chamber [UHPOC] and optical strand burner [OSB]). Although the burning rate of JA2 appears to generally

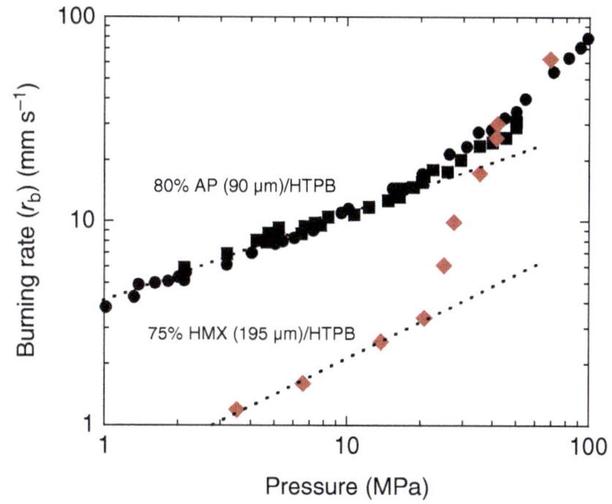

Figure 2.4 Burning rates of two composite propellants consisting of AP/HTPB [9] or HMX/HTPB [10]. The dashed lines are extrapolations from the low-pressure data.

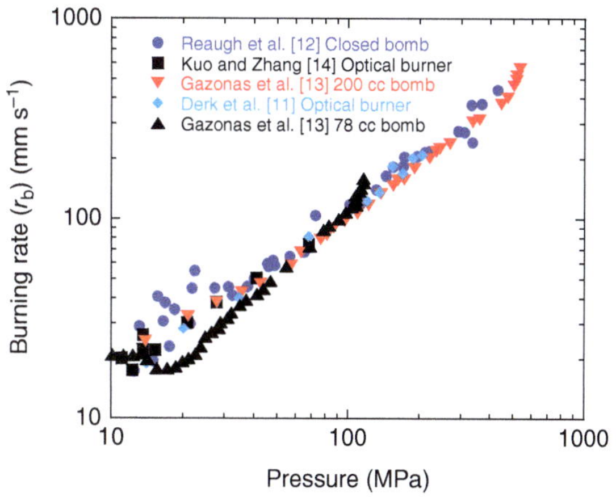

Figure 2.5 Burning rates of JA2 as a function of pressure. Source: Data from Derk et al. [11], Reaugh et al. [12], Gazonas et al. [13], and Kuo and Zhang [14].

follow $r_b = aP^n$ over this pressure range without any significant break points in the pressure exponent, between ~12 and 25 MPa, considerably more scatter is observed in the data. Kuo and Zhang [14] report correlations showing the presence of a slope break in the burning rate for JA2 around 13.8 MPa. For the low-pressure region below 13.8 MPa, a burning rate correlation at room temperature was given as

$$r_b\ (\mathrm{cm\ s^{-1}}) = 1.127[P\ (\mathrm{MPa})/6.894]^{0.63} \text{ for } 0.70 < P < 13.8\ \mathrm{MPa}$$

and for the high-pressure region above 13.8 MPa, the correlation reported was

$$r_b\ (\mathrm{cm\ s^{-1}}) = 5.822[P\ (\mathrm{MPa})/48.26]^{0.97} \text{ for } 13.8 < P < 96.5\ \mathrm{MPa}$$

The burning rates for liquid nitromethane are presented in Figure 2.6. In these experiments, the liquid is contained in tubes, for example, transparent glass or quartz tubes, or paper straws. The tube inner diameters ranged from approximately 5 to 10 mm. Both break wires and visual observation of the regressing surface have been used to obtain the burning rates. As evident from the figure,

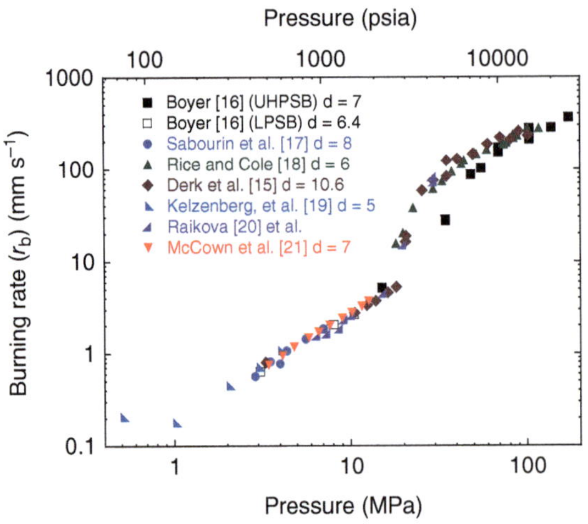

Figure 2.6 Burning rates of nitromethane as a function of pressure. Source: Data from Derk et al. [15], Boyer [16], Sabourin et al. [17], Rice and Cole [18], Kelzenberg et al. [19], Raikova et al. [20] (see [16] for more discussion on the presentation of the Raikova et al. data), and McCown et al. [21]. The listed inner tube diameters "*d*" are in millimeters. The UHPSB (Ultra High-Pressure Strand Burner) of Boyer and Kuo used 6.4 mm inner diameter straws. The LPSB (Liquid Propellant Strand Burner) of Boyer and Kuo used a continuous flow of propellant through a 7 mm inner diameter tube. Source: Derk et al. [15]. Reproduced with permission of Elsevier.

a significant slope break occurs at about 18 MPa, at which point the pressure exponent reaches a value of 21 before reestablishing the low-pressure exponent at pressures above 40 MPa. At the lowest pressures, the low-pressure flammability limit of nitromethane is approximately 3 MPa. The low-pressure data of Kelzenberg et al. [19] were influenced by the presence of oxygen in the surrounding gas enabling the additional heat release from the products of fuel-rich nitromethane combustion with oxygen to stabilize the flame to much lower pressures (much like the counterflow diffusion flames for AP presented earlier).

Figure 2.7 shows the burning rates of liquid hydroxyl ammonium nitrate (HAN)-water solutions and HAN/H_2O/CH_3OH mixtures as a function of pressure. As evident from Figure 2.7, pressure has a significant and unusual effect on the behavior of both the 13 M HAN and HAN/H_2O/CH_3OH burning rates. In the case of 13 M HAN, the burning rates increase and then decrease by nearly 2 orders of magnitude over a change in pressure of 2 orders of magnitude. Over the same pressure range, the burning rates of the HAN/H_2O/CH_3OH propellant increase rapidly by an order of magnitude starting at 1 MPa, then decrease starting at ~8 MPa by about an order of magnitude, and then rapidly increase (again by nearly an order of magnitude) around 13 MPa before decreasing at 30 MPa. The results of Ferguson et al. [23] appear to show a second dip in burning rate to occur near 18 MPa, which begins to recover at 22 MPa. At both pressure extremes, the 13 M HAN and the HAN/H_2O/CH_3OH burning rates are nearly the same.

As evident from most of the burning rate data, the empirical equation $r_b = aP^n$ is valid for only limited pressure ranges. As mentioned previously, most burning

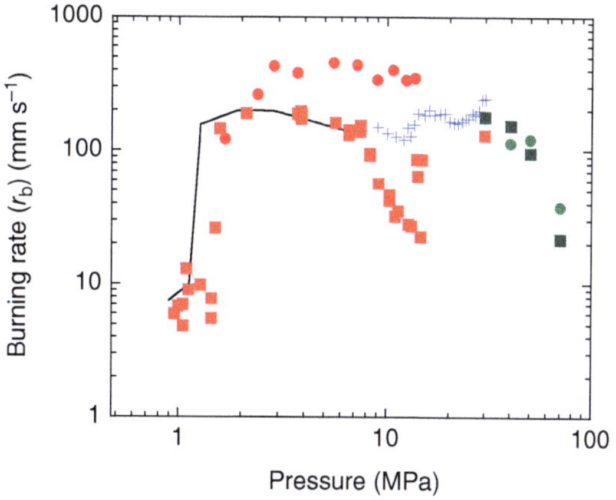

Figure 2.7 Burning rates at ambient temperature of 13 M HAN/water solution (solid red circle and solid green circle) and of a HAN/H_2O/CH_3OH mixture in which methanol was added to 13 M HAN to achieve ~15 wt% CH_3OH (solid black line [22], solid red square, blue cross [23], and solid green square). Source: All other data from Bugay et al. [24].

rate data are limited to pressures below 7 MPa, and for data greater than 50 MPa, observations of the burning processes are not available, and thus the presence of nonideal burning (e.g. burning in cracks and side burning) can only be surmised. Composite propellants are also well known to exhibit slope breaks, burning rate plateaus, and even negative pressure dependencies. In addition, the way that burning rate changes with pressure when the temperature of the energetic material is varied (burning rate temperature sensitivity) can also change the dependence of burning rate on pressure. Understanding of burning rate pressure and temperature sensitivity are key elements of propellant combustion response and stability as evident from a Zel'dovich–Novozhilov analysis [25].

A further examination of the burning rates of the nitramines and nitrate esters reveals an interesting feature between 12 and 25 MPa. In this pressure region, all the burning rates deviate from having a constant pressure exponent, generally showing an increase in pressure exponent followed by a decrease. The deviation is greater for nitromethane than HMX and RDX. For JA2, this region exhibits a slope break near 13.8 MPa where the scatter in data is the largest for the burning rates obtained from closed bomb experiments and the smallest for those obtained from visual observation. The HAN/H_2O/CH_3OH propellant burning rate also exhibits a sudden increase (after a decrease) in this pressure region. Interestingly, methyl nitrite has been measured as a surface product from these propellants. Even AP displays a unique behavior in this pressure region. For the data of Figure 2.2, a decrease in burning rate followed by an increase with a larger pressure exponent is shown. Although these trends may be coincidental, there may be fundamental processes of importance common to all these propellants in this pressure range.

The ability to predict the linear burning rate of an energetic material as a function of pressure and initial temperature is a highly desirable goal. It is important to note that none of the unique pressure dependencies of the above examples have been predicted accurately by detailed combustion models. Furthermore, it is critical to the body of knowledge regarding energetic materials to promote development of next-generation technologies that provide advancements in synthesis of new materials, formulations of propellants, and models for predicting ballistic properties.

The ability to predict linear burning rates of monopropellant deflagration as a function of temperature and pressure is an initial step in this direction. Such predictions depend on understanding of thermochemical equations of state, condensed-phase chemistry, gas-phase chemistry, and the interfacial chemistry that exists between phases. A starting point for construction of these models is a phenomenological picture of the burning process (based on known physics and chemistry), which is also aided by visual observations.

2.3 Visual Observations of Burning Behavior as a Function of Pressure

According to Figure 2.6, liquid nitromethane combustion exhibits a significant slope break at about 18 MPa. Derk et al. [15] have visually observed the combustion process over a pressure range that encompasses the slope break up to pressures as high as 100 MPa as shown in Figure 2.8. The initial experiments were performed by filling glass tubes of 10.6 mm internal diameter with liquid nitromethane and observing the propagation of the flame into the tube. From high-speed photographs at pressures below the slope break, the combustion process is observed to be nearly one dimensional with a sharp transition between the liquid and gas. At the lowest pressure shown (3.3 MPa), little luminosity is produced from the gas-phase flame. At a pressure higher than 12 MPa the gas-phase flame luminosity increased significantly, not from soot, but from chemiluminescence. At pressures just below 18 MPa, occasional bubbles appeared to be ejected from the surface. Independent of the bubble ejection, small surface waves also began to appear. Just above 18 MPa, large hydrodynamic instabilities exist in both longitudinal and radial directions. As the pressure is further increased to above 30 MPa, the large instabilities are replaced by a cellular structure with a slanted propagation front similar to a gas-phase turbulent flame. Under these conditions, a discrete interface is no longer discernable. Correlating the

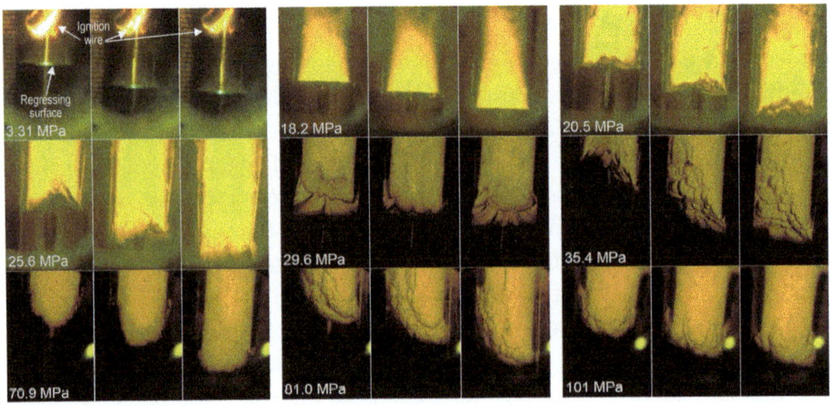

Figure 2.8 Frame images from high-speed photographs of liquid nitromethane burning as a function of pressure. Source: Derk et al. [15]. Reproduced with permission of Elsevier.

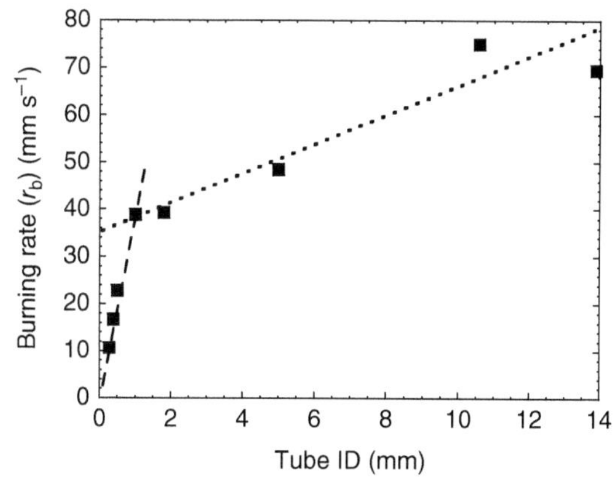

Figure 2.9 Combustion of liquid nitromethane as a function of tube diameter at a pressure of 40 MPa. Source: From Fritz et al. [26].

observations with the burning rate data indicates the sharp slope break occurs with the formation of the large hydrodynamic instabilities and the recovery of a burning rate exponent close to the low-pressure exponent at pressures above 30 MPa occurs with the formation of the turbulent cellular flame structure. The data of Boyer and Kuo (Figure 2.6) conducted in consumable straws were found to have lower burning rates in the transition region where hydrodynamic instabilities occurred for the confined flow but are likely dampened in the unconfined straws provided the consumption rate of the straw is close to the burning rate of the nitromethane.

Recognizing the importance of length scales in turbulent combustion, Derk et al. [15] and Fritz et al. [26] performed experiments of liquid nitromethane burning in tubes of different internal diameters decreasing the diameter to the quenching limit. A representative case is shown in Figure 2.9 for a pressure of 40 MPa. As the tube diameter is decreased, the burning rate decreased at a rate strongly influenced by the scale of the turbulence. For tube diameters less than 1 mm, the rate of decrease in burning rate increased significantly, presumably due to heat loss and radical quenching, until a tube diameter of 125 μm, below which the flame would no longer propagate. Extrapolation of the large tube diameter slope to a diameter of zero gives an indication of the laminar burning rate at 40 MPa of about 35 mm s^{-1}. Placing this value of the burning rate on the plot of Figure 2.6 would still suggest a departure from extrapolation of the low-pressure burning rate by about a factor of 3 at 40 MPa.

The burning rates of HAN-based propellants have received significant attention because of their possible replacement of hydrazine as a monopropellant. Hori et al. [27–30] have studied a wide range of compositions of aqueous HAN and HAN/H_2O/CH_3OH mixtures for pressures as high as 8 MPa in 12 mm inner diameter tubes. Ammonium nitrate (AN) in tiny amounts is often added to these mixtures as a stabilizer. Figure 2.10 illustrates the combustion process of both types of mixtures at two pressures (with AN). The top left row shows combustion of 88 wt% HAN/4.6 wt% AN/7.4% H_2O at pressures of 2 MPa (Figure 2.10a) and at 6 MPa (Figure 2.10b). The bottom left row shows combustion of 73.6 wt% HAN/3.9 wt% AN/6.2 wt% H_2O/16.3 wt% CH_3OH at pressures of 2 MPa (Figure 2.10c) and 6 MPa (Figure 2.10d). On the right side of the figure are the corresponding burning rates.

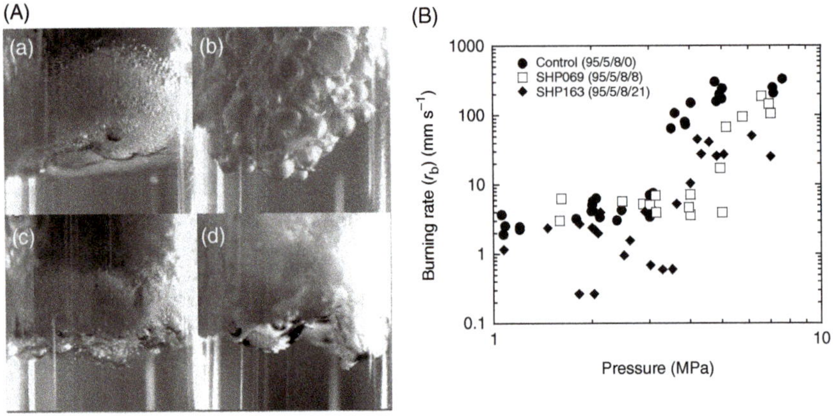

Figure 2.10 ((A): top row) Combustion of 88 wt% HAN/4.6 wt% AN/7.4 wt% H_2O (control) at pressures of 2 MPa (a) and 6 MPa (b); ((A): bottom row) Combustion of 73.6 wt% HAN/3.9 wt% AN/6.2 wt% H_2O/16.3 wt% CH_3OH (SHP163) at pressures of 2 MPa (c) and 6 MPa (d). Source: Katsumi et al. [27]. Reproduced with permission of Springer Nature. (B) Burning rates of HAN-based propellants. Mixture (SHF069) consisted of 81.9 wt% HAN/4.3 wt% AN/6.9 wt% H_2O/6.9 wt% CH_3OH. Source: Katsumi et al. [28]. Reproduced with permission of American Institute of Aeronautics and Astronautics.

Although the trends of the burning rates are qualitatively similar to those shown in Figure 2.7, the burning rate magnitude and pressure at which the slope break occurs are sensitive to the water concentration, the presence of AN, and the concentration of methanol. As shown in the figures, boiling of the components plays a key role in the burning mechanism as do the size of the bubbles, the thickness of the two-phase regime, and the bubble penetration into the liquid [27–30].

Figure 2.11 shows another set of images for the 73.6 wt% HAN/3.9 wt% AN/6.2 wt% H_2O/16.3 wt% CH_3OH mixture at pressures of 1, 3, 4, 6, and 8 MPa. At all these pressures, the bubble process/two-phase zone is clear. The pressure range for these images is below the dip in burning rate that starts at ~8 MPa for the

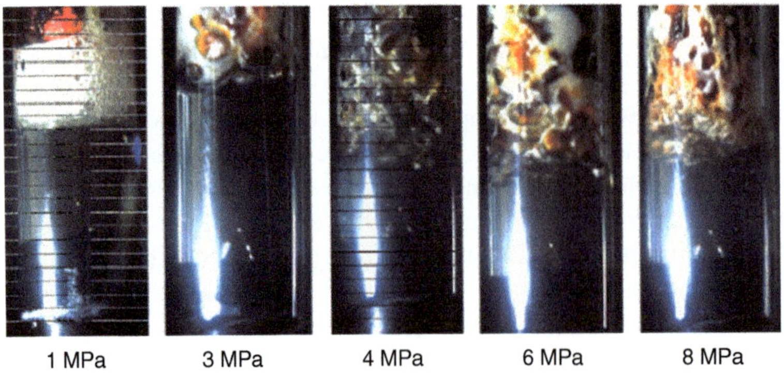

Figure 2.11 Images for combustion of 73.6 wt% HAN/3.9 wt% AN/6.2 wt% H_2O/16.3 wt% CH_3OH at pressures of 1, 3, 4, 6, and 8 MPa. Source: Amrousse et al. [30]. Reproduced with permission of Elsevier.

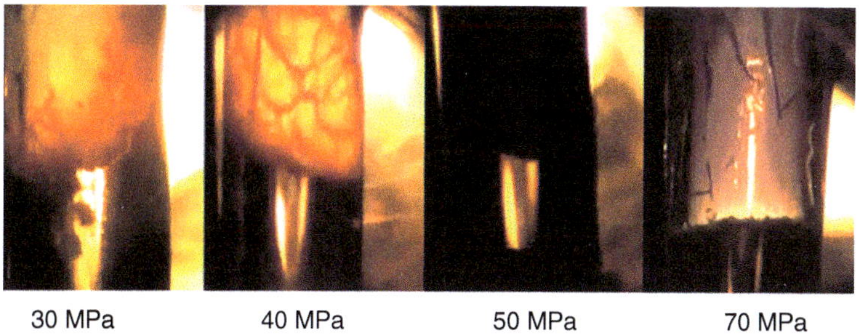

Figure 2.12 Photographs of HAN/H_2O/CH_3OH burning at various pressures in glass tubes of internal diameter 10.6 mm.

CH_3OH mixture and below the fall off in burning rates that start at ~22 MPa for both the aqueous HAN and CH_3OH mixtures presented earlier in Figure 2.7.

Bugay et al. [24] have taken images of the burning process of both a 13 M HAN/H_2O solution and a HAN/H_2O/CH_3OH mixture (13 M HAN with ~15 wt% CH_3OH) at pressures above 30 MPa. Images for the CH_3OH mixture are shown in Figure 2.12 for pressures of 30, 40, 50, and 70 MPa. For pressures of 30 and 40 MPa, the flame structure has characteristics like nitromethane burning at equivalent pressures, that is, a cellular turbulent flame structure. The well-defined bubbles observed at lower pressures are missing, as are large-scale hydrodynamic instabilities. At 50 MPa, the flame luminosity goes dark (which was characteristic of the aqueous HAN mixtures for 30, 40, and 50 MPa), and at 70 MPa, the flame becomes nearly one dimensional with small surface waves. A discrete interface seems to reappear without bubbles, hydrodynamic instabilities, or a turbulent cellular structure. Note that the burning rate at 70 MPa is only about a factor of 2 greater than the burning rate at 0.1 MPa.

The burning rates of nitramines, RDX and HMX, are observed to deviate from a constant pressure exponent at about 12 MPa. Both are known to have melt layers and fizz zones at low pressures. An image of RDX burning at 0.17 MPa is shown in Figure 2.13 [31], from which phenomenological models and subsequently detailed

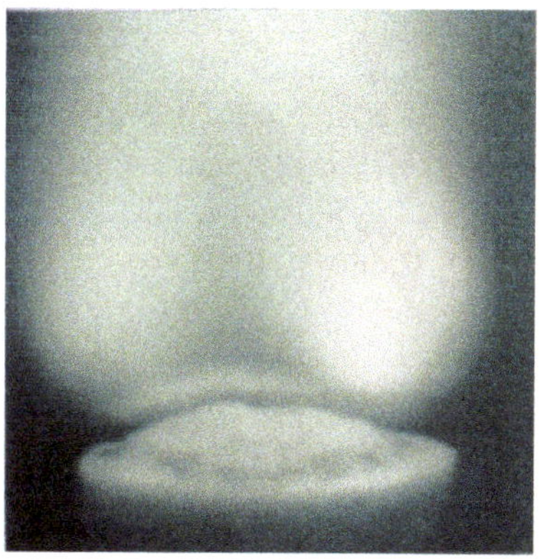

Figure 2.13 Thick foam layer on burning surface of RDX at 0.17 MPa (10 psig). Source: The Pennsylvania State University.

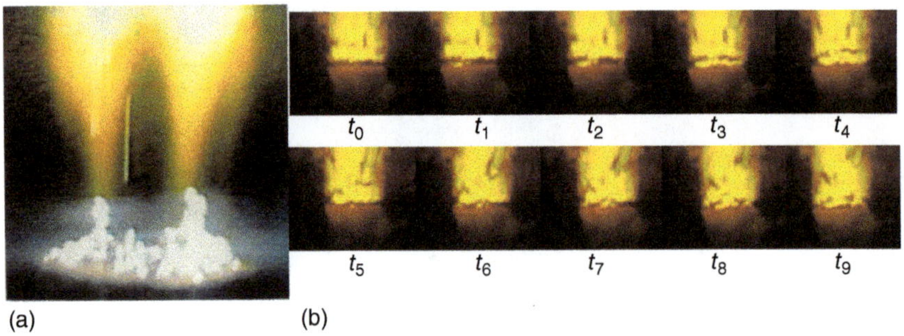

Figure 2.14 (a) Flame attachment to carbonaceous flakes on JA-2 at 1.48 MPa (200 psig) and (b) sequence of burning at 206.8 MPa (30 000 psig). Derk [34] provides additional description of the high pressure burning of JA2. The interval between frames is ~20 ms.

models have been developed [32, 33]. Images of burning nitramines between 12 and 25 MPa have not been reported.

Nitrate ester–based propellants and double base propellants such as JA2 are not known to have significant melt layers. According to the data of Figure 2.5, significant variation in the burning rate has been measured in the vicinity of 12–25 MPa and a slope break has been reported to occur at ~13.8 MPa [14]. An image of JA2 burning at 1.48 MPa is shown in Figure 2.12 where carbonaceous layers are observed to form at the surface.

High-speed video by Derk et al. [11, 34] of JA2 burning from 14 to 200 MPa shows the carbonaceous material to form at all pressures (Figure 2.14). However, at high pressures, the carbon deposits do not reside on the surface but are swept away as they are formed. At the lowest pressures studied (below 14 MPa), the carbon would form a layer on the surface, which would build up, and then periodically fly away from the surface.

AP also exhibits a burning rate slope break around 20 MPa, as well as showing some negative dependence in the vicinity of this pressure. Quenched samples along with burning rate data have been used to interpret possible mechanisms for the change in the burning process [6].

Prior to the negative dependence in the burning rate (Figure 2.2). The surface has been interpreted to have gas entrapped in liquid resulting in froth. As the pressure is increased, the flame remains steady and planar, while the surface structure changes to one with ridges and valleys. In the region of the negative pressure exponent, the flame is intermittent and nonplanar, and needles form on the surface. With a further increase in pressure, the flame becomes steady and planar again and the pressure exponent turns positive with the entire surface covered with needles. The early formation of the needles in the negative dependence region was reported to affect the diffusional processes and consequently the monopropellant flame standoff distance that results in the unsteadiness of the flame [6]. Needles were observed with diameters as small as 5 μm and considerably longer lengths. The effects of needle formation (if it occurs) on the surface of the AP particles in composite propellants (such as those of Figure 2.4) are unknown.

2.4 Discussion

Beginning the discussion with nitromethane combustion, the visual observations indicate the presence of hydrodynamic instabilities that are initiated with the formation of surface waves. Over the range of pressures shown in Figure 2.6, condensed-phase chemistry would be expected to be extremely slow and mostly endothermic. From detailed kinetic modeling calculations, Boyer [35] compared ignition times at the surface temperature to residence times in the thermal wave for nitromethane deflagration over the pressure range from 1 to 6 MPa and found the residence times were 3–4 orders of magnitude smaller than ignition delay times. Consequently, during transcritical combustion of nitromethane, the liquid–gas interface is brought to the critical temperature via conduction from the gas-phase reaction, and not by heating beneath the interface. The presence of the surface waves has been described as a loss of surface tension resulting from the mixture at and above the surface becoming supercritical. Although the critical conditions for pure nitromethane are $T_c = 588$ K and $P_c = 6.6$ MPa (Table 2.1), Derk [34] using the results from numerical modeling calculations [35] and empirical methods [38] estimated a mixture critical pressure of 16 MPa and critical temperature of 485 K. Even though the uncertainty in these methods can be large, the estimated critical pressure is close to the experimental occurrence and appears to be driven by the critical properties of water (Table 2.1). In the supercritical state, a further increase in pressure produces the cellular turbulent flame front whose propagation speed is dependent on the tube diameter, and hence many of the previously reported values of the burning rate at high pressures are not fundamental intrinsic values. Experiments with measurements obtained in smaller tube diameters have enabled an estimate of burning rates at the laminar limit. These results show that the maximum difference between the subcritical burning rate if extrapolated to higher pressures and the supercritical burning rate is about a factor of three, with the difference somewhat decreasing as pressures are further increased [26].

In the case of multicomponent mixtures involving HAN, the critical conditions for HAN were estimated by Kounalakis and Faeth [37] to be $P_c = 7.7$ MPa and

Table 2.1 Properties of major reactants and products.

Species	CO_2	H_2O	N_2	CH_3NO_2	HAN	AN	CH_3OH
Critical properties							
Temperature (K)	304.2	647.3	126.2	588	763	677.4	512.5
Pressure (MPa)	7.4	22.0	3.4	6.3	7.7	7.36	8.08
Volume (cm^3 gmol^{-1})	94	56	89.5	173	196	160	0.2715
Enthalpy of formation at 298.15 K (kcal mol^{-1})	−94.0	−57.8	0.0	−112.6	−95.3	−87.3	−56.98

Source: Data from NIST [36] and Kounalakis and Faeth [37].

T_c = 763 K. Because HAN begins to decompose at 80 °C and attains a maximum rate at 130 °C (with a heating rate of 10 °C min^{-1}), the critical conditions for HAN are not expected to be important to the high-pressure burning of these mixtures. Furthermore, the temperature at the maximum decomposition rate (~130 °C) is not significantly affected by pressure [23]. At the lowest pressures, the condensed-phase reaction is slow and limited by the boiling point temperature of water, which is below the temperature where the maximum decomposition rate occurs. For these conditions, the condensed-phase reaction does not supply enough heat to produce a fast regression rate. As pressure is increased, the boiling point temperature of the water increases beyond the temperature for maximum decomposition rate, and enough heat is supplied to superheat the water producing subsurface boiling in aqueous HAN. The process initiates the hydrodynamic instabilities. Hori and coworkers [27–30] have performed stability analyses to identify the necessary conditions for instability occurrence. The instabilities and bubble formation increase the surface area resulting in a significant increase in burning rate. Different combustion modes consisting of one with a layered structure, one with bubbles in front of the combustion wave, and one where both the layered and bubble structures alternatively appear have been identified below 8 MPa and depend partially on the size of the bubbles and the thickness of the bubble region [27–30]. After the initial increase in burning rate, it remains nearly constant up to about 8 MPa. This pressure range appears to be dominated by bubble formation. Above approximately 22 MPa, the burning rate begins to decrease with further increases in pressure. Visual images of the flame at pressures of 30, 40, and 50 MPa show a dark plume without luminosity and without bubble formation. A sharp interface is not observed suggesting a transition from subcritical to supercritical conditions in the mixture near the interface. Instabilities from bubble formation would also be suppressed due to the high gas densities and the larger amount of heat required from the condensed-phase reaction to raise the temperature to the higher water boiling point temperature with increasing pressure. Heat feedback from an exothermic gas-phase reaction appears to be weakened slowing the reaction. These trends again suggest the importance of the transition from subcritical to critical combustion.

With the addition of methanol, the reaction rate slows relative to aqueous HAN. It is well known that both the concentration of water and methanol affect the reaction rate [27–29, 39]. For pressures less than 8 MPa, the images of burning aqueous HAN methanol mixtures show significant similarity to those of aqueous HAN with bubble formation again playing a significant role in achieving the high burning rates. At 40 MPa, flame propagation of the aqueous HAN methanol mixture is visually like nitromethane combustion showing the absence of bubbles and instabilities and the existence of a possible supercritical cellular turbulent flame structure. Increasing the pressure to 50 MPa shows the luminosity of the gas-phase flame to disappear resulting in a similar flame structure and burning rate as aqueous HAN. A further increase in pressure to 70 MPa decreases the burning rate revealing a nearly one-dimensional interface with waves evident at the interface and luminosity again above it. Because of the condensed-phase HAN reaction, the liquid is continuously heated from beneath the interface to the saturation temperatures of CH_3OH, H_2O, and the decomposition products of HAN. Examination of Table 2.1 reveals the dip in burning rate at about 8 MPa coincides with the critical pressure of pure CH_3OH and the change in burning rate at almost 22 MPa coincides with the critical pressure of pure H_2O. As the reacting system passes through the critical point, significant changes in fluid properties occur, which diminish as the temperature and pressure move further away. In a transcritical reacting system where the temperature is increasing from reactants to final products, a change in heat capacity around the critical point would affect temperatures and consequently reaction rates. In this instance, the large specific heats of CH_3OH and H_2O around their critical points may affect mixture-specific heats to lower the reaction temperature such that reaction rates slow.

Another characteristic of going to pressures beyond the critical point is evident from continuity and the loss of a jump in the gas velocity due to a phase change ($\rho_c r_b = \rho_g v_g$, or $v_g \sim ZP^{n-1}$ where Z is the compressibility factor). That is, the ratio of gas velocity at the surface to the burning rate of the propellant will be close to a maximum at its low-pressure deflagration limit and will decrease up to the critical point. The jump in density around the critical point and the disappearance of the enthalpy requirement for phase change will add further complexity to interpretation of the burning process. These trends affect diffusional gradients and heat transfer necessary for determining the burning rate.

Although images of the burning rate of the solid monopropellants RDX and HMX at the highest pressures are not available, the burning rates do exhibit a slope break around 12 MPa that results in a burning rate increase of a factor of 2–3 above the extrapolated low-pressure burning rate. Both RDX and HMX exhibit condensed-phase chemistry. The major intermediate and final products of each are generally the same as those from nitromethane, and the HAN mixtures with carbon-containing fuels. Table 2.2 provides the species concentrations predicted by detailed modeling calculations for RDX, nitromethane, and JA2. For the RDX calculations, a detailed elementary reaction mechanism was included for both the condensed phase and gas phase with a fizz zone (bubble formation) connecting the two. The nitromethane calculations only included vaporization at the interface, and the JA2 mechanism included a lumped condensed-phase mechanism with a detailed elementary gas-phase mechanism. As evident from the table, there is a subset of approximately 10 species that dominate the mixture composition, although their rank ordering can vary. Hence the critical pressure and temperature for the mixtures at and just above the decomposing interface would be expected to be nearly the same for C–H–O–N propellants varying somewhat with the relative changes in the species concentrations. AP produces hydrogen chloride, which has a critical pressure and temperature of $P_c = 8.3$ MPa and $T_c = 325$ K, in addition to producing water, nitric oxide, oxygen, and nitrogen [42], and therefore it should not be surprising that unusual trends occur in the burning rate in the vicinity of 12–25 MPa as well.

The detailed RDX calculations by Khichar show that for a pressure variation of 0.1–9.1 MPa, the RDX mass fraction at the propellant surface varies from 0.997 to 0.987, with most of this variation occurring below 2 MPa [40]. The calculations also

Table 2.2 Calculated major species in different regions near the condensed- and gas-phase interface for various burning propellants.

	Largest species concentrations within one order of magnitude	Reference
RDX (9.1 atm)		[40]
Liquid-phase melt ($T < 770$ K)	$RDX \gg NO_2 \sim H_2O \sim NO > N_2O > HCN > N_2 > CH_2O > CO > CO_2$ (not including RDX to determine the largest concentrations)	
Gas phase ($T > 770$ K)	$NO_2 > N_2O > H_2O > NO > N_2 > CO \sim H_2 \sim HCN > CO_2 \sim CH_2O$	
Gas phase (~dark zone)	$H_2O > NO > N_2 > CO > N_2O > H_2 \sim HCN > CO_2$	
CH_3NO_2 (6 MPa)		[16]
Condensed phase	Only vaporization considered	
Gas phase ($T_{surface} \sim 620$ K)	$CH_3NO_2 > H_2O > NO > CO > H_2 > CH_4 > N_2 > CO_2$	
Gas phase (~dark zone)	$H_2O > NO > CO > H_2 \sim CH_4 > N_2 > CO_2$	
JA2 (1.6 MPa)		[41]
Condensed phase	$NO_2 > CH_2O > CO > HCO > CH_2 > HONO > H$	
Gas phase ($T > 760$ K)	$CO > CH_2O > NO > HONO > H_2O > CO_2 > CH_2O > NO_2$	
Gas phase (~dark zone)	$CO > H_2O > NO > CO_2 > H_2$	

show that the ratio of the overall heat release in the melt layer to the heat feedback from gas-to-liquid phase varies from less than 0.05% at 0.1 MPa to less than 0.2% at 9.1 MPa. These results differ from those estimated by Zenin using thermocouples to measure temperature and literature estimates for thermal conductivity, specific heat, and enthalpy of vaporization for RDX that show at 9.1 MPa, the ratio approaches five [43]. The modeling results suggest, like nitromethane, that the liquid–gas interface temperature is brought to the critical temperature via conduction from the gas phase. However, the experimental results [43] suggest, like the HAN mixtures, that significant reaction occurs in the condensed phase and the critical conditions of RDX are not important to determine the critical condition of the mixture.

A generalized compressibility chart for a pure substance indicates that at the critical pressure and temperature, the compressibility factor is a minimum (below unity) [44]. Based on density considerations, reaction rates and spatial gradients can therefore increase in the vicinity of the critical point thereby modifying the burning rate. As the reduced pressure ($P_R = P/P_c$) is increased above the critical point, the compressibility factor increases and approaches unity. Above a reduced pressure of approximately 8 (for a reduced temperature, $T_R = T/T_c$, of unity), the compressibility factor continues to increase to values greater than unity. However, an increase in reduced temperature for a given reduced pressure always tends to drive the compressibility factor toward unity above or below $P_R \cong 8$. While mixture behavior can vary from pure substances, these trends would suggest a low-pressure exponent that changes its value at the critical point could eventually reestablish this value at higher pressures. As shown here, the slope breaks that occur between 12 and 25 MPa may be related to the onset of supercritical combustion. It is also evident that this transition could be the initiator of other chemical and physical processes. For the small number of examples described, the burning rate pressure exponents of several energetic materials at high pressures are observed to return close to their low-pressure values (e.g. HMX and CH_3NO_2), whereas others do not (e.g. JA2 and HMX/HTPB and AP/HTPB composites). Additional fundamental data on transcritical and supercritical properties of mixtures will be required to enable detailed modeling of supercritical combustion to further our understanding of slope breaks in burning rates.

The critical combustion conditions of liquid monopropellants have been previously studied by Faeth and co-workers [37, 45]. For liquid ethylene oxide combustion, they found from both experimental and theoretical analyses that transition to critical combustion occurred between pressures of 12–16 MPa, similar to the critical combustion pressure of nitromethane. In the analysis of a HAN-based liquid monopropellant with triethanol ammonium nitrate (TEAN) as the fuel component, the predicted critical pressure was found to be as high as 250 MPa. The predictions were found to be extremely sensitive to the critical temperature of TEAN and the binary interaction parameter between water and TEAN, both of which had large uncertainties in the analysis. However, when interaction parameters were set to zero, the predicted critical pressure dropped to 65 MPa. The strong effects of the interaction parameters were not observed for ethylene oxide combustion or the combustion of other liquid monopropellants. It should also be noted that in the burning rate data of McBratney and Vanderhoff [46], where the HAN/TEAN/H_2O liquid propellant was gelled with Kelzan and studied in small characteristic diameter tubes to suppress the instabilities and turbulence below 60 MPa, no slope breaks were observed, and the pressure exponent was a constant. A slope break was observed at approximately 70 MPa with associated unstable combustion that may be indicative of transition to supercritical combustion. Thus, the pressure range presented in Figure 2.7 may be below a critical combustion pressure for the very complex tri-component ionic HAN mixture with methanol as the fuel. Higher pressures will need to be studied for further understanding.

2.5 Conclusions

Burning rates of energetic materials have long been known to exhibit slope breaks in the burning rate law. The reasons for these slope breaks are rarely known. Visual observations of the burning process along with burning rate measurements are necessary to better understand the combustion process and to understand the validity of the measurements themselves. Extending these measurements and observations to extreme conditions are valuable to understanding phenomena at non-extreme conditions as well as discovery of new combustion behavior. For many burning propellants, transition from subcritical to supercritical conditions appears to occur over a common pressure range from 12 to 25 MPa with the transition resulting in a slope break in the burning rate profile. More studies are needed in the extreme conditions to address many of the questions that were brought to light here.

Acknowledgments

The authors acknowledge support from the Army Research Office under grants W911NF-16-1-0139 and W911NF-17-1-0149. The authors acknowledge the contributions of Gregory Derk, Ryan Fritz, and Ezekiel Bugay in providing some of the experimental results reported in this chapter.

References

1 Zarko, V., Kiskin, A., and Cheremisin, A. (2022). Contemporary methods to measure regression rate of energetic materials: a review. *Progress in Energy and Combustion Science* https://doi.org/10.1016/j.pecs.2021.100980.

2 Fry, R.S., DeLuca, L.T., Gadiot, G. et al. (2012). Solid propellant burning rate measurement methods used within the NATO propulsion community. AIAA 2001-3948.

3 Gupta, G., Jawale, L., Mehilal, D., and Bhattacharya, B. (2015). Various methods for the determination of the burning rates of solid propellants – an overview. *Central European Journal of Energetic Materials* 12 (3): 593–620.

4 Atwood, A., Boggs, T.L., Curran, P.O. et al. (1999). Burning rate of solid propellant ingredients, part 1: pressure and initial temperature effects. *Journal of Propulsion and Power* 15: 740–747.

5 Hightower, J.D. and Price, E.W. (1967). Combustion of ammonium perchlorate. *Symposium (International) on Combustion* 11 (1): 463–472.

6 Boggs, T.L. (1970). Deflagration rate, surface structure, and subsurface profile of self-deflagrating single crystals of ammonium perchlorate. *AIAA Journal* 8 (5): 867–873.

7 Irwin, O.R., Salzman, P.K., and Anderson, W.H. (1963). Deflagration characteristics of ammonium perchlorate at high pressures. *Symposium (International) on Combustion*, 9: 358–364.

8 Johansson, R.H., Connell, T.L. Jr., Risha, R.A. et al. (2012). Investigation of solid oxidizer and gaseous fuel combustion performance using an elevated pressure counterflow experiment for reverse hybrid rocket engine. *International Journal of Energetic Materials and Chemical Propulsion* 11 (6): 511–536.

9 Lengellé, G., Duterque, J., and Trubert, J.F. (2000). Physico-chemical mechanisms of solid propellant combustion. In: *Solid Propellant Chemistry, Combustion, and Motor Interior Ballistics*, Progress in Astronautics and Aeronautics, vol. 185 (ed. V. Yang, T.B. Brill and W.-Z. Ren), 287–334. Reston, VA: American Institute of Aeronautics and Astronautics (AIAA).

10 Beckstead, M.W. (2000). Overview of combustion mechanisms and flame structure for advanced solid propellants. In: *Solid Propellant Chemistry, Combustion, and Motor Interior Ballistics*, Progress in Astronautics and Aeronautics, vol. 185 (ed. V. Yang, T.B. Brill and W.-Z. Ren), 267–285. Reston, VA: American Institute of Aeronautics and Astronautics (AIAA).

11 Derk, G., Risha, G.A., Boyer, J.E., and Yetter, R.A. (2019). High pressure burning rate measurements by direct observation. *International Journal of Energetic Materials and Chemical Propulsion* 18 (3): 213–227.

12 Reaugh, J. E., Maienschein, J.L., and Chandler, J.B. (1997). Laminar burn rates of gun propellants measured in the high-pressure strand burner. *JANNAF, 34th Combustion Subcommittee/Propulsion Systems Hazards Subcommittee and Air-breathing Propulsion Subcommittee Joint Meetings*, Joint Army Navy NASA Air Force Interagency Propulsion Committee, Johns Hopkins University Energetics Research Group, Columbia MD (27–31 October 1997).

13 Gazonas, G.A., Juhasz, A.A., and Ford, J.C. (1996). Strain rate insensitivity of damage-induced surface area in M30 and JA2 gun propellants. *Propellants, Explosives, Pyrotechnics* 21: 307–316.

14 Kuo, K.K. and Zhang, B. (2006). Transient burning characteristics of JA2 propellant using experimentally determined Zel'dovich map. *Journal of Propulsion and Power* 22 (2): 455–461.

15 Derk, G., Boyer, J.E., Risha, G.A. et al. (2021). Experimental and numerical investigation of high-pressure nitromethane combustion. *Proceedings of the Combustion Institute* 38: 3325–3332.

16 Boyer, J.E. (2005). Combustion characteristics and flame structure of nitromethane liquid monopropellant. PhD dissertation. The Pennsylvania State University, University Park, PA.

17 Sabourin, J.L., Yetter, R.A., and Parimi, V.S. (2010). Exploring the effects of nanostructured particles on liquid nitromethane combustion. *Journal of Propulsion and Power* 26 (5): 1006–1015.

18 Rice, T.K. and Cole, Jr., J.B. (1953). Liquid monopropellants burning rate of nitromethane. Naval Ordnance Lab Report NAVORD-2885, White Oak, MD.

19 Kelzenberg, S., Eisenreich, N., Eckl, W., and Weiser, V. (1999). Modelling nitromethane combustion. *Propellants, Explosives, Pyrotechnics* 24: 189–194.

20 Raikova, V.M. (1977). Limit conditions of combustion and detonation of nitroesters and mixtures on their base. PhD thesis. Mendeleev Institute of Chemical Technology, 222 Moscow.

21 McCown, K.W. III, Demko, A.R., and Petersen, E.L. (2014). Experimental techniques to study linear burning rates. *Journal of Propulsion and Power* 30 (4): 1027–1037.

22 Chang, Y. (2002). Combustion behavior of HAN-based liquid propellants. PhD dissertation. The Pennsylvania State University, University Park, PA.

23 Ferguson, R.E., Esparza, A.A., and Shafirovich, E. (2021). Combustion of aqueous HAN/methanol propellants at high pressures. *Proceedings of the Combustion Institute* 18: 3295–3302.

24 Bugay, E., Boyer, J.E., and Yetter, R.A. (2023). Burning behavior of aqueous HAN and aqueous HAN-methanol mixtures at pressures greater than 7 MPa, to be submitted.

25 Novozhilov, B.V. (1992). Theory of nonsteady burning and combustion stability of solid propellants by the Zel'dovich–Novozhilov method, Chapter 15. In: *Non-steady Burning and Combustion Stability of Solid Propellants*, Progress in Astronautics and Aeronautics, vol. 143 (ed. L. De Luca, E.W. Price and M. Summerfield), 601–641. Washington, DC: American Institute of Aeronautics and Astronautics (AIAA).

26 Fritz, R., Boyer, J.E., and Yetter, R.A. (2023). High-pressure burning rates and quenching diameters of nitromethane combustion, to be submitted.

27 Katsumi, T., Kodama, H., Matsuo, T. et al. (2009). Combustion characteristics of a hydroxylammonium nitrate based liquid propellant. Combustion mechanism and application to thrusters. *Combustion, Explosion, and Shock Waves* 45 (4): 442–453.

28 Katsumi, T., Hori, K., Matsuda, R., and Inoue, T. (2010). Combustion wave structure of hydroxylammonium nitrate aqueous solutions. *46th AIAA/ASME/SAE/ASEE Joint Propulsion Conference and Exhibit*, AIAA Paper No. 2010-6900, Nashville, TN.

29 Katsumi, T., Inoue, T., Nakkatsuka, J. et al. (2012). HAN-based green propellant, application, and its combustion mechanism. *Combustion, Explosion, and Shock Waves* 48 (5): 536–543.

30 Amrousse, R., Katsumi, T., Itouyama, N. et al. (2015). New HAN-based mixtures for reaction control system and low toxic spacecraft propulsion subsystem: thermal decomposition and possible thruster applications. *Combustion and Flame* 162: 2686–2692.

31 Boyer, E., Lu, Y.C., Desmarais, K., and Kuo, K.K. (1994). Observation and characterization of burning surface reacting zones of solid propellants. In: *31st JANNAF Combustion Meeting*, Sunnyvale, CA (17–21 October 1994). CPIA Publication 620, vol. II, 211–220.

32 Kuo, K.K. and Acharya, R. (2012). *Applications of Turbulent and Multiphase Combustion*. Wiley.

33 Liau, Y.-C. and Yang, V. (1995). Analysis of RDX monopropellant combustion with two-phase subsurface reactions. *Journal of Propulsion and Power* 11 (4): 729–739.

34 Derk, G. (2019). High-pressure burning rates of JA2 and nitromethane propellants. Master's thesis. The Pennsylvania State University, University Park, PA.

35 Boyer, J.E. and Kuo, K.K. (2007). Modeling of nitromethane flame structure and burning behavior. *Proceedings of the Combustion Institute* 31: 2045–2053.

36 NIST Chemistry WebBook, SRD 69, National Institute of Standards and Technology, US Department of Commerce, https://webbook.nist.gov (accessed 31 March 2022).

37 Kounalakis, M.E. and Faeth, G.M. (1988). Combustion of HAN-based liquid monopropellants near the thermodynamic critical point. *Combustion and Flame* 74: 179–192.

38 Reid, R.C., Prausnitz, J.M., and Poling, B.E. (1987). *The Properties of Gases and Liquids*, 4e. New York: McGraw Hill Book Company.

39 Amariei, D., Courtheous, L., Rossignol, S. et al. (2005). Influence of the fuel on the thermal and catalytic decompositions of ionic liquid monopropellants. *41st AIAA/ASME/SAE/ASEE Joint Propulsion Conference & Exhibit*, Tucson, Arizona (10–13 July 2005), AIAA 2005-3980.

40 Khichar, M. (2021). Thermal decomposition and combustion modeling of RDX monopropellant and RDX-TAGzT pseudo-propellant. PhD dissertation. The Pennsylvania State University.

41 Miller, M.S. and Anderson, W.R. (2004). Burning-rate predictor for multi-ingredient propellants: nitrate–ester propellants. *Journal of Propulsion and Power* 20 (3): 440–454.

42 Tanoff, M.A., Ilincic, N., Smooke, M.D. et al. (1998). Computational and experimental study of ammonium perchlorate combustion in a counterflow geometry. *Proceedings of the Combustion Institute* 2: 2397–2404.

43 Zenin, A., HMX and RDX: combustion mechanism and influence on modern double-base propellant combustion, *Journal of Propulsion and Power* 11 (1995) 752–758. doi:https://doi.org/10.2514/3.23900.

44 Moran, M.J. and Shapiro, H.N. (1996). *Fundamentals of Engineering Thermodynamics*, 3e. New York, NY: Wiley.

45 Chen, L.-D. and Faeth, G.M. (1981). Initiation and properties of decomposition waves inliquid ethylene oxide, *Combustion and Flame* 40: 13–28.

46 McBratney, W.F. and Vanderhoff, J.A., Bum-RateInvestigations of HAN-Based Candidate Liquid Propellants, Technical Report ARL-TR-1927, Army Research Laboratory, Aberdeen Proving Ground, MD 21005-5066, March 1999.

Part II

New Energetic Ingredients

3

Cyclic Nitramines as Nanoenergetic Organic Materials

Alexander A. Larin and Leonid L. Fershtat

Russian Academy of Sciences, N. D. Zelinsky Institute of Organic Chemistry, Laboratory of Nitrogen Compounds, Leninsky Prosp., 47, Moscow 119991, Russia

3.1 Introduction

Nanoenergetic materials are commonly referred to as high-energy materials or high-energy composites, at least one of which is below 100 nm in size. Nanoenergetic materials possess small particle size, large specific surface area, high surface energy, and strong surface activity, so their functional properties are improved compared to energetic materials of conventional size [1]. Strong advantages of nanoenergetics include reduced mechanical sensitivity to impact, friction, and electrostatic discharge, increased sensitivity to short pulses and acceleration of reaction speed or burning speed enabling their potential as solid rocket propellants [2, 3]. Although the friction and even more electrostatic sensitivities are not always reduced for nanoenergetics in comparison to conventional energetic materials, other advantages encourage further research in this field.

Nanoenergetic materials are widely used as new functional components, and many researchers are interested in technologies of production thereof [4, 5]. However, existing synthetic methods for the construction of conventional microscale high-energy materials cannot be directly transferred to the preparation of nanoenergetic materials. For this reason, a group of specific methods is used that can be classified according to various techniques [6, 7]:

- Mechanical grinding and milling;
- Re-precipitation method (cooling crystallization and all types of solvent–anti-solvent processes);
- Spray drying and spray freeze-drying techniques;
- Sol-gel processing;
- Supercritical fluid precipitation processes.

Usually, for the formation of nanoparticles, the method of initial discrete destruction (the method from large to small) or the method of coagulation crystallization (the method from small to large) is used. The peculiarity of this method

Nano and Micro-Scale Energetic Materials: Propellants and Explosives, First Edition.
Edited by Weiqiang Pang and Luigi T. DeLuca.

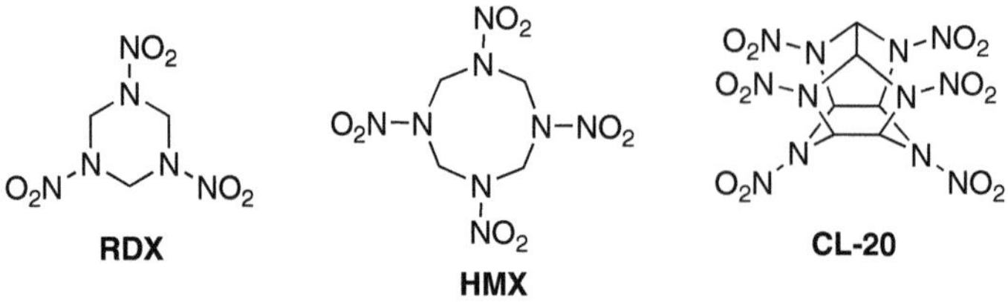

Figure 3.1 Molecular structures of cyclic nitramines used for the preparation of nanoenergetic materials and covered in this chapter. Source: A.A. Larin, L.L. Fershtat.

is processing by means of external fragmentation or microprocessing, which continuously micronizes and disperses common energy particles.

On the other hand, utilization of nanoenergetic materials has a number of disadvantages associated with the methods used for their production. For example, mechanical grinding has a low degree of control over particle size, while wet milling raises a possibility of spontaneous ignition or detonation which is especially relevant for a large-scale manufacturing. Crystallization procedures and sol-gel processing are usually multistep, time-consuming, and produce large amounts of waste. Solvent–antisolvent processes may contaminate crystals with solvent and produce solvates, while supercritical techniques require sophisticated instrumentation dependent from economic factors.

In recent years, various methods for the preparation of energetic materials on nanoscale were developed and the main directions for further application of these substances were proposed [8]. In this regard, main efforts of researchers were directed toward the formulation of nanoenergetic cyclic nitramines (RDX, HMX, CL-20) as the most promising materials for industrial applications [9]. RDX (1,3,5-trinitroperhydro-1,3,5-triazine) and HMX (1,3,5,7-tetranitro-1,3,5,7-tetrazocane) correspond to the most widely utilized energetic materials (Figure 3.1). Both compounds have high detonation performance and their production is cost efficient due to cheap starting reagents used for their synthesis. The latest generation of high-energy materials represented by the caged structure of 2,4,6,8,10,12-hexanitro-2,4,6,8,10,12-hexaazaisowurtzitane (CL-20) delivers more energy but with increased hazards (Figure 3.1) [10, 11]. Therefore, an incorporation of nanotechnology tools to reduce mechanical sensitivities and toxic hazards of cyclic nitramines while retaining their functional properties and application prospective remains highly urgent. In this chapter, we provide a detailed analysis of the state-of-the-art investigations (since 2010) on nanoenergetic RDX, HMX, and CL-20 considering various factors affecting the size and shape of these materials and their influence on physicochemical parameters, detonation performance, and mechanical sensitivity.

3.2 Nanosized RDX

RDX, also referred to as 1,3,5-trinitroperhydro-1,3,5-triazine is one of the benchmark explosives that is known for over a century and is still widely used both in military and civilian technologies. RDX has good detonation velocity (8750 m s^{-1}) and high decomposition temperature (204 °C), which justify its wide utilization in various energetic formulations [12]. In energetic materials science, RDX has been used over the years as a reference for crystallization experiments, aiming to reduce the size of energetic organic crystals below a micrometer. Many crystallization processes inspired from other fields of chemistry, such as polymer crystallization, have been applied but only a few resulted in a significant size reduction [13].

A number of approaches for the preparation of nanosized RDX was proposed so far. One of the classical methods used is the solvent–antisolvent approach. RDX was precipitated into nanosized particles with spherical morphologies using acetone as a solvent and water as the antisolvent [14]. High concentration of RDX solution favors the formation of relatively larger particles while smaller nanoparticles were obtained when the antisolvent temperature was 70 °C. Recently, the micro RDX was comminuted to nano RDX particles by superfine milling [15]. Although according to scanning electron microscopy (SEM) and transmission electron microscopy (TEM) data the size of some of the particles was up to 270 nm, most of the pulverized RDX particles were nanosized (<100 nm). Using Merzhanov's model, a relationship between particle size, pore size, and critical temperature was established. It was concluded that the nanometer particle size led to a very small hot spot size and high critical temperature, which benefited a low sensitivity (IS of micro RDX: H_{50} = 46 cm; IS of nano RDX: H_{50} = 63 cm). At the same time, the activation energy of nanosized RDX is similar to that of micro RDX. In other words, nano RDX showed almost the same reactivity as micro RDX, which ensured the high safety of this nanoenergetics. In small-scale gap test, the shock sensitivity of nano RDX decreased by about 45% compared with that of micro RDX [16]. Interestingly, the values of activation energy obtained by the Kissinger and Ozawa methods for RDX samples with various particle sizes show that the activation energy for micro RDX is 1.5 times higher than that for nano RDX [17]. This fact is attributed to the two reasons: (i) reduction in particle size of RDX powders from micro to nanosize enhances surface area leading to increase in reactivity, and (ii) for the smaller and jagged particles in an explosive less thermal energy is needed to produce localized hot spots, which formed upon ignition of energetic materials. However, the thermolysis characteristics of micro RDX and nano RDX fabricated by wet milling on gram scale did not differ significantly. By comparing the values of activation energy, $\Delta G^{\neq}$, $\Delta H^{\neq}$, and $\Delta S^{\neq}$ between micro and nano RDX, it was shown that a decrease in particle size did not affect the thermal decomposition of the nitramine [16].

Nanosized RDX was also fabricated by evaporation–condensation technique. The average linear diameter of the particles was 35 nm. Thermal stability of nano RDX was comparable to micron-sized nitramine, but the burning rate was increased twofold for nanomaterial. This augmentation of the RDX monopropellant pressed charges is probably caused by the greater porosity for nano RDX [18]. Several factors could be responsible for the observed particle size influence. Atomic force microscopy shows that the RDX nanoparticles within pressed sample are much more accessible for the RDX decomposition gas products, as compared to the micron-sized RDX particles, because of the more developed particle–gas interface. The exothermic reactions of gasification and thermal decomposition of molten RDX on the burn surface are further assisted by inward diffusion of RDX decomposition gas products from the gas phase into the reacting layer resulting in the total acceleration in the heat release process [19].

Using bacterial cellulose (BC) gelatin with a three-dimensional network as a matrix and dimethylformamide (DMF) as a solvent, the RDX nanostructured explosives were prepared through the solvent–antisolvent method. The crystallization of RDX in the solution results from the simultaneous action of the polar DMF and the BC gelatin providing nano RDX with good crystallization and uniformity. RDX particles with diameters of 30–50 nm were coated with the nanofibers of BC gelatin through the hydrogen bonding at high RDX concentrations. The RDX contents in the nanostructured explosives increased with an increase in the concentrations of RDX solutions, and the average content of RDX was not less than 91% at the concentration of $0.30\,g\,ml^{-1}$. An incorporation of BC in the nanostructured explosives was found to be useful to decrease greatly the mechanical sensitivities of nanostructured explosives and improve their safety. When the RDX content was 91%, the impact and friction sensitivities of the nanocomposites were 40% and 48%, respectively, while for raw RDX material these values were 84% and 80%, correspondingly [20]. A similar solution–water suspension method was applied for the preparation of nano RDX-based plastic bonded explosives (PBX) comprising RDX, 2,4-dinitrotoluene, polyvinylacetate, and stearic acid in a mass ratio of 94.5 : 3 : 2 : 0.5. It was found that the particle size of nano RDX-PBX molding powder was smaller than that of micron-sized RDX-based PBX. Main thermolysis parameters (the initial temperature, the termination temperature, the derivative thermogravimetry (DTG) peak temperature, the endothermic peak temperature, and the exothermic peak temperature) of nano RDX-PBX were also lower than those of micro RDX-PBX [21].

Very recently, a solvent–antisolvent method for the synthesis and characterization of nanodiamond-RDX (ND-RDX) composites was employed [22]. Although RDX particles were of micrometer size, an incorporation of ND resulted in a significant desensitization toward impact and friction in comparison to pure nano RDX. The sensitivity thresholds of the composites were raised to 7 J (impact) and 360 N (friction), whereas they were only 4 J and 360 N for the pure material. The desensitization is due to both a lower concentration of defects in a sample on the nanoscale and the heat-dissipating ability of the ND surface and interfaces. Interestingly, it was found that NDs act as seeds for the core–shell growth stimulating the crystallization of RDX on the surface of NDs during the antisolvent process. The RDX layer was about 2–4 nm thick and was assumed to be free from internal inclusions. In Figure 3.2, ND-RDX particles obtained by antisolvent crystallization show an average size of 17 nm with low polydispersity (80% of the population is smaller than 20 nm).

A novel generation of smart energetic materials comprising a combination of nano RDX with nanothermites prepared in turn by mixing aluminum nanoparticles with metal-based oxidizers (WO_3 or $Bi_2(SO_4)_3$) was recently established [23].

Figure 3.2 SEM micrographs of ND-RDX particles obtained by antisolvent crystallization. Source: Guillevic et al. [22]. Reproduced with permission of American Chemical Society.

Such materials were found to be powerful initiating substances. Compared to primary explosives, they have higher reaction heats (3–5 $kJ\,g^{-1}$), similar true densities (2–5 $g\,cm^{-3}$), and release larger amounts of gases. In addition, they can be formulated without using highly toxic heavy metals (Pb, Hg, Co, Ni) and have lower sensitivity levels than primary explosives, making their handling much safer (Table 3.1). The detonation can be induced with small amounts of energetic nanocomposite (about 100 mg), on distances shorter than 2 cm, in a semi-confined configuration. These results are of prime importance for industrial applications for which the initiation of detonation in high explosives without primary explosives has been up till now a major impediment.

Another method for the production of RDX nanoparticles is spray drying. The process sprays a solution containing a dissolved compound or particles in suspension into a hot gaseous stream (air or nitrogen) thus crystallizing into particles and/or drying the granules. This method was employed for the preparation of RDX-based nanocomposites containing polyvinylacetate binder [24]. In-depth investigation of the formation of these nanocomposites showed that the size of RDX crystals decreases radially across the particles from the center to surface, with RDX crystals on the surface generally within the range of hundreds of nanometers, while the binder is distributed uniformly inside the particles. This data suggests that the crystallization of RDX begins with rapid nucleation and the shell formation plays an important role in determining the size distribution of RDX crystals, as the shell seems to stop the crystal growth of RDX on the surface region and promote

Table 3.1 The sensitivity values of RDX-based nanocomposite thermites and their comparison with reference explosives [23].

Energetic material	Composition	Impact sensitivity (J)	Friction sensitivity (N)	Electrostatic discharge (mJ)
Nanothermite	*n*-WO_3/nano Al	49.6	168	0.1
	$Bi_2(SO_4)_3$/nano Al	49.6	168	6.1
Secondary explosives	nano RDX (∼100 nm)	2.1	180	360
Nanocomposite thermites	12.4% nano WO_3/27.6% nano Al/60% nano RDX	1.6	72	16.9
	nano WO_3/nano Al/(10–90%) nano RDX	1.6–34.9	42–216	1.9–120.7
	20% $Bi_2(SO_4)_3$/20% nano Al/60% nano RDX	1.6	60	16.9
Primary explosives	Lead azide	0.6–4	0.3–0.5	6–12
	Lead styphnate	2.2–5	1.5	0.04–0.14

Source: Adapted from Comet et al. [23].

the growth of internal RDX crystals. Therefore, parameters that affect the shell formation and the degree of supersaturation can be used to control the RDX size on the surface region of the particles [24].

According to the spray flash evaporation technique, RDX is dissolved in a volatile solvent (e.g. acetone) and that solution is heated just before being sprayed into vacuum, where the crystallization is triggered by the sudden temperature depression and the solvent evaporation. This method enables a preparation of finely grounded RDX nanomaterial with particle sizes around 100 nm. In contrast to the aforementioned results, as-prepared nano RDX demonstrated increased sensitivity to impact compared to raw material, which sharply contradicts with the concept of desensitization of explosives through nanostructuring. At the same time, the sensitivities of nano RDX to friction and electrostatic discharge were lower than those of micro RDX. To explain such contradictions, it was proposed that sensitivities toward friction and electrostatic discharge are connected with internal defects that cause internal hot spot ignition, whereas sensitivity to impact is effectively influenced by external defects at interfaces that cause external hot spot ignition. X-ray diffraction analysis revealed a higher internal defect density and higher strain content for the micro RDX supporting the proposed theory [25].

The versatility of the spray flash evaporation technique allows the processing of solid (polyvinylpyrrolidone) and liquid (polyethylene glycol) polymers to tune the particle size distribution of the final dried nano, submicron, or micron-sized RDX powder. Polyethylene glycol triggers the early nucleation of RDX with low nucleation rate leading to bigger particles up to 5 μm. On the contrary, polyvinylpyrrolidone acts as a nucleation inhibitor and a growth inhibitor: RDX nuclei are formed in less volume available due to the fission and flashing of droplets, and then the crystal growth is slowed, thus allowing the formation of much smaller particles at 160 nm with a narrow distribution and a spherical shape. Despite a lower thermal stability, the synthesized RDX composites exhibit reduced sensitives in electrostatic discharge, friction, and impact without loss of reactivity [26].

A so-called sol-gel method is used to process energetic materials to achieve their desensitization by being embedded in a matrix, usually SiO_2. These SiO_2-based explosive gels are prepared by dissolving the energetic compound, the silica precursor, and a catalyst in a cosolvent. After the gelification, an antisolvent of the explosive is injected to replace the solvent in the pores and precipitate the explosives in the SiO_2 matrix. By drying with heating or at low pressure, a xerogel with higher density is obtained [27]. The capability of sol-gel method for high solid loading in the porous matrix of SiO_2 was exploited to prepare the RDX/SiO_2 composite xerogels containing up to 90% RDX [28]. The 3D structure of porous gel matrix restricts the growth of RDX particles crystallized out of solvent to nanometer size. The resulted xerogels contained homogeneously dispersed RDX nanoparticles (10–30 nm) with diffuse particle–pore boundaries in the SiO_2 matrix, keeping its basic structural morphology intact. Sensitivity tests performed for several nano RDX samples showed that the less porosity of RDX/SiO_2 composite resulted in its lower sensitivity to impact and friction.

Another nanodispersion process allowing nanostructuration of up to 90% charge of RDX in an organogel matrix was developed. Gel matrix was synthesized from phloroglucinol and nitrophloroglucinol as precursors and formaldehyde as coupling reagent. The impact sensitivity of nanocomposites was equivalent to macroformulations irrespective of the RDX mass content, but the reactivity was lowered as soon as the RDX percentage was below 90% [29].

The sol-gel method was also employed for the preparation of RDX/3,3-bis(azidomethyl)oxetane-tetrahydrofuran (RDX/BAMO-THF) nanocomposites with enhanced thermolysis performance and low sensitivity. The average sizes of RDX particles in RDX/BAMO-THF nanocomposites fell into a nanoscale range, indicating that BAMO-THF gel matrix could effectively inhibit the RDX particles from aggregating. The results of calculated kinetic and thermodynamic parameters indicated that RDX/BAMO-THF nanocomposites present high thermal reactivity and the thermal reactivity of RDX/BAMO-THF nanocomposites decreases with increasing the RDX content. Nevertheless, the impact sensitivities of RDX/BAMO-THF nanocomposites (H_{50} = 24–28 cm) were remarkably lower than those of raw RDX (H_{50} = 17 cm), which showed the opposite trend compared with the thermal reactivity [30].

A similar approach was later used to fabricate RDX-based energetic nanocomposites with glycidyl-azide polymer (GAP) as gel matrix [31]. The average sizes of RDX particles in RDX/GAP composites were at nanoscale, which were much smaller than those of raw RDX. During the vacuum freezing drying process, the dissolved RDX particles were recrystallized and the crystal growth was restrained by GAP gel matrix, resulting in the decrease in RDX particle sizes. The calculated results of kinetic and thermodynamic parameters indicate that RDX/GAP nanocomposites exhibited high thermolysis activities and the thermal stabilities of RDX/GAP nanocomposites increased upon an increase in the RDX content. A modification of this approach toward the construction of RDX/GAP nanocomposites involved a utilization of GAP xerogel with three-dimensional network structure prepared by sol-gel method and formed by polyether type of polyurethane through the curation of hexamethylene diisocyanate [32]. The specific surface area of cured GAP gel was much higher than that of the nanocomposite materials with explosive RDX and the average grain size was about 20–46 nm. Differential scanning calorimetry (DSC) analysis showed thermal decomposition peak temperature of RDX in formed RDX/GAP nanocomposite materials was 33–37 °C lower than that of pure RDX. In contrast, the quantity of the heat released by the nanocomposite materials increased with the increase of the amount of RDX, and it is significantly higher ($Q = 4876\ \text{kJ kg}^{-1}$) than that of the physical blend materials ($Q = 4084\ \text{kJ kg}^{-1}$).

A supercritical modification of the sol-gel method, namely solution enhanced dispersion by supercritical fluids (SEDS) method was applied for the preparation of RDX/SiO_2 nanocomposites [33]. The obtained nanocomposite consisted of RDX nanoparticles (about 60 nm) that were closely coated by more tiny SiO_2 nanoparticles (about 5.4 nm). After fabrication, the crystal phase and molecular structure of RDX did not change. Kinetic evaluation indicated that thermal decomposition of RDX/SiO_2 nanocomposite possessed of much higher activation energy and rate constant than decomposition of raw RDX, arguably due to the introduction of nano-SiO_2. The mechanical sensitivities of RDX/SiO_2 nanocomposite were found to be quite low, much lower than those of raw RDX or of the simple mixture of RDX and SiO_2, which is attributed to the buffer effect of nano SiO_2.

The structure of nano RDX can be stabilized electrostatically. It was shown, that the surface charge of milled RDX in a suspension can vary with electrolyte concentration and pH [34]. The maximum Zeta potential of −39.1 mV, was achieved at a pH = 11 and electrolyte (NaCl) concentration of $0.002\ \text{mol l}^{-1}$. Interestingly, the specific adsorption of ions onto the nano RDX surface, even at low electrolyte concentrations of $0.002\ \text{mol l}^{-1}$, can have a dramatic effect on the zeta potential of the nano RDX particle. According to the Derjaguin–Landau and Verwey–Overbeek (DLVO) theory, 300 nm RDX particles were predicted to be stable at electrolyte concentration of $0.002\ \text{mol l}^{-1}$ and unstable at $0.1\ \text{mol l}^{-1}$. Thus, stabilization could be easily regulated by adjusting the salt concentration of the suspension. Turbidity tests confirmed these results, showing that the stability of the suspension increased significantly with improved zeta potential.

Nano RDX may find application as a part of energetic formulation incorporating other nanoenergetic materials. For example, composition B (Comp B) is an explosive consisting of 59.5% RDX, 39.5% TNT, and 1% wax. As a rule, Comp B is produced by a melt-cast process in which a slurry of RDX crystals dispersed in molten TNT is cast and allowed to solidify into a charge [35]. Nanoenergetics-based Comp B (N-Comp B) consisting of nanoscale RDX and TNT crystals was recently prepared by spray drying and pressing [36]. Compared to other spray-dried energetic materials which are typically spherical, the irregular morphology of N-Comp B molding powder is likely due to the fusion of particles because TNT has a relatively low melting point (81 °C). Structural characterization showed that the majority of the voids inside the formulation were in the nanoscale range but have a large number density. Reduction in shock sensitivity was observed and attributed to the elimination of large voids, and yet the large number density of voids seemed to have constrained the sensitivity decrease. Nano RDX was also embedded in the composite modified double-base propellant either with low or high solid content to increase the maximum tensile strength and to decrease mechanical sensitivities [37, 38].

3.3 Nanosized HMX

HMX, also referred to as 1,3,5,7-tetranitro-1,3,5,7-tetrazocane is a very powerful high-energy material that is widely used in ammunitions, propellants, and energetic formulations [39]. HMX has excellent detonation velocity ($9114\,m\,s^{-1}$) and high decomposition temperature (280 °C) making this compound a golden standard in energetic materials science [12]. HMX exists in four polymorphic forms: α (orthorhombic), β (monoclinic), γ (monoclinic), and δ (hexagonal) [40]. β-HMX has the highest density among all of the known polymorphs and is the thermodynamically most stable polymorph at room temperature. Therefore, it is the desired polymorph to be crystallized and used as an explosive [41].

Nano HMX in a stable β-crystalline form can be easily prepared by an antisolvent precipitation method with an average particle size ranging from 30 to 128 nm under different experimental conditions. A higher temperature (70 °C) of the antisolvent and lower concentration of HMX in acetone (5 mM) favored the formation of smaller particles. Thermal decomposition of thus prepared nano HMX occurred similar to the thermal decomposition of bulk HMX [42].

Ultrafine HMX was prepared by spray drying [43]. An interesting observation by authors was that the in case of production of HMX with incorporation of nanoaluminum, the structure of submicron HMX particles changes with time after production (on the timescale of weeks to months). This effect also propagates in the changes in combustion velocity of pressed charges. The burning rate of HMX of micron- and submicron-sized was found to be equal in 20–100 atm pressure range.

In-depth investigations on the energetics, electronic structure, and vibrational properties of a series of spherical nano β-HMX NPs with the size range 1.4–3.4 nm using a combination of dispersion-corrected DFT and DFTB methods were recently performed [44]. It was found that the HMX nanoparticles with 3.4 nm have higher energy storage in the surface molecules, higher surface tension with great anisotropy, significantly lower enthalpy of sublimation, and lower melting point than the HMX bulk crystal. The melting points of the HMX nanoparticles were less than 400 K, while the nanoparticles with the sizes less than 2.2 nm cannot exist in solid form at room temperature. The electronic structure indicated that the surface molecules of the HMX nanoparticles were polarized due to the quantum confinement, which differed from either gas-phase or solid-phase molecules. The induced surface states remarkably reduced the band gap of the HMX nanoparticles by 0.8–1.5 eV, which can lower the material stability. The surface states on the valence band side originated dominantly from the (100) facets that possessed the highest surface energy among the exposed facets on the nanoparticles. The (100) facets covered by axial NO_2 groups had very high activity in triggering the chemical decomposition of the nanoparticles.

Similar observations were obtained upon DFTB-MD simulations in combination with DFT calculations to study the thermal decompositions of different α-HMX nanoparticles at high temperatures [45]. The global thermal decompositions of the HMX nanoparticles presented great dependence on the temperature and particle size. In the early decomposition stage, the expansion of the HMX nanoparticles was remarkable in competition with the initial chemical decomposition. The smaller NPs had the lower sublimation enthalpies. The following thermal decompositions were greatly dependent on the system temperatures. At low temperatures, the predominating decomposition was found to be the isomerization of the HMX molecule, while at high temperatures, the $N{-}NO_2$ homolysis accompanied by ring cleavage dominated. The concerted ring opening and HONO elimination occurred to a much less extent in the studied temperatures. There were a large number of activated complexes during the low-temperature decomposition channel, which led to more than four times of volumetric expansion. This might be unfavorable in condensed-phase HMX but can be available in real densely packed nanoscale HMX due to their differences in local molecular packing and external conditions. The subsequent expansion and recompression can magnify the local temperature and further accelerate their decompositions. The rate-limiting steps of the dominating decomposition channels at the earliest decomposition stages exhibited a clear second-order reaction character. The global activation energies in the three main channels were calculated to be in the range 22.9–32.6 kcal mol^{-1}.

HMX nanocrystals may also be fabricated using a mechanical ball milling approach. Although according to SEM and TEM data, a significant amount of HMX nanoparticles was presented, a mean size of material was 0.27 μm. As a result, thermal analysis confirmed that there was no distinct difference in thermal decomposition between the raw and the pulverized HMX samples [46].

The decomposition kinetics of bulk and nanoconfined HMX in solid and liquid phases was studied using DSC [47]. Dynamic heating scans showed that HMX rapidly decomposed after melting. For bulk HMX, the decomposition took place in the solid state at the slowest heating rates; started in the solid state, then passed through melting, and completed decomposition in the liquid states for the intermediate heating rates; at the highest heating rates, the decomposition took place only at the liquid state. For nano HMX, the decomposition occurred in a qualitatively similar pattern as in the bulk but the exotherm was shifted to lower temperatures. The activation energy of the bulk and nano HMX decreased upon an increase in the conversion. The reaction kinetics for bulk and nanoconfined HMX were well described by a first-order autocatalytic reaction model with the rate constant in the liquid state being approximately one order of magnitude larger than that in the solid state and with the rate constant increasing as pore size decreased. The decomposition reaction was accelerated for nanoconfined HMX by 1.4 and 1.6 times in the solid state before melting in 50 and 12 nm-diameter pores, respectively, and by 2.5 and 1.7 times in the liquid state in the 50- and 12 nm-diameter pores.

Depending on the initial pressure and temperature parameters, different phases of nano HMX can be obtained by using the spray flash evaporation technique. Low temperature (110 °C, 4 MPa) allows the elaboration of α-HMX and γ-HMX while higher temperature with the same pressure (160 °C, 4 MPa) favors the formation of only γ-HMX and higher pressure (160 °C, 6 MPa) leads to the predominance of the β-HMX phase. The sample synthesized at 110 °C, 4 MPa exhibited a phase transition from γ-HMX to α-HMX probably due to an aging process and the presence of α-HMX seeded in the sample. The sample containing only γ-HMX was rather stable with only a slight part transforming into β-HMX after a very long time. In composites containing HMX and TNT, the HMX also appears in its γ-HMX phase, and no transformation into α- or β-HMX phases occurred with time or heating at 60 °C [48]. However, a complete study with defined conditions of aging is required to give more information about the mechanisms and kinetics of the aging process.

The polymeric stabilization of HMX using polyvinylpyrrolidone (PVP) was investigated by zeta potential and turbidity measurements [49]. The zeta potential values for HMX surface with PVP demonstrated that the shorter PVP polymer chain (lesser molecular weight) adsorbed better than the longer one (greater molecular weight). Interestingly, the solvent dramatically affects the PVP adsorption: no PVP adsorption was observed with 50% ethanol concentration. The turbidity test showed that HMX with lower molecular weight PVP was more stable and stayed suspended for a much longer time compared to the higher one. The turbidity tests also showed that the stability of HMX was enhanced by increasing the ethanol concentration. PVP in 50% water/50% ethanol was able to stabilize HMX by depletion and reduced the size to 180 nm within 10 minutes of milling. Strong aggregation was observed for HMX milled in the absence of PVP (stabilization mechanism).

HMX nanoparticles were successfully prepared via supercritical antisolvent (SAS) process. Taguchi's robust design method was applied to optimize the operation conditions of HMX micronization by this method with aim of preparing HMX nanoparticles. It was found that pressure and temperature of the supercritical CO_2 have no significant effects on the particle size of produced HMX during SAS process; while, other studied variables including HMX concentration, solution flow rate, solvent identification, and flow rate of CO_2 have a considerable role in determination of product particles in this procedure. Under optimized conditions of SAS process (3.5 mol l^{-1} HMX concentration, 3 ml min^{-1} solution flow rate, cyclohexanone as solvent with 70 ml min^{-1} flow rate), HMX nanoparticles with average particle size of 56 nm were prepared [50].

HMX/GAP nanocomposites with high reactivity and low sensitivity were recently prepared by a sol-gel-supercritical method [51]. The average particle sizes of HMX particles in HMX/GAP nanocomposites were estimated at nanoscale, which indicated that HMX particles were much smaller than those of raw HMX particles and GAP gel matrix could effectively prevent the HMX particles from aggregating. Thus prepared HMX/GAP nanocomposites exhibited high thermal reactivity, and the onset decomposition temperature gradually decreased (by 75–80 °C) with the decrease of the content of HMX in HMX/GAP nanocomposites. The thermal reactivity of HMX/GAP nanocomposites increased as the content of HMX decreased in HMX/GAP nanocomposites. Nevertheless, the mechanical sensitivities of HMX/GAP nanocomposites were lower than those of raw HMX, which showed the opposite trend in comparison with the thermal reactivity.

A new insensitive explosive with HMX crystals, nanosized graphene oxide (GO) sheets, and viton was prepared by a solvent–slurry process. According to SEM and XPS analysis, the content of 4% viton, 1% GO, and 95% HMX provided a successful cladding layer. DSC data demonstrated that the GO sheets enhance the thermal stability of the HMX crystals. The drop height (H_{50}) of raw HMX was 19.6 cm, which indicated that HMX is sensitive to impact stimuli. After coating with viton, the impact sensitivity of HMX/viton significantly decreased, with H_{50} of 47.7 cm. In addition, with the introduction of graphite and GO, the impact sensitivity of the composites further decreased, and GO sheets exhibited lower impact sensitivity than graphite sheets toward HMX crystals. This is attributed to the fact that nano GO with a large specific surface area can form a dense layer on the HMX surface and transmit heat between each HMX crystal under impact stimuli. Therefore, fewer hot spots are generated, leading to highly reduced impact sensitivity. On the other hand, the HMX/viton/GO composites demonstrated the lowest shock sensitivity, with a small clapboard amount of 4.66 mm, evidently below that of HMX/viton (9.25 mm) and HMX/viton/graphite (7.49 mm). This implies that the GO sheet is an efficient absorber and diverter under the shock wave stimuli [52].

Very recently, a novel HMX@dual shell composite comprising HMX as a core, adhesive polydopamine (PDA) as an inner shell, and hexagonal boron nitride nanosheets (hBNNS) as an external shell was prepared through a facile bioinspired strategy of PDA in situ polymerization. The thermal behavior results indicated that the formed PDA shell could increase the $\beta \rightarrow \delta$ phase transition temperature and decrease the thermal decomposition temperature of HMX crystal due to the heat-resistant coating and the endothermic effect of PDA chain segment thermal motion, while the hBNNS could reduce the phase transition temperature but increase the thermal decomposition temperature owning to its excellent thermal conductivity and stability. The impact and friction sensitivities of the core@dual shell composites could be reduced significantly compared with the raw HMX and HMX@PDA composite, which should be attributed to the isolation and buffering effect, and outstanding thermal conductivity of soft hBNNS (Table 3.2) [53].

A novel core@double-shell (CDS) HMX-based energetic composite was constructed with an inner nano 1,3,5-triamino-2,4,6-trinitrobenzene (nano TATB) shell and outer PDA shell fabricated via a facile ultrasonic method and a simple immersion method, respectively. The inner nano TATB shell reduced the sensitivity of HMX while maintaining explosion performance; the outer PDA shell enhanced the interfacial interaction between explosive crystals and polymer binder. According to the DSC data, the $\beta \rightarrow \delta$ phase transition temperature shifted from 197.0 °C for pure HMX to 212.8 °C for HMX@TATB@PDA hybrid particles [54]. As shown in Figure 3.3, the as-prepared HMX particles have flat surfaces and sharp edge angles. After PDA coating, a relatively rough surface made up of a compact PDA shell was observed on the HMX surface. The self-polymerization of dopamine resulted in the deposition of PDA on the surface of nano-TATB and the filling of the nanovoids. The

Table 3.2 Comparison of mechanical sensitivities of HMX and HMX-based core@dual shell composites [53].

Energetic material	Shell content (%)	IS (H_{50}) (cm)	FS (*P*) (%)
HMX	—	38	100
HMX@PDA	PDA, 2.8	46	60
HMX@PDA@hBNNS	PDA, 2.8; hBNNS, 0.6	63	57
HMX@PDA@hBNNS	PDA, 2.8; hBNNS, 1.2	68	55
HMX@PDA@hBNNS	PDA, 2.7; hBNNS, 2.3	72	52

Source: Adapted from Zhang et al. [53].

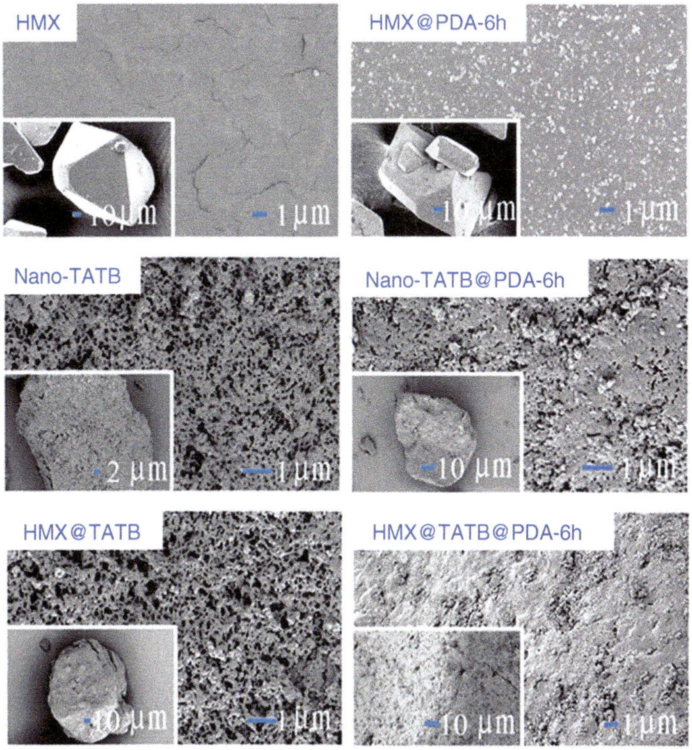

Figure 3.3 SEM images for HMX, nano-TATB, HMX@TATB, and their corresponding PDA-coated particles for six hours. Source: Lin et al. [54]. Reproduced with permission of American Chemical Society.

HMX@TATB particles exhibited the typical feature of well-developed core–shell structure. The uniform and continuous porous surface confirmed that nano-TATB was compactly coated on HMX crystals with almost full coverage. After nano-TATB coating, the edges and corners of HMX crystals became blurred. As for CDS particles, the PDA coating formed the second shell with a similar morphology of nano-TATB@PDA.

A series of energetic cocrystals of HMX and 2,6-diamino-3,5-dinitropyridine-1-oxide with different molar ratios under the constraint of 2D triaminoguanidine-glyoxal polymer has been prepared by a solvent–antisolvent method. Obtained cocrystals were of irregular polyhedron shape, whereas the obtained two types of cocrystals were plate-like with smooth and integrated surfaces. The phase and thermal stabilities of the new cocrystal were better than those of HMX due to an increased hydrogen bonding. The detonation velocity and detonation pressure of the prepared cocrystal were calculated to be 9.82 km s^{-1} and 46.2 GPa, respectively, exceeding those of raw HMX [55].

3.4 Nanosized CL-20

2,4,6,8,10,12-Hexanitro-2,4,6,8,10,12-hexaazaisowurtzitane (CL-20 or HNIW) is the top high-energy material known so far (Figure 3.4). It is a highly fusible explosive, that is more powerful than RDX and HMX [56]. Its high detonation rate and pressure make the CL-20 a suitable candidate for a wide range of military and commercial purposes. Under normal conditions, the four experimentally isolated polymorphs of CL-20 are known: α-, β-, γ-, and ε- phases. The stability of the isolated polymorphic forms decreased in the following order: $\varepsilon > \gamma > \alpha > \beta$. The ε-phase is the most attractive one because of its higher density (2.04 g·cm^{-3}) and lower sensitivity compared to other three phases [57].

The impact and friction sensitivity changes for HMX and CL-20 with particle size were given in [58]. The impact sensitivity is not affected by the decreasing of the particle sizes, while the friction sensitivity reduces. The authors suggested that the effect is partly caused by the measurement technique (the height of grooves on porcelain plate where the sample is poured, becomes greater that the particle size [59]), in part by the other crystal modifications present in submicron nitramines.

Historically, various methods for the preparation of nano CL-20 using either dissolution/recrystallization approach or crushing the explosive were employed [60]. During grinding, the structural elements of the solid are destroyed mechanically, increasing the dispersion of the mass. Compared to grinding, crystallization procedures for the deposition of explosive particles have significant advantages. Crystals can be grown slowly and with clear confidence that after the crystallization process is completed, they will have a well-defined crystal shape and structure. However, in recent years, all known methods for production of nanoenergetic

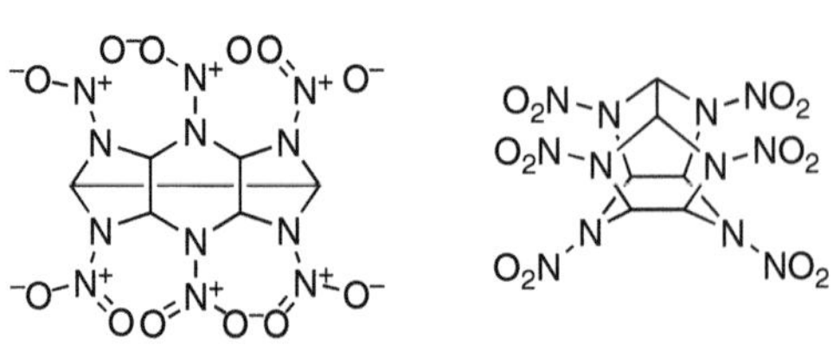

Figure 3.4 Molecular structure of CL-20 in 2D (left) and in 3D format (right). Source: A.A. Larin, L.L. Fershtat.

Table 3.3 Critical thermal explosion and impact sensitivity of raw and ultrafine CL-20 prepared by ultrasound and spray-assisted precipitation.

Energetic material	Shape	Particle size	H_{50} (cm)	T_c (°C)
Raw CL-20	Polyhedral	30–100 μm	13	235.6
Ultrafine CL-20	Spherical	470 nm	38	229.8

materials, such as grinding [61], recrystallization by solvent–insoluble type [62], microemulsion [63], and rapid expansion of supercritical solution [64] were successfully transferred to the preparation of nano CL-20.

3.4.1 Ultrasound- and Spray-Assisted Precipitation of Ultrafine CL-20

The first attempts to prepare nanoparticles of CL-20 were performed a decade ago [65]. However, it was not possible to produce the desired nanoparticles. As becomes obvious from SEM images raw CL-20 has a polyhedral morphology and its particle size is about 30–100 mm. An ultrasound- and spray-assisted recrystallization method [62] was explored to prepare spherical ultrafine CL-20 particles with a narrow particle size distribution and a mean diameter of 470 nm. Comparing functional properties, the drop height (H_{50}) as a value of impact sensitivity increased for ultrafine CL-20 from 13 to 38 cm (Table 3.3). On the contrary, the critical explosion temperature of ultrafine CL-20 decreased from 235.6 to 229.8 °C, which suggests that the thermal stability of ultrafine CL-20 is lower than that of raw CL-20.

3.4.2 Preparation of CL-20 Nanoparticles Via Oil in Water Microemulsions

CL-20 nanoparticles were also prepared by the oil in water microemulsion method [66]. In this study, various factors that affect the limpid and thermodynamic stability of microemulsion formation were optimized via experimental microemulsion method for micronization. The studies on various parameters illustrated that the nature of the organic solvent has the most significant effect on microemulsion formation, in which *n*-butyl acetate was the best as an immiscible organic solvent. The weight ratio of water/organic phase played no significant role at the investigated levels. Under the optimal conditions for the microemulsion method proposed by the Taguchi method, β-HNIW nanoparticles with an average size of 80 ± 10 nm were prepared.

3.4.3 Production of Nanoscale CL-20 Using Ultrasonic Spray-Assisted Electrostatic Adsorption Method (USEA)

A new method for the production of nano CL-20 using ultrasonic spray-assisted electrostatic adsorption method (USEA) was recently developed [67]. Various experimental conditions affecting safety factors and the crystallization process were studied. Thus prepared nano CL-20 showed a wide size distribution in the range from 150 to 600 nm with an average value of 270 nm. The exothermic peak of ultrafine CL-20 increased by 12 °C compared to the traditionally manufactured raw CL-20 up to 232.9 °C. There was no obvious aggregation of nano CL-20 particles in the SEM images. It was also observed that chaotic and complicated aerosol size distribution created by a piezoelectric converter during the aerosol production process affects the size of the aerosol and its distribution. Thus, droplets of small aerosol size crystallize small particles, and droplets of large aerosol size crystallize large particles at the crystallization stage. The repulsive force existing in the electrostatic field between the nanoparticles prevents the crystals from sticking together, while the ultrasound effect can also reduce the size to nanoscale. It is expected that such a simple and continuous approach will become an attractive potential for the manufacture of other organic nanoparticles.

3.4.4 Preparation of Nano CL-20 Via Sonocrystallization

Sonocrystallization is the use of powerful ultrasound to control the crystallization process and is used as a mechanical means for mixing and cavitating the medium during the salting process. Crystallization by ultrasound is an important method of controlling the crystallization process. Ultrasound is mainly used to influence the process of nucleation and orderly control of crystallization [68]. Ultrasound creates cavitation in liquid media, and each cavitation event consists of first opening a small (several tens of micrometers) gas or steam void, followed by a strong collapse cavitation as a means of generating nuclei and (due to high local supersaturation conditions) allows crystals to form and grow. At high intensity, ultrasound can be used instead of seed crystals and/or to start nucleation at a lower degree of volume supersaturation than usual.

Sonocrystallization is a safe method of using ultrasound to control the point of nucleation and the total number of nuclei formed to obtain the best crystal composition of products. This process provides homogeneous, non-agglomerated crystals of nanoscale CL-20 [60]. Nanoparticles of explosives have a very large surface area and a better ability to heat transfer, and therefore they are more insensitive to stimulating the formation of hot spots due to impact, friction, or even impact. This can probably be explained by the smaller number of occlusions and dislocations present in these particles.

CL-20 nanoparticles were obtained by precipitation crystallization via spraying of a prepared solution of CL-20 in ethyl acetate as a solvent with isooctane as an antisolvent [60]. The results showed that nano CL-20 particles have the shape of spheres or ellipsoids with an average size of 95 nm. Due to their small diameter and high surface energy, the particles tended to agglomerate. CL-20 sample showed random crystals with a high degree of agglomeration and CL-20 particle size obtained at 600 nm. Application of ultrasonication provided the correct uniform shape of a crystal with an average particle size of 95 nm and a suitable particle size distribution. The data obtained during the study of CL-20 sensitivity to impact were improved

Table 3.4 Effect of particle size on the sensitivity.

Average particle size (μm)	Shape	Friction sensitivity (kg)	Electrostatic discharge (J)	Impact sensitivity (cm)
15	Spheres or ellipsoids	6.4	45	25
4		8	49	32
1		No reaction	49	40
0.095		No reaction	60	55

Table 3.5 Thermal behavior of bulk, micro- and nanoscale CL-20.

Type of particles	Average particles size (μm)	*T* (phase transition), (°C)	T_{max} (dec.) (°C)
Raw	40	171	244.7
Micro	0.5	156	235.5
Nano	0.26	151	232.9

from 25 cm for a particle size of 15 μm to 55 cm for an average particle size of 95 nm (Table 3.4) [60].

3.4.5 Preparation of Micro-Sized Particles and their Comparison with Nanoscale CL-20

The comparison of thermal transitions of micro and nano CL-20 prepared by ultrasonic spray-assisted electrostatic adsorption was considered recently [67]. Thermal transitions of bulk, micro, and nano CL-20 were investigated using DSC and TG measurements. As it turned out, the thermal behavior of CL-20 involves an endothermic process with a solid–solid phase transition. The endothermic reaction indicates that a phase transition is taking place in the CL-20 molecule. Three endothermic peaks were observed at 171, 156, and 151 °C, which corresponded to the phase shift temperature ε→γ for bulk CL-20, micro CL-20, and nano CL-20, respectively (Table 3.5).

The crystal phase transition is associated with the transformation of N–NO_2 groups of CL-20. It should be noted that the temperature of the nano CL-20 phase transition is lower than that of bulk and micro CL-20. It is believed that a possible reason for this is the difference between the particle size distribution of nano CL-20 and micro CL-20. Small nano CL-20 crystals have better reactivity, which leads to a low phase transition temperature. The DSC data for nano CL-20 showed the process of exothermic decomposition in the temperature range 220–250 °C. However, for raw and micro CL-20 materials, the maximum exothermic decomposition was 244.7 and 235.5 °C, respectively. This value is higher than that of nano CL-20, namely, there is a shift of 12 and 3 °C to a lower temperature.

3.4.6 Method for Production of Nano CL-20 in Supercritical CO_2 and 1,1,1,2-Tetrafluoroethane (TFE)

The preparation of microcrystals of the polymorphic composition of CL-20 was carried out using supercritical methods of SAS, GAS (gaseous antisolvent), and RESS (rapid expansion of supercritical solution) [69]. The use of solvents containing carbon dioxide and 1,1,1,2-tetrafluoroethane can be effectively controlled by both the micronization method and the media used. In the first case, when liquefied or supercritical CO_2 is used as an antisolvent, the SAS and GAS methods allow using only α-CL-20 in the form of a solvate with CO_2, regardless of the process parameters. In the second case, the use of TFE as an active medium for the GAS method allows selectively obtaining homogeneous ultrafine particles of ε-CL-20 under mild conditions, while the use of RESS leads mainly to the formation of β-phase crystals. Unlike known analogs, the TFE-based method developed by GAS for the formation of ε-CL-20 is characterized by resistance to the initial quality of the substrate (untreated CL-20), increased environmental and industrial safety, ease of product separation, and the ability to control the morphology of particles and the crystalline phase.

3.4.7 Creation of Nano CL-20 Particles by the Method of Bidirectional Rotary Mill

Nano CL-20 samples were produced by a bidirectional rotary mill and were successfully pulverized to be pseudospheres with an average size of 200 nm [61]. The thermal decomposition peak temperature of nano CL-20 was 239.6 °C lower than that of micrometer-sized CL-20 at a heating rate of 15 °C min^{-1}. The test results revealed that the impact and friction sensitivities of the CL-20 nanocrystals were reduced by 116.2% and 22%, respectively (Table 3.6).

3.4.8 Electrospray of CL-20 Particles

Spherical particles of nano and microscale CL-20 can be continuously formed by electrospray [57]. The morphology and particle size mainly depend on the manufacture of nano- and micron-sized CL-20 spheres by electrospray properties of the precursor solution. Solid particles of several hundred nanometers in size could be obtained if acetone or a mixture of acetone and a small amount of ethyl acetate were used as a solvent. However, when ethyl acetate was used as a solvent, hollow layered spherical particles ranging in size from a submicron to several microns were obtained. The layers of hollow spheres were a cluster of primary particles with a

Table 3.6 Sensitivities of raw and nano CL-20.

CL-20	T_{p0} (°C)	T_b (°C)	H_{50} (cm)	P (%)
Raw	223	232	14	82
Nano	215	224	29	66

Table 3.7 Sensitivities of raw, micro, and nano CL-20 prepared by electrospray technique.

CL-20	Density (g cm^{-3})	Shape	H_{50} (cm)	P (%)
Raw	1.97		15	76
Nano	1.96	Nano-sized hollow sphere	24	60
Micro	2.03	Micro-sized hollow sphere	20	66

size of nanometers. The sprayed samples had the form of β-polymorph. Thus, the density of the sprayed CL-20 was 1.97 g cm^{-3}, which was lower than that of the untreated CL-20. Impact sensitivity and friction sensitivity decreased by 59.2% and 21.1%, respectively (Table 3.7), for smaller solid particles compared to untreated CL-20. The activation energy for thermal decomposition of sprayed CL-20 was 175.66 and 189.01 kJ·mol^{-1} for untreated CL-20, which shows that sprayed CL-20 decomposes more easily than raw CL-20.

The densities of nano and micro CL-20 were nearly the same, 1.97 and 1.96 g cm^{-3}, respectively. This also indirectly proved that the hollow layered micron CL-20 consisted of primary particles, which were also solid particles. Compared with the raw CL-20 density (2.03 g cm^{-3}), the decrease in density values was mainly caused by changes in the crystalline phase from ε to β. The change in sensitivity values with decreasing particle size can be explained by the theory of hot spots. The smaller particle size led to better mechanical sensitivity, since the phase change from ε to β increased mechanical sensitivity. Consequently, the change in particle size had a greater impact than the change in the crystal phase for nano and micro CL-20. The difference in sensitivity between raw and micro materials was mainly due to the grinding process. The sharp edges of the broken particle would create hot spots under the influence of mechanical stimuli.

3.4.9 Production of Sub-Micro CL-20-Based Energetic Polymer Composite Ink

Sub-micro CL-20 material was prepared by the ball milling method and then mixed with an energetic binder consisting of GAP (glycidyl azide polymer) and polyisocyanate to study properties of polymer composite ink [70]. Sub-micro CL-20-based ink showed wettability and good uniformity without cracks, porosity, and voids. The sub-micro CL-20 and CL-20/GAP samples exhibited different impact sensitivities (H_{50}). The obtained CL-20 samples displayed reduced sensitivity in contrast to refined CL-20; the H_{50} value increased from 21.1 to 37.2 cm. A possible reason for this phenomenon is that the binder reduces the friction between the particles of CL-20, thus reducing the phenomenon of stress concentration and also partially absorbing the heat, preventing self-heating of the CL-20 to a certain extent. The thermal properties of micro CL-20 and CL-20/GAP samples showed that the raw CL-20 and sub-micro CL-20 exhibited exothermic decomposition processes in the temperature range from 230 to 250 °C and from 220 to 260 °C, respectively. The

Table 3.8 Sensitivities of CL-20-based nanoscale composites produced by ultrasound-assisted emulsion.

CL-20 based composite	H_{50}, (cm)	Shape	Particle size (μm)
CL-20/DNB	55	Polyhedron	500
CL-20/TNT	30	Prism	300
CL-20/PNCB	63	Sphere	5–10
CL-20 and PNCB mixture	31	Polyhedron	2–8

reason the T_{max} value is lower than that for raw CL-20 is attributed to the high surface energy at micro diameters [70].

3.4.10 Nanoscale Composites Based on CL-20

Using the ultrasound-assisted emulsion (UAE) method a micron-scale monodisperse spherical energy composite consisting of CL-20 and *p*-nitrochlorobenzene (PNCB) was manufactured [71]. The surface of the composite was smooth, and the average particle size was about 5–8 μm. UAE is a convenient, environmental friendly, and fast process. Thus obtained CL-20/PNCB composite showed better thermodynamic stability. The drop height corresponding to an explosion probability of 50% showed that the microscale spherical composite CL-20/PNCB was more insensitive compared to other energetic materials composites based on CL-20 (Table 3.8).

Figure 3.5 shows the morphologies of the raw materials and the as-prepared CL-20/PNCB composite. The raw CL-20 possessed a fusiform microstructure (Figure 3.5a), and the PNCB was an irregular prismatic crystal (Figure 3.5b). In contrast, the shape of the CL-20/PNCB composite was spherical with good dispersibility, as shown in Figure 3.5c. The surface of the composite was smooth, and the average particle size was about 5–8 μm, which was calculated by taking statistics from more than 200 particles, as shown in Figure 3.5d. The parameters of the CL-20 composite, such as its smooth spherical surface and monodisperse size distribution, present advantages for reducing its sensitivity.

Sensitivity depends on the structure, shape, and size of materials. The newly formed structure and the interaction between molecules in the finished composite can increase the stability of the crystal structure and contribute to a decrease in sensitivity. Similarly, the extremely small-sized CL-20/PNCB composite also exhibited size-dependent properties, including thermal decomposition, performance, and especially sensitivity. It was found, that the average diameter of small particles was too small to become a hot spot during exposure, which led to a decrease in impact sensitivity. Meanwhile, when faced with external stimuli, spherical CL-20/PNCB composite material can significantly improve insensitivity compared to nonspherical crystals by reducing friction between neighboring particles. For the above reasons, the impact sensitivity of the finished spherical composite CL-20/PNCB showed satisfactory results.

Figure 3.5 SEM micrographs of (a) raw CL-20 crystal, (b) raw PNCB crystal, (c, d) micro-sized target spherical CL-20/PNCB composite. Source: Zhu et al. [71]. MDPI, CC BY 4.0.

3.4.11 Comparison of the Detonation Performance of Micro–/Nanoscale High-Energy Materials

Considering the main energy materials (RDX, HMX, CL-20), it is impossible not to compare their detonation and physicochemical characteristics for micro and nanoscale sizes. The average particle sizes of micron-sized RDX, HMX, and CL-20 are 83.64, 120.36, and 49.29 mm with wide size distributions, and the average sizes of the nano-sized particles are 0.16, 0.16, and 0.18 mm with narrow size distributions, respectively. The micron-sized particles are polyhedral, irregular, and heterogeneous, while the nano-sized particles are semispherical and homogeneous. As shown in Table 3.9, the mean explosion probability of micro-sized RDX, HMX, and CL-20 was 80%, 86%, and 88%, respectively; however, under the same experimental conditions, the same values for nano-sized RDX, HMX, and CL-20 were 50%, 58%, and 66%, which means friction sensitivities were decreased by 30%, 28%, and 22%, respectively. As listed in Table 3.3, compared with the micron-sized samples, the gap thicknesses (δ) of the nano-sized samples are 9.21, 7.88, and 25.41 mm thinner, which means that the shock sensitivities of the nano-sized samples are decreased by 59.9%, 56.4%, and 58.1%, respectively. Furthermore, the standard deviations (S_{dev}) of the nano-sized samples are smaller than those of the micron-sized samples, which states that the size and morphology of the particles are close to each other. As a result, the shock initiation probabilities of the samples are close to each other. The characteristic heights for 50% probability of initiation (H_{50}) of micro-sized RDX, HMX, and CL-20 were 49.8, 44.1, and 13.6 cm using 2.5 kg drop hammer, respectively. However, the H_{50} of nano-sized RDX, HMX, and CL-20 was increased to 99.1, 63, and 29.4 cm, which means that the impact sensitivities were reduced by 99%, 42.8%, and 116.2%, respectively. Thus, nanoscale energetic materials can

Table 3.9 The size, friction, and impact sensitivities of RDX, HMX, and CL-20.

Energetic material	Average size (μm)	Thickness (δ)	S_{dev}	P (%)	H_{50} (cm)
Micro-RDX	83.64	15.38	0.41	80	49.8
Nano RDX	0.16	6.17	0.32	50	99.1
Micro-HMX	120.36	13.96	0.40	86	44.1
Nano-HMX	0.16	6.08	0.32	58	63
Micro-CL-20	49.29	43.72	0.42	88	13.6
Nano-CL-20	0.18	18.31	0.35	66	29.4

Table 3.10 Mechanical properties of raw and nanoscale CL-20 and PETN.

Energetic material	H_{50} (cm)		P (%)		Shape	Particle size (nm)
	2.5 kg hammer	5 kg hammer	66 °C, 2.45 MPa	80 °C, 2.45 MPa		
Raw CL-20	18.2	8.3	100	100	Sphere	
Nano CL-20	37.6	15.4	100	100	Sphere	116
Raw PETN	24.5	10.1	100	69	Sphere	
Nano PETN	31.2	17.5	90	63	Sphere	446

be used for applications that require less sensitive energetic materials with high detonation performance [72].

The special properties of nanoenergetic CL-20 and pentaerythritol tetranitrate (PETN) were researched using the mechanical grinding method. Although the d_{90} values for crushed CL-20 and PETN were only 111.6 and 446.7 nm, respectively, instead of traditional nanoparticles smaller than 100 nm, the authors still called them nano CL-20 and nano PETN. Analyses such as XRD, IR, and XPS showed that the crushed samples had the same crystalline phase, molecular structure, and surface elements as the raw explosives [73]. Nevertheless, it was found that thus prepared CL-20 and PETN showed parameters close to those of raw CL-20 and PETN, respectively. DSC-IR assay was used to investigate the decomposition mechanisms of nano CL-20 and nano PETN. For nano CL-20, the main products were CO_2 and N_2O. Authors attributed this to redox reactions between the cleaved NO_2 radicals and the C–NO_2 fragments. For nano PETN, the main product has changed to NO_2. In addition, the mechanical and thermal properties of raw CL 20, nano CL-20, raw PETN, and nano PETN were tested (Table 3.10). Undoubtedly, nanoexplosive substances were more insensitive than micro-explosive substances. According to the theory of hot spots, this feature was explained by a significant reduction in the size of pores and voids, which was the result of a significant reduction in the particle size of explosives.

Coarse CL-20 and PETN are very sensitive to mechanical stress. Due to its high sensitivity, PETN was commonly used as a booster explosive and even as a primary explosive in main charges. Nano PETN has less mechanical sensitivity than raw PETN. The results for CL-20 samples were similar to the results for PETN samples. In fact, explosions during mechanical sensitivity tests were also initiated by a heating element of a certain type, which is also a thermochemical problem. Theoretically, a random explosion was initiated by the formation of hot spots that formed when the explosive charge was stimulated by impact, friction, or shock wave. After they have been initiated, the hot spots will grow and spread, which will determine whether the charge will explode. The precursors of hot spots are pores or voids embedded in the explosive charge. When exposed to mechanical action or a detonation wave, pores and voids were compressed in an adiabatic process, because heat was not dissipated in such a short time, which led to a sharp increase in temperature inside the pores [74]. The micron-scale morphologies of nano CL-20 and nano PETN were determined with a field emission scanning electron microscope, and the resulting SEM images showed that particles were with sphere-like morphologies. The results of size particles were not as expected in that the d_{90} values of milled CL-20 and PETN were only 111.6 and 446.7 nm, respectively, instead of traditional nanoparticles with sizes less than 100 nm.

3.4.12 Preparation of Nano-Sized CL-20/NQ Co-Crystal Via Vacuum Freeze Drying

One of the strategies that is currently pursued to retain the performance of explosives while significantly reducing their sensitivity is to employ their cocrystals in formulations, as cocrystals can have distinct properties compared to the corresponding coformer crystals [75]. Using a vacuum freeze-drying method, nano-sized energetic cocrystal consisting of CL-20 and a typical insensitive explosive used in propellants nitroguanidine (NQ) was prepared [76].

The sensitivity results are listed in Table 3.11. As shown in the table, the friction sensitivity of the cocrystal decreased to 52% from the significantly higher friction sensitivity of 100% for neat CL-20. As for the impact sensitivity, the result of the cocrystal was 36% that was evidently lower than for neat CL-20, even at the same level with NQ. It is conjectured that the intermolecular hydrogen bond increases the stability of molecular system of the cocrystal. Moreover, as it was mentioned above, nano-sized particles also have lower sensitivities. NQ represents needle crystal or crystalline powder. ε-CL-20 particles are irregular polyhedrons with uneven size and a very wide distribution range. Most of the particles are 50 mm or more in diameter. CL-20/NQ nano cocrystals are regularly spherical-shaped particles with a narrow size distribution. Most of the particle sizes are less than 500 nm. According to the hot spot theory, the spherical shape will help to uniformly spread heat to the interior of the particle and to form less amount of hot spots. Therefore, the mechanical sensitivity can be greatly reduced.

Freeze vacuum drying is an efficient way to prepare nano-sized energetic cocrystals, which also has the potential for batch preparation. Therefore, a production

Table 3.11 Results of sensitivity test for NQ, CL-20, 1 : 1 mixture, and CL-20/NQ cocrystal.

Energetic material	H_{50}, (cm)	*P* (%)	Shape
NQ	0	0	Needle crystal
CL-20	100	100	Polyhedron
CL-20/NQ mixture	76	80	Sphere
Nano CL-20/NQ	36	52	Sphere

of other energetic cocrystals for various applications using this approach can be achieved in the near future.

3.4.13 Nanoscale 2CL-20·HMX High Explosive Cocrystal Synthesized by Bead Milling

The 2CL-20·HMX cocrystal was first produced by Bolton et al. by a solution precipitation route [77] and has since been made by solvent drop grinding using resonant acoustic mixing [78]. Recently, an energetic nanoscale cocrystal of CL-20 and HMX in a 2 : 1 M ratio, namely 2CL-20·HMX, was prepared by a novel method of bead milling an aqueous suspension of ε-CL-20 and β-HMX [79]. The conversion of the coformers to the cocrystal form was monitored by powder X-ray diffraction and scanning electron microscopy of specimens sampled at various milling times. Complete conversion to the cocrystal form was achieved after 60 minutes of milling. Rounded 2CL-20·HMX cocrystal particles with a mean size below 200 nm were produced. As an inherently safe manufacturing method, the aqueous bead milling process has great potential in advancing cocrystal research and applications in the field of energetic materials.

Figure 3.6 shows scanning electron microscopy images of raw CL-20, raw HMX, and CL-20/HMX cocrystal explosive. Figure 3.6c displays spherical-shaped microparticles with sizes ranging from 0.5 to 5 μm. As indicated in Figure 3.6d, CL-20/HMX microparticles are composed of numerous tiny cocrystals, which are plate-like in shape. Plate-like cocrystals have thicknesses below 100 nm [80].

Another recent research [81] deals with CL-20/HMX cocrystals, which were prepared with high purity, uniform morphology, well-proportioned size distribution, compact internal structure, and reduced sensitivity through a solvent-induced self-assembling approach using corresponding nanoparticles as the basic units. Nanoparticle self-assembly is a novel strategy to obtain CL-20/HMX cocrystals, indicating that it is a promising CL-20/HMX cocrystals after assembly exhibit a desensitization effect. For HMX obtained by recrystallization, the impact energy was 5.5 J, and the sensitivity value for recrystallized CL-20 with size of about 50 μm was 3 J. In contrast, the impact energy of the CL-20/HMX cocrystal corresponding to 4.5 J was measured [82]. Thus, the impact energy of the CL-20/HMX cocrystal is significantly higher than that of raw CL-20, indicating that energy close to that of HMX was gained, while the sensitivity was reduced by the self-assembly method.

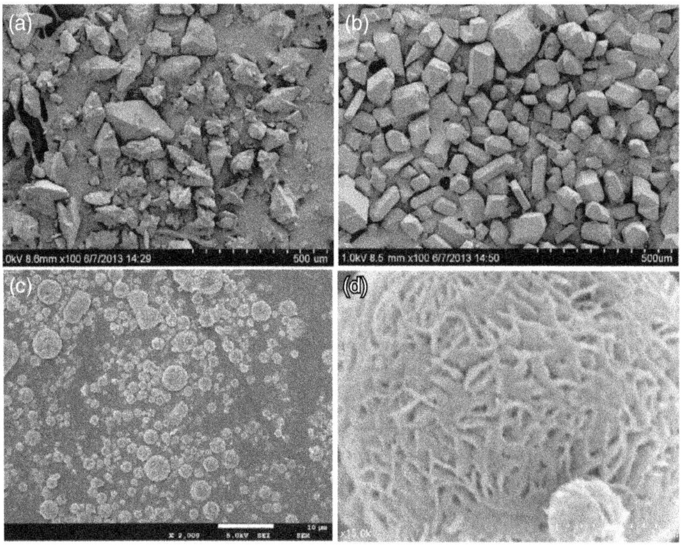

Figure 3.6 SEM images of explosive samples: (a) raw CL-20; (b) raw HMX; and (c, d) CL-20/HMX cocrystal explosive. Source: Zhu et al. [71]. MDPI, CC BY 4.0.

Moreover, uniform morphology and few defects may enhance the impact energy of the CL-20/HMX cocrystal. Thus, the impact energy of the CL-20/HMX cocrystals increases due to interfacial self-assembly strategy, providing a facile and appreciable advantage in the field of nanoscale energy materials.

3.4.14 Mechanochemical Fabrication and Properties of CL-20/RDX Nano Co/Mixed Crystals

It was found, that by milling CL-20 and RDX together, a nano CL-20/RDX co/mixed crystal explosive with a mean particle size of 141.6 nm with near-spherical shape was formed [83]. To evaluate the safety of CL-20/RDX, tests on the impact, friction, and thermal sensitivities were performed, with the results presented in Table 3.12. Raw CL-20 and raw RDX are of the mean size of 59 and 88 μm, respectively. The sample "CL-20/RDX blend" is a simple mixture of raw CL-20 and raw RDX. After milling, the impact and friction sensitivities of the nano CL-20/ RDX co/mixed crystal significantly decreased. The characteristic heights (H_{50}) of raw CL-20 and raw RDX were 15 and 42 cm, respectively, while the same value of the nano co/mixed crystal explosive was 51 cm, which was higher than the 36 cm value of raw CL-20 and 10 cm of raw RDX. In addition, the explosion probability of friction (P) was 56% lower than that of raw CL-20, raw RDX, and their blends. However, the thermal sensitivity test results are contrary to the mechanical sensitivity test results. The burst point (T_b) of CL-20/RDX was only 243 °C, which means that it has the highest thermal sensitivity compared to those of the raw materials. CL-20/RDX is not suitable for use as a heat-resistant explosive.

After milling, the morphology of the co/mixed crystal explosive was near-spherical, and the particle size revealed a normal distribution. The milled sample

Table 3.12 Mechanical and thermal sensitivity of raw CL-20, raw RDX, CL-20/RDX blend, and CL-20/RDX co/mixed crystal prepared by mechanochemical fabrication.

Energetic material	Shape	Particle size	H_{50} (cm)	P (%)	T_b (°C)
Raw CL-20		59 μm	15	100	284
Raw RDX		88 μm	42	74	301
CL-20/RDX blend	Near-spherical	141.6 nm	30	100	—
CL-20/RDX co/mixed crystal	Near-spherical		51	56	244

Table 3.13 Special height of raw material and nano CL-20/HMX cocrystal prepared by milling method.

Energetic material	H_{50} (cm)	P (%)	T_b (°C)
Raw CL-20	15	100	283.57
Raw HMX	27	74	301.20
CL-20/HMX blend	20	100	—
CL-20/HMX co/mixed crystal (particle size 81.6 nm)	33	56	243.51

showed the same molecular structure and surface elements as the raw materials, but the XRD test shows that CL-20/RDX has a new crystal phase and the Raman and IR spectra gave a supplementary confirmation for the existence of a cocrystal phase in the milled sample. The activation energy of the thermal decomposition of CL-20/RDX was 206.49 kJ mol^{-1} which is higher than that of raw RDX.

3.4.15 Preparation of Nano CL-20/HMX Cocrystal by Milling Method

Owing to milling CL-20 and HMX together, nano-Cl-20/HMX cocrystal with mean particle size of 81.6 nm was prepared under the conditions of the 0.3 mm diameter of milling balls [84]. The micromorphology of the explosives was found to be near-spherical and the particle size revealed normal distribution. Before and after milling, the element composition and molecular structure of the CL-20/HMX did not change in comparison to the raw CL-20 and HMX. In impact sensitivity tests, with 5 kg hammer, the characteristic height (H_{50}) of CL-20/HMX was 33 cm. Compared with the starting raw materials and simple mixture, impact sensitivity of CL-20/HMX cocrystal was much better (Table 3.13).

3.4.16 Synthesis of Nano CL-20/HMX Co-Crystals by Ultrasonic Spray-Assisted Electrostatic Adsorption Method

In another research [85] a facile, large-scale synthesis of an energetic nano cocrystal composed of a 2 : 1 M ratio of CL-20 and HMX via the USEA method was performed. The mean particle size was 50 nm, and the particle size distribution

Table 3.14 Sensitivity of raw CL-20, raw TNT, CL-20/TNT mixture, and CL-20/TNT co-crystal prepared by mechanical ball-milling method.

Energetic material	Impact sensitivity (H_{50}, cm)	Friction sensitivity (P, %)
Raw CL-20	13	100
Raw TNT	92	4
CL-20/TNT cocrystal	39	68
CL-20/TNT mixture	18	92

was relatively narrow. The cocrystal produced was stable because of a number of C—H hydrogen bonds, each involving nitro group oxygen atoms. The density of the nanosized CL-20/HMX cocrystal was 1.945 g cm^{-3} at room temperature and 2.006 g cm^{-3}at 100 K. Its detonation velocity was 9480 m s^{-1}, which is higher than that of pure HMX explosive and comparable with that of pure CL-20. The cocrystal CL-20/HMX exhibited high energy-release efficiency and a unique exothermic peak at 243.5 °C. The peak shifts are left by 40 °C from that of pure HMX. This strategy not only combined the advantages of cocrystals and nanoeffects but also opens up new perspectives and advances in development in the science and technology of organic cocrystal materials. The method may be employed in industrial production on a large scale for preparing high-performance materials with excellent properties.

3.4.17 Preparation of Nano-CL-20/TNT Cocrystal Explosives by Mechanical Ball-Milling Method

Nano CL-20/TNT cocrystal explosive was successfully prepared by mechanical ball milling with 0.38 mm grinding beads [86]. The proposed mechanical ball-milling method is a green cocrystal preparation technique. Using this method, Zhao et al. prepared nanoscale CL-20/HMX cocrystal explosives [84] and Song et al. prepared nanoscale CL-20/RDX cocrystal explosives [83].

Impact and friction sensitivities of CL-20, TNT, the CL-20/TNT mixture, and thus obtained nano CL-20/TNT cocrystal explosives are summarized in Table 3.14. According to this data, nano CL-20/TNT cocrystal exhibited lower impact and friction sensitivity than the raw materials or CL-20/TNT mixture. Compared with CL-20 and the CL-20/TNT mixture, the characteristic drop height of nano CL-20/TNT cocrystal explosives prepared by the mechanical ball-milling method increased by 26 and 22 cm, respectively, indicating that mechanical ball-milling can effectively reduce the sensitivity of CL-20.

Both impact and friction sensitivities of the nano-CL-20/TNT cocrystal are lower than those of the raw materials or CL-20/TNT mixture, demonstrating improved safety of the nano CL-20/TNT cocrystal explosives. By combining nanocrystallization and cocrystal formation, this study provides an alternative route for industrial preparation of cocrystal explosives. Additionally, nano-CL-20/TNT cocrystal explosives are promising substitutes for CL-20.

3.4.18 Preparation of Nanoscale CL-20/Graphene Oxide by One-Step Ball Milling

The nanoscale CL-20-based/graphene oxide composite was first prepared in 2018 [87]. A one-step production method involved exfoliating graphite materials (GIMs) off into graphene materials (GEMs) in aqueous suspension of CL-20 and forming CL-20/graphene materials (CL-20/GEMs) composites by using ball milling. Mechanical ball milling is a desirable choice because it is suitable for massive and continuous preparation of uniform morphology crystals that maintained the original crystal form. The impact sensitivities of CL-20/GEMs composites were contrastively investigated. The impact sensitivities of CL-20 with different content of GEMs are lower than that of milling CL-20. The reduced impact sensitivities of CL-20/GEMs were supposed from the excellent lubrication and heat conduction of GEMs, which could reduce the internal folding dislocations and hot spots [88]. Moreover, the impact sensitivities were reduced with the increase of GEM content.

3.4.19 Preparation and Properties of CL-20 Based Composite by Direct Ink Writing

Direct ink writing (DIW) is the state-of-the-art method for energetic materials loading [89, 90]. Compared with other conventional charging styles for energetic materials, DIW offers an attractive alternative for depositing functional materials in complex structures due to its 3D motion and precision. Despite the particle size of raw CL-20 ranges from 50 to 200 mm, the mean thickness of 100 layers of composite is 0.24 mm. The thickness of single film is 2.4 mm. The density of composite stays at 1.70 g cm^{-3}, while the theoretical density is 1.97 g cm^{-3}. Furthermore, the inks prepared by energetic materials are safe. An application of DIW to the preparation of CL-20-based composite resulted in a formation of energetic material with a drop height higher than that of raw ε-CL-20, since the particle size of CL-20 reduced after recrystallization and the binders coated in the surface reduce the appearance rate of hot spots [91]. Thus, the impact sensitivity decreased to a certain degree though the morphology of CL-20 changed (Table 3.15).

CL-20-based all-liquid ink was deposited into glass to obtain sub-micro composite. The properties of composite were characterized by SEM, XRD, drop height, and DSC. Thickness of single layer is 2.4 mm and density of composite approaches 86% of theoretical density. The morphology phase of CL-20 changes for the addition of binders, nevertheless, the impact sensitivity is reduced. Critical size of detonation is around 13 × 0.4 mm, which contributed to detonation transformation under

Table 3.15 Impact sensitivity of CL-20-based composite and raw CL-20.

Energetic material	H_{50} (cm)	Average size particles (μm)	Density (g cm^{-3})
Raw ε-CL-20	13	50–200	1.93
CL-20 based composite	23	0.05–0.1	1.70

Table 3.16 Impact sensitivity of CL-20 based composite, sub-micro CL-20 and raw CL-20.

Energetic material	Impact sensitivity, H_{50} (cm)
Raw εCL-20	13
Sub-micro CL-20	34
CL-20 based composite	40

micro dimension. DSC results indicated that thermal stability of composite slightly decreased compared to raw ε-CL-20.

3.4.20 CL-20 Based Explosive Ink of Emulsion Binder System for Direct Ink Writing

A method for the preparation of emulsion, the high-density viton A and polymer binder mixed together formed a uniformly binder system of explosive ink, sub-micro CL-20 as body explosive, formulated into explosive ink [88]. As is shown in Table 3.16, the particle size of the raw CL-20 is about 400 mm. The particle size of sub-micro CL-20 powder refined by ball milling are all below 1 mm with no obvious edges and the particles are spherical in shape, which meets the preparation conditions of explosive ink. Combined DIW technology was used to load explosive ink in the condition of micro size to obtain CL-20-based composite. Interestingly, the morphology of CL-20 did not change even if ethyl acetate was used as a solvent for viton A in the binder system, the CL-20-based composite had less internal defects and the impact sensitivity was reduced. The drop height of sub-micro CL-20 was much higher than that of raw ε-CL-20, suggesting that sub-micro CL-20 is more difficult to explode under impact stimulus. Comparison of sub-micro CL-20 and CL-20 based composite showed that the H_{50} value increases from 34 to 40 cm. A possible reason for this phenomenon is that the binder reduces the friction between the particles of CL-20 and leads to a reduction in the probability to form hot spots [92]. Critical size of detonation of the prepared CL-20-based composite was around 13 × 0.17 mm and detonation velocity was 8079 m s^{-1}.

3.5 Conclusions and Future Outlook

Covered in the present chapter nanoscale cyclic nitramines correspond to the highly energetic materials which are applied in wide area of industrial and military fields. In recent decade, numerous approaches toward the preparation of nanosized RDX, HMX, and CL-20 energetic materials were developed. The main goal of these investigations was directed not only to enhance the explosive performance, but also to achieve the necessary safety concerns arose from the production, transportation, and storage of energetic materials. The safety of energetic materials is defined in terms of the sensitivity to external stimuli, such as impact, friction, and thermal energy. Desensitizing energetic materials for propellants or explosives is still of great importance because of the reduced shock sensitivity, resulting in lessened explosion risks.

New ideas for surface modification of energetic crystals and fabrication of high-performance energetic materials using general, facile, and scalable approaches based on sol-gel processing, spray drying, and supercritical technologies were proposed. In addition, it has been proved that in situ polymerization became an important and matured way to construct polymer-coated core–shell energetic materials. Nano cocrystallization not only combined the advantages of cocrystals and nanoeffects, but also unveiled new perspectives and advances in development in the science and technology of organic cocrystal materials. These methods may be employed in industrial production on a large scale for preparing high-performance materials with excellent properties. There is no doubt that the construction of novel nanoenergetic composites, explosives, and propellants will be in the spotlight of energetic chemistry and engineering in next decades and will provide new fundamental and applied perspectives for advanced materials.

Declaration of Originality

We confirm that this article has not been published elsewhere and it has not been submitted for publication elsewhere.

References

1 Ma, X., Li, Y., Hussain, I. et al. (2020). Core–shell structured nanoenergetic materials: preparation and fundamental properties. *Advanced Materials* 32: 2001291. https://doi.org/10.1002/adma.202001291.

2 DeLuca, L.T. (2019). Nanoenergetic ingredients to augment solid rocket propulsion, Chapter 6. In: *Nanomaterials in Rocket Propulsion Systems* (ed. Q.-L. Yan, G.-Q. He, P.-J. Liu and M. Gozin), 177–261. Amsterdam: Elsevier.

3 Gao, B., Qiao, Z., and Yang, G. Review on nanoexplosive materials, Chapter 2. In: *Nanomaterials in Rocket Propulsion Systems* (ed. Q.-L. Yan, G.-Q. He, P.-J. Liu and M. Gozin), 31–79. Amsterdam: Elsevier.

4 Patel, V.K., Joshi, A., Kumar, S. et al. (2021). Molecular combustion properties of nanoscale aluminum and its energetic composites: a short review. *ACS Omega* 6: 17–27. https://doi.org/10.1021/acsomega.0c03387.

5 Rossi, C. (2014). Two decades of research on nano-energetic materials. *Propellants, Explosives, Pyrotechnics* 39: 323–327. https://doi.org/10.1002/prep.201480151.

6 Huang, B., Cao, M.-H., Nie, F. et al. (2013). Construction and properties of structure- and size-controlled micro/nano-energetic materials. *Defence Technology* 9: 59–79. https://doi.org/10.1016/j.dt.2013.10.003.

7 Van der Heijden, A. (2018). Developments and challenges in the manufacturing, characterization and scale-up of energetic nanomaterials – a review. *Chemical Engineering Journal* 350: 939–948. https://doi.org/10.1016/j.cej.2018.06.051.

8 Zeng, G., Zhao, L., Zenget, X. et al. (2018). Preparation progress of micro/nano-energetic materials. *IOP Conference Series: Materials Science and Engineering* 382: 022030. https://doi.org/10.1088/1757-899X/382/2/022030.

9 Kuchurov, I.V., Zharkov, M.N., Fershtat, L.L. et al. (2017). Prospective symbiosis of green chemistry and energetic materials. *ChemSusChem* 10: 3914–3946. https://doi.org/10.1002/cssc.201701053.

10 Simpson, R.L., Urtiew, R.A., Ornellas, D.L. et al. (1997). CL-20 performance exceeds that of HMX and its sensitivity is moderate. *Propellants, Explosives, Pyrotechnics* 22: 249–255. https://doi.org/10.1002/prep.19970220502.

11 Fedyanin, I.V., Lyssenko, K.A., Fershtat, L.L. et al. (2019). Crystal solvates of energetic 2,4,6,8,10,12-hexanitro-2,4,6,8,10,12-hexaazaisowurtzitane molecule with [bmim]-based ionic liquids. *Crystal Growth & Design* 19: 3660–3669. https://doi.org/10.1021/acs.cgd.8b01835.

12 Klapötke, T.M. (2018). *Energetic Materials Encyclopedia*. Berlin: DeGruyter.

13 Pessina, F. and Spitzer, D. (2017). The longstanding challenge of the nanocrystallization of 1,3,5-trinitroperhydro-1,3,5-triazine (RDX). *Beilstein Journal of Nanotechnology* 8: 452–466. https://doi.org/10.3762/bjnano.8.49.

14 Kumar, R., Siril, P.F., and Soni, P. (2014). Preparation of nano-RDX by evaporation assisted solvent-antisolvent interaction. *Propellants, Explosives, Pyrotechnics* 39: 383–389. https://doi.org/10.1002/prep.201300104.

15 Wang, Y., Song, X., Song, D. et al. (2015). Foci for determining the insensitivity features of nanometer RDX: nanoscale particle size and moderate thermal reactivity. *Central European Journal of Energetic Materials* 12 (4): 799–815.

16 Wang, Y., Jiang, J., Song, D. et al. (2013). A feature on ensuring safety of superfine explosives. The similar thermolysis characteristics between micro and nano nitroamines. *Journal of Thermal Analysis and Calorimetry* 111: 85–92. https://doi.org/10.1007/s10973-011-2191-4.

17 Fathollahi, M., Mohammadi, B., and Mohammadi, J. (2013). Kinetic investigation on thermal decomposition of hexahydro-1,3,5-trinitro-1,3,5-triazine (RDX) nanoparticles. *Fuel* 104: 95–100. https://doi.org/10.1016/j.fuel.2012.09.075.

18 Frolov, Y.V., Pivkina, A.N., Zav'yalov, S.A. et al. (2010). Physicochemical characteristics of the components of energetic condensed systems. *Russian Journal of Physical Chemistry B* 4 (6): 916–922. https://doi.org/10.1134/S1990793110060072.

19 Pivkina, A., Ulyanova, P., Frolov, Y. et al. (2004). Nanomaterials for heterogeneous combustion. *Propellants, Explosives, Pyrotechnics* 29 (1): 39–48. https://doi.org/10.1002/prep.200400025.

20 Luo, Q., Pei, C., Liu, G. et al. (2015). Insensitive high cyclotrimethylenetrinitramine (RDX) nanostructured explosives derived from solvent/nonsolvent method in a bacterial cellulose (BC) gelatin matrix. *NANO: Brief Reports and Reviews* 10: 1550033. https://doi.org/10.1142/S1793292015500332.

21 Liu, J., Bao, X. -z., Rong, Y. -b. et al. (2018). Preparation of nano-RDX-based PBX and its thermal decomposition properties. *Journal of Thermal Analysis and Calorimetry* 131: 2693–2698. https://doi.org/10.1007/s10973-017-6731-4.

22 Guillevic, M., Pichot, V., Cooper, J. et al. (2020). Optimization of an antisolvent method for RDX recrystallization: influence on particle size and internal defects. *Crystal Growth & Design* 20: 130–138. https://doi.org/10.1021/acs.cgd.9b00893.

23 Comet, M., Martin, C., Klaumünzer, M. et al. (2015). Energetic nanocomposites for detonation initiation in high explosives without primary explosives. *Applied Physics Letters* 107: 243108. https://doi.org/10.1063/1.4938139.

24 Qiu, H., Stepanov, V., di Stasio, A.R. et al. (2015). Investigation of the crystallization of RDX during spray drying. *Powder Technology* 274: 333–337. https://doi.org/10.1016/j.powtec.2015.01.032.

25 Klaumünzer, M., Pessina, F., and Spitzer, D. (2017). Indicating inconsistency of desensitizing high explosives against impact through recrystallization at the nanoscale. *Journal of Energetic Materials* 35: 375–384. https://doi.org/10.1080/07370652.2016.1199610.

26 Pessina, F., Schnell, F., and Spitzer, D. (2016). Tunable continuous production of RDX from microns to nanoscale using polymeric additives. *Chemical Engineering Journal* 291: 12–19. https://doi.org/10.1016/j.cej.2016.01.083.

27 Reiser, J.T., Ryan, J.V., and Wall, N.A. (2019). Sol–gel synthesis and characterization of gels with compositions relevant to hydrated glass alteration layers. *ACS Omega* 4: 16257–16269. https://doi.org/10.1021/acsomega.9b00491.

28 Ingale, S.V., Sastry, P.U., Wagh, P.B. et al. (2013). Preparation of nano-structured RDX in a silica xerogel matrix. *Propellants, Explosives, Pyrotechnics* 38: 515–519. https://doi.org/10.1002/prep.201100160.

29 Wuillaume, A., Beaucamp, A., David-Quillot, F. et al. (2014). Formulation and characterizations of nanoenergetic compositions with improved safety. *Propellants, Explosives, Pyrotechnics* 39: 390–396. https://doi.org/10.1002/prep.201400021.

30 Chen, T., Li, W., Jiang, W. et al. (2018). Preparation and characterization of RDX/BAMOTHF energetic nanocomposites. *Journal of Energetic Materials* 36: 424–434. https://doi.org/10.1080/07370652.2018.1473900.

31 Chen, T., Gou, B., Hao, G. et al. (2019). Preparation, characterization of RDX/GAP nanocomposites, and study on the thermal decomposition behavior. *Journal of Energetic Materials* 37: 80–89. https://doi.org/10.1080/07370652.2018.1539786.

32 Li, G., Liu, M., Zhang, R. et al. (2015). Synthesis and properties of RDX/GAP nano-composite energetic materials. *Colloid and Polymer Science* 293: 2269–2279. https://doi.org/10.1007/s00396-015-3620-x.

33 Zhang, J., Liu, Y., Zhang, X. et al. (2017). Thermal decomposition and sensitivities of RDX/SiO_2 nanocomposite prepared by an improved supercritical SEDS method. *Journal of Thermal Analysis and Calorimetry* 129: 733–741. https://doi.org/10.1007/s10973-017-6210-y.

34 Doukkali, M.A., Patel, R.B., Stepanov, V. et al. (2017). The effect of ionic strength and pH on the electrostatic stabilization of NanoRDX. *Propellants, Explosives, Pyrotechnics* 42: 1066–1071. https://doi.org/10.1002/prep.201700096.

35 Smith, D.L. and Thorpe, B.W. (1973). Fracture in the high explosive RDX/TNT. *Journal of Materials Science* 8: 757–759. https://doi.org/10.1007/BF00561228.

36 Qiu, H., Stepanov, V., Patel, R.B. et al. (2017). Preparation and characterization of nanoenergetics based composition B. *Propellants, Explosives, Pyrotechnics* 42: 1309–1314. https://doi.org/10.1002/prep.201700165.

37 Liu, J., Ke, X., Xiao, L. et al. (2018). Application and properties of nano-sized RDX in CMDB propellant with low solid content. *Propellants, Explosives, Pyrotechnics* 43: 144–150. https://doi.org/10.1002/prep.201700211.

38 Yu, H., Sun, S., Gao, J. et al. (2021). Application of nano-sized RDX in CMDB propellant with high solid content. *Propellants, Explosives, Pyrotechnics* 46: https://doi.org/10.1002/prep.202100076.

39 Qiu, H., Stepanov, V., Chou, T. et al. (2012). Single step production and formulation of HMX nanocrystals. *Powder Technology* 226: 235–238. https://doi.org/10.1016/j.powtec.2012.04.053.

40 Lewis, J.P., Sewell, T.D., Evans, R.B. et al. (2000). Electronic structure calculation of the structures and energies of the three pure polymorphic forms of crystalline HMX. *Journal of Physical Chemistry B* 104: 1009–1013. https://doi.org/10.1021/jp9926037.

41 Brand, H.V., Rabie, R.L., Funk, D.J. et al. (2002). Theoretical and experimental study of the vibrational spectra of the α, β, and δ phases of octahydro-1,3,5,7-tetranitro-1,3,5,7-tetrazocine (HMX). *Journal of Physical Chemistry B* 106: 10594–10604. https://doi.org/10.1021/jp020909z.

42 Kumar, R., Siril, P.F., and Soni, P. (2015). Optimized synthesis of HMX nanoparticles using antisolvent precipitation method. *Journal of Energetic Materials* 33: 277–287. https://doi.org/10.1080/07370652.2014.988774.

43 Muravyev, N., Frolov, Y., Pivkina, A. et al. (2010). Influence of particle size and mixing technology on combustion of HMX/Al compositions. *Propellants, Explosives, Pyrotechnics* 35 (3): 226–232. https://doi.org/10.1002/prep.201000028.

44 Liu, Z., Zhu, W., and Xiao, H. (2016). Surface-induced energetics, electronic structure, and vibrational properties of β-HMX nanoparticles: a computational study. *Journal of Physical Chemistry C* 120: 27182–27191. https://doi.org/10.1021/acs.jpcc.6b09795.

45 Liu, Z., Zhu, W., Ji, G. et al. (2017). Decomposition mechanisms of α-octahydro-1,3,5,7-tetranitro-1,3,5,7-tetrazocine nanoparticles at high temperatures. *Journal of Physical Chemistry C* 121: 7728–7740. https://doi.org/10.1021/acs.jpcc.7b01136.

46 Wang, Y., Jiang, W., Song, X. et al. (2013). Insensitive HMX (octahydro-1,3,5,7-tetranitro-1,3,5,7-tetrazocine) nanocrystals fabricated by high-yield, low-cost mechanical milling. *Central European Journal of Energetic Materials* 10 (2): 277–287.

47 Bari, R., Koh, Y.P., McKenna, G.B. et al. (2020). Decomposition of HMX in solid and liquid states under nanoconfinement. *Thermochimica Acta* 686: 178542. https://doi.org/10.1016/j.tca.2020.178542.

48 Pichot, V., Seve, A., Berthe, J.-E. et al. (2017). Study of the elaboration of HMX and HMX composites by the spray flash evaporation process. *Propellants, Explosives, Pyrotechnics* 42: 1418–1423. https://doi.org/10.1002/prep.201700171.

49 Doukkali, M.A., Gauthier, E., Patel, R.B. et al. (2019). Polymeric stabilization of nano-scale HMX suspensions during wet milling. *Propellants, Explosives, Pyrotechnics* 44: 557–563. https://doi.org/10.1002/prep.201800153.

50 Bayat, Y., Pourmortazavi, S.M., Iravani, H. et al. (2012). Statistical optimization of supercritical carbon dioxide antisolvent process for preparation of HMX nanoparticles. *Journal of Supercritical Fluids* 72: 248–254. https://doi.org/10.1016/j.supflu.2012.09.010.

51 Chen, T., Jiang, W., Du, P. et al. (2017). Facile preparation of 1,3,5,7-tetranitro-1,3,5,7-tetrazocane/glycidylazide polymer energetic nanocomposites with enhanced thermolysis activity and low impact sensitivity. *RSC Advances* 7: 5957–5965. https://doi.org/10.1039/c6ra27780b.

52 Wang, J., Ye, C., An, C. et al. (2016). Preparation and properties of surface-coated HMX with viton and graphene oxide. *Journal of Energetic Materials* 34: 235–245. https://doi.org/10.1080/07370652.2015.1053016.

53 Zhang, S., Gao, Z., Jia, Q. et al. (2020). Bioinspired strategy for HMX@hBNNS dual Shell energetic composites with enhanced desensitization and improved thermal property. *Advanced Materials Interfaces* 7: 2001054. https://doi.org/10.1002/admi.202001054.

54 Lin, C., Huang, B., Gong, F. et al. (2019). Core@Double-Shell structured energetic composites with reduced sensitivity and enhanced mechanical properties. *ACS Applied Materials & Interfaces* 11: 30341–30351. https://doi.org/10.1021/acsami.9b10506.

55 Xue, Z.-H., Zhang, X.-X., Huang, B. et al. (2021). Assembling of hybrid nano-sized HMX/ANPyO cocrystals intercalated with 2D high nitrogen materials. *Crystal Growth & Design* 21: 4488–4499. https://doi.org/10.1021/acs.cgd.1c00386.

56 Mao, X., Jiang, L., Zhu, C. et al. (2018). Effects of aluminum powder on ignition performance of RDX, HMX, and CL-20 explosives. *Advances in Materials Science and Engineering* 2018: 5913216. https://doi.org/10.1155/2018/5913216.

57 Yan, S., Li, M., Sun, L. et al. (2018). Fabrication of nano- and micron- sized spheres of CL-20 by electrospray. *Central European Journal of Energetic Materials* 15: 572–589. https://doi.org/10.22211/cejem/100682.

58 Muravyev, N.V., Meerov, D.B., Monogarov, K.A. et al. (2021). Sensitivity of energetic materials: evidence of thermodynamic factor on a large array of CHNOFCl compounds. *Chemical Engineering Journal* 421: 129804. https://doi.org/10.1016/j.cej.2021.129804.

59 Radacsi, N., Bouma, R.H.B., Krabbendam-la Haye, E.L.M. et al. (2013). On the reliability of sensitivity test methods for submicrometer-sized RDX and HMX particles. *Propellants, Explosives, Pyrotechnics* 38 (6): 761–769. https://doi.org/10.1002/prep.201200189.

60 Bayat, Y. and Zeynali, V. (2011). Preparation and characterization of nano-CL-20 explosive. *Journal of Energetic Materials* 29: 281–291. https://doi.org/10.1080/07370652.2010.527897.

61 Guo, X., Ouyang, G., Liu, J. et al. (2014). Massive preparation of reduced-sensitivity nano CL-20 and its characterization. *Journal of Energetic Materials* 33: 24–33. https://doi.org/10.1080/07370652.2013.877102.

62 Wang, J., Li, J., An, C. et al. (2012). Study on ultrasound- and spray-assisted precipitation of CL-20. *Propellants, Explosives, Pyrotechnics* 37: 670–675. https://doi.org/10.1002/prep.201100088.

63 Mandal, A.K., Thanigaivelan, U., Pandey, R.K. et al. (2012). Preparation of spherical particles of 1,1-diamino-2,2-dinitroethene (FOX-7) using a micellar nanoreactor. *Organic Process Research & Development* 16: 1711–1716. https://doi.org/10.1021/op200386u.

64 Stepanov, V., Krasnoperov, L.N., Elkina, I.B. et al. (2005). Production of nanocrystalline RDX by rapid expansion of supercritical solutions. *Propellants, Explosives, Pyrotechnics* 30: 178–183. https://doi.org/10.1002/prep.200500002.
65 Wang, P.Y., Shi, C.H., He, Q.Y. et al. (2009). Preparation and characterization of ultrafine CL-20. *Initiators & Pyrotechnics* 1: 19–21.
66 Bayat, Y., Zarandi, M., Khadiv-Parsi, P. et al. (2015). Statistical optimization of the preparation of HNIW nanoparticles via oil in water microemulsions. *Central European Journal of Energetic Materials* 12: 459–472.
67 Wang, D., Gao, G., Yang, G. et al. (2016). Preparation of CL-20 explosive nanoparticles and their thermal decomposition property. *Journal of Nanomaterials* 5462097. https://doi.org/10.1155/2016/5462097.
68 Suslick, K.S. (1995). Applications of ultrasound to materials chemistry. *Materials Research Society* 20: 29–34. https://doi.org/10.1557/S088376940004464X
69 Zharkov, M.N., Kuchurov, I.V., and Zlotin, S.G. (2020). Micronization of CL-20 using supercritical and liquefied gases. *CrystEngComm* 22: 7549–7555. https://doi.org/10.1039/d0ce01167c.
70 Wang, D., Zheng, B., Guo, C. et al. (2016). Formulation and performance of functional sub-micro CL-20-based energetic polymer composite ink for direct-write assembly. *RSC Advances* 6: 112325–112331. https://doi.org/10.1039/C6RA22205F.
71 Zhu, Y., Lu, Y., Gao, B. et al. (2018). Synthesis, characterization, and sensitivity of a CL-20/PNCB spherical composite for security. *Materials* 11: 1130. https://doi.org/10.3390/ma11071130.
72 Liu, J., Jiang, W., Yang, Q. et al. (2014). Study of nano-nitramine explosives: preparation, sensitivity and application. *Defence Technology* 10: 184–189. https://doi.org/10.1016/j.dt.2014.04.002.
73 Song, X., Wang, Y., and An, C. (2018). Thermochemical properties of nanometer CL-20 and PETN fabricated using a mechanical milling method. *AIP Advances* 8: 1–14. https://doi.org/10.1063/1.5030155.
74 Wang, J.Y., Li, H.Q., and An, C.V. (2015). Preparation and characterization of ultrafine CL-20/TNT cocrystal explosive by spray drying method. *Chinese Journal of Energetic Materials* 23: 1103–1106. https://doi.org/10.16251/j.cnki.1009-2307.2015.11.013.
75 Millar, D.I., Maynard-Casely, H.E., Allan, D.R. et al. (2012). Crystal engineering of energetic materials: co-crystals of CL-20. *CrystEngComm* 14: 3742–3749. https://doi.org/10.1039/C2CE05796D.
76 Gao, H., Du, P., Ke, X. et al. (2017). A novel method to prepare nano-sized CL-20/NQ co-crystal: vacuum freeze drying. *Propellants, Explosives, Pyrotechnics* 42: 889–895. https://doi.org/10.1002/prep.201700006.
77 Bolton, O., Simke, L.R., Pagoria, P.F. et al. (2012). High power explosive with good sensitivity: a 2:1 cocrystal of CL-20:HMX. *Crystal Growth & Design* 12: 4311–4314. https://doi.org/10.1021/cg3010882.
78 Anderson, S.R., am Ende, D.J., Salan, J.S. et al. (2014). Preparation of an energetic-energetic cocrystal using resonant acoustic mixing. *Propellants, Explosives, Pyrotechnics* 39: 637–640. https://doi.org/10.1002/prep.201400092.
79 Qiu, H., Patel, R.B., Damavarapu, R.S. et al. (2015). Nanoscale 2CL-20·HMX high explosive cocrystal synthesized by bead milling. *CrystEngComm* 17: 4080–4083. https://doi.org/10.1039/C5CE00489F.

80 An, C., Li, C., Ye, B. et al. (2017). Nano-CL-20/HMX cocrystal explosive for significantly reduced mechanical sensitivity. *Journal of Nanomaterials* 3791320. https://doi.org/10.1155/2017/3791320.

81 Zhang, M., Tan, Y., Zhao, X. et al. (2020). Seeking a novel energetic co-crystal strategy through the interfacial self-assembly of CL-20 and HMX nanocrystals. *CrystEngComm* 22: 61–67. https://doi.org/10.1039/c9ce01447k.

82 Liu, Y., Niu, S., Lai, W. et al. (2019). Crystal morphology prediction of energetic materials grown from solution: insights into the accurate calculation of attachment energies. *CrystEngComm* 21: 4910–4917. https://doi.org/10.1039/C9CE00848A.

83 Song, X., Wang, Y., and Zhao, S. (2018). Mechanochemical fabrication and properties of CL-20/RDX nano co/mixed crystals. *RSC Advances* 8: 34126–34135. https://doi.org/10.1039/C8RA04122A.

84 Zhao, S., Song, X., Wang, Y. et al. (2018). Characterization of nano-CL-20/HMX cocrystal prepared by mechanical milling method. *Journal of Solid Rocket Technology* 41: 479–482. https://doi.org/10.7673/j.issn.1006-2793.2018.04.014.

85 Gao, B., Wang, D., Zhang, J. et al. (2014). Facile, continuous and large-scale synthesis of CL-20/HMX nano co-crystals with high-performance by ultrasonic spray-assisted electrostatic adsorption method. *Journal of Materials Chemistry A* 2: 19969–19974. https://doi.org/10.1039/C4TA04979A.

86 Hu, Y., Yuan, S., Li, X. et al. (2020). Preparation and characterization of nano-CL-20/TNT cocrystal explosives by mechanical ball-milling method. *ACS Omega* 5: 17761–17766. https://doi.org/10.1021/acsomega.0c02426.

87 Ye, B., An, C., Zhang, Y. et al. (2018). One-step ball milling preparation of nanoscale CL-20/graphene oxide for significantly reduced particle size and sensitivity. *Nanoscale Research Letters* 13: 42. https://doi.org/10.1186/s11671-017-2416-y.

88 Li, Q., An, C., Han, X. et al. (2018). CL-20 based explosive ink of emulsion binder system for direct ink writing. *Propellants, Explosives, Pyrotechnics* 43: 533–537. https://doi.org/10.1002/prep.201800064.

89 Nellums, R.R., Son, S.F., and Groven, L.J. (2014). Preparation and characterization of aqueous nanothermite inks for direct deposition on SCB initiators. *Propellants, Explosives, Pyrotechnics* 37: 670–675. https://doi.org/10.1002/prep.201400013.

90 Wang, J., Xu, C., and An, C. (2017). Preparation and properties of CL-20 based composite by direct ink writing. *Propellants, Explosives, Pyrotechnics* 42: 1139–1142. https://doi.org/10.1002/prep.201700042.

91 An, C.W., Li, H.Q., Geng, X.H. et al. (2013). Preparation and properties of 2,6-diamino-3,5-dinitropyrazine-1-oxide based nanocomposites. *Propellants, Explosives, Pyrotechnics* 38: 172–175. https://doi.org/10.1002/prep.201200164.

92 Mellor, A.M., Wiegand, D.A., and Isom, K.B. (1995). Hot spot histories in energetic materials. *Combustion and Flame* 101: 26–28. https://doi.org/10.1016/0010-2180(94)00171-NL.

4

Clathrates of CL-20: Thermal Decomposition and Combustion

Valery P. Sinditskii, Nikolay V. Yudin, Nikita A. Kostin, Sergei A. Filatov, and Valery V. Serushkin

Mendeleev University of Chemical Technology, Chemical Engineering Department, 9 Miusskaya Square, 125047 Moscow, Russia

4.1 Introduction

High-energy-density materials (HEDMs) are attracting wide attention in recent decades [1]. The performance of HEDMs is defined by parameters such as density, enthalpy of formation, and oxygen balance (OB). The best energy performance is achieved in a balanced substance. However, the traditional strategy of integrating oxidizing groups into an organic molecule is limited by the structure of the molecule. Recently, there is a new strategy for constructing advanced HEDMs by an intermolecular host–guest inclusion [2, 3]. The driving forces for the self-assembly of two individual molecules are usually intermolecular interactions – van der Waals interactions, hydrogen bonds, etc. As is well known, 2,4,6,8,10,12-hexanitro-2,4,6,8,10,12-hexaazaisowurtzitane (CL-20, HNIW) is one of the most powerful explosives [4] due to the good OB, a positive heat of formation ($\Delta H_f^0 = 374.9 - 377.4$ kJ mol^{-1}) [4, 5] and high density ($\rho = 2.04\,\text{g}\,\text{cm}^{-3}$) [6]. One of the features of CL-20 is its high tendency to form crystallites with other substances [7] and solvents [8]. In one of its modifications (α), CL-20 forms an aqueous hemihydrate [4a, 9]. In this crystalline solvate, water molecules only occupy structural cavities in the crystal, and the places occupied by guest molecules are taken up only partially. This type of solvate has been called a pseudopolymorphic solvate [10].

The inclusion of water molecules into the lattice of an energetic material does not lead to an increased OB. However, if we introduce molecules similar in size to a water molecule, but possessing oxidizing properties, into the cavities of CL-20 crystals, we could improve the performance of energetic material. Recently, Matzger's group has synthesized two polymorphic solvates of CL-20 with hydrogen peroxide [11], which is a liquid oxidizer. The resulting solvates have high crystallographic densities (1.966 and 2.033 g cm^{-3}) and sensitivity to mechanical stimuli similar to that of ε-CL-20. The DSC traces show that hydrogen peroxide solvates lose H_2O_2 molecules at higher temperatures (165 and 190 °C) compared to the evaporation of

Nano and Micro-Scale Energetic Materials: Propellants and Explosives, First Edition.
Edited by Weiqiang Pang and Luigi T. DeLuca.

water from the α-CL-20 (158 °C), which is associated with stronger hydrogen bonds of the guest molecule with nitro groups of CL-20. Detonation parameters (velocity and pressure) of the two solvates are predicted to surpass the performance of all known forms of HMX and all low-density forms of CL-20, with the orthorhombic hydrogen peroxide solvate expected to exceed the properties of even ε-CL-20 [11].

Along with the guest molecules, which in the initial state are liquids, it turned out that this approach allows the synthesis of clathrates of CL-20 even with such gaseous molecules as CO_2 [12] and N_2O [13]. The electron paramagnetic resonance (EPR) measurements of the gentle decomposition of CL-20 indicate that NO_2 and NO radicals are trapped by solid crystals and these compounds remain stable for a long time at room temperature [14]. The long-term stability of the NO_x products formed during α-CL-20 decomposition strongly suggests that these radicals may be trapped within the clathrate cages. Indeed, recently a polymorphic solvate of CL-20 with liquid oxidizer N_2O_4 has been synthesized [15].

Our work describes the synthesis and determination of the structure of new CL-20 clathrates with liquid oxidizer N_2O_4 and a gaseous oxidizer N_2O, thereby improving the OB of the material. To evaluate the perspectives of practical use of these clathrates, thermal stability, vapor pressure over solid substances, combustion behavior, and energetic performance have been investigated.

4.2 Host–guest Energetic Material Based on CL-20 and Nitrogen Oxides

4.2.1 Synthesis and Determination of the Structure of New Clathrates

The preparation of the new solvate of CL-20 and liquid N_2O_4 was carried out by the precipitation method, using ε-CL-20 or β-CL-20 as the starting materials [15]. Acetonitrile or nitromethane was used as a "strong" solvent, and liquid N_2O_4 and its mixture with CCl_4 were used as a precipitant. The amount of precipitant was 5–10 times bigger than the amount of solvent. After mixing the CL-20 solution with the precipitant an almost white, crystalline powder (Figure 4.1) is obtained, which upon storage became yellowish brown. It is well known, that N_2O_4 dissociates with the formation of NO_2 even at room temperature. Therefore, the synthesized solvate is the first example of explosive composition containing stable free radicals.

The clathrate of CL-20 and a gaseous oxidizer N_2O was obtained by bubbling a current of N_2O into a solution of ε-CL-20 or β-CL-20 in a mixture of ethyl acetate–heptane (in a volume ratio of 1 : 2) at a temperature of 15–20 °C. During bubbling, the solvents were distilled off, and a white crystalline powder was formed in the flask (Figure 4.2). The formation of CL-20/N_2O host–guest complex was synthesized in work [13] by a solvent/anti-solvent method. The acetone/CCl_4 solution was saturated with N_2O gas through continuous aeration.

The resulting clathrates were analyzed by powder X-ray diffraction (PXRD), IR-, and Raman spectroscopy. X-ray diffraction patterns were recorded at 298 K using a guiner imaging plate G 670 camera with CuKα1 radiation ($\lambda = 1.540598$ Å). Infrared

Figure 4.1 Optical microscope image of CL-20·¼N_2O_4 solvate. Source: Yudin et al. [15]/John Wiley & Sons.

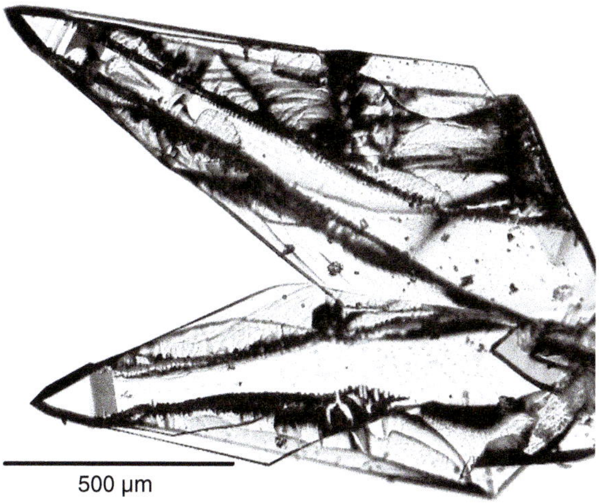

Figure 4.2 Optical microscope image of CL-20·0.5N_2O clathrate.

spectra in the region of 4000–450 cm^{-1} were measured by KBr pellet method or ATR (diamond) with a thermo scientific nicolet iS50 FTIR and Avatar 360 FTIR spectrometers.

The IR spectra of the CL-20/N_2O_4 solvate shows a new band at 1749 cm^{-1}, which characterizes N_2O_4 [16] in the cavity of the crystal lattice. The IR spectrum cannot allow identifying the presence of NO_2 radical in the new compound, since the NO_2 vibrational band (1610–1617 cm^{-1} [17]) is overlapped by the absorption of the CL-20 nitro groups. The IR and Raman spectra of the new solvate (with the exception of vibrational band related to the C—H bond) coincide with those of α-CL-20·0.5H_2O and α-CL-20·0.5CO_2. This indicates that the CL-20 molecule in the new solvate is in the same modification as in the solvates with H_2O and CO_2. The ratio of the peak areas of C—H (3030 cm^{-1}) and N_2O_4 (1749 cm^{-1}) vibrations depends on content of N_2O_4 in the solvate. The relative vibration intensity for solvate precipitated with pure N_2O_4 at 1749 cm^{-1} is 1.3 times higher than intensity for solvate precipitated by solution of N_2O_4 in CCl_4. According to thermogravimetry, first compound contains up to 0.35 moles of N_2O_4 but quickly loses excess during storage with the formation of a more stable form of CL-20·¼N_2O_4.

In the IR spectrum of the CL-20/N_2O clathrate, a new band appears at 2245 cm^{-1} with shoulder at 2227 cm^{-1}. The new strong peak could be ascribed to the asymmetric stretching vibration of N_2O, which shifted to a higher peak by 20 cm^{-1} compared with the free gas molecule N_2O [18]. This difference indicates the existence of host–guest interactions and that the N_2O molecule, according to X-ray diffraction analysis, remains linear within the solvate. The profile and position of other vibrational bands in the IR and Raman spectra of the clathrate coincide with those of α-CL-20·0.5H_2O and α-CL-20·0.5CO_2.

The freshly prepared solvate of CL-20 with N_2O_4 is almost white, but when stored for several days, it acquires a tan color, indicating partial dissociation of N_2O_4 to NO_2. EPR spectroscopy was used to confirm the presence of paramagnetic NO_2 particles located in cavities of CL-20 lattice. EPR spectrum was recorded at Bruker EMX 200 spectrometer at 9.3 GHz. The g-values were measured using a standard containing Mn^{2+} ions doped into MgO (g = 1.985). A typical spectrum of the sample is shown in Figure 4.3.

An ESR spectrum shows a triplet pattern that is due to the hyperfine interaction of the lone unpaired electron with nucleus ^{14}N ($I = 1$). The triplet pattern is centered at g = 2.0016 and has a hyperfine coupling of AN = 60.6 G. The shape and parameters of the EPR spectrum characterize the NO_2 radical, trapped in a solid matrix [19, 20].

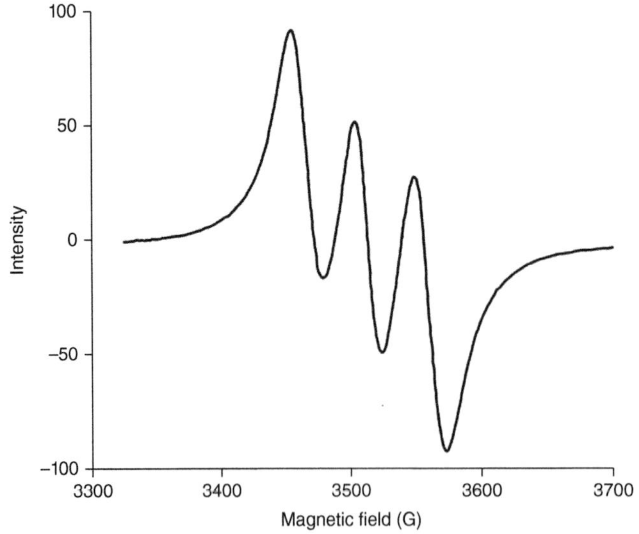

Figure 4.3 ESR spectrum of NO_2 in CL-20 matrix.

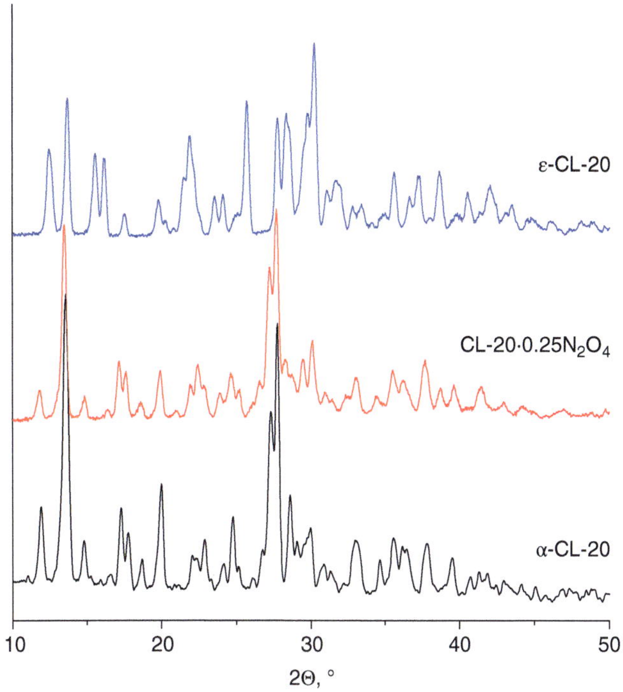

Figure 4.4 Powder patterns of CL-20·¼N_2O_4 in comparison with α-CL-20 and ε-CL-20.

The powder pattern of the CL-20 solvate with N_2O_4 [15] obtained by PXRD is very different from that of starting ε-CL-20 and indistinguishable from that of α-CL-20 (Figure 4.4), which suggests that both solvates are the isostructural materials and differ only in guest molecules (N_2O_4 and water, respectively). The density for the solvate (2.019 g cm^{-3}) obtained from PXRD data at room temperature is higher than the density determined by pycnometry (1.981 ± 0.003 g cm^{-3}). This is due both to the presence of defects in the crystal lattice and to the loss of N_2O_4 during pycnometric measurements, as well as due to part of the cavities contains NO_2 molecules rather than N_2O_4 molecules.

The X-ray diffraction powder pattern of the CL-20 clathrate with N_2O also demonstrates that the new compound is isostructural to α-CL-20 [13]. The crystal structure of the N_2O host–guest complex was elucidated with help of X-ray diffraction analysis and it was found that the clathrate has a 2 : 1 molar ratio of CL-20 to N_2O [13]. The CL-20·0.5N_2O clathrate has a high crystallographic density of up to 2.038 g cm^{-3} at 293 K, which is close to that of ε-CL-20 (2.044 g cm^{-3}) and higher than that of α-CL-20·0.5H_2O (1.97 g cm^{-3}).

It is interesting to note that X-ray diffraction analysis of the CL-20 clathrates with nitrogen oxides shows that their structure is formed due to hydrogen bonds between the oxygen of "guest" molecule and the hydrogen of "host" molecule. The situation is different for the CL-20 hydrate and CL-20 solvate with hydrogen peroxide, where there is a hydrogen bond between the oxygen of the "host" nitro group and the hydrogen of the "guest" molecules.

4.2.2 Thermal Stability of the New Clathrates

Thermal stability of the new clathrates was initially evaluated with help of differential scanning calorimetry (DSC) and thermogravimetric analysis (TGA) in nonisothermal conditions. DSC and TGA were performed with a DSC 822e Mettler Toledo in the temperature range of 25–300 °C at different heating rates.

DSC analysis of the α-CL-20·¼N_2O_4 solvate has shown an endothermic peak at 165 °C with the heat effect of 26.2 J g^{-1} (10 °C per min, Figure 4.5). At close temperatures, the $\alpha \rightarrow \gamma$ phase transition was observed in CL-20·0.5H_2O [21], CL-20·0.5H_2O_2 [11a] and CL-20·0.5CO_2 [14], which was accompanied by the loss of the guest molecules. According to the thermogravimetry of CL-20·¼N_2O_4, the mass loss at this stage is 4.7%, which corresponds to 0.25 mole of N_2O_4 (4.99 %). DSC of anhydrous α-CL-20 shows that the modification transition needs 12.5 J g^{-1} [21]. Hence the remaining heat at 165 °C (13.7 kJ g^{-1}) is spent on the evaporation of the N_2O_4 molecules (~25.3 kJ per mole of N_2O_4 or ~12.6 kJ per mole of NO_2). After evaporation of guest molecules, the CL-20 decomposes at a temperature of 255 °C with a heat release of 3086 J g^{-1}, which is typical for the decomposition of pure CL-20 (3000 ± 200 J g^{-1} [22]).

DSC analysis of the CL-20 clathrate with N_2O has shown an endothermic peak at 182 °C with the heat effect of 34 J g^{-1} (Figure 4.6). The mass loss in the region 165–185 °C (4.58%) corresponds to one half of a mole of N_2O (4.78%). The remaining CL-20 decomposes at a temperature of 248 °C (2595 J g^{-1}). Taking into account the heat, which is spent on the $\alpha \rightarrow \gamma$ phase transition, the evaporation of N_2O accounts for 19.8 kJ per mole of N_2O.

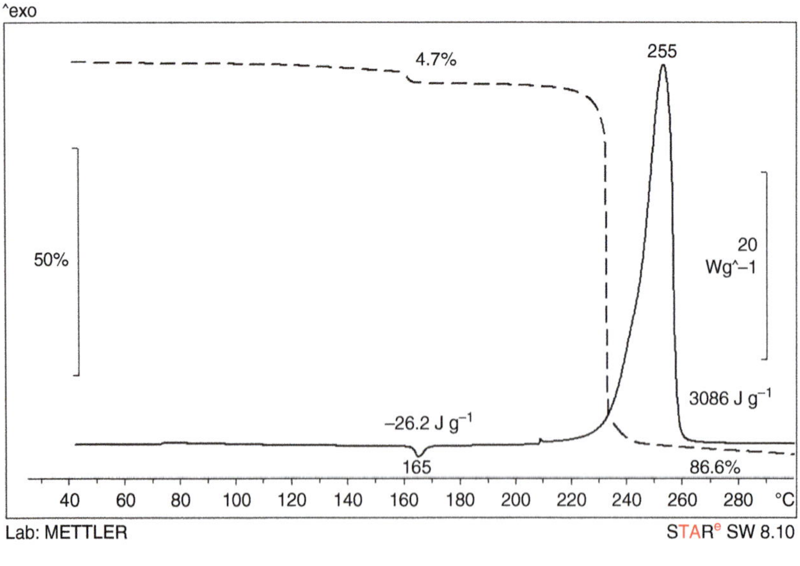

Figure 4.5 TG and DSC curves of CL-20·¼N_2O_4 solvate at a heating rate 10 °C/min.

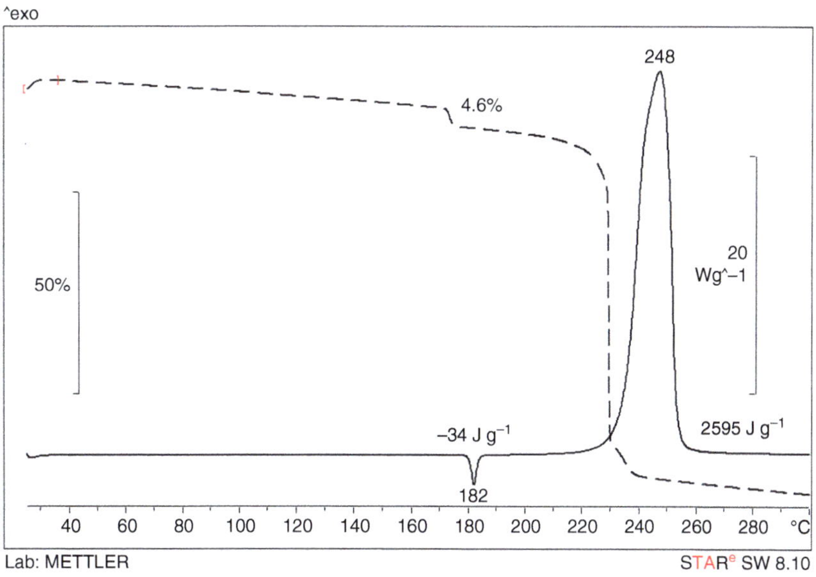

Figure 4.6 TG and DSC curves of CL-20·0.5N_2O clathrate at a heating rate 10 °C/min.

The non-isothermal decomposition kinetics of the CL-20·¼N_2O_4 solvate was calculated with the help of the Kissinger's method [23] from DSC data obtained at different heating rates: k, s^{-1} = 3.72·10^{15}·exp(21270/T), E_a = 176.8 kJ mol^{-1}. The rate constants of CL-20·0.5N_2O decomposition coincide with those of CL-20·¼N_2O_4 (Figure 4.7).

Study on the thermal decomposition of CL-20·¼N_2O_4 and CL-20·0.5N_2O under isothermal conditions was carried out in thin-walled glass manometers of compensation type in the temperature range of 180–210 °C. The gas evolution curves are similar to those for CL-20 decomposition, which is known [24] proceeds with acceleration in time. In the case of new clathrates the gas evolution curves have the initial pressure shock, which is due to the rapid release of NO_2 and N_2O. The maximum volumes of gaseous decomposition products of solvates are 695–705 cm^3 g^{-1}. These amounts are greater than can be expected upon decomposition of pure CL-20 taking into account its fraction in the clathrates (~600 cm^3) and gas volume of 0.5 NO_2 or 0.5N_2O (~30 cm^3). The larger amounts of gaseous products indicate the interaction of oxidizers NO_2 and N_2O with CL-20.

The rate constants were calculated using the first-order reaction model with self-acceleration [25]. This model describes the process up to 25–30% of decomposition then gas release slows down. Study under isothermal conditions shows that the decomposition of both clathrates does not obey the first order, so the data obtained under non-isothermal conditions DSC are formal and cannot be used to predict the stability of the compound.

The kinetics of the decomposition of CL-20·¼N_2O_4. CL-20·0.5N_2O and the 2CL-20/HMX co-crystal are shown in Figure 4.7. Our previous study shows that 2CL-20/HMX co-crystal [25] contains CL-20 as a mixture of α and β conformers, and the second component HMX does not affect its decomposition kinetics. As can

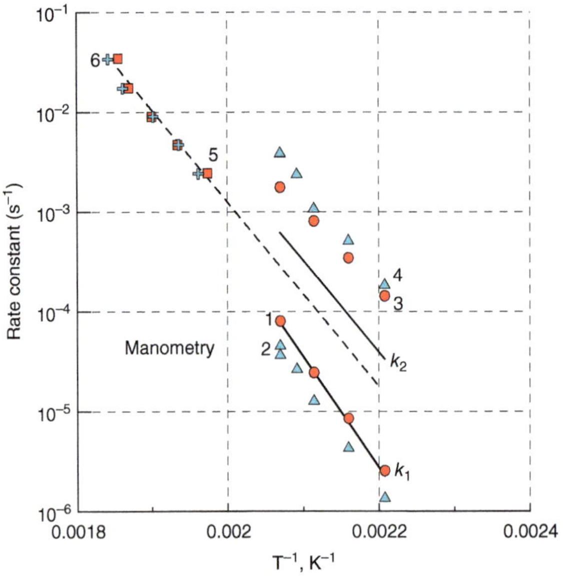

Figure 4.7 Comparison of rate constants of decomposition of CL-20·¼N_2O_4 (red circles 1,3) and CL-20·0.5N_2O (blue triangles 2,4) clathrates and CL-20 in the composition of the 2CL-20/HMX co-crystal [25] under isothermal (solid lines k_1 and k_2) conditions, as well as rate constants of decomposition of CL-20·¼N_2O_4 (5 and dashed line) and CL-20·0.5N_2O (6) under non-isothermal conditions (DSC).

be seen from Figure 4.7, the rate constants of the non-catalytic decomposition stage of CL-20·¼N_2O_4 solvate [15] completely coincide with those for CL-20 (k_1). These rate constants characterize the strength of broken bonds in the CL-20 molecule in the solid state ($k_1 = 1.11 \cdot 10^{19} \cdot \exp(-25860/T)$, $E_a = 215.0\,\text{kJ mol}^{-1}$). At the same time the decomposition at the acceleration stage ($k_2 = 1.39 \cdot 10^{16} \cdot \exp(-21540/T)$, $E_a = 179.1\,\text{kJ mol}^{-1}$) in the presence of NO_2 proceeds much faster than in CL-20. A similar situation is observed in the case of decomposition of CL-20·0.5N_2O, and, despite the fact that N_2O, unlike NO_2, is not a radical, the rate of acceleration of CL-20·0.5N_2O decomposition is higher. Kinetic data obtained are shown in Table 4.1.

Table 4.1 Kinetic parameters of decomposition of the host–guest materials under isothermal conditions.

Compound	Stage	Temperature interval, °C	Log A	E_a (kJ/mol)	Coef. of determination
CL-20·¼N_2O_4	Initial	180–210	19.43	207.6	0.999
	Acceleration		13.74	152.4	0.999
CL-20·0.5N_2O	Initial	180–210	18.21	208.7	0.996
	Acceleration		14.44	157.3	0.993

4.2.3 Vapour Pressure Above the New Clathrates

The thermal stability of the energetic materials is a mandatory characteristic that determines the possibility of using the new material. In the case of solvates, an important characteristic is also the ability of the substance to retain the guest molecule. On the one hand, such information is provided by DSC data on the evaporation temperature of the guest molecule, and on the other hand, knowledge of vapor pressure above the new clathrates can be useful.

The glass Bourdon gauge was used to measure vapor pressures above CL-20·¼N_2O_4 and CL-20·0.5N_2O. Figure 4.8 represents the temperature dependencies of vapor pressure for CL-20·¼N_2O_4 [15], CL-20·0.5N_2O, NO_2 [26], N_2O [27], and $Ca(NO_3)_2$·4H_2O [28]. As can be seen from Figure 4.8, the vapor pressure above CL-20·¼N_2O_4 is more than 2 orders of magnitude lower than the vapor pressure above liquid N_2O_4 [26] and lower than the vapor pressure of water above a typical crystalline hydrate, $Ca(NO_3)_2$·4H_2O [28]. The difference in vapor pressure over CL-20·0.5N_2O and solid N_2O is even greater. Nevertheless, according to the obtained equation in the range of 100–200 °C (ln$p$ (atm) = −2075/T + 1.83), the NO_2 pressure over the CL-20·¼N_2O_4 solvate crystals will be 4.5 mm Hg at 25 °C [15]. According to the data obtained, the heat of vaporization is 17.2 kJ per mole of NO_2. Taking into account the experimental error, this value agrees with the estimated value of the heat of NO_2 vaporization obtained from the DSC data (12.6 kJ per mole of NO_2).

In the case of α-CL-20·0.5N_2O, the vapor pressure was measured in the range 130–210 °C: ln$p$ (atm) = −2755/T + 2.4). According to the data obtained, the N_2O pressure over the α-CL-20·0.5N_2O clathrate crystals will be 0.8 mm Hg at 25 °C, and the heat of vaporization is 22.9 kJ per mole of N_2O, which is in good agreement with the estimate of the vaporization heat obtained in the DSC experiments (19.8 kJ per mole of N_2O).

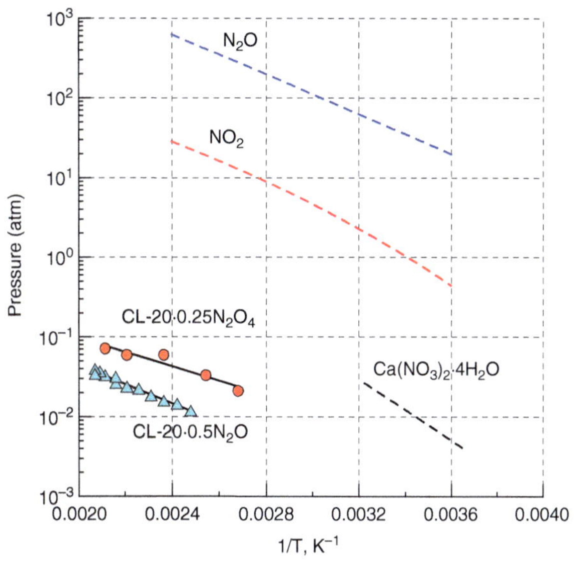

Figure 4.8 The vapor pressure over CL-20·¼N_2O_4 (red circles) and CL-20·0.5N_2O (blue triangles) clathrates, in comparison of vapor pressure over liquid N_2O_4, solid N_2O and $Ca(NO_3)_2$·4H_2O.

It should be noted that the precisely measured enthalpy of sublimation N_2O is $25.1 \pm 0.4\,kJ\,mol^{-1}$ [29]. Similar data on the vapor pressure over solid NO_2 (just over NO_2, not N_2O_4) and its values of the enthalpy of sublimation are absent in the literature.

A vacuum stability test (100 °C/48 hours) carried out in [13] showed satisfactory stability of the α-CL-20·0.5N_2O solvate. The gases released over 48 hours amounted to $0.72\,ml\,g^{-1}$, while in the case of ε-CL-20, only $0.04\,ml\,g^{-1}$ was released. It is obvious that the differences in the amount of released gases are primarily due to the evaporation of N_2O rather than the decomposition of the host molecule, the thermal stability of which has not changed. Apparently, it will be more difficult for the CL-20·¼N_2O_4 solvate, which has a higher vapor pressure, to pass the vacuum stability test.

4.2.4 Combustion Behaviors of the New Clathrates

An improvement in the OB of explosive leads to an increase in its energetic characteristics. It was interesting to see how this would affect the burning rate of the new clathrates.

Both clathrates in the form of pressed samples into transparent acrylic tubes of 4 mm internal diameter burn faster than ε-CL-20 [30]. Since CL-20 in the clathrates is in the α-form, combustion of α-CL-20 in the range of 0.5–10 MPa was also studied (Figure 4.9). As can be seen from Figure 4.9, the burning rate of CL-20·¼N_2O_4 and CL-20·0.5N_2O ($r_b = 6.3p^{0.72}$ for both) exceeds not only the burning rate of α-CL-20 ($r_b = 4.3p^{0.82}$), but also ε-CL-20 [30b,c]. Thus, the replacement of water with

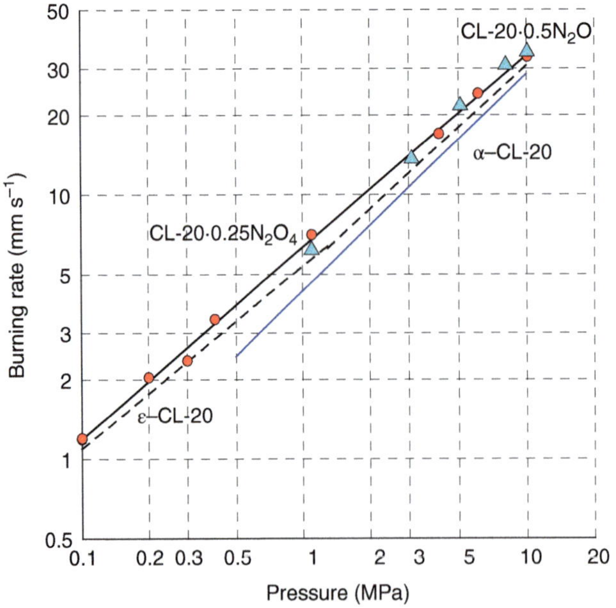

Figure 4.9 Comparison of the burning rates of CL-20·¼N_2O_4 (circles), CL-20·0.5N_2O (triangles) clathrates, ε-CL-20 (dashed line), and α-CL-20·0.5H_2O (solid line).

N_2O_4 and N_2O in the molecule of CL-20 solvate leads to a visible increase in the burning rate.

It can be assumed that the increase in the burning rate of clathrates is explained by the increase in their energetic characteristics. To confirm this assumption, it is necessary to obtain the temperature distribution in the combustion wave of the clathrates. Thin (5–7 μm) tungsten–rhenium thermocouples were used for study of the flame structure of CL-20·¼N_2O_4 at low pressures. Combustion of CL-20·¼N_2O_4 solvate [15] as well as ε-CL-20 [30b,c] proceeds at atmospheric pressure in a flameless mode and is accompanied by the formation of white smoke. The temperature of this process does not exceed 700 °C, which means that the additional oxidizer in CL-20 molecule does not increase the heat release of the first stage of CL-20 decomposition. The heat flux from the gas phase to the surface of the burning substance also remains practically unchanged. A bright luminous flame appears at a distance of 1 mm from the combustion surface of CL-20·¼N_2O_4 as the ambient pressure increases to 0.4 MPa. Despite the appearance of a flame, a heat feedback from the gas remains very low; this indicates that reactions in the condensed phase have the leading role during combustion.

The temperature profiles of ε-CL-20 [30b], α-CL-20 [15], and the CL-20·¼N_2O_4 solvate [15] show a characteristic bend in the area of 170–200 °C caused by the phase transition to the γ-conformation. Thus, regardless of the initial conformation of the studied substances, the same γ-conformer undergoes sublimation, melting, evaporation, and decomposition at the burning surface. However, it turned out that the surface temperatures (T_S) of the α-conformers (α-CL-20 and α-CL-20·¼N_2O_4) are definitely lower than those of the ε-conformer (Figure 4.10).

The surface temperatures for α-CL-20 and CL-20·¼N_2O_4 measured in [15] are compared with those for ε-CL-20 [30b, 31] in Figure 4.11. Based on the fact that at low pressures the surface temperature of ε-CL-20 does not reach the melting point, data of thermocouple-aided measurements [30b, 31] were used in [32] to estimate

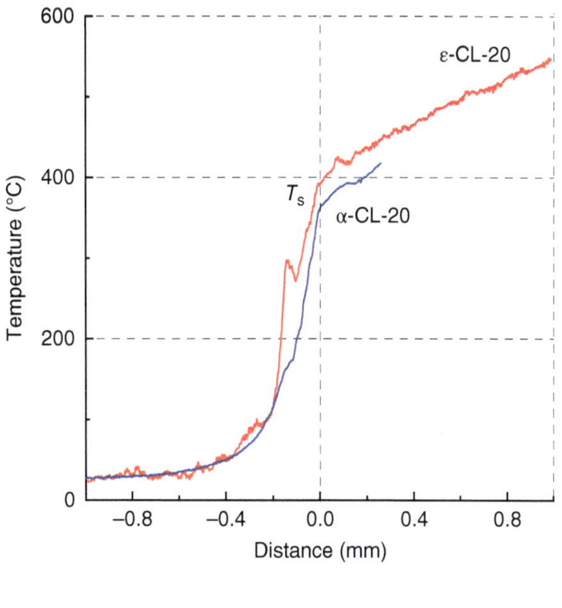

Figure 4.10 Comparison of temperature profiles for combustion of ε-CL-20 and α-CL-20 at a pressure of 0.05 MPa. T_S is the surface temperature.

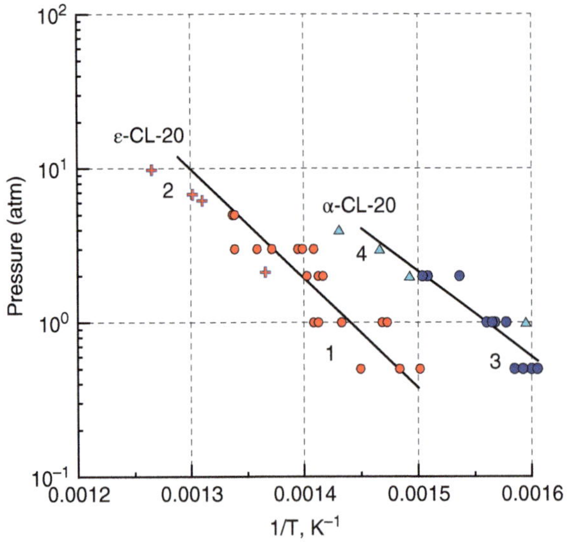

Figure 4.11 Comparison of pressure dependencies of surface temperatures for ε-CL-20 (1[30b], 2[31]), α-CL-20·0.5H_2O (3), and CL-20·¼N_2O_4 (4) solvates. Lines are a description of data (1) and (3, 4).

the enthalpy of sublimation (147.3 kJ mole^{-1}). This value within measurement error is in good agreement with the calculated enthalpy of CL-20 sublimation (157.6 kJ mole^{-1} [5b]). The surface temperatures of the two solvates of α-CL-20 (CL-20·0.5H_2O and CL-20·¼N_2O_4) in the pressure range of 0.05–0.4 MPa are described by the equation $\ln p = -12750/T + 19.9$ (Figure 4.11), the slope of which gives a significantly lower enthalpy of evaporation (106.0 kJ mole^{-1}).

Such a large difference in the enthalpies of evaporation of two conformers can be explained by assuming that the γ-form of CL-20 in the combustion wave "remembers" the history of its occurrence. Indeed, the process of formation of a new conformation, like formation of a new crystal lattice, takes time [33, 34]. It is surprising, however, that the difference in evaporation temperatures of the two conformers is so significant. So, the surface temperature of ε-CL-20 and α-CL-20 at atmospheric pressure differs by more than 50 degrees.

Another reason for the different surface temperatures of CL-20 conformers may be differences in their degradation products. That is, the surface temperature can be determined by the evaporation temperature of the more volatile decomposition product rather than the boiling temperature of the starting material. One of the well-known examples of this phenomenon is the burning of ammonium dinitramide (ADN), the surface temperature of which is determined by the evaporation of ammonium nitrate, which is formed in the melt at the surface as intermediate product of the ADN decomposition [35].

It is well known that the α-conformer of CL-20, as well as the γ-conformer obtained from it, has lower thermal stability compared to the ε-form [24]. Like other nitramines, the initial step of CL-20 decomposition includes the cleavage of N–NO_2. Conformers differ by the positions of NO_2 groups in space, so we can assume that the α-conformer is characterized by a decomposition channel, which is minor for the ε-conformer. For example, based on the B3LYP level of density functional theory it was suggested in [36] that after the elimination of two nitro groups from five-member rings, the C–C connecting two five-member rings breaks to form a stable 1,4,5,8-tetranitro-1,3a,4,4a,5,7a,8,8a-octahydrodiimidazo[4,5-b:4',5'-e]pyrazine. Possibly, it is the evaporation of this intermediate that determines the surface temperature of the α-conformer. In order for the evaporation of an intermediate compound to form a discontinuity in the condensed phase, which the thermocouple will record as a break in the temperature profile (surface temperature), at least 10% of this compound must be formed.

Comparing the kinetics of the leading combustion reaction of explosive with the kinetics of its decomposition allows checking the correctness of the surface temperature determination. The thermocouple study of CL-20·¼N_2O_4 shows that the reactions in the condensed phase control the solvate burning rate. Under these conditions, the decomposition reaction of explosive is usually the limiting one. The combustion of substances, which obey the condensed-phase combustion model, is well described by the classical Zel'dovich equation [37, 30b]. Using this equation and knowing the dependences of the burning rate and surface temperature on pressure, it is possible to derive the kinetic parameters of the reaction that controls the burning rate.

The specific heat, c_p, as 1.0 kJ·kg^{-1}·K^{-1} [31] and density of pressed samples of CL-20·¼N_2O_4 solvate, ρ, as 1.88 g cm^{-3} was used for calculation. The thermal diffusivity of the condensed phase, χ, (1.2·10^{-3} cm^2·s^{-1}) was obtained from thermocouple-aided measurements [15]. The reaction heat effect (695 J g^{-1}) was estimated using the value of first flame temperature and average specific heat of gas phase. Similar parameters were used to calculate the rate constants of the leading reaction of α-CL-20·0.5H_2O, except for the density (1.84 g cm^{-3}).

The comparison shows (Figure 4.12) that the rate constants of the acceleration stage of the CL-20·¼N_2O_4 solvate decomposition, k_2, agree well with the

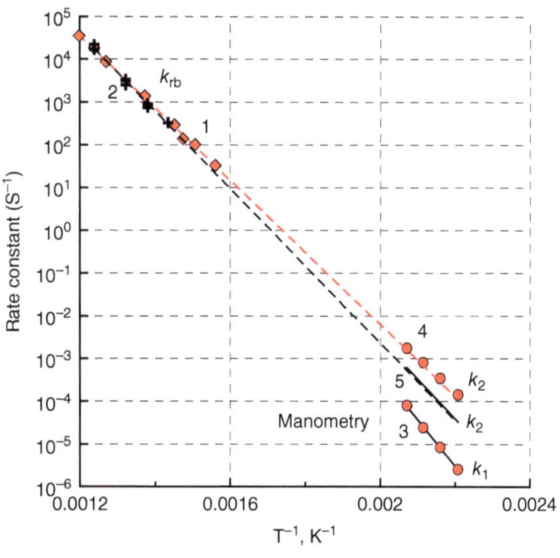

Figure 4.12 Comparison of the rate constants of the leading combustion reaction (k_{rb}) of CL-20·¼N_2O_4 solvate (1, red squares and dashed line) and α-CL-20 (black crosses and dashed line) with kinetics of CL-20·¼N_2O_4 solvate (red circles, 3 and 4, k_1 and k_2) and α-CL-20 (5, k_2, solid line) decomposition.

kinetics of the leading combustion reaction: k_{rb}, s^{-1} = 4.67·10^{14}·exp(−19410/T), E_a = 161.4 kJ mol^{-1}. The rate constants extracted from the combustion wave of α-CL-20·0.5H_2O (k_{rb}, s^{-1} = 2.8·10^{15}·exp(−20820/T), E_a = 173.1 kJ mol^{-1}) are also consistent with the acceleration kinetics of its decomposition (k_2, s^{-1} = 7.77·10^{15}·exp(−21240/T), E_a = 176.6 kJ mol^{-1}).

Thus, the kinetics of self-acceleration reaction of the decomposition at surface temperature determines the burning rate of CL-20·¼N_2O_4 solvate as well as CL-20·0.5H_2O solvate. It was assumed in the decomposition study of various CL-20 co-crystals [25] that the acceleration during decomposition of CL-20 is not only due to autocatalysis by the decomposition products. A topochemical character of decomposition associated with destruction of the crystal lattice and the possible dissolution of CL-20 in its liquid decomposition products may be of certain importance. The rate constants of the acceleration stage of the decomposition of CL-20 solvates are approaching the rate constants of the decomposition of CL-20 in the liquid phase [38]. The assumption that the surface temperature of α-CL-20 solvates is determined by the evaporation of the decomposition intermediate is in good agreement with the fact that, in contrast to ε-CL-20, the burning rate of solvates even at low pressure is determined by the kinetics of its self-acceleration reaction of the decomposition, that is, practically by the kinetics of decomposition in the liquid phase. It is the formation of these intermediates that leads to the dissolution of CL-20. Earlier it was discovered in the study of the combustion of co-crystals CL-20 [39], that in some cases even a two-fold dilution of CL-20 with the second compound practically does not result in decreasing combustion rate. The reason for such unusual behavior is presumed to lie in the amorphous state of CL-20, which forms after the evaporation of the second volatile component. In case of clathrates, the crystal lattice of CL-20 after evaporation of the guest molecule remains unchanged, but the formation of decomposition intermediates leads to the dissolution of CL-20 and the combustion rate is determined by the kinetics of decomposition in the liquid phase.

4.2.5 Energetic Performance of the New Clathrates

To estimate the performance of energetic material, it is necessary to know the enthalpy of formation of the compound. The enthalpy of formation of ε-CL-20 was determined in many works, the following $\Delta H_f^o = 374.9 - 377.4$ kJ mol^{-1} [5] can be considered the most reliable values. As shown by X-ray diffraction studies, the molecule of CL-20 in solvates is in α-modification. The experimentally determined enthalpy of formation of this modification $\Delta H_f^o = 338.99$ kJ mol^{-1} [40] is significantly lower than the enthalpy of formation of ε-CL-20. However, the measured phase transitions between the modifications $\varepsilon \rightarrow \gamma$ (8.4–10 kJ mol^{-1} [41]) and $\alpha \rightarrow \gamma$ (2.8 kJ mol^{-1} [41]) show that the difference in the enthalpies of formation of CL-20 modifications does not exceed 10 kJ mol^{-1}. Most likely, the underestimated value of the enthalpy of formation of α-modification CL-20, obtained in [40], is due to the presence of an admixture of difficult-to-remove water in α-CL-20.

In this regard, we used the most reliable values of the enthalpy of formation ε-CL-20 and the enthalpy of formation of guest molecules in the solid state to

Table 4.2 Detonation Characteristics of Various CL-20 Clathrates in Comparison with ε-CL-20 and 2CL-20/HMX.

Compound	Density ($g\,cm^{-3}$)	Enthalpy of formation ($kJ\,mol^{-1}$)	Heat of explosion ($kJ\,mol^{-1}$)	Detonation velocity ($m\,s^{-1}$)	Detonation pressure (GPa)
ε-CL-20	2.04	377.4	6555	9480	46.75
2CL-20/HMX	1.945	861.9	6477	9297	42.65
α-CL-20·0.5H_2O	1.973	231.8	6363	9265	42.74
CL-20·0.5H_2O_2	2.033	276.2	6575	9510	46.82
CL-20·0.5H_2O_2	1.966	276.2	6560	9315	43.18
CL-20·¼N_2O_4	2.019	368.6	6635	9475	46.43
CL-20·0.5NO_2	2.019	385.3	6670	9485	46.56
CL-20·0.5N_2O	2.038	405.9	6525	9515	47.48

calculate the enthalpy of formation of CL-20 clathrates. Most likely, the real enthalpies of formation of clathrates will be somewhat lower than those values due to the appearance of hydrogen bonds between the components. The resulting values are shown in Table 4.2.

Energetic characteristics of clathrates (heat of explosion, detonation velocity, and detonation pressure) were calculated using computer package "Shock and detonation general kinetics and thermodynamics in reactive systems." [42] The obtained parameters are shown in Table 4.2. To evaluate the energy levels of CL-20 clathrates, parameters of ε-CL-20 and one of the most promising energetic co-crystal 2CL-20/HMX [7c, 25] have been added to Table 4.2.

It should be noted that bomb calorimetry experiments performed in [11b] gave high value of the enthalpy of formation of CL-20·0.5H_2O_2 ($374 \pm 64\,kJ\,mol^{-1}$), although due to the large measurement uncertainty, the minimum experimental value is close to that given in Table 4.2. The thermochemical prediction shows that even with an enthalpy of formation of $276.2\,kJ\,mol^{-1}$ the CL-20·0.5H_2O_2 solvate has larger detonation velocity than ε-CL-20 at the theoretical maximum density, which was confirmed experimentally [11b].

The enthalpy of formation of the CL-20·0.5N_2O clathrate [13] was obtained by using the atomization method with PM6 calculations. The transition of the enthalpy of formation from a gaseous state to a solid one was carried out taking into account the lattice energy, which is calculated by using COMPASS field with the field assigned charges, Ewald and atom-based summation methods for electrostatic and van der Waals interactions. The calculated value obtained ($407.4\,kJ\,mol^{-1}$) is in good agreement with the sum of the enthalpies of formation of the components.

For calculation of the enthalpy of formation of CL-20·¼N_2O_4, the value for solid N_2O_4 ($-35.05\,kJ\,mol^{-1}$[43]) was used. However, as was shown experimentally, during storage of the solvate, the N_2O_4 contained in it dissociates with the formation of the NO_2 radicals. There are no data for solid NO_2 since NO_2 dimerizes to N_2O_4 in the condensed state. It doesn't happen in the solvate, since NO_2 radicals are isolated in the cavities of the host molecule. The enthalpy of the NO_2 radical in the solid state can be calculated based on the known value for the gaseous state (33.1 kJ mol^{-1} [43]) and the value of the heat of sublimation obtained in our experiments (17.2 kJ mol^{-1}). As can be seen from Table 4.2, the solvate with completely dissociated N_2O_4, having a higher enthalpy of formation, differs only slightly from the solvate of CL-20·¼N_2O_4 in terms of energetic performance.

Both CL-20·0.5N_2O and orthorhombic CL-20·0.5H_2O_2 have predicted detonation velocities and pressures that slightly outperform those of ε-CL-20 and one of the most promising energetic co-crystal 2CL-20/HMX. Solvate CL-20·¼N_2O_4, having a lower density, is comparable in energy characteristics with ε-CL-20. Similar results for solvates with H_2O_2 were obtained in [11], but the value for CL-20·0.5N_2O given in [13] is somewhat overestimated in our opinion.

The most important characteristic of explosives is the sensitivity of energetic materials to initiation by external stimuli, such as impact, friction, and electrostatic shocks, without the knowledge of which it is impossible to discuss the application of new compounds. Such studies were carried out in [11, 13, 15] in which CL-20 solvates with oxidizers were first obtained. Solvates of CL-20 with H_2O_2 and CL-20·¼N_2O_4 have sensitivities slightly below or similar to that of ε-CL-20, yet with an increase in the overall OB of the system. These materials can be classified as sensitive secondary explosives. According to the data of [13], which are consistent with our data, the CL-20·0.5N_2O clathrate is less sensitive than ε-CL-20 in impact sensitivity and is similar to ε-CL-20 in friction sensitivity. The electrostatic spark sensitivity of CL-20·0.5N_2O is also lower than the sensitivity of ε-CL-20 [13].

Another necessary characteristic of energetic materials, without which a practical application is impossible, is thermal safety prediction. Self-accelerating decomposition temperature (SADT) is an important parameter for evaluating the thermal hazards during storage and transportation. According to Ref. [44], SADT is defined as the lowest temperature at which an overheating in the middle of specific packaging exceeds 6 °C after a lapse of the period of seven days. The kinetics obtained allows calculating SADT under adiabatic conditions. Based on the kinetics we can calculate the degree of decomposition, and the heat effect obtained in DSC may be used for calculation of the temperature rise.

The SADTs for CL-20·¼N_2O_4 and CL-20·0.5N_2O, from which the storage temperatures of the solvates can be deduced, is 129.3 and 133.1 °C, respectively, which exceeds the usual storage temperatures of explosives and close to SADT of such co-crystals as CL-20/HMX [25, 45].

4.3 Conclusion Remarks

The physicochemical properties of the new CL-20 clathrates with liquid oxidizer N_2O_4 and a gaseous oxidizer N_2O such as thermal stability and combustion behaviors are investigated.

The thermal stability was studied by isothermal and non-isothermal methods. Evaporation of NO_2 and N_2O from the clathrates, as well as the solvates with water and hydrogen peroxide, do not disturb the α-CL-20 crystal lattice, resulting in unchanged thermal stability. However, under closed conditions, NO_2 and N_2O increase the rate of the autocatalytic decomposition reaction in comparison with neat CL-20. Nevertheless, the SADTs for CL-20·¼N_2O_4 and CL-20·0.5N_2O, from which the storage temperatures of the clathrates can be deduced, exceed the typical storage temperatures of explosives.

The temperature dependencies of vapor pressure over solid CL-20·¼N_2O_4 and CL-20·0.5N_2O crystals were determined for the first time. It has been shown that the vapor pressures of NO_2 and N_2O above the clathrate crystals are similar to the pressure above conventional crystalline hydrates. The obtained enthalpies of sublimation are consistent with the DSC data on evaporation of NO_2 and N_2O from clathrates. The vapor pressure over the clathrates is most likely associated with their evaporation temperatures in DSC, which increase in the series CL-20·0.5H_2O < CL-20·¼N_2O_4 (CL-20·0.5NO_2) ~ CL-20·0.5H_2O_2 < CL-20·0.5N_2O. The vapor pressure over the clathrates has an effect on the vacuum stability test, however, as shown in [13], the α-CL-20·0.5N_2O clathrate passes this test satisfactorily.

The combustion behaviors of new clathrates were determined using a constant pressure device. It turned out that the replacement of water with N_2O_4 and N_2O in the molecule of CL-20 solvate resulted in a visible increase in the burning rate. Thermocouple measurements have revealed that reactions in the condensed phase play the leading role during combustion. It turned out that the surface temperatures of the α-conformers (α-CL-20 and α-CL-20·¼N_2O_4) are lower than those of the ε-conformer. Moreover, the surface temperatures of the two clathrates of α-CL-20 in the pressure range of 0.05–0.4 MPa are described by the equation with a significantly lower enthalpy of evaporation (106.0 kJ $mole^{-1}$) than that for ε-CL-20 (147.3 kJ $mole^{-1}$). It has been suggested that the different surface temperatures of CL-20 conformers may be explained by formation of different degradation products. In this case, the surface temperature can be determined by the evaporation temperature of the more volatile decomposition product rather than the boiling temperature of the initial material. The kinetic parameters of the reaction that controls the burning rate were obtained using the burning rate and surface temperature dependences on pressure. It appeared that the kinetics of the leading combustion reaction of the CL-20·¼N_2O_4 solvate agree well with the rate constants of its self-acceleration stage of the decomposition. This most likely indicates that the formation of decomposition intermediates leads to the dissolution of CL-20 and the rate of combustion is determined by the kinetics of decomposition in the liquid phase. A similar effect can have the destruction of the crystal lattice and the subsequent decomposition of CL-20 in amorphous form.

Thus, the strategy for the preparation of novel host–guest energetic materials using the incorporating oxidizer N_2O_4 and N_2O into the crystal lattice cavity of CL-20 really makes it possible to obtain, promising from a practical point of view, energetic materials superior in energetic characteristics to such a powerful explosive as ε-CL-20.

The resulting compounds are of interest as powerful explosives and energetic fillers for rocket propellants.

Acknowledgments

The authors are grateful to Dr. N.N. Kondakova and Dr. N.N. Ilyicheva for carrying out DSC and TGA measurements, student M.V. Ivanyan for her help in conducting experiments on thermal decomposition. Density measurements were performed on equipment of the Center for collective use of the Mendeleev University of Chemical Technology.

References

1 (a) Sikder, A.K. and Sikder, N.A. (2004). Review of advanced high performance, insensitive and thermally stable energetic materials emerging for military and space applications. *Journal of Hazardous Materials* 112: 1–15; (b) Klapötke, T.M. (ed.) (2007). *High Energy Density Materials*, vol. 125, 286. Springer; (c) Chavez, D.E., Parrish, D.A., Mitchell, L., and Imler, G.H. (2017). Azido and tetrazolo-1,2,4,5-tetrazine N-oxides. *Angewandte Chemie, International Edition* 56 (13): 3575–3578; (d) Yu, Q., Yin, P., Zhang, J. et al. (2017). Pushing the limits of oxygen balance in 1, 3, 4-oxadiazoles. *Journal of the American Chemical Society.* 139 (26): 8816–88194; (e) Klenov, M.S., Guskov, A.A., Anikin, O.V. et al. (2016). Synthesis of tetrazino-tetrazine-1,3,6,8-tetraoxide (TTTO). *Angewandte Chemie, International Edition* 55 (38): 11472–11475.

2 (a) McDonald, K.A., Bennion, J.C., Leone, A.K., and Matzger, A.J. (2016). Rendering non-energetic microporous coordination polymers explosive. *Chemical Communications* 52: 10862–10865. (b) Kent, R.V., Vaid, T.P., Boissonnault, J.A., and Matzger, A.J. (2019). Adsorption of tetranitromethane in zeolitic imidazolate frameworks yields energetic materials. *Dalton Transactions* 48: 7509–7513.

3 Wang, Y., Song, S., Huang, C. et al. (2019). Hunting for advanced high-energy-density materials with well-balanced energy and safety through an energetic host–guest inclusion strategy. *Journal of Materials Chemistry A* 7 (33): 19248–19257.

4 (a) Simpson, R.L., Urtiew, P.A., Ornellas, D.L. et al. (1997). CL-20 performance exceeds that of HMX and its sensitivity is moderate. *Propellants, Explosives, Pyrotechnics* 22 (5): 249–255. (b) Nielsen, A.T., Chafin, A.P., Christian, S.L. et al. (1998). Synthesis of polyazapolycyclic caged polynitramines. *Tetrahedron* 54: 11793–11812.

5 (a) Miroshnichenko, E.A., Kon'kova, T.S., Matyushin, Y.N., and Inozemtsev, Y.O. (2009). Bond dissociation energies in nitramines. *Russian Chemical Bulletin* 58 (10): 2015–2019. (b) Dorofeeva, O.V. and Suntsova, M.A. (2015). Enthalpy of formation of CL-20. *Computational & Theoretical Chemistry* 1057: 54–59.

6 Nielsen, A.T. (1997). Caged polynitramine compound. US Patent 5693794, publ. December 2.

7 (a) Bolton, O. and Matzger, A.J. (2011). Improved stability and smart-material functionality realized in an energetic cocrystal. *Angewandte Chemie, International Edition* 50 (38): 8960–8963. (b) Millar, D.I., Maynard-Casely, H.E., Allan, D.R. et al. (2012). Crystal engineering of energetic materials: Co-crystals of CL-20. *CrystEngComm* 14 (10): 3742–3749. (c) Bolton, O., Simke, L.R., Pagoria, P.F., and Matzger, A.J. (2012). High power explosive with good sensitivity: a 2:1 cocrystal of CL-20:HMX. *Crystal Growth & Design* 12 (9): 4311–4314. (d) Yang, Z., Li, H., Zhou, X. et al. (2012). Characterization and properties of a novel energetic–energetic cocrystal explosive composed of HNIW and BTF. *Crystal Growth & Design* 12 (11): 5155–5158.

8 (a) Aldoshin, S.M., Aliev, Z.G., Goncharov, T.K. et al. (2011). New conformer of 2,4,6,8,10,12-hexanitro-2,4,6,8,10,12-hexaazaisowurtzitane (CL-20). Crystal and molecular structures of the CL-20 solvate with glyceryl triacetate. *Russian Chemical Bulletin* 60 (7): 1394–1400. (b) Goncharov, T.K., Aliev, Z.G., Aldoshin, S.M. et al. (2015). Preparation, structure, and main properties of bimolecular crystals CL-20—DNP and CL-20—DNG. *Russian Chemical Bulletin* 64 (2): 366–374.

9 Lempert, D.B. and Chukanov, N.V. (2014). The outlook for the use of pseudopolymorphic solvates in energetic materials. *Central European Journal of Energetic Materials* 11 (2): 285–294.

10 Bernstein, J. (2002). *Polymorphism in Molecular Crystals*, 410. Oxford: Clarendon Press.

11 (a) Bennion, J.C., Chowdhury, N., Kampf, J.W., and Matzger, A.J. (2016). Hydrogen peroxide solvates of 2,4,6,8,10,12-hexanitro-2,4,6,8,10,12-hexaazaisowurtzitane. *Angewandte Chemie, International Edition* 55 (42): 13118–13121. (b) Vuppuluri, V.S., Bennion, J.C., Wiscons, R.A. et al. (2019). Detonation velocity measurement of a hydrogen peroxide solvate of CL-20. *Propellants, Explosives, Pyrotechnics* 44 (3): 313–318.

12 Saint Martin, S., Marre, S., Guionneau, P. et al. (2010). Host–guest inclusion compound from nitramine crystals exposed to condensed carbon dioxide. *Chemistry - A European Journal* 16 (45): 13473–13478.

13 Xu, J., Zheng, S., Huang, S. et al. (2019). Host–guest energetic materials constructed by incorporating oxidizing gas molecules into an organic lattice cavity toward achieving highly-energetic and low-sensitivity performance. *Chemical Communications* 55 (7): 909–912.

14 Pace, M.D. (1992). Free radical mechanisms in high density nitrocompounds: hexanitroisowurtzitane, a new high-energy nitramine. *Molecular Crystals and Liquid Crystals Science and Technology. Section A. Molecular Crystals and Liquid Crystals* 219 (1): 139–148.

15 Yudin, N.V., Sinditskii, V.P., Filatov, S.A. et al. (2020). Solvate of 2,4,6,8,10,12-hexanitro-2,4,6,8,10,12-hexaazaisowurtzitane (CL-20) with both N_2O_4 and stable NO_2 free radical. *ChemPlusChem* 85 (9): 1994–2000.

16 Bibart, C.H. and Ewing, G.E. (1974). Vibrational spectrum, torsional potential, and bonding of gaseous N_2O_4. *The Journal of Chemical Physics* 61 (4): 1284–1292.

17 St. Louis, R.V. and Crawford, B. Jr., (1965). Infrared spectrum of matrix-isolated NO_2. *The Journal of Chemical Physics* 42 (3): 857–864.

18 Zakirov, V., Sweeting, M., Lawrence, T., and Sellers, J. (2001). Nitrous oxide as a rocket propellant. *Acta Astronautica* 48 (5–12): 353–362.

19 Kasai, P.H., Weltner, W. Jr., and Whipple, E.B. (1965). Orientation of NO_2 and other molecules in neon matrices at 4 K. *The Journal of Chemical Physics* 42 (3): 1120–1121.

20 Schwartz, R.N., Clark, M.D., Chamulitrat, W., and Kevan, L. (1986). Electron-spin resonance of NO_2 trapped in SiO_2 thin solid films. *Journal of Applied Physics* 59 (9): 3231–3234.

21 Foltz, M.F., Coon, C.L., Garcia, F., and Nichols, A.L.I.I.I. (1994). The thermal stability of the polymorphs of hexanitrohexaazaisowurtzitane, part II. *Propellants, Explosives, Pyrotechnics* 19 (3): 133–144.

22 Turcotte, R., Vachon, M., Kwok, Q.S. et al. (2005). Thermal study of HNIW (CL-20). *Thermochimica Acta* 433 (1): 105–115.

23 Kissinger, H.E. (1957). Reaction kinetics in differential thermal analysis. *Analytical Chemistry* 29 (11): 1702–1706.

24 Nedelko, V.V., Chukanov, N.V., Raevskii, A.V. et al. (2000). Comparative investigation of thermal decomposition of various modifications of hexanitrohexaazaisowurtzitane (CL-20). *Propellants, Explosives, Pyrotechnics* 25 (5): 255–259.

25 Sinditskii, V. P., Yudin, N. V., Fedorchenko, S. I., Egorshev, V. Y., Kostin, N. A., Gezalyan, L. V., Zhang, J. G. (2020). Thermal decomposition behavior of CL-20 co-crystals. *Thermochimica Acta*, 691, 178703, DOI https://doi.org/10.1016/j.tca.2020.178703.

26 Stull, D.R. (1947). Vapor pressure of pure substances. Organic and inorganic compounds. *Industrial and Engineering Chemistry* 39 (4): 517–540.

27 Blue, R.W. and Giauque, W.F. (1935). The heat capacity and vapor pressure of solid and liquid nitrous oxide. The entropy from its band spectrum. *Journal of the American Chemical Society* 57 (6): 991–997.

28 Ewing, W.W. (1927). Calcium nitrate. II. The vapor pressure-temperature relations of the binary system calcium nitrate-water. *Journal of the American Chemical Society* 49 (8): 1963–1973.

29 Bryson, C.E., Cazcarra, V., and Levenson, L.L. (1974). Sublimation rates and vapor pressures of water, carbon dioxide, nitrous oxide, and xenon. *Journal of Chemical & Engineering Data* 19 (2): 107–110.

30 (a) Atwood, A.I., Boggs, T.L., Curran, P.O. et al. (1999). Burning rate of solid propellant ingredients, part 1: Pressure and initial temperature effects. *Journal of Propulsion and Power* 15 (6): 740–747. (b) Sinditskii, V.P., Egorshev, V.Y., Berezin, M.V. et al. (2003). Combustion behavior and mechanism of high-energy caged nitramine hexanitrohexaazaisowurtzitane. *Zhurnal Fizicheskoi Khimii* 22 (7): 69–74. (c) Sinditskii, V.P., Egorshev, V.Y., Serushkin, V.V. et al. (2010).

Combustion of energetic materials governed by reactions in the condensed phase. *International Journal of Energetic Materials and Chemical Propulsion* 9 (2): 147–192.

31 Hommel, J. and Trubert, J.F. (2002). Study of the condensed phase degradation and combustion of two new energetic charges for low polluting and smokeless propellants- HNIW and ADN. In: *Proc. 33 Inter. Ann. Conf. ICT*, paper V10, 1–17.

32 Sinditskii, V.P., Egorshev, V.Y., Serushkin, V.V. et al. (2009). Evaluation of decomposition kinetics of energetic materials in the combustion wave. *Thermochimica Acta* 496 (1–2): 1–12.

33 Chukanov, N.V., Dubovitskii, V.A., Zakharov, V.V. et al. (2009). Phase transformations of 2,4,6,8,10,12-hexanitrohexaazaisowurtzitane: the role played by water, dislocations, and density. *Russian Journal of Physical Chemistry B* 3 (3): 486–493.

34 Chukanov, N.V., Zakharov, V.V., Korsunskii, B.L. et al. (2009). The kinetics of the polymorphic transition of the α-Form of 2,4,6,8,10,12-hexanitrohexaazaisowurtzitane. *Russian Journal of Physical Chemistry A* 83 (1): 29–33.

35 (a) Sinditskii, V.P., Egorshev, V.Y., Levshenkov, A.I., and Serushkin, V.V. (2006). Combustion of ammonium dinitramide, part I: burning behavior. *Journal of Propulsion and Power* 22 (4): 769–776. (b) Sinditskii, V.P., Egorshev, V.Y., Levshenkov, A.I., and Serushkin, V.V. (2006). Combustion of ammonium dinitramide, part 2: combustion mechanism. *Journal of Propulsion and Power* 22 (4): 777–785.

36 Okovytyy, S., Kholod, Y., Qasim, M. et al. (2005). The mechanism of unimolecular decomposition of 2, 4, 6, 8, 10, 12-hexanitro-2, 4, 6, 8, 10, 12-hexaazaisowurtzitane. A computational DFT study. *The Journal of Physical Chemistry. A* 109 (12): 2964–2970.

37 Zeldovich, Y.B. (1942). On the combustion theory of powders and explosives. *Journal of Experimental and Theoretical Physics* 12: 498–510.

38 Korsounskii, B.L., Nedel'ko, V.V., Chukanov, N.V. et al. (2000). Kinetics of thermal decomposition of hexanitrohexaazaisowurtzitane. *Russian Chemical Bulletin* 49 (5): 812–818.

39 Sinditskii, V.P., Chernyi, A.N., Egorshev, V.Y. et al. (2019). Combustion of CL-20 cocrystals. *Combustion and Flame* 207: 51–62.

40 Golfier, M., Graindorge, H., Longevialle, Y., and Mace, H. (1998). New energetic molecules and their applications in energetic materials. In: *Proc. of 29th Inter. Annual Conf. of ICT*DWS Werbeagentur und Verlag GmbH, Fraunhofer Karlsruhe, Germany, 30 June–3 July, paper V3, 1–18.

41 Golovina, N.I., Utenyshev, A.N., Bozhenko, K.V. et al. (2009). The energy parameters of 2,4,6,8,10,12-hexanitro-2,4,6,8,10,12-hexaazaisowurtzitane polymorphs and their phase transitions. *Russian Journal of Physical Chemistry A* 83 (7): 1153–1159.

42 (a) Kondrikov, B.N. and Sumin, A.I. (1987). Equation of state for gases at high pressure. *Combustion, Explosion, and Shock Waves* 23 (1): 105–113. (b) Sumin, A.I., Gamezo, V.N., Kondrikov, B.N., and Raikova, V.M. (2000). Shock and detonation general kinetics and thermodynamics in reactive systems computer package. In: *Proc. 11th Int. Detonation Symposium*, USA. Bookcoomp, Ampersand, 30–35.

43 Chase, M.W. Jr., (1998). NIST-JANAF thermochemical tables, Fourth Edition. *Journal of Physical and Chemical Reference Data Monographs* 9: 1–1951.

44 Recommendations on the Transport of Dangerous Goods. (2003). Manual of Tests and Criteria, 4 revised ed., United Nations, ST/SG/AC.10/11/Rev. 4, United Nations, New York and Geneva.

45 Zhao, L., Yin, Y., Sui, H. et al. (2019). Kinetic model of thermal decomposition of CL-20/HMX co-crystal for thermal safety prediction. *Thermochimica Acta* 674: 44–51.

5

HMX and CL-20 Crystals Containing Metallic Micro and Nanoparticles

Alexander B. Vorozhtsov[1], Georgiy Teplov[2], and Sergei D. Sokolov[1]

[1]*National Research Tomsk State University, Laboratory for High Energy and Special Materials, Lenin Ave., 36, Tomsk 634050, Russia*
[2]*Federal Research and Production Center "Altay", Socialistitcheskaya St., 1, Byisk 659322, Russia*

5.1 Introduction

Currently, cyclic nitramines such as 1,3,5,7-tetranitro-1,3,5,7-tetraazacyclooctane (octogen, HMX) and 2,4,6,8,10,12-hexanitro-2,4,6,8,10,12-hexaazaisowurtzitane (CL-20) are the basic components of creating solid propellants for various applications [1–4]. It is caused by their high brisance and chemical stability in comparison to nitrate ester explosives of similar energy.

However, using cyclic nitramines in solid propellants is difficult due to their high solubility in many polar solvents including, plasticizers and polymers used in fuel preparation. It results in a decrease of polymorphic transition energy of nitramines as well as uncontrolled reactions of complex formation with other mixture components. As a result, the specific impulse of the finished product may decrease due to nitramine density reduction, the mechanical vulnerability may increase as well as fracturing of fuel and mechanical durability reduction.

The agglomeration of metal particles is another reason hindering the realization of the energy potential of most modern solid propellants. They are main combustible component of most high-energy materials. To reduce agglomeration, material must be optimized for rapid ignition. Obviously, it is necessary to develop an integrated approach to solve current problems with solid propellants and use effective cyclic nitramines in fuel compositions. This approach should take into account possible complex formulation with other mixture components, blocking the polymorphic transitions, and reducing existing energy losses. On the one hand, this may allow to ensure the participation of the main components in the formation of a complex propellant morphology when solving the issues of compatibility of cyclic nitramines with other fuel components. On the other hand, this may allow the development of an effective method for modifying the characteristics of existing explosives and the creation of new composite materials.

It is known that metal particles become the main combustible components of many high-energy materials [5]. Their processes of oxidation and combustion cause

Nano and Micro-Scale Energetic Materials: Propellants and Explosives, First Edition.
Edited by Weiqiang Pang and Luigi T. DeLuca.

a sharp increase in the volumetric specific impulse. Aluminum powder is commonly used as the metallic fuel in solid propellants. The aluminum content can be up to 5–15 mass% of solid propellant. The particle size is about 5–50 μm. However, the possibility of using nanoparticles with a size of 10–100 nm is being actively studied [6–8]. The chemical and physical characteristics of high-energy materials containing nanopowder (nMe) are very different from micro powder (μMe). It makes nMe a promising material for solid propellants, explosives, and other compounds [5, 7, 9, 10]. Partial or complete replacement of μMe by nMe leads to a decrease of ignition temperature and an increase in burning rate of the mixture, as well as a reduced agglomeration in the near-surface zone [6, 11, 12].

Despite the benefits of using nMe, there are some negative effects that prevent their use. For example, nMe is more susceptible to the influence of an oxidizing or aggressive environment [13]. Also, it has low chemical compatibility with some fuel components and low thermal stability in the complete formulations [14–18]. The main reason for these negative effects is the large specific surface of nMe.

The manuscript covers the research for CL-20 and HMX crystal synthesis with additive compounds such as μAl (ASD-6), nAl (Alex), Fe_2O_3, $B_{amorphous}$, AlB_2, and TiB_2. Fe_2O_3 is used as a catalyst for solid propellants [19, 20].

5.2 Research on High-Energy Cyclic Nitramines HMX and CL-20

High-energy cyclic nitramines such as HMX and CL-20 were chosen as study materials (Table 5.1). They have high thermal stability and relatively low sensitivity to impact and friction compared to nitrate ester compositions.

Table 5.1 Structure and characteristics of cyclic nitramines used for research purposes.

Cyclic nitramines	Structural formula	Enthalpy of formation (kJ mol^{-1})	ρ (g cm^{-3})	u_D (m s^{-1})	References
HMX	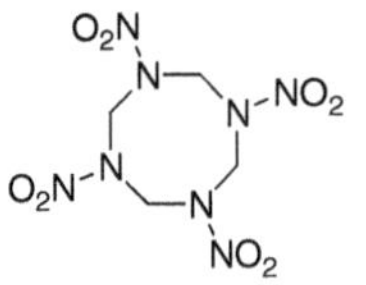	187.9	1.91 (β-form)	9100	[21]
CL-20	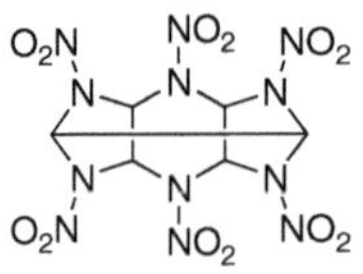	377.4	2.04 (ε-form)	9660	[21]

Table 5.2 Composition, physical, and chemical characteristics of powders.

Characteristics	Powders					
	μAl	nAl	Fe_2O_3	$B_{amorphous}$	AlB_2	TiB_2
Average particles size	3 μm	100 nm	90 nm	100 nm	80 nm	60 nm
Apparent density ($g\,cm^{-3}$)	2.7	1.0–1.2	1.03	2.46	2.12	4.52
Specific surface, ($10^3\,m^2\,kg^{-1}$)	0.5–0.65	15.5	10	—	5.42	45
Melting point (°C)	660	640	1550	2400	—	—
Powder color	Gray	Gray	Brown	Black	Black	Black

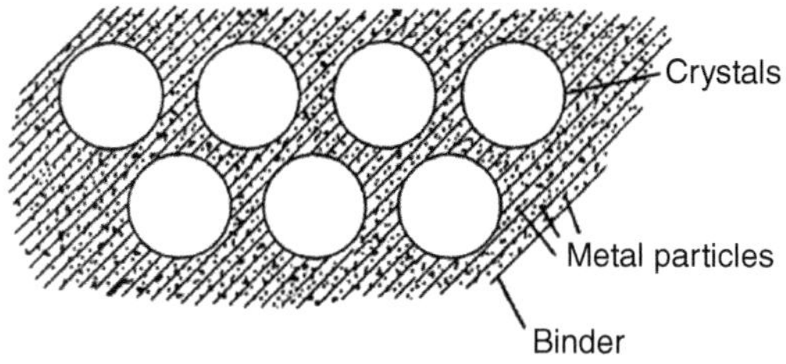

Figure 5.1 Scheme of separation of solid-propellant components.

CL-20 is the most powerful and commercially available explosive with high density, formation heat, and detonation velocity. However, a strong tendency to complex with other fuel components and increased sensitivity to mechanical stress limit the use of CL-20.

Aluminum powders (μAl and nAl), ferric oxide (III) (Fe_2O_3), amorphous black boron ($B_{amorphous}$), aluminum boride (AlB_2), and titanium boride (TiB_2) were chosen to be embedded into cyclic nitramine crystals. Powder characteristics according to the producer and research results are provided in Table 5.2.

5.2.1 Synthesis of HMX and CL-20 Crystals with Inclusion of Metal Particles

The main volume during the preparation of the high-energy materials is filled by dispersed oxidizing crystals (ammonium perchlorate [AP], ammonium nitrate, and others) and different energetic additives (HMX, CL-20, and others). The metal particles with high concentration fill in the intercrystalline space left (Figure 5.1). This leads to their high aggregation and close contact with each other. It is the necessary condition of agglomeration processes influencing negatively the solid-propellant characteristics.

The contacting metal particles in the condensed phase can form clusters during heating. The cluster size significantly depends on the size and volumetric concentration of particles in the fuel. If the volume concentration of aluminum in the solid-propellant matrix exceeds the percolation limit ($\nu_{cr} = 0.15$–0.17), then clusters are created that may unite into larger agglomerates or drops during the combustion process [22, 23].

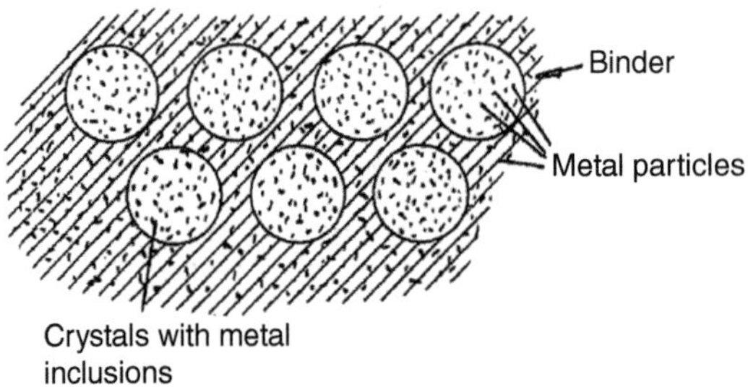

Figure 5.2 Scheme of the distribution of metal particles with inclusion of modified components.

As noted above, one of the negative aspects of the use of metal nanopowder is their low chemical compatibility with some fuel components, as well as the low thermal stability of finished compositions. In the case of aluminum, the catalytic activity of a nanopowder depends on the activity and content of γ-Al_2O_3. Metal nanopowders actively interact with some main components during long-term storage. Therefore, it is inappropriate to use chemical compounds containing nitrate ester, nitrile, and several other active groups when developing compositions for a blended energy material (BEM) with metal nanopowders and their oxides. The most stable compounds in mixtures with aluminum nanopowder are the nitrates and perchlorates, as well as the substances containing such active fragments (i.e. C–NO_2, N–NO_2) and hydrocarbon compounds [14].

There are other difficulties connected with the use of metal. For instance, the excessively high system ductility, the fragility of finished compositions, and the low combustion effectiveness. These difficulties arise from large the specific surface of the metal particles. In addition, they are the result of the high hygroscopicity toward liquid binder components of fuel.

One of the solutions to these difficulties is the encapsulation and uniform distribution of metallic fuel throughout the volume. It results in a reduction of agglomeration processes and metal particle contacts with functional groups of fuel components. It also facilitates the increase in ballistic efficiency of finished products and ensures acceptable operational characteristics of BEM. Therefore, it is proposed to partially remove the active metal from the intercrystalline space of fuel into the cyclic nitramine crystals (Figure 5.2).

5.3 Production of Cyclic Nitramine Crystals with Metal Inclusions

5.3.1 Production of CL-20 Crystals with Metal Inclusions

The method for production of CL-20 crystals with metal inclusions is based on the classical crystallization and precipitation process [24]. The method includes the dissolution of a calculated amount of CL-20 in acetone followed by adding metal particles and the precipitator, namely *o*-Xylene. The precipitator mass fraction increases during the distillation of the solvent, thus, increasing the cyclic nitramine

Table 5.3 The conditions for CL-20 crystallization with metal inclusions.

Initial concentration of CL-20 (%)	Dissolvent/seeding agent ratio	Stirring rate (rpm)	Crystallization temperature (°C)	Period of crystallization (h)
50	1/2	450–600	45–60	1–2

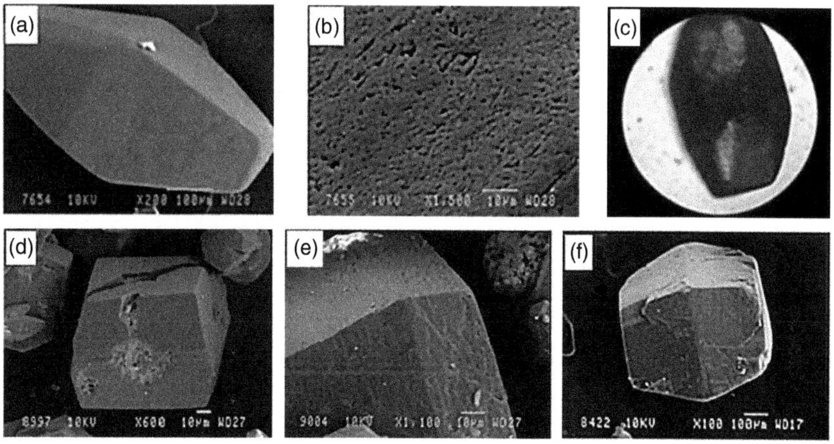

Figure 5.3 Microstructure of the crystals: CL-20/μAl (a–c), CL-20/nAl (d, e) and CL-20/Fe_2O_3 (f).

concentration. The nucleation of crystals and their growth are observed upon reaching the critical concentration.

One way to influence the size, morphology, and polymorphous modification of CL-20 crystals is by adding a seeding agent [25]. Thus, the hypothesis was formulated. The metal nanopowders (nMe) may be used as a seeding agent as a starting point for crystal formation. Further crystallization leads to the capture of more metal particles inside the nitramine crystal.

It has been established that the stirring rate during the distillation of the dissolvent has a determining influence on the granulometric composition of the crystals and the mass concentration of metal in the crystals.

The values of the main crystallization parameters for production of CL-20 crystals with metal inclusions are provided in Table 5.3. The microstructure of CL-20 crystals with μAl, nAl, and Fe_2O_3 inclusions at a stirring speed of 450 rpm is shown in Figure 5.3.

At the first stage, CL-20 crystals with inclusions of spherical dispersed μAl were produced at a stirring speed of 450 rpm. The general appearance of the CL-20/μAl crystals is provided in Figure 5.3a. A large amount of micropores are observed on the surface of the CL-20/μAl crystals (Figure 5.3b). These micropores contained particles of aluminum, which was removed by treatment with ethyl alcohol. Figure 5.3c shows the illumination of the crystal by transmitted light. Agglomerates of aluminum fines are visible on the surface in the center of the crystal. It may be explained by the gradual extraction of aluminum during the whole processing of the sample dissolution by acetone. The particles produced by such a process were quantitatively collected and dried. Their mass made up 5% (450 rpm) and 9% (600 rpm) of the dissolved sample. The subsequent increase in the stirring speed leads to a decrease in the surface concentration of metal particles, which reduces the metal content in the final CL-20 crystals. The average size of CL-20/μAl crystals was 300–500 μm. The pycnometer method was used to measure the density of crystals. It has been established that the density of CL-20/μAl (5%) crystals is 2.07 $g\,cm^{-3}$, and CL-20/μAl (9%) is 2.1 $g\,cm^{-3}$. This data correlates well with calculations of the density of the additive based on the compositions of 5% μAl/95% ε-CL-20 and 9% μAl/91% ε-CL-20, respectively.

Also, CL-20 crystals with inclusions of nAl (BET = 15 $m^2\,g^{-1}$) and Fe_2O_3 (BET = 9.4 $m^2\,g^{-1}$) were studied. Both nAl and Fe_2O_3 particles are trapped inside the crystal during the crystallization of CL-20 in the acetone/*o*-Xylene system. The surface of CL-20/nAl and CL-20/Fe_2O_3 crystals is shown in Figure 5.3d–f. The average size of CL-20/nAl and CL-20/Fe_2O_3 crystals was 50–100 and 500–600 μm, respectively. The shape of the crystals is triangular pyramids. This is probably due to the fact that the metal particles act as inoculating crystals.

It should be noted that the size and shape of CL-20/μAl and CL-20/nAl crystals are very different despite similar conditions of the experiment. CL-20/nAl (Figure 5.3d, e) and CL-20/Fe_2O_3 crystals (Figure 5.3f) have bevel faces and numerous surface defects. It seems to be connected with influence of the metal particles that serve as a seeding agent for crystal formation.

The CL-20/$B_{amorphous}$ and CL-20/AlB_2 crystals are similar to CL-20/μAl crystals. In both cases, the crystals have bipyramidal form with bevel tops (Figure 5.4). In the case of CL-20/$B_{amorphous}$ crystals, the maximum fraction of boron inside CL-20

Figure 5.4 Microstructure of CL-20/$B_{amorphous}$ (a–c) and CL-20/AlB_2 (d–f) crystals.

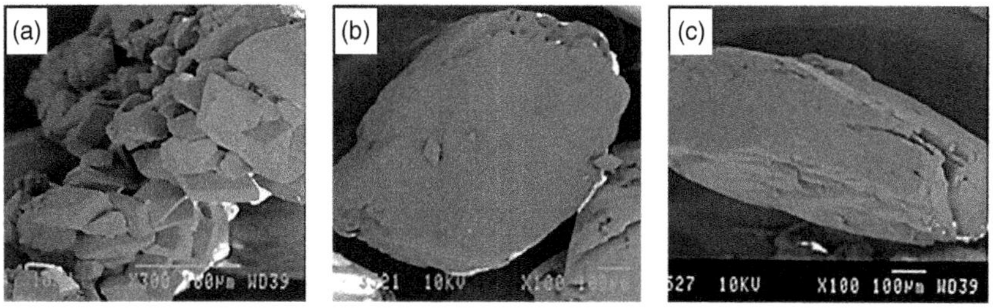

Figure 5.5 Microstructure of CL-20/TiB_2 crystals at magnifications ×300 (a) and ×100 (b, c).

crystals did not exceed 7%, while for CL-20/AlB_2 it was 8%. There is a significant difference in the density of crystals with inclusions: 2.018 g cm^{-3} (CL-20/$B_{amorphous}$) and 2.119 g cm^{-3} (CL-20/AlB_2).

TiB_2 is the densest (exceeds 4.5 g cm^{-3}) among the powders of the investigated intermetallic compounds. It was impossible to create a homogenous suspension of metal particles in the reaction environment due to the extremely high density of TiB_2 powder. Increasing the stirring rate and replacement of the propeller stirrer by a turbine or magnet stirrer did not lead to the formation of a suspension. TiB_2 particles always fell out at the bottom of the device and did not participate in CL-20 crystallization. Many crystal faces had not enough time to finish their growth. Thus, many defects are observed (Figure 5.5). After CL-20 crystallization a small amount of metal is precipitated on its surface. This metal is easily removed by ethanol.

As noted above, the effective delivery of metal particles during the growth of CL-20 crystal is the necessary requirement for production of the target product. However, the excessive intensiveness of stirring leads to an increase in the number of crystal defects. Therefore, CL-20 crystal production with inclusions of this metal during evaporating–precipitating crystallization is impossible. This is due to the physical properties of TiB_2.

5.3.2 Production of HMX Crystals with Metal Inclusions

Modification of the physicochemical and explosive properties of HMX by metal inclusions can lead to an increase in the specific impulse in already existing solid-propellant compositions based on it. Several systems for HMX crystallization are known in the literature [26, 27]. This is due to the low solubility of HMX in most organic solvents.

HMX crystallization was conducted with two methods:

1) A calculated amount of HMX was dissolved in 10% ε-caprolactam (CPL) solution ($T_{heating}$ = 100–120 °C) followed by cooling down to 60 °C. The mass ratio of CPL to HMX was 5 to 1. Then the solution was cooled down to 15 °C and the precipitated crystals were filtered out. A detailed description is presented in [26]. HMX crystal produced according to method 1 is shown in Figure 5.6a.

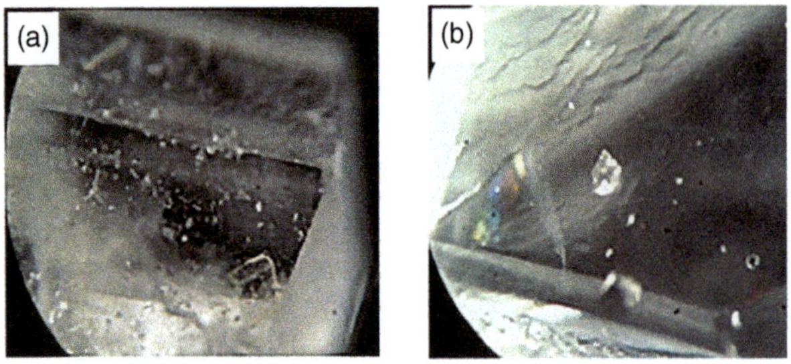

Figure 5.6 HMX crystals produced according to method 1 (a) and method 2 (b).

2) A calculated amount of HMX was dissolved in a mixture of acetone and water at 50 °C. The ratio of HMX to acetone/water was 9 to 1. The produced saturated solution was evaporated down to half of the initial volume and the precipitated crystals were filtered out. A detailed description is presented in [27]. HMX crystal produced according to method 2 is shown in Figure 5.6b.

The metal nanoparticles (20 mass%) were added to the mixtures after HMX dissolution stage. To avoid local overheating and overcooling during crystallization, it is necessary to ensure effective mixing of the reaction mass. The design of the stirring device was the main tool for effective stirring. Thus, the blade stirring, turbine stirring, and magnet stirring were used to create different flows of the liquid. As a result of numerous experiments on HMX crystallization with various rates and types of stirring devices, optically transparent crystals without any visible inclusions of metal particles were always produced (Figure 5.6).

There are difficulties in obtaining suspensions with inclusions of TiB_2 particles. TiB_2 particles fell out at the bottom of the device and did not participate in HMX crystallization. This is due to their high density.

Unexpected results were achieved in the experiments with $B_{amorphous}$ and AlB_2. To use these substances, a suspension was produced with a uniform distribution of solid particles in the entire volume of reaction mass as was the case for CL-20 crystallization. However, with the growth of HMX crystals both in CPL and acetone/*o*-Xylene, the particles were not captured. An increase in the content of particles up to 100% compared to HMX_{raw} as well as the use of various stirring devices did not lead to a crystal production with metal inclusions. The best result with HMX crystals was achieved with μAl. Nevertheless, the mass concentration of metal in the nitramine crystals did not exceed 2%.

It has been found that the stirring rate strongly influences the shape and size of the HMX and CL-20 crystals. HMX crystals produced at different stirring rates are shown in Figure 5.7.

A low stirring rate leads to the creation of polycrystalline structures with a ramified surface and many defects (Figure 5.7a). At the same time, an increase in the stirring rate three times leads to crystal formation that does not have enough time to create the final shape (Figure 5.7c). The most regular shape of crystals with average size of 250–300 μm was produced in the experiments at a stirring rate of 400–450 rpm (Figure 5.7b).

Figure 5.7 Microstructure of HMX/μAl crystals produced at stirring rates of 200 rpm (a), 450 rpm (b), and 600 rpm (c).

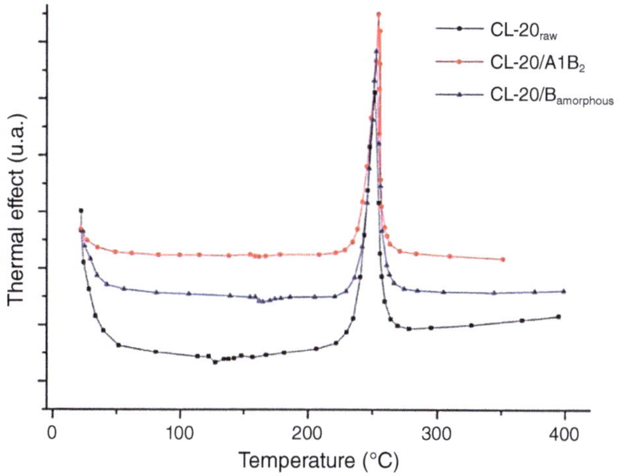

Figure 5.8 DTA curves for the CL-20_{raw}, CL-20/$B_{amorphous}$, and CL-20/AlB_2 crystals.

5.4 Research on the Physicochemical and Explosive Characteristics of CL-20 and HMX Crystals with Metal Inclusions

Calorimetric studies of the materials were performed using differential thermal analysis (DTA), thermogravimetric analysis (TGA) and differential scanning calorimetry (DSC). The DTA was conducted for CL-20_{raw}, CL-20/$B_{amorphous}$, and CL-20/AlB_2 crystals (Figure 5.8). One exothermic peak at ~255 °C and one endothermic peak at ~135 °C (CL-20_{raw}) and ~165 °C (CL-20/AlB_2 and CL-20/$B_{amorphous}$) were observed for all considered materials. This reaction corresponds to the polymorphic transformation from ε- to γ-form. Apparently, the intermetallic particles due to their higher thermal conductivity compared to pure CL-20, absorb part of the heat and thereby increase the minimum temperature to overcome the energy barrier of the polymorphic transition.

An increase in the specific surface of metal particles leads to a greater shift of the peaks (Figure 5.9): for μAl (5–7 °C), Fe_2O_3 (8–10 °C), and nAl (10–12 °C). At

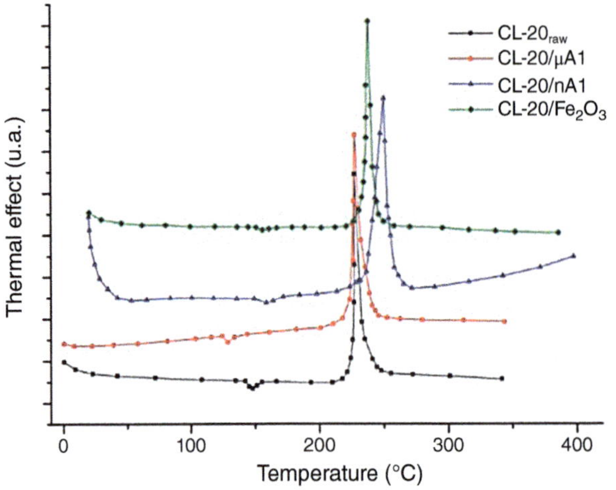

Figure 5.9 DTA curves for $CL\text{-}20_{raw}$, CL-20/μAl, CL-20/Alex, and $CL\text{-}20/Fe_2O_3$ crystals.

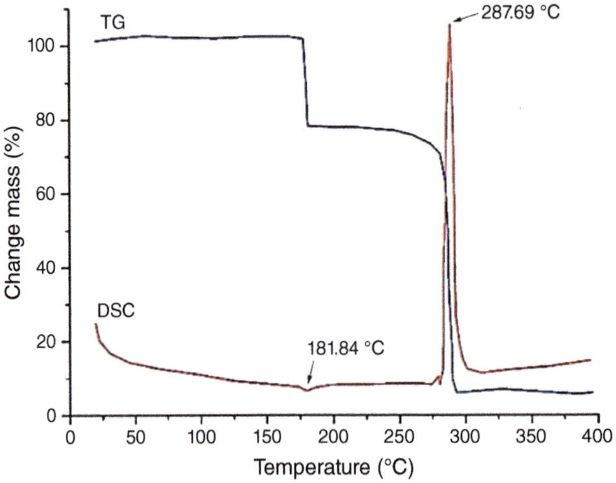

Figure 5.10 TG and DSC curves of HMX/μAl crystals.

a temperature close to the HMX polymorphic transition temperature (~170 °C), a sharp abrupt decrease of mass appears (Figure 5.10). However, this effect does not appear for HMX_{raw}.

A detailed observation of HMX/μAl crystals shows that at the temperature of 170–180 °C the crystals start hopping and even leave the heating surface. Aluminum forms stretched structures inside the crystal and local overheating of aluminum-containing zones occurs with a temperature increase. It is connected with large coefficient of thermal expansion of aluminum in comparison to HMX.

A similar effect at the heated surface is not observed for CL-20 crystals with metal inclusions. The mass of samples at T = 130–155 °C does not change. This

Table 5.4 Results of comparative tests to determine electrical resistance and minimum ignition energy of CL-20 and CL-20/μAl.

Crystal	Environmental conditions	ρ_v (Ohm cm)	ρ_s (Ohm)	W_{min} (mJ)	Result
CL-20_{raw}	$\varphi = 46\%$; $t = 22\,°C$	$>10^{13}$	$>10^{13}$	10	Ignition
CL-20/μAl	$\varphi = 46\%$; $t = 22\,°C$	$>10^{13}$	$>10^{13}$	3.5	Ignition

Notation: φ – air humidity, ρ_v – volume resistance, ρ_s – surface resistance, W_{min} – minimum ignition energy.

Table 5.5 Volume and composition of released gases according to ACM data.

Crystal	Volume of released gases ($cm^3\ g^{-1}$)	Gas composition (%)
CL-20/$B_{amorphous}$	0.004	84 (N_2) 16 (CO_2)
CL-20/AlB_2	0.012	39 (NO) 20 (N_2) 41 (CO_2)
CL-20/μAl	0.004	70 (N_2) 30 (CO_2)
CL-20/nAl	0.056	19 (NO) 12 (N_2) 12 (CO_2) 46 (N_2O)

is important when using such crystals in mixtures of energy materials, during assembly and operation of which the temperature can reach 100 °C.

The effect of metal on electrical resistivity ρ_v and minimum ignition energy W_{min} was studied on CL-20_{raw} and CL-20/μAl crystals (Table 5.4).

It has been established that the minimum energy of an electric spark required to ignite a dust-air mixture of the CL-20/μAl complex is 3.5 mJ, and for CL-20_{raw} this parameter is 10 mJ. The decrease in the minimum ignition energy in the case of a complex with inclusions may be due to the influence of spherical aluminum. It is more sensitive to the action of an electric spark compared to CL-20. It should be noted that in both cases there is a flash and complete combustion of the dust-air mixture of the samples. According to their electrical properties, the CL-20_{raw} and CL-20/μAl crystals are dielectrics.

The chemical resistance of the samples was evaluated by the amount of released gases at $T = 80\,°C$ for 24 hours using the ampoule chromatographic technique (ACM) and the stability control equipment "Vulkan." The induction period was about three hours. Subsequent exposure occurs without gas release. The obtained results show the high chemical resistance of the studied objects. The volumes and compositions of the released gases are presented in Table. 5.5.

The total volume of released gases with inclusions of nAl and AlB_2 particles is an order of magnitude higher than for CL-20/μAl and CL-20/AlB_2 crystals. This may be due to the large catalytic effect of the decomposition of cyclic nitramine under the action of particles with a larger specific surface area. According to ACM for CL-20/$B_{amorphous}$ and CL-20/μAl, 0.004 $cm^3\ g^{-1}$ of gaseous products were

Table 5.6 Sensitivity to friction (according to Russian standard 50874-96).

Characteristics	Results		
	$CL\text{-}20_{raw}$	$CL\text{-}20/AlB_2$	CL-20/μAl
Bottom sensitivity limit (MPa)	118	118	88

Table 5.7 Sensitivity to impact (according to Russian Standard B 84-892-74).

Characteristics	Load weight (kg)	Results		
		$CL\text{-}20_{raw}$	$CL\text{-}20/AlB_2$	CL-20/μAl
Bottom sensitivity limit (mm); device №2	10	—	>50	—
	2	120	200	100
Explosion frequency ratio (%); device №1, $H = 250$ mm	10	100	100	100
	2	—	0	84

released. The total volume of released gases for CL-20/nAl was 0.056 $cm^3\ g^{-1}$, and for $CL\text{-}20/AlB_2$ it was 0.012 $cm^3\ g^{-1}$.

The sensitivity characteristics to external mechanical impact were conducted to evaluate the safety of production and operation of CL-20, CL-20/μAl, and $CL\text{-}20/AlB_2$ crystals. The results of sensitivity to friction (according to GOST 50874-96) and to impact (according to GOST B 84-892-74) are presented in Tables 5.6 and 5.7, respectively.

It was revealed that $CL\text{-}20/AlB_2$ crystals are less sensitive to impact than $CL\text{-}20_{raw}$, while inclusion of μAl into the crystalline structure of the nitramine decreases the sensitivity levels insignificantly in both devices №1 and №2.

It may be that the inclusion of intermetallic compounds such as AlB_2 results in partial dispersion of the energy introduced into the sample by impact or friction. It reduces the possibility of triggering reaction via the mechanism of so-called "hot spots."

5.5 Research on the Combustion of Fuel Samples Based on CL-20 Crystals with Metal Inclusions

One of the key stages in the study of the characteristics of CL-20 and HMX crystals with metal inclusions was testing them in solid fuel mixtures close to standard ones. According to [28], the choice of fuel-binder substance without considering the high probability for complex formation of CL-20 and HMX can lead to co-crystallization that influences negatively the mechanical, rheological, and ballistic characteristics of the fuels. It has been established that CL-20 and HMX may co-crystallize with plasticizer components (i.e. nitrate ester, linear

Table 5.8 The components of high-energy compositions with μAl.

	The content of components, mass. %					
Sample	**SKDM-80**	**μAl**	**AP**	**$CL\text{-}20_{raw}$**	**CL-20/μAl (5%)**	**CL-20/μAl (9%)**
№1-1	15	15	30	40	—	—
№1-2	15	12	30	—	43	—
№2-1	15	20	15	50	—	—
№2-2	15	15	15	—	—	55

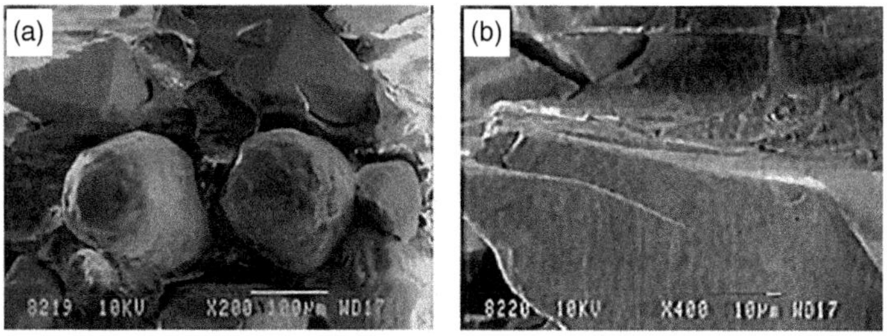

Figure 5.11 Microstructure of sample №1-2 surface.

nitramines, triacetin, and other substances) without any additional extraction agents [29, 30].

In this research, the samples (groups №1 and №2) contained the following: divinyl rubber based on butadiene rubber, which was plasticized with transformer oil in a ratio of 15/85 (SKDM-80) as a combustible binder; bimodal ammonium perchlorate (AP) with fractions of 160–315 μm and less than 50 μm at the ratio of 60/40 as an oxidizing agent; aluminum (μAl); $CL\text{-}20_{raw}$ and CL 20/μAl crystals. The main difference was in applying CL-20/μAl crystals in compositions №1-2 instead of $CL\text{-}20_{raw}$. The content of high-energy compositions is presented in Table 5.8.

The microstructure of the sample №1-2 is shown in Figure 5.11a,b. A uniform distribution of CL-20/μAl crystals and AP over the fuel volume is observed. A good adhesion of the components to the binder is also visible. It is important that there is no free aluminum on the sample surface. The same result was observed for composition №2-2.

To conduct research on the burning rate, the samples of uncured high-energy compositions were prepared. The samples with a diameter of 14 mm and a height of 10–30 mm were used. The device (Crawford bomb) of constant pressure was used to study burning rates. The research was conducted in nitrogen environment. The error did not exceed 3% at a confidence level of 0.95. The coefficients (b, ν) of the power dependence of burning rate (u) on pressure (p): $u(p) = bp^{\nu}$ have been determined for all the compositions (Figure 5.12).

Different b and ν coefficients for all the samples are shown on Figure 5.12. The b and ν coefficients from the burning rate equation for sample №1-1 are 3.32 and 0.41, respectively. There is a simultaneous increase in both b (up to 3.58) and ν

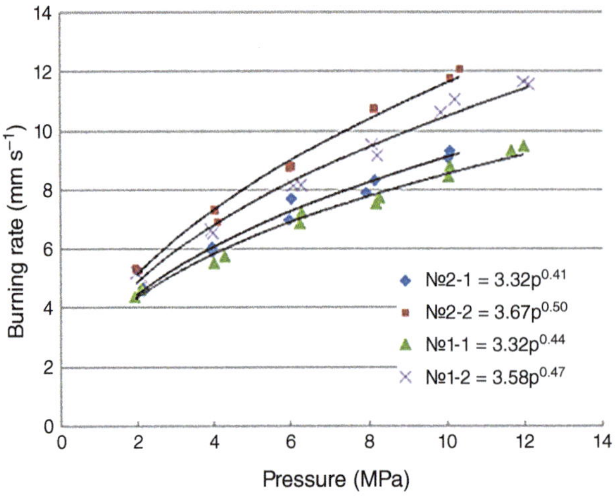

Figure 5.12 Dependence of burning rates on pressure for samples of groups №1 and №2.

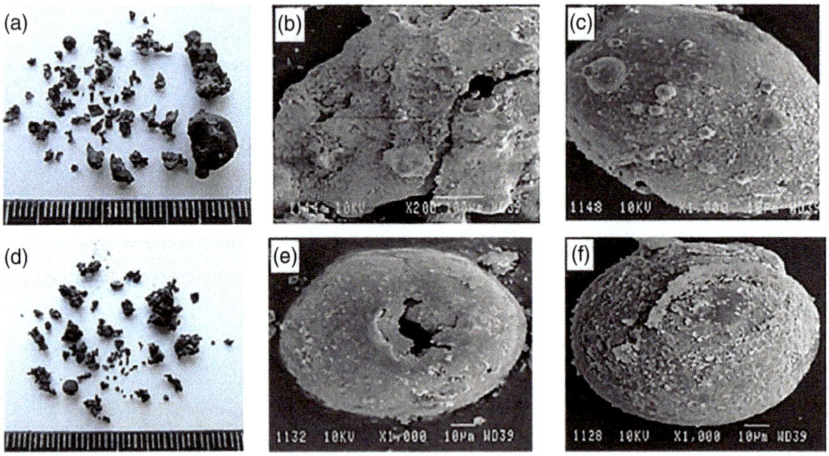

Figure 5.13 Microstructure and morphology of agglomerates created as a result of combustion of samples №2-1 (a–c) and №2-2 (d–f).

(up to 0.47) coefficients in the case of replacing a part of aluminum and CL-20_{raw} by CL-20/μAl (5%) crystals. It may be connected with the increased combustion efficiency of aluminum particles inside CL-20 crystals. The addition of aluminum particles also causes catalytic effects facilitating thermal decomposition of CL-20 itself.

The evaluation of agglomerates and slag size created during combustion of high-energy compositions is an important parameter (Figure 5.13).

The size of the agglomerates is 10–100 times greater than the size of the initial aluminum particles, due to which a size equal to 1000 μm is achieved (sample №2-1). The main reason for agglomerates formation is heating in high-volume mode as well as melting and coagulation of initial aluminum particles. The agglomerated particles size decreases significantly when CL-20 crystals are replaced

Table 5.9 The components of high-energy compositions with Fe_2O_3.

Sample	The content of components (mass%)						
	SKDM-80	AP	μAl	$CL\text{-}20_{raw}$	CL-20/ Fe_2O_3 (10%)	CL-20/ Fe_2O_3 (15%)	Fe_2O_3
№3-1	15	40	40	27	3		—
№3-2	15	40	40	27		3	—
№3-3	15	40	40	30	—	—	—
№3-4	15	40	40	30	—	—	0.4 over

Figure 5.14 Dependence of burning rates on pressure for samples group №3.

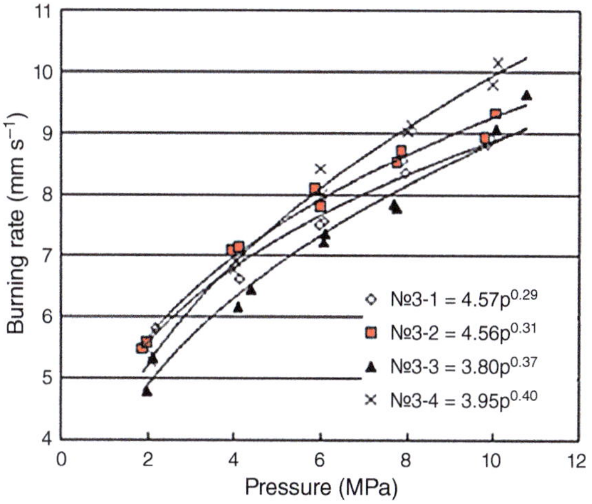

by CL-20/μAl (9%) crystals. Heating at *k*-phase (sample №2-2) causes both thermal decomposition of binder and dispersed nonmetallic particles as well as sintering of contacting aluminum particles with formation of relatively long-lasting coral-like conglomerates. It can be concluded that CL-20/μAl crystals inclusion causes significant reduction of agglomeration processes due to uniform distribution of aluminum particles. There is a large amount of heat energy released in the combustion process, and the coefficient of fuel composition efficiency increases. On the other hand, it may be related to the rate of heat energy release, which means that the same amount of heat is released in a shorter time.

In case of ferric oxide, for the samples №3-1 and №3-2, CL-20/Fe_2O_3 crystals containing 10 and 15 mass% Fe_2O_3, respectively, were added at 3 mass%. The basic compositions for comparison were samples №3-3 and №3-4. The content of AP and μAl for all compositions was the same and amounted to 40% and 15%, respectively. List the components of the investigated high-energy compositions is presented in Table 5.9.

The curves of high-energy compositions showing burning rate vs. pressure are provided in Figure 5.14.

For sample №3-3 (without Fe_2O_3) *b* and *v* coefficients are equal to 3.80 and 0.37 respectively. Addition of Fe_2O_3 in pure kind (sample №3-4) results in the expected increase in coefficient *b* to 3.95 and a slight growth of coefficient *v* up to 0.40. This effect may be explained by the reactions in the condensed phase and ammonium perchlorate decomposition catalysis. Moving to samples with CL-20/Fe_2O_3 crystals inclusion, coefficient *v* drops to 0.29–0.31, while coefficient *b* increases up to 4.56–4.57. It seems that for samples №3-1 and №3-2, the portions of reactions in condensed phase increases but the influence of Fe_2O_3 changes. Thus, it provides active decomposition of CL-20 that it contacts closely inside the CL-20/Fe_2O_3 complex.

5.6 Conclusions

The work covers research in the synthesis of high-energy cyclic nitramine crystals of CL-20 and HMX with additives of μAl (ASD-6), nAl (Alex), Fe_2O_3, $B_{amorphous}$, AlB_2, and TiB_2. The influence of metal inclusions micro and nanoparticles and their oxides on physical and chemical characteristics of energetic materials was revealed. The following conclusions may be drawn upon the research conducted:

1) It has been established that among the main parameters affecting the particle size and the number of inclusions, the stirring rate of the reaction mixture during the distillation of the solvent is of decisive importance. A low stirring rate led to the formation of polycrystalline structures with a branched surface and multiple defects.
2) It was found that the size and shape of the produced CL-20/μAl and CL-20/nAl crystals differ a lot despite similar conditions of experiment. The general kind of CL-20/$B_{amorphous}$ and CL-20/AlB_2 crystals is similar to CL-20/μAl.
3) It has been established that the density of CL-20/μAl (5%) crystals is 2.07 g cm^{-3}, and CL-20/μAl (9%) is 2.1 g cm^{-3}. The maximum proportion of boron inside CL-20 crystals did not exceed 7% (CL-20/$B_{amorphous}$) and 8% (CL-20/AlB_2). There is a significant difference in the density of crystals: 2.018 g cm^{-3} (CL-20/$B_{amorphous}$) and 2.119 g cm^{-3} (CL-20/AlB_2).
4) It was observed that CL-20/AlB_2 crystals are less sensitive to impact in comparison to CL-20_{raw}. CL-20/μAl crystals have insignificantly lower sensitivity limit.
5) It has been established that the minimum energy of an electric spark required to ignite a dust-air mixture of the CL-20/μAl complex is 3.5 mJ, and for CL-20_{raw} it is 10 mJ.
6) According to ACM for CL-20/$B_{amorphous}$ and CL-20/μAl, 0.004 $cm^3\ g^{-1}$ of gaseous products were released. The total volume of released gases for CL-20/nAl was 0.056 $cm^3\ g^{-1}$, and for CL-20/AlB_2 it was 0.012 $cm^3\ g^{-1}$.
7) It was detected that the replacement of CL-20_{raw} crystals by its crystals with metal particles inclusion results into higher burning rate and a reduction of slag and agglomeration processes.

Funding

This study was supported by the Tomsk State University Development Program (Priority-2030). This work was carried out with financial support from the Ministry of Education and Science of the Russian Federation (State assignment No. FSWM-2020-0028).

Acknowledgments

The SEM researches were carried out with the equipment of Tomsk Regional Core Shared Research Facilities Center of National Research Tomsk State University (Grant of the Ministry of Science and Higher Education of the Russian Federation no.075-15-2021-693 (no. 13.RFC.21.0012)).

References

1 Zhang, J. and Guo, W. (2021). The role of electric field on decomposition of CL-20/HMX cocrystal: a reactive molecular dynamics study. *Journal of Computational Chemistry* 42: 2202–2212.

2 Sergienko, A.V., Popenko, E.M., Slyusarsky, K.V. et al. (2019). Burning characteristics of the HMX/CL-20/AP/polyvinyltetrazole binder/Al solid propellants loaded with nanometals. *Propellants, Explosives, Pyrotechnics* 44: 217–223.

3 Chiquete, C. and Jackson, S.I. (2021). Detonation performance of the CL-20-based explosive LX-19. *Proceedings of the Combustion Institute* 38: 3661–3669.

4 Li, Y., Li, B., Zhang, D., and Xie, L. (2021). Theoretical studies on CL-20/HMX based energetic composites under external electric field. *Chemical Physics Letters* 778: 138806.

5 Dreizin, E.L. (2009). Metal-based reactive nanomaterials. *Progress in Energy and Combustion Science* 35: 141–167.

6 DeLuca, L.T., Galfetti, L., Severini, F. et al. (2005). Burning of nano-aluminized composite rocket propellants. *Combustion, Explosion and Shock Waves* 41: 680–692.

7 Sakovich, G.V., Arkhipov, V.A., Vorozhtsov, A.B. et al. (2010). Investigation of combustion of HEM with aluminum nanopowders. *Nanotechnologies in Russia* 5: 91–107.

8 Ivanov, Y.F., Osmonoliev, M.N., Sedoi, V.S. et al. (2003). Productions of ultra-fine powders and their use in high energetic compositions. *Propellants, Explosives, Pyrotechnics: An International Journal Dealing with Scientific and Technological Aspects of Energetic Materials* 28: 319–333.

9 Meda, L., Marra, G., Galfetti, L. et al. (2005). Nano-composites for rocket solid propellants. *Composites Science and Technology* 65: 769–773.

10 Vorozhtsov, A., Lerner, M., Rodkevich, N. et al. (2020). Preparation and characterization of Al/HTPB composite for high energetic materials. *Nanomaterials* 10: 2222.

11 Puri, P. and Yang, V. (2007). Effect of particle size on melting of aluminum at nano scales. *The Journal of Physical Chemistry C* 111: 11776–11783.

12 Sossi, A., Duranti, E., Manzoni, M. et al. (2013). Combustion of HTPB-based solid fuels loaded with coated nanoaluminum. *Combustion Science and Technology* 185: 17–36.

13 Vorozhtsov, A.B., Lerner, M., Rodkevich, N. et al. (2016). Oxidation of nano-sized aluminum powders. *Thermochimica Acta* 636: 48–56.

14 Popok, V.N. and Vdovina, N.P. (2017). The interaction of metal nanopowders with components of the blend compositions. *Butlerov Communications* 8: 147–154.

15 Popok, V.N., Pivovarov, Y.A., and Bychin, N.V. (2015). The influence of micro- and nanosized powders of aluminum and passivating additives on the curing rheokinetics of 1,2-dinitrile-2,4,6-thriethylbenzene and mechanical characteristics of compositions based on the base of rubber SKD. *Butlerov Communications* 44: 120–131.

16 Stepanov, A.M., Zyablitskiy, S.A., Popok, V.N., and Pevchenko, B.V. (2010). Study into the chemical stability of HEMs with different lots of aluminum nanopowder. *Polzunovskiy vestnik* 4-1: 120–124.

17 Popok, V.N. and Vdovina, N.P. (2009). Issledovanie sovmestimosti nanoporoshkov s komponentami vysokojenergeticheskih materialov [The study into the compatibility of nanodispersed powders with high energy materials] (in Russian). *Izvestija vuzov* 52: 99–101.

18 Popok, V.N., Vdovina, N.P., and Bychin, N.V. (2013). Compatibility of nanodispersed powders of metals and their oxides with components of mixed energy materials. *Nanotechnologies in Russia* 8: 87–93.

19 Chakravarthy, S.R., Price, E.W., and Sigman, R.K. (1997). Mechanism of burning rate enhancement of composite solid propellants by ferric oxide. *Journal of Propulsion and Power* 13: 471–480.

20 Ma, Z., Li, F., and Bai, H. (2006). Effect of Fe_2O_3 in Fe_2O_3/AP composite particles on thermal decomposition of AP on burning rate of the composite propellant. *Propellants, Explosives, Pyrotechnics* 31: 447–451.

21 Lide, D.R. (ed.) (2004). *CRC Handbook of Chemistry and Physics*, vol. 85. CRC press.

22 Rashkovskii, S.A. (1999). Struktura geterogennyh kondensirovannyh smesej [Structure of heterogenous condensed mixtures] (in Russian). *Fizika goreniâ i vzryva* 35: 65–74.

23 Rashkovskii, S.A. (2002). Role of the structure of heterogeneous condensed mixtures in the formation of agglomerates (in Russian). *Fizika goreniâ i vzryva* 38: 65–76.

24 Sysolyatin, S.V., Lobanova, A.A., Chernikova, Y.T., and Sakovich, G.V. (2005). Methods of synthesis and properties of hexanitrohexaazaisowurtzitane. *Russian Chemical Review* 74: 829–839.

25 Sanderson, A.J., Hamilton, R.S., and Warner, K.F. (2002). Crystallization of 2,4,6,8,10,12-hexanitro-2,4,6,8,10,12-hexaazatetracyclo[5.5.0.05,903,11]-dodecane. Patent US6350871B1, filed by 21 March 2001.

26 Savinova, M.A., Pechenev, Y.G., Fyrne, V.V., and Dmitrieva, T.A. (1994). Method of octogen crystallization. Patent RU2024495C1, filed by 24 April 1990.

27 Basu, S., Gawande, N.M., Apte, M.E., and Narasimhan, V.L. (2004). Crystallization of HMX in acetone-water system. *Indian Journal of Chemical Technology* 11: 575–581.

28 Popok, V.N. (2012). Vlijanie strukturnogo faktora na parametry gorenija smesevyh jenergeticheskih materialov [Influence of structural factor on the energy materials burning parameters] (in Russian). *Butlerov Communications* 32: 75–87.

29 Zhou, S., Pang, A., and Tang, G. (2017). Crystal transition behaviors of CL-20 in polyether solid propellants plasticized by nitrate esters with co-existence of HMX and CL-20. *New Journal of Chemistry* 41: 15064–15071.

30 Popok, V.N. and Zharkov, A.S. (2015). Harakteristiki termicheskogo razlozhenija i teplovogo vzryva nekotoryh komponentov smesevyh jenergeticheskih materialov [Characteristics of thermal decomposition and heat explosion of some blended energy materials] (in Russian). *Butlerov Communications* 42: 11–16.

6

Effects of TKX-50 on the Performance of Solid Propellants and Explosives

WeiQiang Pang[1,2], Thomas M. Klapötke[3], Luigi T. DeLuca[4], Dihua OuYang[5], Zhao Qin[2], YiWen Hu[1], XueZhong Fan[1], and FengQi Zhao[2]

[1] *Xi'an Modern Chemistry Research Institute, Zhangba East Road No. 168, Xi'an 710065, China*
[2] *Science and Technology on Combustion and Explosion Laboratory, Xi'an 710065, China*
[3] *University of Munich (LMU), Energetic Materials Research, Department of Chemistry, Butenandtstr. 5 – 13 (D), 81377 Munich, Germany*
[4] *Politecnico di Milano, Space Propulsion Laboratory (SPLab), Aerospace Science and Technology Dept., Campus Bovisa Sud 34 Via La Masa, I-20156 Milan, Italy*
[5] *Xi'an University of Architecture and Technology, College of Resource Engineering, No. 13, middle section of Yanta Road, Beilin District, Xi'an 710068, China*

6.1 Introduction

The search for novel, high-energy-density compounds (HEDCs) with low sensitivity, is the eternal goal in energetic materials (EMs). Ammonium perchlorate (AP), is normal choice for use in solid rocket and missile propellants. However, the use of AP causes various environmental concerns due to the release of perchlorate into groundwater, and in addition, the generation of hydrochloric acid causes the depletion of the ozone layer as well as leading to high concentrations of acid rain. In recent years, considerable efforts have been devoted to developing solid propellants that use green oxidizers, which are less hazardous and are environmental friendly, and which produce only chlorine-free combustion products. Moreover, with the service of high-value weapon platforms, the demand for insensitive munitions (IMs) is increasing every day [1, 2]. The design and synthesis of insensitive EMs have become one of the hot topics of research in EMs worldwide. It is known that the gravimetric specific impulse and/or density specific impulse (I_{sp}) of propellants can be increased by the inclusion of specific high-energy ingredients [3, 4]. However, not only are EMs with increased energy contents needed, but they must also show decreased sensitivity toward external stimuli such as thermal, friction, and impact. In this context, many efforts are ongoing to design and synthesize new high-energy explosives with improved performance, but with low sensitivity. Dihydroxylammonium 5,5′-bistetrazole-1,1′-diolate (TKX-50, Figure 6.1), is an ionic energetic compound with nitrogen-rich tetrazole groups, a high enthalpy of formation, high nitrogen content and good thermal stability. This compound

Nano and Micro-Scale Energetic Materials: Propellants and Explosives, First Edition.
Edited by Weiqiang Pang and Luigi T. DeLuca.

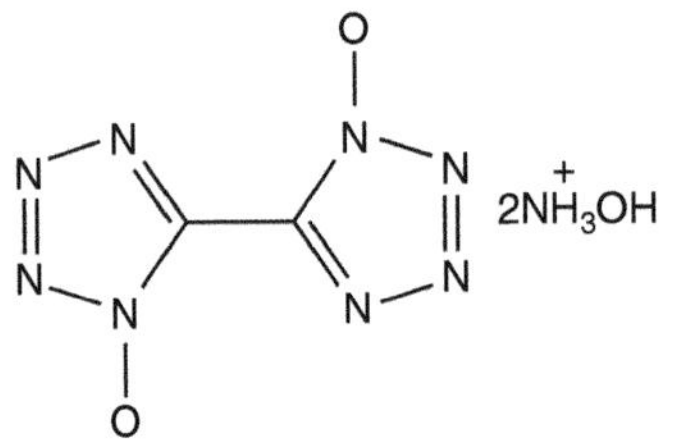

Figure 6.1 Structure of TKX-50.

was synthesized in 2012 and has proven to be of great interest for researchers ever since [5–7]. Table 6.1 lists the physicochemical properties of TKX-50, which are compared with those of other EMs. It was found that the detonation velocity of TKX-50 is higher than that of 2,4,6,8,10,12-hexanitro-2,4,6,8,10,12- hexaaza-isowurtzitane (ε-CL-20), the detonation pressure is 42.4 GPa, and the impact sensitivity (20 J) and friction sensitivity (120 N) of TKX-50 are similar to that of 1,3,5-trinitro-1,3,5-triazinane (RDX), and much lower than that of ε-CL-20 [18]. Moreover, investigations on the thermal decomposition, non-isothermal decomposition kinetics, calculated detonation velocity, and safety evaluation of TKX-50 show that TKX-50 is an example of a promising EM with good thermal performance and sensitivity [8, 19–21]. With respect to the effect of TKX-50 on solid propellants, theoretical calculations show that TKX-50 should be a good replacement for RDX in composite modified double-base (CMDB) propellants. Substitution of 1,3,5,7-tetranitro-1,3,5,7-tetrazocane (HMX) by TKX-50 in nitrate ester plasticized polyether (NEPE) propellants results in a two to five seconds increase of the gravimetric specific impulse (I_{sp}) [22, 23]. Furthermore, explosives are some of the most important charge methods of warhead systems. In particular, 2,4,6-trinitrotoluene (TNT) is used as the molten carrier in melt cast explosives that have been widely used in industrial explosives and military explosives all over the globe. However, nowadays, the increasing demand for explosives with low sensitivities for use as IMs has kick-started extensive research into high explosive formulations [24, 25]. TKX-50, however, also has drawbacks. For example, the use of lithium hydroxide in the synthesis of TKX-50 is expensive, which increases the preparation cost [26–28].

In the quest for safer and cheaper explosives with higher performance, TKX-50 has been developed. It is a new generation high explosive with low sensitivity and is highly cost-effective [29, 30]. On account of its good performance and sensitivity, TKX-50 is currently one of the most promising EMs. To explore the possible future applications of TKX-50 in solid propellants and explosives, its safety and performance properties are of great interest to researchers. Thus, the purpose of this paper is to review the compatibility of TKX-50/CL-20, TKX-50/HMX, TKX-50/pentaerythritol tetranitrate (PETN) co-crystals, as well as of TKX-50/glycidyl azide polymer (GAP), TKX-50/AP and TKX-50/graphene oxide (GO) mixtures, and their performance was emphatically analyzed by thermal decomposition techniques. The aim of this work is mainly to explore and clarify the influence of TKX-50 on the combustion, stability, and detonation performance of solid propellants and explosives.

Table 6.1 Physicochemical properties of TKX-50.

Compounds	Chemical formula	N (%)	Ω (%)	ΔH_f (kJ mol^{-1})	ρ (g cm^{-3})	D (m s^{-1})	P (GPa)	H_{50} (cm)	F (%)
TKX-50	$C_2H_8N_{10}O_4$	59.3	−27.10	446.6	1.918 (100 K); 1.887 (298 K)	9698	42.4	100 [8]	24 [8]
RDX	$C_3H_6N_6O_6$	37.8	−21.61	70.70	1.818	8750	38.0	38 [9]	88 [9]
HMX	$C_4H8N_8O_8$	37.8	−21.61	75.02	1.905	9221	41.5	32 [10], 7.5 J [11]	100 [10], 112 N [11]
FOX-7	$C_2N_4O_4H_4$	37.8	−18.50	−134.0	1.885	8870	33.96	72 [12], 25 J [13]	240 N [13]
TATB	$C_6N_6O_6H_6$	32.6	−145.1	−87.0	1.857	7606	24.40	>320 [12], 103 J [14]	360 N [11]
TNT	$C_7H_5N_3O_6$	18.5	−73.96	−55.5	1.648	7459	23.5	15 J [15]	353 N [15]
Ammonium dinitramide (ADN)	$N_4H_4O_4$	45.2	+25.8	−150	1.82	8100	23.72	33.9 [16]	20 [16]
ε-CL-20	$C_6H_6N_{12}O_{12}$	38.3	−10.95	416.0	2.04	9455	46.7	18 [17], 4 J [11]	100 [17], 48 N [11]

Note: N = nitrogen content, %; Ω = oxygen balance, %; ΔH_f = enthalpy of formation, kJ mol^{-1}; ρ = density, g cm^{-3}; H_{50} = impact sensitivity, characteristic drop height, cm; F = friction sensitivity, %; D = detonation velocity, m s^{-1}; P = detonation pressure, GPa.

6.2 Physicochemical Properties of TKX-50

In both military and civilian utilizations, the highest performing solid propellants, and explosives take advantage of the same strategy: caged and cyclic nitramines [1]. The high-energy compounds with high detonation performance are often extremely sensitive due to their unprecedented energetic content. To improve and balance the energetic and safety features of innovative EMs, TKX-50 was prepared and sensitivity tests were performed [18]. The impact sensitivity of TKX-50 (20 J) is much lower than that of RDX, HMX, and ε-CL-20 (from 4 to 7.5 J) suggesting that TKX-50 can be used without desensitization. The friction sensitivity of TKX-50 (120 N) is comparable or lower than that of RDX, HMX, or ε-CL-20, which is important in an industrial context. Moreover, the electrostatic sensitivity of TKX-50 (0.1 J) is far higher than what human body generates (25 mJ). The thermal stability of TKX-50 is comparable to that of RDX, with TKX-50 showing a decomposition onset at 221 °C. The EC_{50} (concentration for 50% of maximal effect) of TKX-50 (130 ppm) lies above RDX (91 ppm), indicating that TKX-50 shows a lower toxicity to *Vibrio fischeri*, etc.

The scanning electron microscope (SEM) of TKX-50 showed in Figure 6.2. The grain size distributions of well-dried TKX-50 particles prepared in the Xi'an Modern Chemistry Research Institute (MCRI) compared with AP (105–147 μm), which was used and replaced by TKX-50 in the composition, are showed in Table 6.2. TKX-50 crystals prepared by the direct method are long flakes with large aspect ratios, uneven particle size, and a smooth surface (Figure 6.2a). The morphology of TKX-50 crystals can be improved using crystal control technology, which results in the ratio of length to diameter showing a significant decrease, and it shows a trend toward spheroidization under the effect of the crystal control agent (Figure 6.2b). TKX-50 particles are regular polyhedra with smooth surfaces and few defects, which may be one of the reasons for their low mechanical sensitivity. The average particle size of TKX-50 is 233.7 μm, which is extremely higher than that of the first mode AP particles, and the width of TKX-50 (1.31) is much larger than that of AP particles (0.686).

Figure 6.2 SEM photograph of TKX-50 crystal prepared by different methods. (a) Direct preparation. (b) Crystal control technology. Source: Ren et al. [31], Reproduced with permission from Chinese journal of energetic materials.

Table 6.2 Characteristics of AP and TKX-50 particles.

Items	Unit	AP (105–147 μm)	TKX-50
d_{10}	μm	111.5	116.9
d_{50}	μm	155.8	233.7
d_{90}	μm	218.4	423.2
Span	—	0.686	1.31
Density	$kg{\cdot}m^{-3} \times 10^3$	1.94	1.88
Relative atomic mass	—	117.5	236.15
Specific surface area (SSA)	$m^2\ g^{-1}$	1.13	0.03
Oxygen balance	%	34.04	−27.10

Note: d_{10} = particle diameter corresponding to 10% of the cumulative undersize distribution, μm; d_{50} = median particle diameter, μm; d_{90} = particle diameter corresponding to 90% of the cumulative undersize distribution, μm; width = $(d_{90} - d_{10})/d_{50}$; and specific surface area refers to the particle size distribution determined by Malvern Mastersizer. It should be noted that the numbers shown in the table are not very reliable for nonspherical particles.
Source: Reproduced from [32], with permission from Propellants Explos. Pyrotech., Wiley, 2018.

At the same time, pretreatment of TKX-50 at different times by ball milling method was introduced (Figure 6.3) [33]. It was found that the particle distribution of TKX-50 was changed after using the ball milling method as the pretreatment process. The experimentally determined mechanical sensitivity results show that ball milling pretreatment is successful in lowering the mechanical sensitivity. Moreover, the detonation velocities of TKX-50-based explosives (TKX-50/energetic thermoplastic elastomer [ETPE]/others = 95.5/3/1.5) measured using the probe method (copper method) increased from 8699 to 9037 m s^{-1} on increasing the tube diameter from φ20 mm × 20 mm to φ60 mm × 60 mm.

Not only has the stability of solid TKX-50 attracted attention considerably, but also the stability of TKX-50 in various solvents has also been carried out. For example, TKX-50 is stable in dimethyl sulfoxide (DMSO), deionized water, ethyl acetate, methanol, ethanol, petroleum ether, and hexane at both 25 and 50 K, but is unstable in *N*,*N*-dimethylformamide (DMF). TKX-50 reacts with DMF in various ways over the temperature range (25–150 K) [34]. It was also found that a shorter heating time and lower temperature are required for the complete conversion of TKX-50 into diammonium 5,5′-bistetrazole-1,1′-diolate (ABTOX) in the liquid phase with respect to the solid phase.

Additionally, the mechanical sensitivity of TKX-50 can be reduced by polymer coating. For example, TKX-50 was coated with a fluororubber or ETPE, and the sensitivity and formability were investigated [31]. It was found that the friction sensitivity and impact sensitivity of coated explosives were lower than that of TKX-50 slightly, suggesting that ETPE or fluororubbers could be used to coat TKX-50 to increase its safety (Table 6.3). The reasons for the lowering of the sensitivity may be because ETPE or the fluororubber binder fills the voids of TKX-50 with fewer "hot spots" during mechanical impact and friction stimuli.

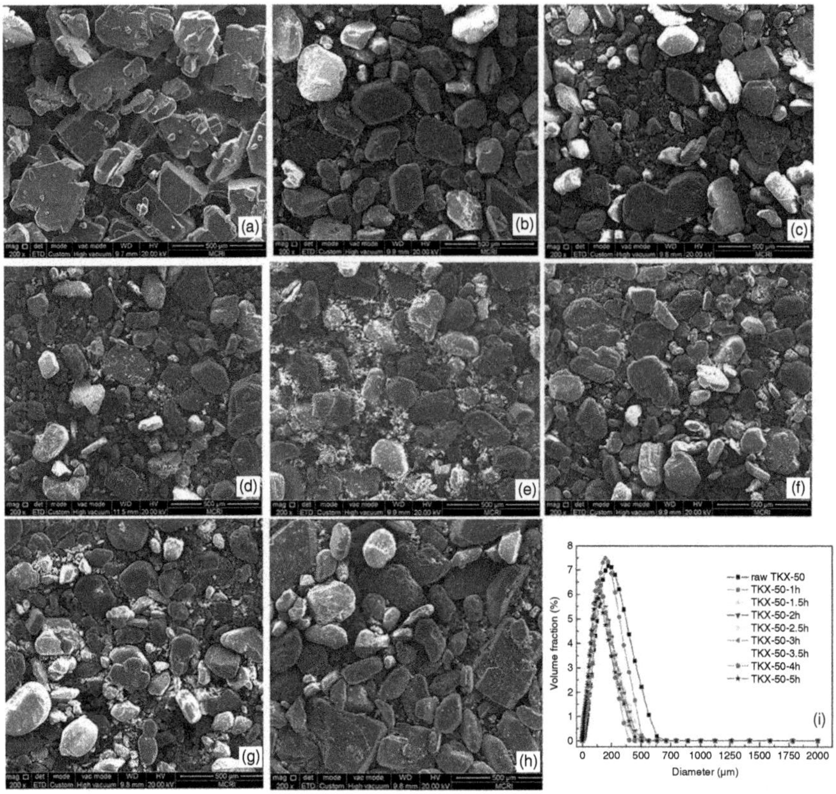

Figure 6.3 SEM and particle size distribution of TKX-50 after different milling time durations. (a) Raw TKX-50; (b) 1 hour; (c) 1.5 hour; (d) 2 hours; (e) 2.5 hours; (f) 3 hours; (g) 4 hours; (h) 5 hours; (i) particle size distribution curves. Source: Xing et al. [33], Reproduced with permission from John Wiley & Sons.

Table 6.3 Mechanical sensitivity data of TKX-50 and of TKX-50 coated with different binders.

Samples	Impact sensitivity (%)	Friction sensitivity (%)
TKX-50	38	40
TKX-50@Fluororubber	28	24
TKX-50@ETPE	32	30

6.3 Interactions Between TKX-50 and EMs

In contrast to TKX-50 with high energetic storage, low mechanical sensitivity, and low toxicity, several other high-energy materials, such as ε-CL-20, HMX, and PETN possess high energy and high density, however, their high mechanical sensitivity is a disadvantage, which has limited their application. Many effective methods have been developed to decrease their sensitivities, such as surface coating [35, 36], crystal quality improving [37, 38], and granularity decreasing [39, 40]. However, these methods still have some problems, such as complicated processes and cost. Combining these high-energy compounds with TKX-50 to form co-crystals or mixtures is a novel approach for the preparation of insensitive EMs.

6.3.1 TKX-50/EMs Co-crystals

Co-crystals can improve the stability and compatibility of compounds due to a new structure formation. The formation and properties of co-crystals have been widely studied with respect to pharmaceuticals, various areas of chemistry, and EMs [41, 42]. The co-crystallization of high EMs has attracted enormous interest since it can mediate the power – safety contradiction to a certain extent. In near recent years, quite lots of co-crystal EMs have been simulated and prepared, and some excellent properties were observed [43–47]. TKX-50, an ionic salt, contains two hydroxylammonium cations, it can form hydrogen bonds with other EMs easily. For example, a hydrogen bond can be formed between hydrogen atoms of the H_3NOH^+ cation in TKX-50, and an oxygen atom of the $-NO_2$ group in ε-CL-20. Co-crystals such as TKX-50/CL-20 can be investigated as possible solutions to combat the problems of the use of single-compound explosives in applications, such as the sensitivity of ε-CL-20, and thereby greatly enlarge the possible applications range of TKX-50 and ε-CL-20. RDF spectra were used to show that both hydrogen bonding and van der Waals forces are present in the TKX-50/CL-20 co-crystal. Furthermore, TKX-50/CL-20 co-crystal has been determined to possess excellent properties and possess both TKX-50 and ε-CL-20 advantages. It has been suggested that this new co-crystal could reduce the limitations of TKX-50 and ε-CL-20 in applications by reducing the disadvantageous properties each individual compound possesses [48]. The TKX-50/CL-20 co-crystal was prepared by using the solvent–non-solvent method [49]. The crystal morphology of the individual components ε-CL-20 and TKX-50 are near spherical in shape, and the average particle size is 1 μm with uniform size distribution. In contrast, the morphology of the TKX-50/CL-20 co-crystal is that of a long and thin lamellar structure, and the average particle size of TKX-50/CL-20 co-crystal is 10 μm, indicating that the co-crystal process not only changes the morphology of the original particles, but also results in the formation of a new type of crystal (Figure 6.4). The differential scanning calorimeter (DSC) curves (Figure 6.5) of ε-CL-20, TKX-50, CL-20/TKX-50 mixture, and ε-CL-20/TKX-50 co-crystal showed that there were two exothermic decomposition peaks in the thermal decomposition process of TKX-50/CL-20 mixture, which indicate that the thermal decomposition process of the TKX-50/CL-20 mixture is just an overlaying of the individual ε-CL-20 and TKX-50 components thermal decomposition processes. For the TKX-50/CL-20 co-crystal, the first exothermic decomposition peak appears at 171.6 °C, which is lower than that of the TKX-50/CL-20 mixture (229.85 °C), due to the hydrogen bond destruction in the co-crystal structure and a small number of TKX-50 decomposition. Subsequently, a large amount of co-crystals begin to decompose, the second exothermic thermal

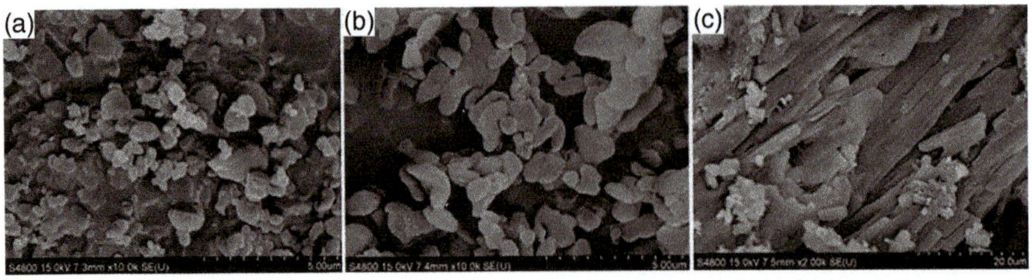

Figure 6.4 SEM photographs of ε-CL-20, TKX-50 and CL-20/TKX-50 co-crystals. (a) Raw ε-CL-20; (b) raw TKX-50; (c) TKX-50/CL-20 co-crystals. Source: Yuan et al. [49], Reproduced with permission from Chinese journal of energetic materials.

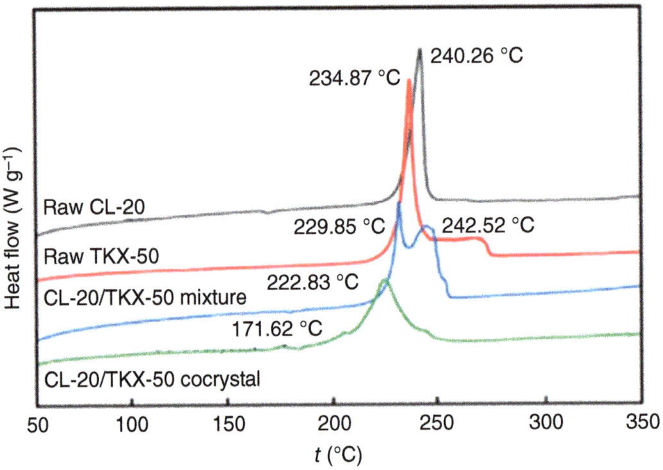

Figure 6.5 DSC curves of ε-CL-20, TKX-50, TKX-50/CL-20 mixture and CL-20/TKX-50 co-crystal. Source: Reproduced from [49], with permission from *Chin. J. Explos. Propellants*, (Huo Zha Yao Xue Bao), 2020.

decomposition peak appears at 222.8 °C, which indicates the hydrogen bond and a new structure between TKX-50/CL-20 co-crystal molecules formation [49].

The cohesive energy density (CED) of the TKX-50/HMX co-crystal was calculated by means of molecular dynamics (MD) [50]. It was found that the CED value decrease with an increase in temperature gradually. The CED value of TKX-50 is much higher than that of HMX, which indicates that TKX-50 possesses far lower sensitivity than HMX (Figure 6.6). Furthermore, the CED value of the TKX-50/HMX co-crystal is between that of TKX-50 and HMX, and it is far higher than that of HMX, which indicates that the co-crystal structure has resulted in a great improvement with respect to the high sensitivity problems of HMX, and in addition, it also shows better thermal stability (Figure 6.6a). Moreover, the modulus of elasticity (E), shear modulus (G), and bulk modulus (K) values showed a downtrend with an increase in temperature, and the fracture strength, hardness, and rigidity of the TKX-50/CL-20 co-crystal decreased (Figure 6.6b). Additionally, the E, K, and G values of TKX-50/CL-20 co-crystal were found to be lower than those of the individual TKX-50 and HMX components, the K/G and

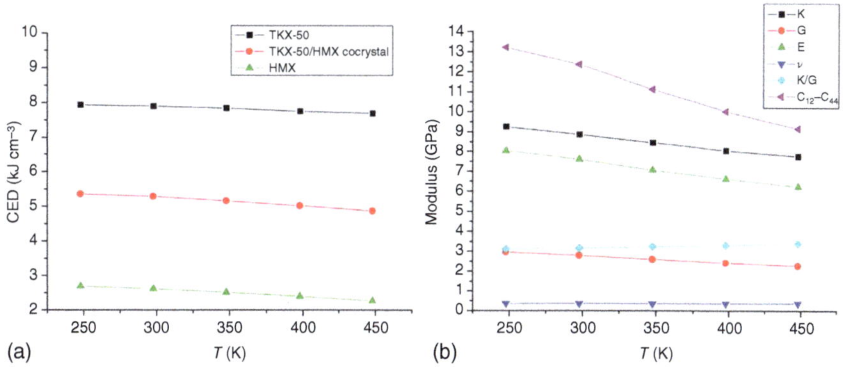

Figure 6.6 CED spectra and mechanical properties of TKX-50/HMX co-crystal. Source: Xiong et al. [50]/The Royal Society of Chemistry/CC BY 3.0.

Figure 6.7 SEM images of (a) raw PETN, (b) raw TKX-50, (c) PETN/TKX-50 co-crystal. Source: Xiao et al. [52], Reproduced with permission from Royal Society of Chemistry.

C_{12}–C_{44} values were found also to be larger than those of individual TKX-50 and HMX component at the same temperature, which indicate that the rigidity and hardness of the co-crystal were lower and the ductility higher in the co-crystal. Hence, the co-crystal structure shows improved mechanical properties in comparison with pure TKX-50 or HMX, making it much more attractive for widespread application.

As is well-known, the detonation velocity ($8310\,m\,s^{-1}$) and density ($1.77\,g\,cm^{-3}$) of PETN are high. However, PETN has not been widely used for its poor chemical stability in vacuum stability tests (VST) and also because of its high mechanical sensitivity [51]. Intermolecular hydrogen bonds in the TKX-50/PETN co-crystal could be formed between the $-NO_2$ in PETN and the NH_3OH^+ cation in TKX-50. Therefore, the TKX-50/PETN co-crystal was prepared by using the solvent/non-solvent method. It can be seen that original PETN represents crystal structure with a wide range of granularity distribution from 100 to 500 mm (Figure 6.7a). Raw TKX-50 shows irregular polyhedron structures with short rod shapes (Figure 6.7b). The image of TKX-50/PETN co-crystals (Figure 6.7c) shows a different particle shape. The co-crystal morphology was uniform, and the average particle size was 1 mm with integrated and smooth surfaces. Thus, such co-crystals would be expected to show advantageous properties in terms of a lower mechanical sensitivity, as well as excellent formability in comparison with those of TKX-50 and PETN.

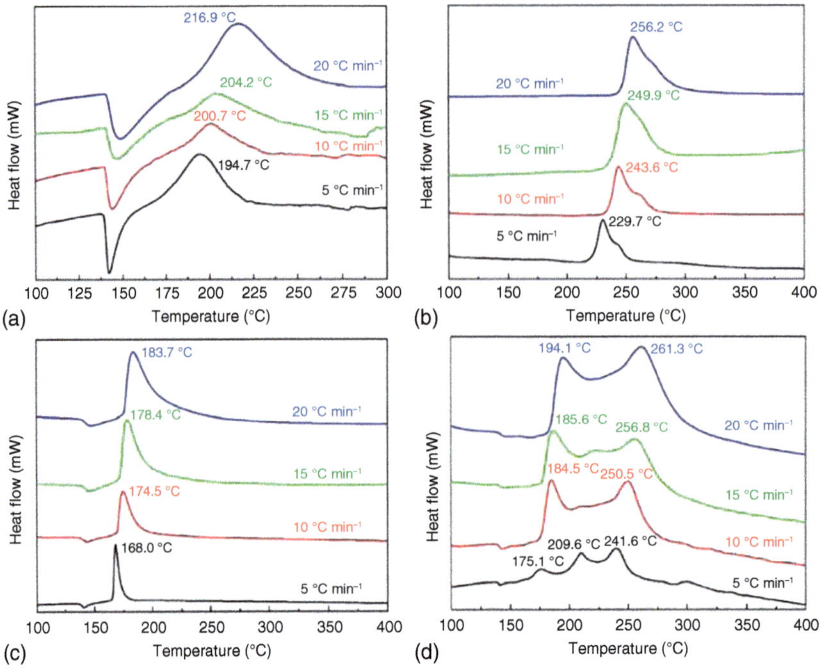

Figure 6.8 DSC curves of (a) raw PETN, (b) raw TKX-50, (c) PETN/TKX-50 mixture and (d) PETN/TKX-50 co-crystal at different heating rates (5, 10, 15, and 20 K min^{-1}). Source: Reproduced from [52], with permission from *RSC Adv.*, 2019.

The thermal decomposition of the TKX-50/PETN co-crystal was also investigated and compared with that of the component compounds, as well as with that of the TKX-50/PETN mixture. The DSC traces of these four samples at various heating rates (5, 10, 15, and 20 K min^{-1}) showed that the thermal behavior of the co-crystal was different from those of the original compounds as well as of the mixture. The thermogravimetry (TG) curve of the co-crystal was found to be similar to those of both TKX-50 and the mixture, with all showing two mass loss steps. In contrast, PETN showed only one mass loss step (Figure 6.8).

The high mechanical sensitivity of ε-CL-20, HMX, and PETN is a major limitation for their widespread application, and this is the fundamental problem that the use of co-crystal technology aims to solve. The impact and friction sensitivities of raw TKX-50, ε-CL-20, HMX, PETN, TKX-50/CL-20, TKX-50/HMX, and TKX-50/PETN co-crystals were tested (Table 6.4). It can be clearly seen that the mechanical impact sensitivity decrease, and the characteristic drop height (H_{50}) increases from 14.5 cm for PETN to 42.2 cm for TKX-50/PETN co-crystal, indicating for the co-crystal a decrease of about 191% in impact sensitivity compared to raw PETN. The friction sensitivity of the co-crystal was decreased from 68% for PETN to 12% for the co-crystal, which was even lower than that of TKX-50 (28%). This reason for the lower sensitivity of the co-crystal could be due to the intermolecular hydrogen bond formed between TKX-50 and PETN [53].

Table 6.4 The mechanical sensitivity of various EMs as well as of related co-crystals.

	Typical compounds				Co-crystals	
Samples	**TKX-50**	**ε-CL-20**	**HMX**	**PETN**	**TKX-50/CL-20**	**TKX-50/PETN**
H_{50} (cm)	55.4 [53]	15.0	13.9	14.5 [53]	34.0 [52]	42.2 [53]
F (%)	28 [53]	100	84	68 [53]	—	12 [53]

6.3.2 TKX-50/EMs Mixtures

TKX-50, as an organic explosive with ionic structure, is an innovative EM that makes different interactions with interfaces, such as HMX or RDX. For example, to reduce the sensitivity of the high-energy solid propellant, TKX-50 was added to the GAP propellant instead of HMX. Experiments and simulations were carried out on the interfacial interaction between GAP and TKX-50 with the mass ratio of 10/90, with the number of GAP, HMX, and TKX-50 molecules being 1, 48, and 60, respectively. A stronger adhesion energy was found for TKX-50/GAP mixture based on the static and dynamic mechanical properties and interfacial tension studies [54]. It was found from the contact angle tests of GAP, HMX, TKX-50 that the surface tension of HMX and TKX-50 were 50.33 and 70.13 $mN\,m^{-1}$, respectively, while the surface tension of the GAP polymer was 13.97 $mN\,m^{-1}$, which may be due to the polar force. Based on the spreading coefficient and adhesion energy of the TKX-50/GAP and HMX/GAP mixtures, the adhesion energy between GAP and TKX-50 could be determined to be 62.61 $mN\,m^{-1}$, which is larger than that of the GAP/HMX mixture (57.51 $mN\,m^{-1}$), meaning more energy is required to separate GAP and TKX-50 than for GAP/HMX, demonstrating the stronger interaction between GAP and TKX-50 than that of GAP and HMX.

The interaction between TKX-50 and AP is much important for designing and judging the feasibility of TKX-50-based composite propellants. Therefore, the interaction mechanism of TKX-50/AP was investigated by using theoretical calculations and experimental methods [55]. The results showed that the superimposition of positive charges in TKX-50 and negative charges in AP makes the electrostatic potential distribution of TKX-50/AP compounds different from that of the individual TKX-50 and AP components. The interaction between TKX-50 and AP in TKX-50/AP mixture depends on hydrogen bonds and van der Waals forces, whereby the strength and number of the hydrogen bonds are significantly larger than that of van der Waals forces.

In another example, the TKX-50/GO composite was prepared by means of the solvent–non-solvent method [56]. As is showed in Figure 6.9, TKX-50 can be described as being a tabular crystal with different particle sizes bearing a high draw ratio (Figure 6.9a). In contrast, crystals of the TKX-50/GO composite appeared to be polyhedrons with every crystal face (Figure 6.9b), which may be ascribed to the phenol, hydroxyl, epoxide, carboxylic, etc. functional groups of GO restrained the fast-growing crystal face of TKX-50 growth in the coating process. As shown in Figure 6.9c,d, the TKX-50 crystals surface was smooth and clean, and

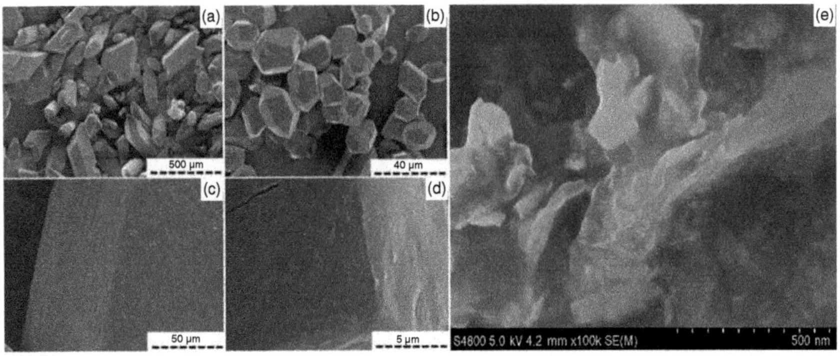

Figure 6.9 SEM images of TKX-50 and TKX-50/GO. (a, b) Low-magnification; (c, d) high-magnification; (e) GO. Source: Wang et al. [56], Reproduced with permission from John Wiley & Sons.

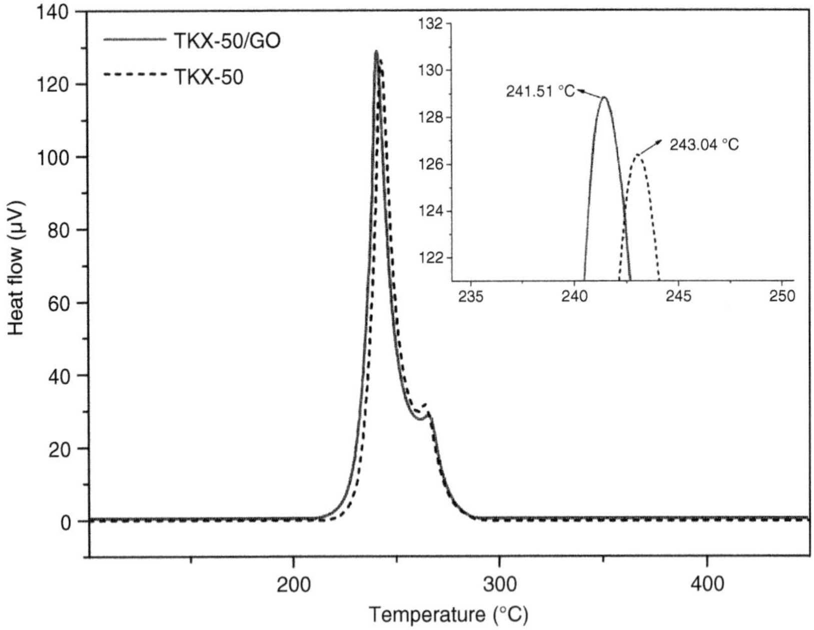

Figure 6.10 DTA curves of TKX-50 and TKX-50/GO at 10 °C m^{-1}. Source: Reproduced from [56], with permission from *Propellants Explos. Pyrotech.*, Wiley, 2017.

a large amount of wrinkles could be observed easily on the composite surface, the TKX-50/GO composite exhibited a uniform particle size distribution in 20 mm, owing to the surface of TKX-50 was coated by GO [57].

Since it is known that GO has a considerable effect on the thermal performance of TKX-50, the thermal decompositions of the TKX-50/GO composite and TKX-50 were investigated and compared (Figure 6.10). The first exothermic thermal decomposition peak temperatures of TKX-50/GO composite and TKX-50 were

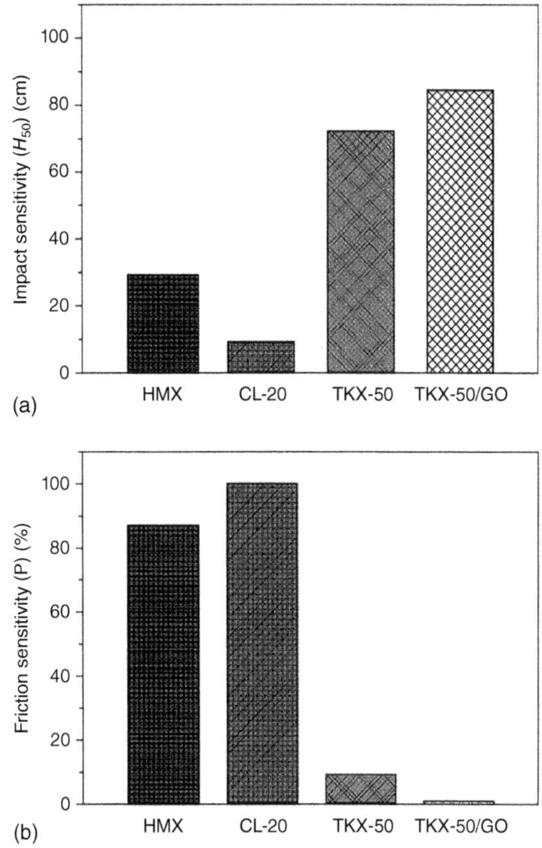

Figure 6.11 The impact and friction sensitivity of TKX-50/GO composite. Source: Reproduced from [56], with permission from *Propellants Explos. Pyrotech.*, Wiley, 2017.

241.51 and 243.04 °C, respectively at the heating rate of 10 °C min^{-1}. The only small forward shift in the peak temperatures of 2 °C and the similar shape of the thermal decomposition peaks indicate that the thermal stability of TKX-50 is not obviously reduced by being coated with GO.

Since the sensitivity of the TKX-50/GO composite is hugely important for applications, the impact sensitivity and friction sensitivity of the TKX-50/GO composite were tested and compared with that of HMX, ε-CL-20, and TKX-50 under the same conditions. The results are showed in Figure 6.11. It was found that TKX-50 (72.1 cm [IS], 9% [FS]) showed a lower mechanical sensitivity than either HMX (29.1 cm, 87%) or ε-CL-20 (8.9 cm, 100%). In addition, compared to TKX-50, the impact sensitivity and friction sensitivity of TKX-50/GO composite (84.4 cm, 1%) showed a reduction of 17% and 8%, respectively, indicating that the sensitivity of TKX-50 toward impact and friction was lower after coating with GO. Furthermore, the impact sensitivity and friction sensitivities of the TKX-50/GO composite were compared with those of other composites as summarized in Table 6.5. From these results, it could be concluded that the TKX-50/GO composite is less mechanical sensitive than the other tested composites.

Table 6.5 The impact and friction sensitivity of TKX-50/GO and other composites.

Samples	Content of carbon materials (wt%)	Impact sensitivity		Friction sensitivity P_F (%)
		P_I (%)	H_{50} (cm)	
TKX-50/GO	0.5	8	84	1
HMX/GO-1 [58]	1.0	70	—	40
HMX/GO-2 [58]	2.0	10	43	32
HMX/CNT [59]	20	—	45	24
HMX/Vition/G [60]	1.0	—	49	—
HMX/Vition/GO [60]	1.0	—	66	—
HMX/[60]Fullerene3 [60]	1.0	60	—	70
CL-20/551glue/G [61]	0.5	20	—	22
CL-20551glue/GO [61]	0.5	25	—	30
CL-20/551glue/rGO [61]	0.5	22	—	28

A further valuable idea that was investigated is the possible use of TKX-50 in polymer-bonded explosives (PBXs), with the aim of improving the performance of, and range of possible applications for TKX-50. As well-known, polymer binders are often used to improve explosive brittleness, which is important in terms of safety, since the use of amorphous polymers can result in PBXs with better flexibility or compressibility toward external mechanical stimuli, that is, it desensitizes [62, 63]. In this context, to improve the possible practical applications of TKX-50, polyvinylidene difluoride (PVDF) and polychlorotrifluoroethylene (PCTFE) were combined to TKX-50, forming different PBXs. The mechanical and interfacial properties of these PBXs were carried out by using the MD method [64]. The results showed that the binding energy of TKX-50/PVDF is in general higher than that of TKX-50/PCTFE, which may be ascribed to the larger potentials and smaller interatomic distance between the TKX-50 and PVDF surfaces. The modulus of HMX or RDX is lower than that of TKX-50, but the modulus decrease after TKX-50 mixed with polymers, with a minimum being achieved in the TKX-50/PCTFE. The ductility of TKX-50 is preferable to that of RDX, but poor slightly to that of HMX, PVDF, or PCTFE addition to TKX-50 improves its ductility significantly. At the same time, PVDF and PCTFE decrease the compression sensitivity of TKX-50, whereby the desensitizing effect of the latter is much more obvious and is due to the superior compressibility and greater stress reduction that results in TKX-50/PCTFE system. In addition, the interactions between TKX-50 and F_{2311}, F_{2314}, F_{2611}, F_{2614} were investigated using MD [65]. Negative ΔE_{bind} values were found for all four systems, attributing to the good compatibility of TKX-50 with the fluoropolymers. The interaction of fluoropolymers/TKX-50 is dominated by the electrostatic energies between TKX-50 and the fluoropolymers are higher than the van der Waals energies. Moreover, the CEDs of TKX-50 and the four systems mentioned vs. temperature are shown in Figure 6.12. The CED values of the

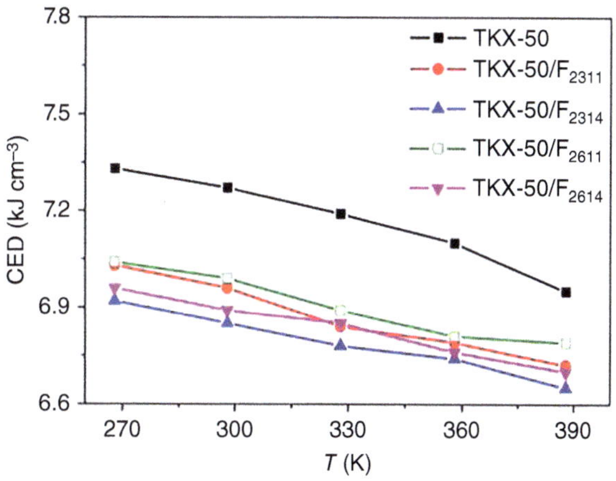

Figure 6.12 CED versus temperature of TKX-50 and of the four systems. Source: Reproduced from Song et al. [65], with permission from IOP Publishing.

four systems and of TKX-50 gradually decrease with increasing temperature, which indicates that the energy required to overcome the intermolecular forces declines, showing that the sensitivity of the system decreases gradually as the temperature decreases [66, 67]. The CED value of TKX-50 is higher than that of the four systems, which may be due to the regularity and integrity of the ion arrangement in TKX-50 crystals. The CED values of the four mixed explosives with TKX-50 are reduced by the addition of a small amount of fluorine polymer, the stiffness of the four mixed explosives is weakened, and the elasticity is enhanced.

The mechanical properties are important for the preparation, processing, and application of EMs. For example, Figure 6.13 illustrates the mechanical properties of TKX-50 and four fluoropolymers over the temperature range from 268 to 388 K. On increasing the temperature, the E, G, and K values of TKX-50 and the TKX-50/fluoropolymers are gradually decreased, which indicates that the fracture strength and stiffness of these systems decrease.

The ratio of bulk modulus to shear modulus (K/G) and Poisson's ratio (ν) can be used to characterize the ductility of EMs. The lower the values of K/G and ν are, the worse the ductility of EMs. In addition to the fluorine-containing polymers mentioned above, polyethylene glycol (PEG), and ethylene vinyl acetate (EVA) copolymer were also used to simulate their interactions with TKX-50 (Figure 6.14). In comparison with TKX-50, the modulus of elasticity (E), G, and K of the four TKX-50/ fluoropolymers systems are all small at the same temperature, which indicates that the stiffness of TKX-50/ fluoropolymers and the resistance to deformation is weakened (Figure 6.14a). The K/G and ν values of all four TKX-50/fluoropolymers systems are larger than those of TKX-50, indicating the ductility of the TKX-50/fluoropolymers systems is superior than that of TKX-50. Additionally, obvious differences were observed in the ductility of the TKX-50/fluoropolymers systems (Figure 6.14b). The plasticity of TKX-50 can be improved by adding small amounts of polymer binders. The effectiveness was found

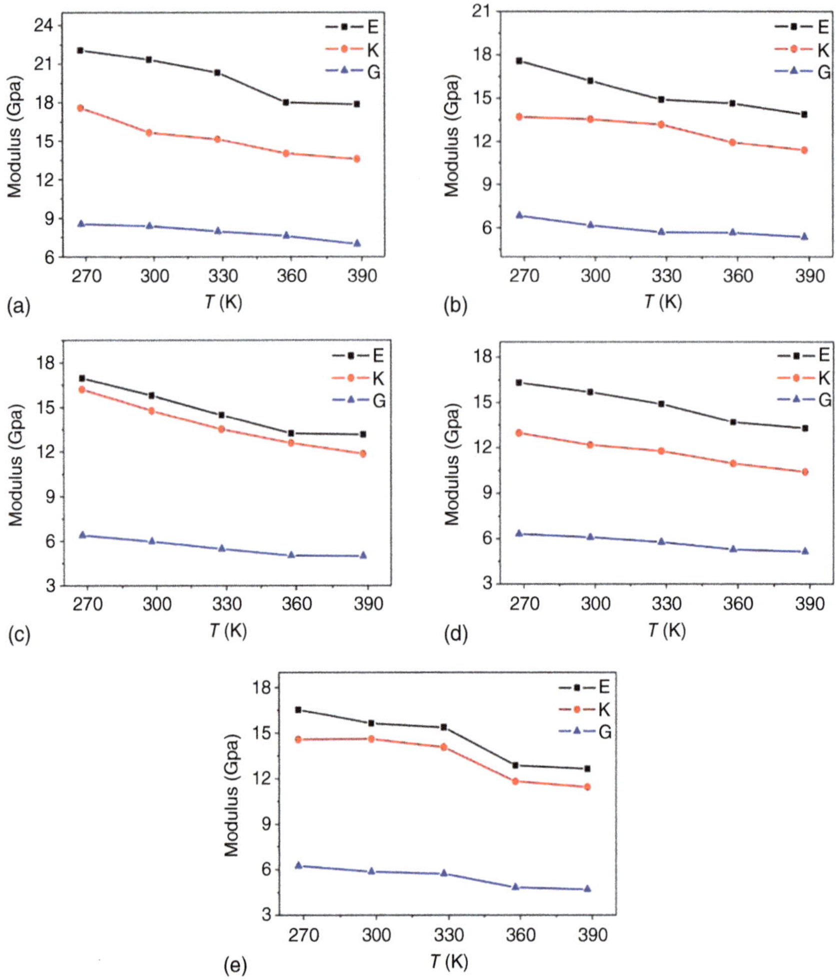

Figure 6.13 Mechanical properties of TKX-50 and of TKX-50/fluorine-containing polymer composites at different temperatures. (a) TKX-50; (b) TKX-50/F_{2311}; (c) TKX-50/F_{2311}; (d) TKX-50/F_{2611}; (d) TKX-50/F_{2614}. Source: Reproduced from Song et al. [65], with permission from IOP Publishing.

to increase in the order [PEG] > [EVA] > [F_{2641}] = [F_{2311}]. The overall performance of the four TKX-50/fluoropolymers systems suggests that PEG can be regarded to be the best binder choice for TKX-50 mechanical properties improvement (Figure 6.14c).

In addition, the detonation performances of four TKX-50/fluoropolymers were investigated (Table 6.6). The theoretical detonation velocity and detonation pressure of TKX-50 are higher than those of the four systems, which may be attributed to the fact that fluoropolymers are not explosives. However, the TKX-50/fluoropolymers are still high-energy EMs with superior performance.

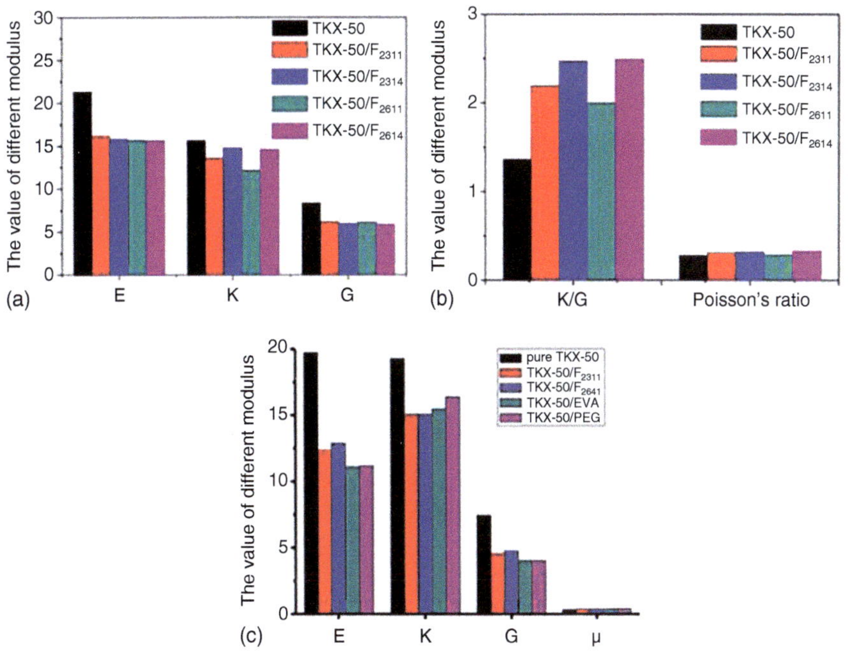

Figure 6.14 Mechanical properties (a), quotient K/G and Poisson's ratio of TKX-50 and four systems (b); TKX-50 based PBXs (c) at 298 K. Source: Reproduced from Refs. [65, 68], with permission from IOP Publishing, Reproduced with permission from RSC Adv., 2016.

Table 6.6 Detonation properties of TKX-50 and four composites at 298 K.

Properties	TKX-50	TKX-50/F_{2311}	TKX-50/F_{2314}	TKX-50/F_{2611}	TKX-50/F_{2614}
D (m s^{-1})	9680.00[a)]	9457.48	9452.10	9461.78	9465.00
P (GPa)	42.40[a)]	41.95	42.48	41.25	42.06

a) From [18].
Note: D-detonation velocity, m s^{-1}; P-detonation pressure, GPa.
Source: Reproduced from Song et al. [65], with permission from IOP Publishing.

6.4 Performance of Nano-sensitized TKX-50

EMs with nano-sized structures show great potential with respect to safety and detonation performance, due to their unique physicochemical properties [69–71]. In recent years, nano-sized EMs such as ε-CL-20, FOX-7, and HNS have been prepared that show enhanced performance [72–74]. For instance, nano-sized sensitized TKX-50 was prepared by using a combination of ball milling technology [75]. The process used for the preparation is shown in Figure 6.15. The suspension of TKX-50 nanoparticles was filtered and separated by using ethanol as the liquid medium.

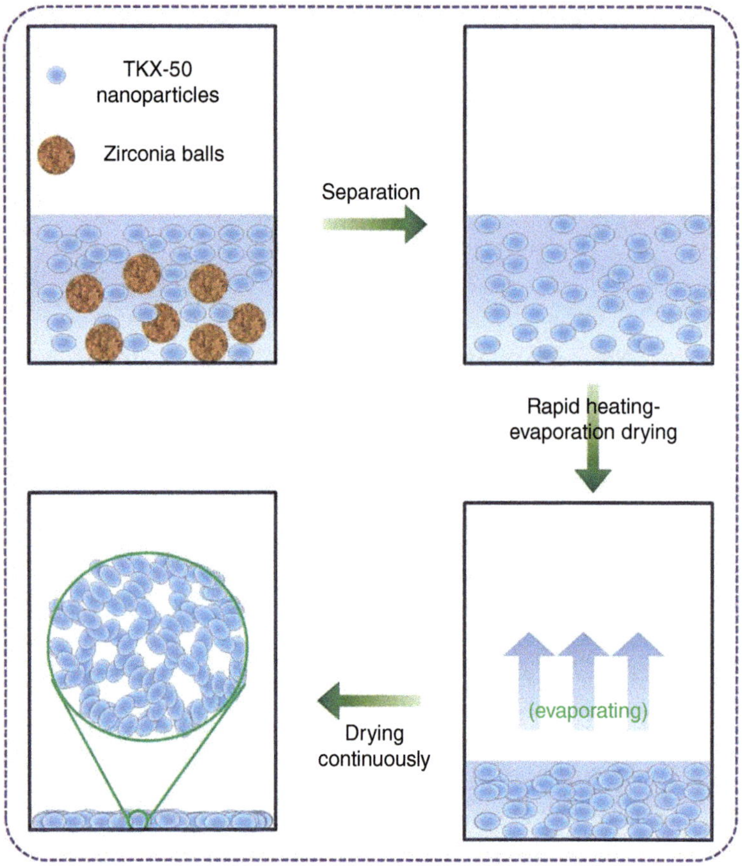

Figure 6.15 Scheme showing the formation of nano-sensitized TKX-50. Source: Reproduced from Deng et al. [75], with permission from Taylor & Francis.

Micro/nanostructures composed of nano-sized sensitized TKX-50 particles with the specific surface area (SSA) being up to 3.528 $m^2\ g^{-1}$. The growth and formation mechanism were described in Ref. [76].

The nano-sized sensitized TKX-50 showed an "ellipsoid-like" morphology with smoother particle surfaces, in contrast to that of raw TKX-50 which shows a micro-sized irregular prism morphology (Figure 6.16a,b). The nano-sized sensitized TKX-50 obtained possesses less particle sizes, with narrow granularity distributions of 100–300 nm, obtained by fast heating evaporation in the drying process (Figure 6.16c–e).

Furthermore, nano-sized sensitized TKX-50 showed similar thermal decomposition processes (Figure 6.17a) as that of raw TKX-50, which shows exothermic thermal decomposition peaks at 250.5 and 273.0 K. The nano-sized sensitized TKX-50 showed a forward shift in these thermal decomposition temperatures to lower temperatures by 10.3 and 2.2 K, respectively. The activation energy (E_a) of the primary thermal decomposition process of TKX-50 was calculated to be

Figure 6.16 SEM images of (a, b) raw TKX-50, and (c–e) the nano-sensitized TKX-50. Source: Deng et al. [75], Reproduced with permission from Taylor & Francis Group.

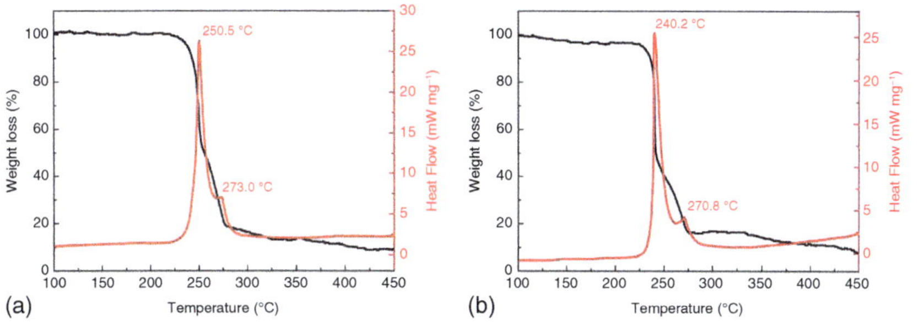

Figure 6.17 TG-DSC curves of raw TKX-50 (a) and nano-sensitized TKX-50 (b). Source: Reproduced from Deng et al. [75], with permission from Taylor & Francis.

220.07 kJ mol^{-1}, while the E_a of nano-sized sensitized TKX-50 prepared by ball milling treatments was found to be 116.37 kJ mol^{-1} [77]. Furthermore, the characteristic drop height (H_{50}) of raw TKX-50 was 65 cm and the friction sensitivity is only 25%, while nano-sized sensitized TKX-50 showed a higher mechanical sensitivity with an H_{50} of 15 cm and friction sensitivity of 100%.

6.5 Application in Solid Propellants

6.5.1 Ideal Energetic Performance

The specific impulse and density are two properties that play an essential role in developing high-energy solid propellants and explosives. The theoretical specific impulses of composite solid propellants with TKX-50 were calculated by using "Energy Calculation Star (ECS)" software [14] based on the minimum free-energy program and compared with corresponding ones without TKX-50.

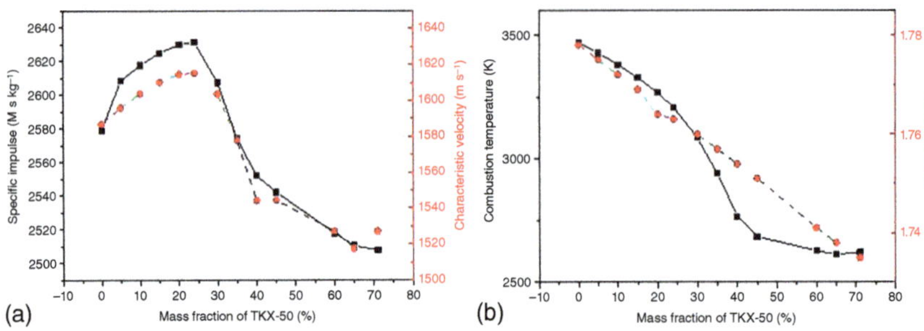

Figure 6.18 Effect of TKX-50 content on the energetic performance of HTPB propellant. Source: Reproduce from [23], with permission from *Chinese Journal of Energetic Materials* (Han Neng Cai Liao), 2014.

For all calculations, the enthalpy of formation of TKX-50 446.6 kJ mol^{-1} was used unless otherwise specified.

6.5.1.1 HTPB/TKX-50

The theoretical specific impulse and characteristic velocity of HTPB solid propellant were calculated for increasing replacement of AP by TKX-50 (Figure 6.18). The mass fractions of the reference formulation were as follows: HTPB 10%, AP 71%, Al 16%, and other additives 3%.

The theoretical specific impulse and characteristic velocity were found to initially increase with an increase in mass fraction of TKX-50, and then to decrease, while the combustion temperature and density decreased with an increase in mass fraction of TKX-50. Compared to the reference composition without TKX-50, the specific impulse was increased by 34.3 N s kg^{-1} (3.5 seconds) at 24% content of TKX-50. The reason for this may be that the oxygen coefficient of the formulation decreases with an increase in mass fraction of TKX-50 due to the negative oxygen coefficient of TKX-50. Moreover, AP addition to the formulation increases the hydrogen chloride (HCl) content and the average molecular weight of the combustion gas.

The reported standard enthalpy of formation of TKX-50 (446.6 kJ mol^{-1} [18]) is higher than the experimentally determined value (210 kJ mol^{-1}), which is similar to that reported by Nicolich (193 kJ mol^{-1}) [76]. Figure 6.19 shows a comparison of the theoretical performance of different formulations in which the same mass fraction of TKX-50 replaces RDX: the oxygen coefficient of RDX is higher than that of TKX-50 (Table 6.1), and the values of 446.6 and 210 kJ mol^{-1} for the standard enthalpy of formation for TKX-50 were used. It can be seen that when the 210 kJ mol^{-1} standard enthalpy of formation of TKX-50 is used, the theoretical specific impulse of the formulation reaches a maximum when a 15% TKX-50 content is used, while the specific impulse of the RDX formulation is higher than that of TKX-50 in the mass fraction range of 5–28%. However, if a value of 446.6 kJ mol^{-1} is used for the enthalpy of formation of TKX-50, the theoretical specific impulse for the formulation reaches a maximum for a mass fraction of TKX-50 of 24%, and the specific impulse of the TKX-50 formulation is much higher than that of RDX over the mass fraction range of 5–28% (Figure 6.19a,b). When the value used for the standard enthalpy of formation

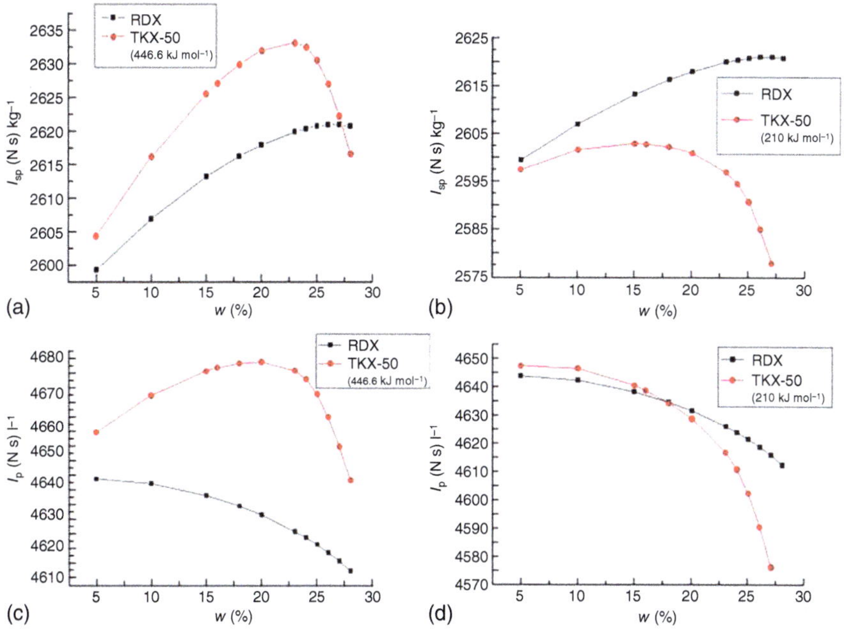

Figure 6.19 Graphs showing the theoretical performance of HTPB/TKX-50 (RDX) formulations. (a,b) Theoretical specific impulse, formulation; (c,d) theoretical density specific impulse. Source: Reproduced from [77], with permission from *Journal of Solid Rocket Technology* (Gu Ti Huo Jian Ji Shu), 2018.

for TKX-50 is 446.6 kJ mol^{-1}, the density-specific impulse of the TKX-50 formulation is much higher than that of RDX. However, when a value of 210 kJ mol^{-1} is used as the standard enthalpy of formation of TKX-50, the density-specific impulse of the TKX-50 formulation is higher than that of RDX only when the mass fraction of TKX-50 is less than 18% (Figure 6.19c,d).

Meantime, according to the report in Ref. [32], the gravimetric specific impulse of HTPB-based propellants with TKX-50 first increases with increasing mass fraction of TKX-50 particles in the compositions and then decrease. While for the heat of explosion, the measured heat of explosion of HTPB-based propellant containing TKX-50 is lower than that of propellant without TKX-50, attributing may be to TKX-50 is an ionic salt structure without energetic group $-NO_2$. The energy release method is high-energy bonds broken instead of redox reactions, bringing the calculated energetic characteristics such as enthalpy of formation, heat of detonation, and heat of combustion have a large difference with the test results. Thus, the real enthalpy of formation of TKX-50 needs to be investigated further.

The density and heat of explosion of the HTPB propellant containing various mass fractions of TKX-50 particles were investigated. Since the density of TKX-50 is lower than that of AP (Table 6.1), replacing AP in the solid-propellant composition by TKX-50 leads to a lower propellant density. Furthermore, increasing the propellant ingredients with high density and solid particle percentages with smaller particle size lead to an increase in propellant density.

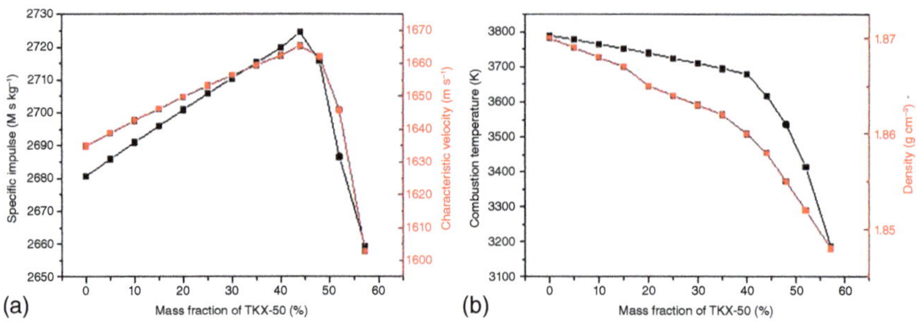

Figure 6.20 Effect of TKX-50 content on the energetic performance of GAP propellant. Source: Reproduced from [23], with permission from *Chinese Journal of Energetic Materials* (Han Neng Cai Liao), 2014.

6.5.1.2 GAP/TKX-50

The theoretical specific impulse of GAP solid propellant was calculated for increasing replacement of HMX and AP by TKX-50 (Figure 6.20). The mass fractions of the reference formulation were as follows: GAP 7.6%, AP 17.2%, HMX 40%, Al 18%, NG/BTTN 17.2%. It was found that with increasing TKX-50 content, the theoretical specific impulse and characteristic velocity of the propellants first increased linearly and then decreased. The theoretical specific impulse reached the maximum value when the TKX-50 content was 44%, increasing by 43.9 $N\,s\,kg^{-1}$ (4.48 seconds), while the density and combustion temperature decreased. The decrease in the density of the propellant is attributed to the density of TKX-50 being lower than that of HMX and AP.

Meantime, the theoretical specific impulse of GAP solid propellants with TKX-50 (GAP 10%, nitroglycerin [NG] 15%, and aluminum [Al] 18%, with the other 57% consisting of energetic fillers [45% HMX + 12% AP, 54% ε-CL-20 + 3% AP or 41% TKX-50 + 16% AP]) was evaluated and compared with that of the HTPB propellant (70% AP, 16% Al, 12% HTPB, and 2% epoxy), the detailed calculations were showed in Ref. [78]. It was found that the I_{sp} of the GAP composite propellant based on TKX-50 is the highest, exceeding those of the corresponding HMX (274 seconds), CL-20 (275 seconds), and HTPB (264 seconds) formulations. However, it must be pointed out that TKX-50 is less impact sensitive and friction sensitive than ε-CL-20, and it can contribute to a considerable increase in the safety during propellant operations in comparison with the corresponding one with ε-CL-20 [79, 80].

6.5.1.3 NEPE/TKX-50 System

The ideal energetic performance of the NEPE solid propellant was calculated for increasing replacement of AP and HMX by TKX-50 (Figure 6.21). The mass fractions of the reference formulation were as follows: PEG 7.0%, AP 42%, HMX 28%, Al 5%, NG/BTTN 18%. It was found that when the AP mass fraction replaced by TKX-50 reaches 20%, the theoretical specific impulse of the propellant reaches a maximum, increasing by 20.4 $N\,s\,kg^{-1}$ (2.1 seconds). It was also found that the theoretical specific impulse was the smallest when the AP content in the formulation

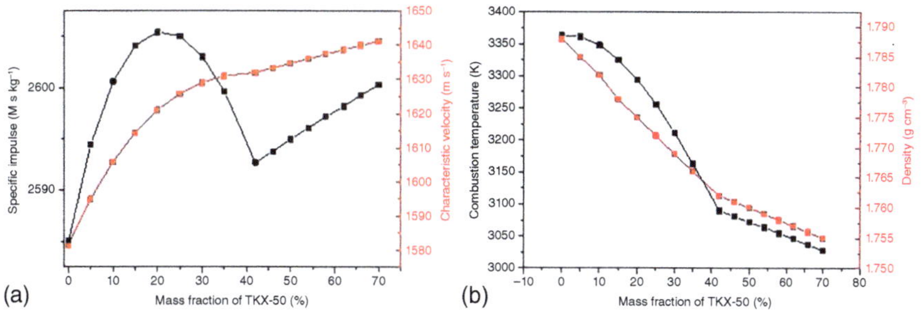

Figure 6.21 Effect of TKX-50 content on the energetic performance of the NEPE propellant. Source: Reproduced from [23], with permission from *Chinese Journal of Energetic Materials* (Han Neng Cai Liao), 2014.

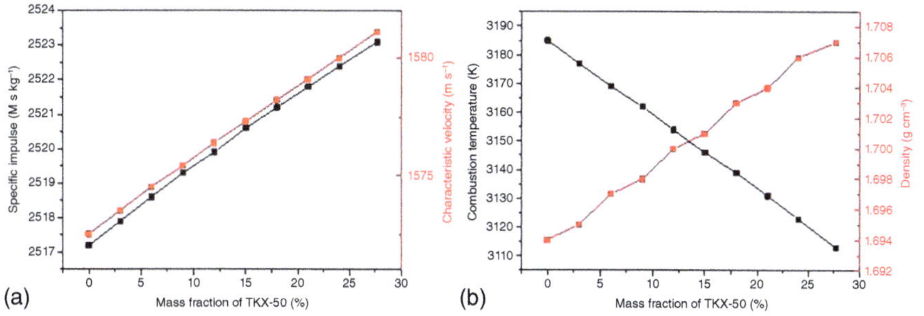

Figure 6.22 Effect of TKX-50 content on the energetic performance of the CMDB propellant. Source: Reproduced from [23], with permission from *Chinese Journal of Energetic Materials* (Han Neng Cai Liao), 2014.

was reduced to zero, which indicates that there is an optimal value of TKX-50 added to NEPE propellant in terms of propellant performance.

6.5.1.4 CMDB/TKX-50 System

The ideal energetic performance of the CMDB solid propellant was calculated for increasing replacement of RDX by TKX-50 (Figure 6.22). The mass fractions of the reference formulation were as follows: NC 25%, NG 33%, RDX 27.7%, Al 5%, and other additives (DINA, C_2, V, etc.) 9.3%. It was found that the theoretical specific impulse, characteristic velocity, and density increase with an increase in the mass fraction of TKX-50 in the formulation, while the combustion temperature decreases, indicating that the addition of TKX-50 gives a positive contribution to the energetic features of the CMDB propellant.

6.5.2 Combustion Features

6.5.2.1 Combustion Behavior of TKX-50

Concerning the environmental impact of EMs has grown in recent years, resulting in a demand for new, green EMs, especially with respect to clean combustion products. Since TKX-50 is a nitrogen-rich compound, its thermal decomposition

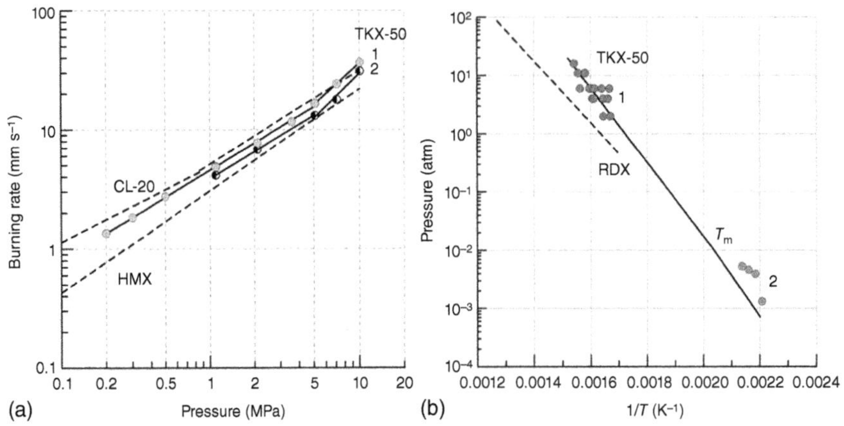

Figure 6.23 A comparison of burning rate *vs.* pressure dependencies for ε-CL-20, HMX, and TKX-50 (1-pure substance, 2-covered with 3% of a fluorelastomer) (a). A comparison of the pressure dependencies of the surface temperatures for TKX-50 and RDX (1-surface temperatures at different pressure, 2-sublimation pressures at different temperatures) (b). Source: Reproduced from Sinditskii et al. [81], with permission of Elsevier.

releases N_2 without smoke or solid residues. Therefore, the combustion behavior of TKX-50 was investigated [81]. The results showed that the burning rate of TKX-50 is larger than that of HMX, and approaches that of ε-CL-20. There is a slight break at 5 MPa for the burning rate on pressure dependence (Figure 6.23a). In the combustion of TKX-50, the major reaction is in the condensed phase. The thermal decomposition rates of TKX-50 in both liquid and solid phases are certified to be slightly larger than the corresponding one of RDX. The surface temperature of TKX-50 in the 0.2–1.6 MPa pressure interval increases from 312 to 376 °C, but are slightly lower than the surface temperatures of RDX (Figure 6.23b).

6.5.2.2 Combustion Behavior of Solid Propellants Containing TKX-50

Various factors, such as solid particle size and mass fraction, oxidizer, pressure, and temperature, affect the burning rate of solid propellants. The burning rates of HTPB composite propellants with and without TKX-50 (the mass fraction of TKX-50 is 0%, 5%, 10%, 20%, 30%, and 40% for PHT-1 to PHT-6, respectively) were determined (Figure 6.24). The results show that the burning rates of the tested propellants increased with an increase in pressure. The increasing degree with 40% mass fraction of TKX-50 over the pressure range of 1–15 MPa is obviously less than those of the other compositions, and the pressure exponent is 0.245, which is the lowest one among all the tested samples. The tested burning rate and calculated pressure exponent of the HTPB composite propellant with TKX-50 decreased with an increase in mass fraction of TKX-50 in the composition. Furthermore, the particle size of TKX-50 has much influence on the burning rate of propellant, it was found that the burning rates of formulation increase effectively with decreasing TKX-50 particle size, also the pressure exponent over the tested pressure range 1–15 MPa increase when adding TKX-50 with smaller particle size to the propellant formulation (Figure 6.24b).

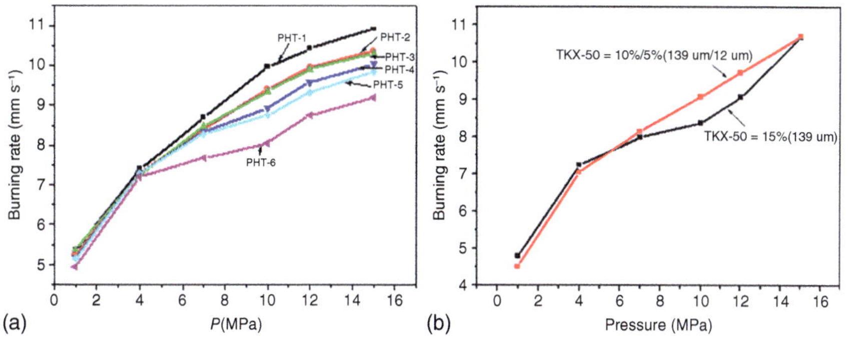

Figure 6.24 The burn rate of the HTPB composite propellant containing TKX-50. Effect of TKX-50 mass fraction ranging from 0% (PHT-1) to 40% (PHT-6) (a); effect of particle size (b). Source: Reproduced from [32], with permission from *Propellants Explos. Pyrotech.*, Wiley, 2016.

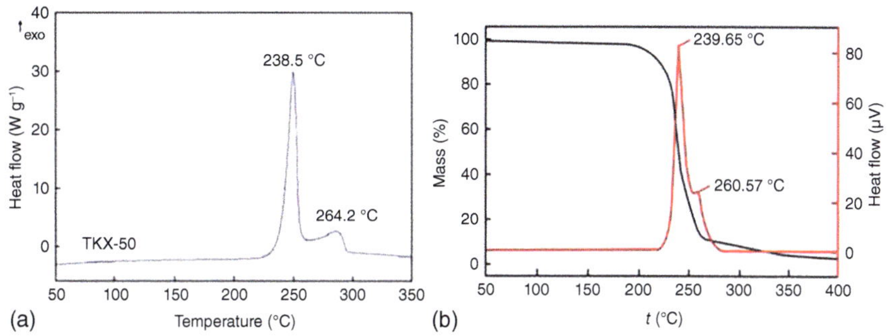

Figure 6.25 DSC (a) and TG–DTG (b) curves of TKX-50. Source: Reproduced from [83], with permission from *Chin. J. Explos. Propellants* (Huo Zha Yao Xue Bao), 2015.

6.5.3 Thermal Decomposition

As known to us, the lower heating rate is, the lower the peak temperature is, the smaller the peak area is, which is closely related to the type of sample and the type of thermal transformation. Even though the heating rate of thermal decomposition with DSC is much slower than that of propagation of EMs, it is one of the important experimental techniques for evaluation performance of EMs [82]. The differential scanning calorimetry (DSC) and thermogravimetry–differential thermogravimetry (TG–DTG) curves at the heating rate of $10\,\mathrm{K\,min^{-1}}$ of TKX-50 are showed in Figure 6.25. It can be seen that the exothermic thermal decomposition peak temperatures of TKX-50 are 238.5 and 264.2 °C, only single visible exothermic peak appeared in the two thermal decomposition steps, and the mass loss in this process is 84.2%.

The kinetic parameters of the thermal decomposition were investigated at the heating rates of 1, 5, 10, and $20\,\mathrm{K\,min^{-1}}$ by the Ozawa and Kissinger [83]. The results show that the activation energy and pre-exponential factor for the first thermal decomposition stage are $147.05\,\mathrm{kJ\,mol^{-1}}$ and $10^{12.91}\ \mathrm{s^{-1}}$, respectively.

The reaction mechanism for the first exothermic decomposition stage obeys the Jander equation with $n = 1/2$, and the kinetic equation can be expressed as: $\frac{d\alpha}{dT} = \frac{10^{12.91}}{\beta} \times 4(1-\alpha)^{\frac{1}{2}}\left[1-(1-\alpha)^{\frac{1}{2}}\right]^{\frac{1}{2}} e^{-\frac{14\,705}{RT}}$.

The safety aspects of ammunition have gained increased attention in recent times [84]. The pressure properties and safety performance of TKX-50 thermal decomposition were investigated in a confined space using the accelerating rate calorimeter (ARC) method, and the effect of the heating rate and the sample mass on the safety properties were studied [85]. The experimental results showed that owing to the kinetic effect, the ignition temperature (T_i) decreased when the heating rate was reduced due to the kinetic influence. The ignition pressure (P_i) was found to increase as the heating rate decreased. HMX exhibited worse safety performance than TKX-50 after ignition even though the high heating rate is detrimental to the safety of TKX-50. The pressure increase rate *vs.* temperature of TKX-50 and HMX is showed in Figure 6.26. The ignition pressure of the explosives increased with an decreased in the heating rate. The pressure decreased at the same pressure when the heating rate was reduced after ignition, indicating that the thermal decomposition of TKX-50 at a lower heating rate will be less violent and safer for the shell (Figure 6.26a). A lower dP/dt_{max} value was obtained when the sample was heated at a lower rate, indicating that a low heating rate was beneficial to the safety of HMX (Figure 6.26b). In the case of the HMX/TKX-50 mixture (50 : 50, wt/wt), no deviation in the thermal decomposition peak temperature of HMX was observed, while TKX-50 shows a decrease by 4 and 15 K for the peak temperatures which are now observed at 506 and 529 K, respectively. When HMX is mixed with TKX-50, rapid thermal decomposition of TKX-50 occurs at the beginning of the thermal decomposition process. It should be noted that the onset temperature (T_{onset}) of the thermal decomposition changed little when TKX-50 and HMX were mixed and decomposed. The two-step thermal decomposition reaction of TKX-50 has a mass loss of 54%, but for HMX it is only 35%. The mass loss ratio of HMX/TKX-50 mixture is significantly higher than that of HMX, which indicates that some HMX also decomposes in the TKX-50 thermal decomposition process (Figure 6.26c).

To obtain reliable kinetic data for EMs, it is crucial to choose appropriate experimental conditions. For a more comprehensive survey of the thermal decomposition kinetics and mechanism of TKX-50, a variety of thermo-analytical tests were carried out in different conditions. Some thermo-analytical techniques (such as Kissinger, iso-conversional, and formal kinetic approaches) were compared to the ARC values independently [87]. The thermal stability of TKX-50 and RDX are similar on the basis of DSC data, while the ARC results of hexogen (RDX), which are more germane to real experimental application, were found to be slightly higher than that of TKX-50, but similar to that of ε-CL-20 [18, 88].

Furthermore, reactive force field (ReaxFF-lg) in the MD simulations was conducted to explore the oxidation and thermal decomposition of normal TKX-50 (N-TKX-50) and twinned TKX-50 (T-TKX-50) at 1000, 2000, and 3000 K, respectively (Figure 6.27) [89]. Results indicate that the potential energy of both N-TKX-50 and T-TKX-50 decreases with the energy released from the stable products formation.

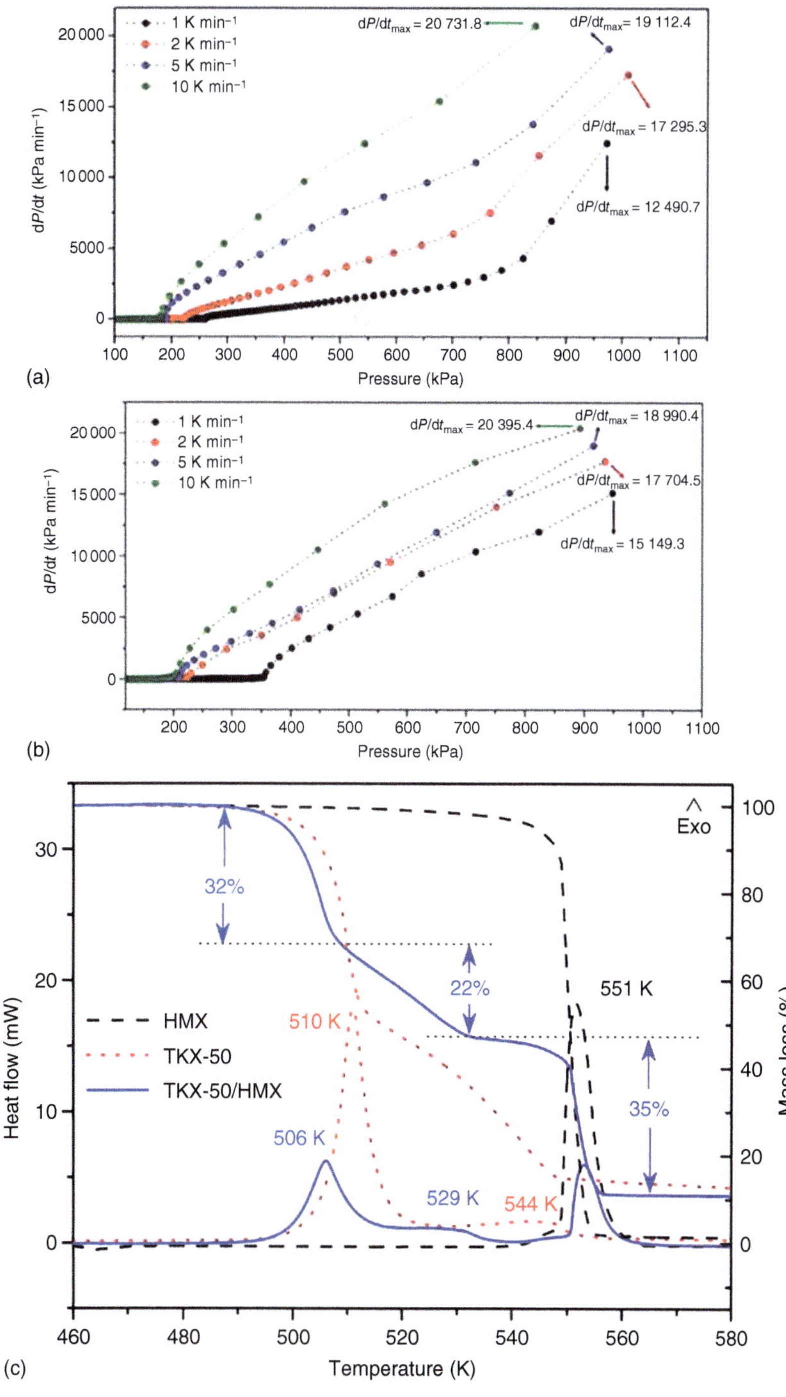

Figure 6.26 Plots of pressure rise rate *vs.* pressure of TKX-50 (a) and HMX (b, 72.5 mg) and TKX-50/HMX mixture (1 : 1) decomposition at different heating rates. Source: Reproduced from Refs. [85, 86], with permission of American Chemical Society.

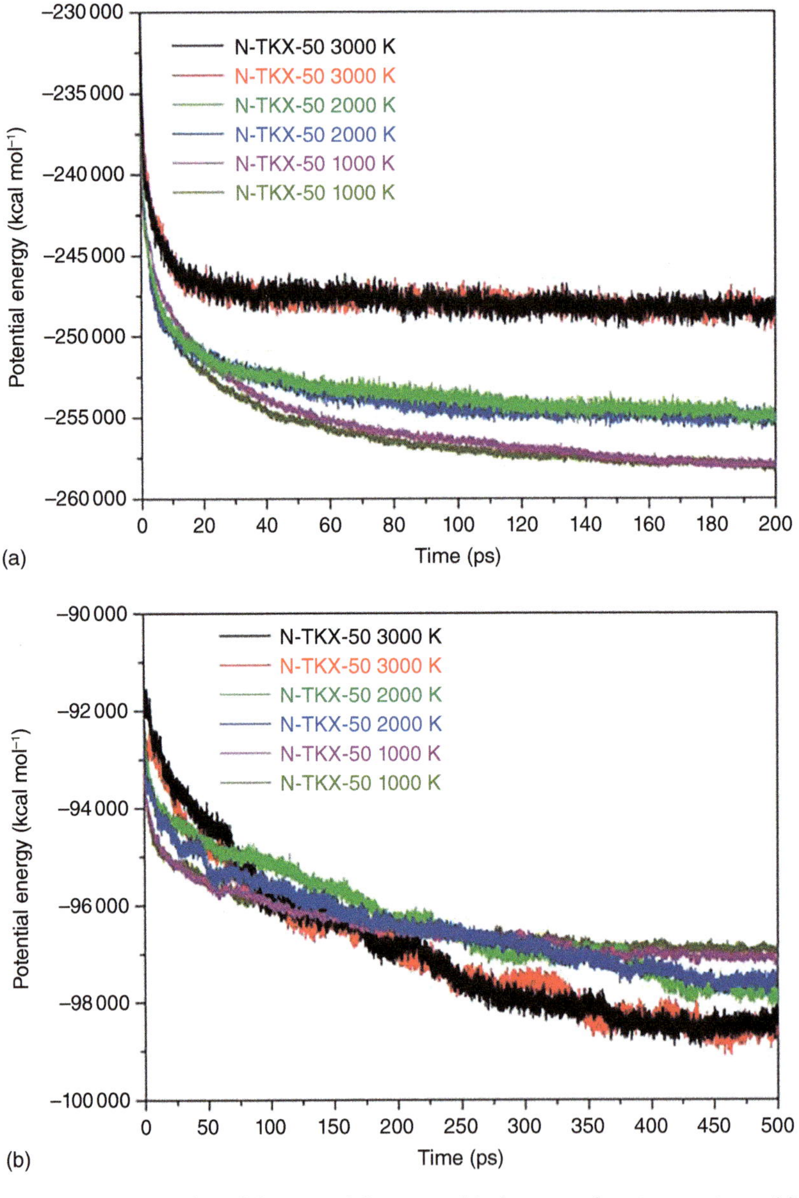

Figure 6.27 Evolution of the potential energy with time at various temperatures. (a) Thermal decomposition; (b) oxidation. Source: Reproduced from Li et al. [89], with permission from The Royal Society of Chemistry, 2020.

The decrease rate in the potential energy decreases with the thermal decomposition proceeding and eventually comes to plateaus (Figure 6.27a). Similar to the thermal decomposition process, the potential energies of both N-TKX-50 and T-TKX-50 decrease as the oxidation reactions proceed, and the potential energy of each system eventually reaches to stable. Unlike the thermal decomposition, the final stable

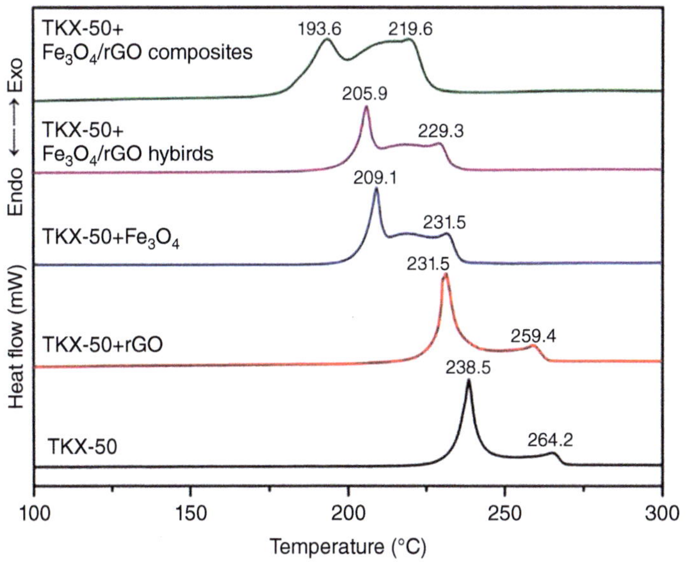

Figure 6.28 DSC curves of pure TKX-50 and TKX-50 mixed with different catalysts. Source: Reproduced from Zhang et al. [90], with permission from Springer Nature, 2020.

data of both N-TKX-50 and T-TKX-50 potential energy increase with an decrease in temperature (Figure 6.27b). It is rational that high temperature promotes combustion and oxidation, and produces stable end products.

Based on the theoretical and experimental investigations of pristine TKX-50, it is crucial to explore the catalytic effect of positive combustion catalysts (such as metal oxides, transition metal oxides, and reduced GO) on the thermal decomposition of TKX-50. For example, DSC analyses at a heating rate of 10 K min^{-1} were carried out to estimate the thermal decomposition peak temperature and catalytic effect of Fe_3O_4, reduced graphene oxide (rGO), Fe_3O_4/rGO hybrid, and Fe_3O_4/rGO nanocomposite on TKX-50 (Figure 6.28) [90]. The DSC curves of TKX-50 show two continuous steps along with a strong exothermic peak at 238.5 °C, followed by a weak exothermic peak at 264.2 °C [91]. Addition of rGO barely moves the first and second exothermic peaks to lower temperatures by 7.0 and 4.8 K, respectively. When Fe_3O_4 is introduced to TKX-50, the two exothermic peaks move to the lower temperatures of 209.1 and 231.5 °C. Adding Fe_3O_4/rGO hybrid to TKX-50 reduces the two exothermic peaks to 205.9 and 229.3 °C. Importantly, with the addition of the Fe_3O_4/rGO nanocomposite, the two exothermic peaks show a larger movement to the lower temperatures of 193.6 and 219.6 °C. Out of the samples which were tested, the Fe_3O_4/rGO nanocomposite shows the best catalytic behavior, by markedly reducing the decomposition temperature of TKX-50.

Additionally, bimetallic iron oxides (BIO, such as $NiFe_2O_4$, $ZnFe_2O_4$, and $CoFe_2O_4$) were introduced and the catalytic behavior of the BIO on the thermal decomposition of TKX-50 was investigated using DSC and TG–DTG (Figure 6.29) [92]. SEM images showed that the MFe_2O_4 (M = Ni, Zn, and Co) exhibits spherical

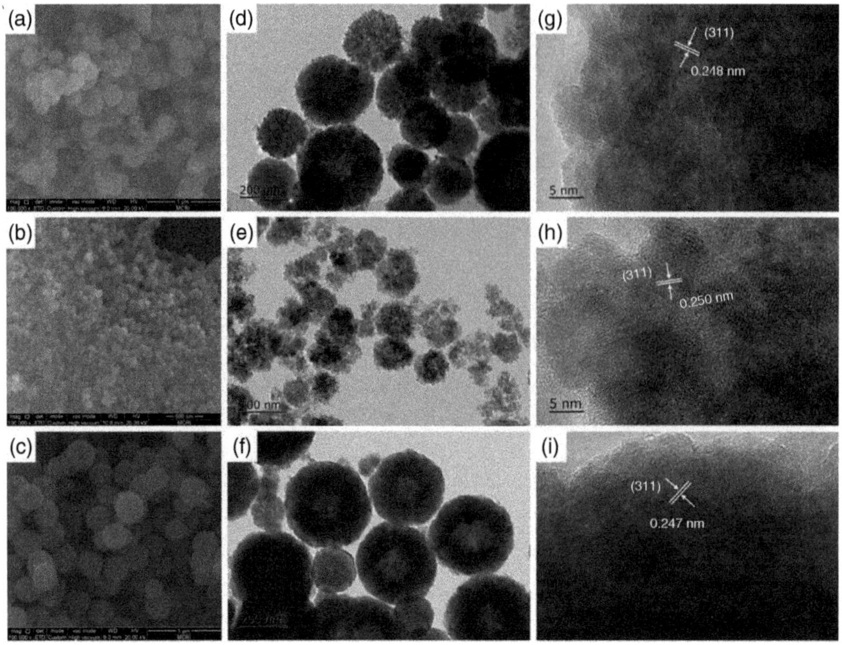

Figure 6.29 SEM (a–c), TEM (d–f) and HRTEM (g–i) images of $CoFe_2O_4$ (a, d, g), $NiFe_2O_4$ (b, e, h) and $ZnFe_2O_4$ (c, f, i). Source: Reproduced from Zhang et al. [92], Reproduced with permission from Springer Nature.

morphology with uniform granularity distribution. The $NiFe_2O_4$, $ZnFe_2O_4$, and $CoFe_2O_4$ samples show rough surfaces with particle sizes of 70, 360, and 220 nm, respectively (Figure 6.29a–c). Transmission electron microscope (TEM) images (Figure 6.29d–f) indicate that MFe_2O_4 samples were composed of agglomerated hollow powders, fitting with the SEM surface images. The high-resolution transmission electron microscope (HRTEM) images of $NiFe_2O_4$, $ZnFe_2O_4$, and $CoFe_2O_4$ samples (Figure 6.29g–i) show clear lattice fringes with inter-planar distances of 0.250, 0.247, and 0.248 nm, respectively. Additionally, the Brunner–Emmet–Teller (BET) data indicate that the SSAs of the $NiFe_2O_4$, $ZnFe_2O_4$, $CoFe_2O_4$, and Fe_2O_3 samples are 105.1, 34.68, 42.07, and 21.26 $m^2\ g^{-1}$, respectively.

The TG–DTG and DSC curves of TKX-50 mixed with Fe_2O_3 and MFe_2O_4 (M = Ni, Zn, and Co) at a heating rate of 10 $K\ min^{-1}$ were investigated and compared with those TKX-50 (Figure 6.30). The DSC results indicate that the high thermal decomposition peak temperature (T_{HDP}) and low thermal decomposition peak temperature (T_{LDP}) of TKX-50 is evidently reduced after being mixed with $NiFe_2O_4$, $ZnFe_2O_4$ or $CoFe_2O_4$. The T_{LDP} of TKX-50 mixed with $NiFe_2O_4$, $ZnFe_2O_4$, and $CoFe_2O_4$ was lowered by 29.3, 38.0, and 39.2 K, respectively. Additionally, the T_{HDP} of TKX-50 mixed with $NiFe_2O_4$, $ZnFe_2O_4$, and $CoFe_2O_4$ were lowered by 39.9, 47.0, and 51.2 K, respectively, in comparison with the T_{LDP} and T_{HDP} of pristine TKX-50 (239.9 and 268.0 °C). The reduced peak temperature of TKX-50 showed the excellent catalytic feature of MFe_2O_4. Furthermore, the MFe_2O_4 (M = Zn and Co) samples are more

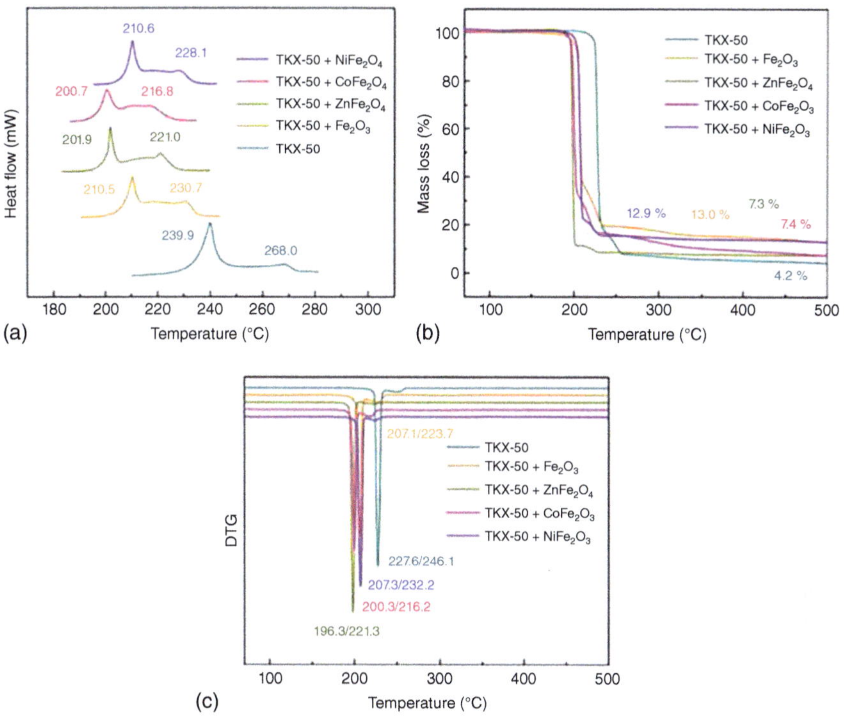

Figure 6.30 DSC and TG–DTG curves of TKX-50 before and after being mixed with different catalysts at 10 K min^{-1}. (a) DSC curves, (b) TG curves, (c) DTG curves. Source: Reproduced from Zhang et al. [92], with permission from Springer Nature, 2020.

effective at reducing the decomposition temperature than either Fe_2O_3 or $NiFe_2O_4$. The TG results indicate that the mass loss of pristine TKX-50 after the first thermal decomposition step is 81.80%, and the total mass loss reach to 450 °C is 95.20%. After mixing with positive combustion catalysts, the residual weight of TKX-50 is showed in Figure 6.30b. The reduction in the weight of TKX-50 residue confirms the superior catalytic effect of $ZnFe_2O_4$ and $CoFe_2O_4$. Additionally, the DTG peak temperatures were significantly lowered indicating the excellent catalytic activity of $ZnFe_2O_4$ and $CoFe_2O_4$ with respect to that of TKX-50 indicating that $CoFe_2O_4$ and $ZnFe_2O_4$ have a better catalytic effect than $NiFe_2O_4$ or Fe_2O_3 with respect to that of TKX-50.

As reported in the literature, MXene is an innovative class of two-dimensional (2D) nanomaterial, and knowing the catalytic thermal decomposition and combustion is essential for practical applications. For example, two-dimensional Ti_3C_2 MXene nano-sized sheets were combined for the thermal decomposition of TKX-50 [93]. The microstructured image of Ti_3C_2 MXene is showed in Figure 6.31a. MXene is composed of some nano-sized sheets with a thickness of 40 ± 10 nm, which stack together at unequal intervals, like the folded bellows of an accordion. Figure 6.31b shows the TEM image of the separated Ti_3C_2 MXene nano-sized sheet. The thermal decompositions of MXene with different mass fractions of Ti_3C_2 MXene were

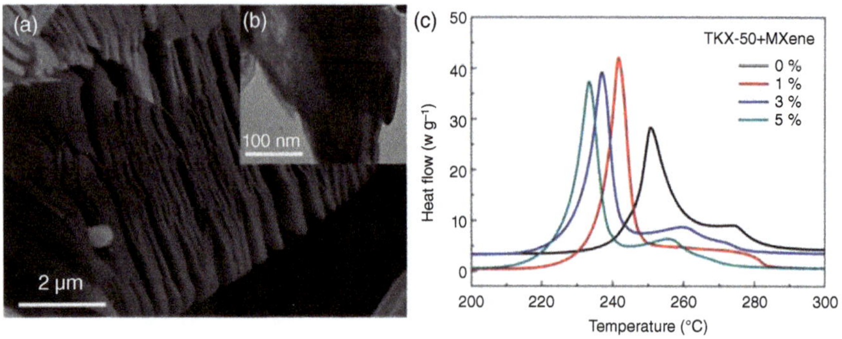

Figure 6.31 SEM (a), TEM (b) images of MXene and DSC curves of TKX-50 and MXene mixtures (c) at a heating rate of 10 K min^{-1}. Source: Reproduced from Zhu et al. [93], Reproduced with permission from Canadian Science Publishing.

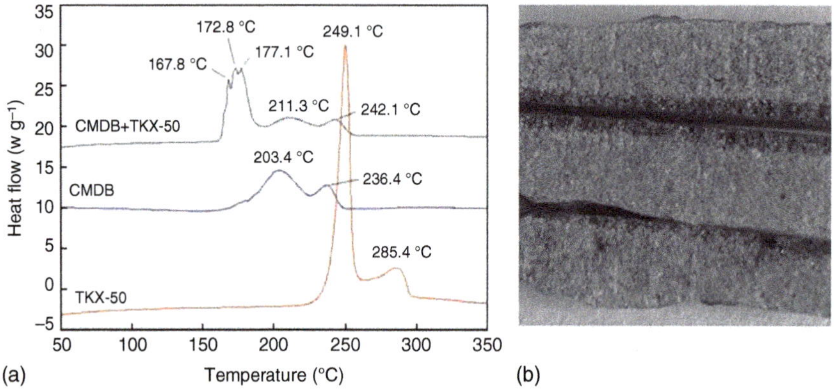

Figure 6.32 DSC curves and solidified surface of CMDB propellant containing TKX-50. (a) DSC; (b) Sldified CMDB-TKX-50 propellant. Source: Reproduced from Bi et al. [27], United States Environmental Protection Agency.

studied by DSC, as showed in Figure 6.31c. The first thermal decomposition peak temperatures of TKX-50 containing 1%, 3%, and 5% of Ti_3C_2 MXene decreased by 8.8 K, 13.5 K, and 17.2 K, respectively, indicating the catalytic effect of MXene decreased with a decrease in mass fraction. Furthermore, the heat release of TKX-50 with 5% Ti_3C_2 MXene (2907 J g^{-1}) is obviously larger than that of TKX-50 (2197 J g^{-1}), showing MXene enhances TKX-50 thermal decomposition.

For assessing its practical applications, TKX-50 was added to a CMDB propellant having the following composition (mass fractions): RDX 30–33%, NC 22–25%, NG 30–33%, Al 4%, DINA 3.5–6.5%, NTO-Pb 3% and additives 1.5%. The mass ratio of slurry and TKX-50 is 10 : 3. Solidification was performed at 70 °C for three days. Two thermal decomposition stages for the CMDB propellant are observed at 203.4 and 236.4 °C (Figure 6.32). The exothermic peak temperature was reduced to 167.8, 172.8, and 177.1 °C on addition of 30% TKX-50. The oxidation–reduction reaction of NO_2 produced by self-decomposition of TKX-50, and the exothermic decomposition and decomposition products of TKX-50 accelerate the thermal decomposition of NC, RDX, and other compounds. It can be concluded that TKX-50 did not react with the components of the CMDB propellant during the solidification process.

6.6 Application in Explosives

TKX-50 is an innovative EM with high-energy storage, high detonation velocity, low sensitivity, and low toxicity, which can not only be added to solid propellants but can also be used in explosives. Polyglycidyl nitrate (PGN), GAP, PEG, and polytetrahydrofuran (poly-THF) were introduced to TKX-50 with ranging from 1% to 12% in mass fraction to form different types of explosives. The microscopic deformation behavior of TKX-50/polymers was carried out by using nonequilibrium molecular dynamics (NEMD) simulations. The dependence of the tensile strength and Young's modulus on the mass fraction was explored. The simulation results showed that the elastic moduli of nanocomposite TKX-50/polymers decreased with increasing polymer concentration. PGN and GAP self-crimp on the TKX-50 surface, while PEG and poly-THF are extended on the TKX-50 surface, indicating that polymers improve the mechanical properties of TKX-50 in the following order: [GAP] < [PGN] < [poly-THF] ≈ [PEG] [94].

For applications, the safety and risk performance of using TKX-50 in explosives have to be considered. For instance, to optimize TKX-50 and TNT-based melt cast compositions, and to promote its application in IMs, the thermal decomposition, mechanical sensitivity as well as detonation velocity (VOD) were investigated (Table 6.7). It was found that the RDX-based composition (RT) was less friction insensitive than the TKX-50-based composition (TT), and TKX-50-based melt cast composition (TT) was more impact sensitive than the RDX-based composition (RT). At the same time, the TTA composition is less sensitive than the RTA composition, and the thermal stability of TKX-50/TNT (30/70) is better than that of the RDX/TNT (30/70) composition. The TTA composition was found to show a larger thermal decomposition temperature than that of TNT and RDX. Similarly, TKX-50/TNT

Table 6.7 The melt cast explosive composition and properties of formulations.

Samples	Mass fraction of components (%)				Percent TMD	VOD ($m\,s^{-1}$)	CJ pressure (Pa)	Mechanical sensitivity	
	RDX	TKX-50	TNT	Al				H_{50} (cm)	F (kg)
TT	—	30	70	—	1.72	7921	2.92	82	36
RT	30	—	70	—	1.71	7844	2.66	104	30
TTA	—	25	60	15	—	—	—	98	36
RTA	25	—	60	15	—	6908	—	>170	36

Note: TKX-50-based non-aluminized and aluminized melt cast formulations, namely TT and TTA; RDX-based aluminized and non-aluminized melt cast formulations, namely RT and RTA, respectively; H_{50}-characteristic drop height, cm; F-friction sensitivity, kg.
Source: Adapted from Badgujar et al. [95].

composition shows larger thermal stability than that of the RDX/TNT one, indicating that the addition of TKX-50 in melt cast formulations could be advantageous for future warhead applications [95].

Furthermore, the performance of TKX-50-based composites was investigated by using TNT/wax or TNT as the carrier, and compared with that of HMX, 3-nitro-1,2,4-triazol-5-one (NTO), and hexahydro-1,3,5-trinitro-1,3,5-s-triazine (RDX)-based compositions. TKX-50, HMX, NTO, and RDX (main explosive) in the composites (60%) and TNT (molten carrier) (40% or 30%) with the addition of wax (0% or 10%) in mass fraction were combined (Figure 6.33) [96]. The SEM image of the TKX-50 compositions showed that TKX-50 was well coated by the molten TNT, and the coating effect of wax on the TKX-50 surface was better by the observation of no crystals were stretched out. The tensile strength of the TKX-50 and NTO formulations columns was decreased on addition of wax. In contrast, the enhancement in the tensile strength of HMX and RDX columns on addition of wax may be attributable to wax being a good polymer for HMX and RDX, and improving the resistance of HMX and RDX to stretch. The addition of wax could make the most efficient decrease the mechanical sensitivity of the compositions and increase the safety performance of the composites. At the same time, the safety performance of TKX-50-based melt cast explosives was studied and compared to that for HMX-based melt cast compositions [97]. The results indicated that weaker attractive forces are present in HMX–HMX than in TKX-50/TKX-50 pairs, but unsimilarly in TKX-50/TNT than in HMX/TNT, revealing that the safety properties of TKX-50-based compositions is better than that of HMX-based compositions.

6.7 Conclusions

In the near future, propellant formulations utilizing high-energy materials with insensitive features are expected to advance the current state-of-the-art in solid rocket propellants and reduce the environmental concerns caused by the use of AP. TKX-50, is an example of a promising EM with good thermal performance and sensitivity, which has attracted great interest from researchers worldwide and shows potential for use in solid rocket propellants. The interaction of TKX-50/CL-20, TKX-50/HMX, TKX-50/PETN co-crystals and TKX-50/GAP, TKX-50/AP, TKX-50/GO mixtures were reviewed and the performances were analyzed in detail. Furthermore, the energetic properties and combustion performance of solid propellants containing different mass fractions of TKX-50 particles were evaluated, and several analytical techniques were introduced to monitor the burning rate and pressure exponent. In addition, the advantages and disadvantages of TKX-50 have also been discussed.

To overcome the disadvantages of present and potential EMs such as ε-CL-20 and HMX, some methods such as coating and co-crystallization were developed to reduce sensitivity, increase stability and performance, and decrease costs. Moreover, the application of co-crystals, such as TKX-50/CL-20 or TKX-50/HMX that should show reduced environmental impact has been shown to be a good method

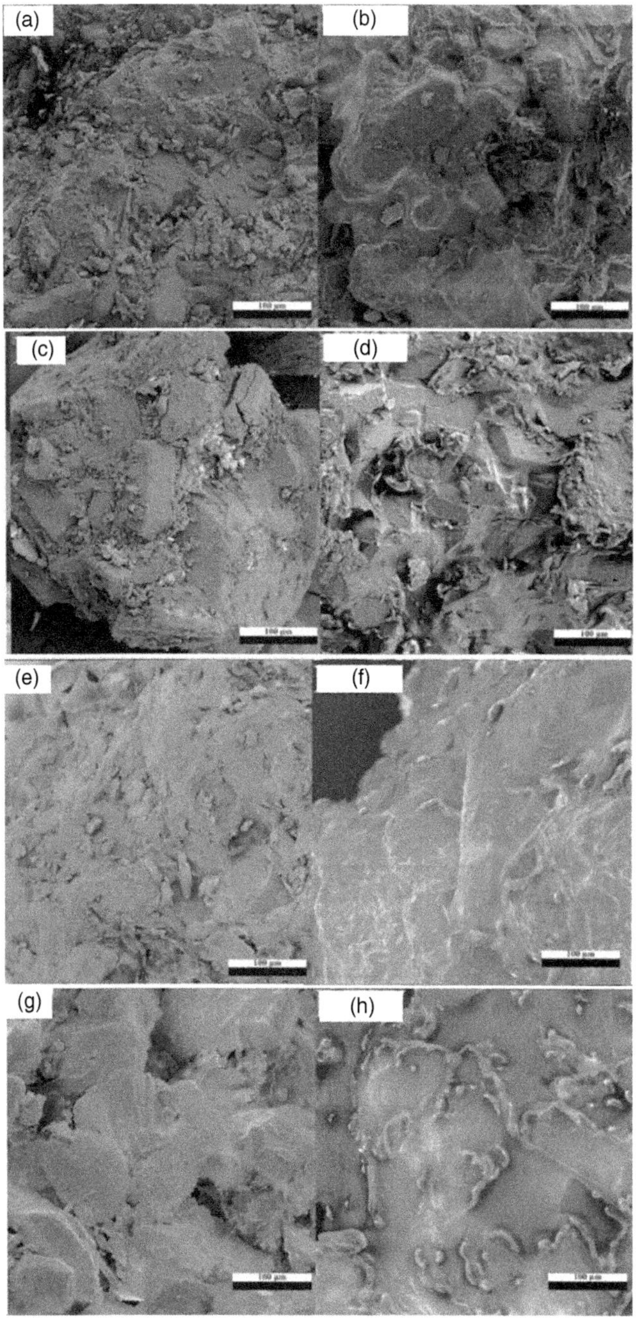

Figure 6.33 SEM images of different composite explosives. (a) TKX-50 60%, TNT 40%, H_{50} = 63.7 cm; (b) TKX-50 60%, TNT 30%, wax 10%, H_{50} = 68.0 cm; (c) HMX 60%, TNT 40%, H_{50} = 40.2 cm; (d) HMX 60%, TNT 30%, wax 10%, H_{50} = 46.8 cm; (e) NTO 60%, TNT 40%, H_{50} = 64.2 cm; (f) NTO 60%, TNT 30%, wax 10%, H_{50} = 79.3 cm; (g) RDX 60%, TNT 40%, H_{50} = 58.8 cm; (h) RDX 60%, TNT 30%, wax 10%, H_{50} = 66.2 cm. Source: Zhou et al. [96], Reproduced with permission from Taylor & Francis Group.

for increasing the performance of solid propellants and explosives. It has been demonstrated that the advantages of TKX-50 can promote the development of solid propellant and explosives with insensitive performance, and it is also trusted that the work discussed in this review will enhance the researchers' interest with respect to innovative EMs with low sensitivity for solid propellants and explosives, as well as offer a fundamental overview of promising ingredients for insensitive ammunitions.

References

1 Trache, D., Klapötke, T.M., Maiz, L. et al. (2017). Recent advances in new oxidizers for solid rocket propulsion. *Green Chemistry* 20: https://doi.org/10.1039/C7GC01928A.

2 Pang, W.Q., DeLuca, L.T., Gromov, A. et al. (2020). *Innovative Energetic Materials: Properties, Combustion and Application*. Singapore: Springer.

3 Huang, H.F., Shi, Y.M., Yang, J. et al. (2015, 2015). Compatibility study of dihydroxylammonium 5,5′-bistetrazole-1,1′-diolate (TKX-50) with some energetic materials and inert materials. *Journal of Energetic Materials* 33: 66–72.

4 Wang, J.Y., Ye, B.Y., An, C.W. et al. (2016). Preparation and properties of surface-coated HMX with viton and graphene oxide. *Journal of Energetic Materials* 34: 235–245.

5 Churakov, A.M. and Tartakovsky, V.A. (2004). Progress in 1,2,3,4-tetrazine chemistry. *Chemical Reviews* 104: 2601–2616.

6 Gobel, M., Karaghiosoff, K., Klapçtke, T.M. et al. (2010). Nitrotetrazolate-1-*N*-oxides and the strategy of *N*-oxide introduction. *Journal of the American Chemical Society* 132: 17216–17226.

7 Kröber, H. and Teipel, U. (2008). Crystallization of insensitive HMX. *Propellants, Explosives, Pyrotechnics* 33: 33–36.

8 Huang, H.F., Shi, Y.M., and Yang, J. (2015). Thermal characterization of the promising energetic material TKX-50. *Journal of Thermal Analysis and Calorimetry* 121: 705–709.

9 Liu, J., Li, Q., Zeng, J.B. et al. (2013). Mechanical pulverization for the production of sensitivity reduced nano-RDX. *Explosive Materials* 42 (3): 1–5.

10 Song, X.L., An, C.W., Guo, X.D. et al. (2008). Effect of preparation methods on mechanical sensitivity and thermal decomposition of HMX. *Chinese Journal of Energetic Materials* 16 (6): 698–702.

11 Smirnov, A., Lempert, D., Pivina, T. et al. (2011). Basic characteristics for estimation polynitrogen compounds efficiency. *Central European Journal of Energetic Materials* 8 (4): 233–247.

12 Liu, J.P., Liu, L.L., and Liu, X.B. (2020). Development of high-energy-density materials. *Science China Technological Sciences* 63 (2): 195–213.

13 Zhou, C., Huang, X.P., Zhou, Y.S. et al. (2007). Crystal structure and thermal decomposition of FOX-7. *Chinese Journal of Explosives & Propellants* 30 (1): 60–63.

14 Ji, G.F., Xiao, H.M., Ju, X.H. et al. (2003). Periodic DFT studies on the structure of crystalline TATB. *Acta Chimica Sinica* 61 (8): 233–247.

15 Miao, C.C., Ji, Y.X., Qian, L. et al. (2015). Research progress of novel bistetrazole-type energetic material TKX-50. *Chemical Propellants & Polymeric Materials* 13 (5): 7–12.

16 Gettwert, V., Franzin, A., Bohn, M.A. et al. (2017). Ammonium dinitramide/glycidyl azide polymer (ADN/GAP) composite propellants with and without metallic fuels. *International Journal of Energetic Materials and Chemical Propulsion* 16 (1): 61–79.

17 Bayat, Y. and Zeynali, V.J. (2011). Preparation and characterization of nano-CL-20 explosive. *Journal of Energetic Materials* 29: 281–291.

18 Fischer, N., Fischer, D., Klapotke, T.M. et al. (2012). Pushing the limits of energetic materials – the synthesis and characterization of dihydroxylammonium 5,5′-bistetrazole-1,1′-diolate. *Journal of Materials Chemistry* 22: 20418–20422.

19 Niu, H., Chen, S.S., Jin, S.H. et al. (2016). Thermolysis, nonisothermal decomposition kinetics, calculated detonation velocity and safety assessment of dihydroxylammonium 5, 5′-bistetrazole-1, 1′-diolate. *Journal of Thermal Analysis and Calorimetry* 126: 473–480.

20 Xu, Y.B., Tan, Y.X., Cao, W.G. et al. (2020). Thermal decomposition characteristics and thermal safety of dihydroxylammonium 5,5′-bistetrazole-1,1′-diolate based on microcalorimetric experiment and decoupling method. *Journal of Physical Chemistry C* 124: 5987–5998.

21 Xiao, L.B., Zhao, F.Q., Luo, Y. et al. (2016). Thermal behavior and safety of dihydroxylammonium 5,5′-bistetrazole-1,1′-diolate. *Journal of Thermal Analysis and Calorimetry* 123: 653–657.

22 Sinditskii, V.P., Filatov, S.A., Kolesov, V.I. et al. (2015). Dihydroxylammonium 5,5′-bistetrazole-1,1′- diolate (TKX-50): physico-chemical properties and mechanism of thermal decomposition and combustion. *2015 International Autumn Seminar on Propellants, Explosives & Pyrotechnics (IASPEP)*, QingDao China, pp. 221–238.

23 Li, M., Zhao, F.Q., Luo, Y. et al. (2014). Energetic characteristics computation of propellants containing dihydroxylammonium 5,5′-bistetrazole-1,1′-diolate (TKX-50). *Chinese Journal of Energetic Materials* 22 (3): 286–290.

24 Oxley, J.C., Smith, J.L., Donnelly, M.A. et al. (2016). Thermal stability studies comparing IMX-101 (dinitroanisole/nitroguanidine/NTO) to analogous formulations containing dinitrotoluene. *Propellants, Explosives, Pyrotechnics* 41: 98–113.

25 Provatas, A. and Wall, C. (2016). Ageing of Australian DNAN based melt-cast insensitive explosives. *Propellants, Explosives, Pyrotechnics* 41: 555–561.

26 Bi, F.Q., Xiao, C., Xu, C. et al. (2014). Synthesis and properties of dihydroxylammonium 5,5′-bistetrazole- 1,1′-diolate. *Chinese Journal of Energetic Materials* 22 (2): 272–273.

27 Bi, F.Q., Fan, X.Z., Fu, X.L. et al. (2014). Interaction of dihydroxylammonium 5,5′- bistetrazole-1,1′- diolate with CMDB propellant components. *Journal of Solid Rocket Technology* 37 (2): 214–218.

28 Yao, L.N., Wang, C.L., Han, Z.X. et al. (2020). Effects of binders on mechanical sensitivity and molding properties of TKX-50-based pressed explosive. *Explosive Materials* 49 (6): 36–41.

29 Chen, W.T., Chen, W.C., You, M.L. et al. (2015). Evaluation of thermal decomposition phenomenon for 1,1-bis(tert-butylperoxy)-3,3,5-trimethylcyclohexane by DSC and VSP2. *Journal of Thermal Analysis and Calorimetry* 122: 1125–1133.

30 Yuan, B., Yu, Z.J., and Bernstein, E.R. (2015). Initial mechanisms for the decomposition of electronically excited energetic salts: TKX-50 and MAD-X1. *The Journal of Physical Chemistry. A* 119 (12): 2965–2981.

31 Ren, X.T., Zhang, G.T., He, J.X. et al. (2016). Calculation and control of crystal morphology of dihydroxylammonium 5,5′-bistetrazole-1,1′-diolate. *Chinese Journal of Explosives and Propellants* 39 (2): 6–71.

32 Pang, W.Q., Li, J.Q., Wang, K. et al. (2018). Effects of dihydroxylammonium 5,5′-Bistetrazole-1,1′-diolate on the properties of HTPB based composite solid propellant. *Propellants, Explosives, Pyrotechnics* 43: 1013–1022.

33 Xing, X.L., Zhao, S.X., Wang, X.F. et al. (2019). The detonation properties research on TKX-50 in high explosives. *Propellants, Explosives, Pyrotechnics* 44: 408–412.

34 Jia, J.H., Xu, J.J., Cao, X. et al. (2019). Stability of dihydroxylammonium 5,5′-bistetrazole-1,1′-diolate (TKX-50) in solvents. *Propellants, Explosives, Pyrotechnics* 44: 989–999.

35 Nandi, A.K., Ghosh, M., Sutar, V.B. et al. (2012). Surface coating of cyclotetramethylenetetranitramine (HMX) crystals with the insensitive high explosive 1,3,5-Triamino-2,4,6-trinitrobenzene (TATB). *Central European Journal of Energetic Materials* 9: 119–130.

36 Ma, Z.G., Gao, B., Wu, P. et al. (2015). Facile, continuous and large-scale production of core-shell HMX@TATB composites with superior mechanical properties by a spray-drying process. *RSC Advances* 5: 21042–21049.

37 Doherty, R.M. and Watt, D.S. (2008). Relationship between RDX properties and sensitivity. *Propellants, Explosives, Pyrotechnics* 33: 4–13.

38 Borne, L., Mory, J., and Schlesser, F. (2008). Reduced sensitivity RDX (RS-RDX) in pressed formulations: respective effects of intra-granular pores, extra-granular pores and pore sizes. *Propellants, Explosives, Pyrotechnics* 33: 37–43.

39 Radacsi, N., Stankiewicz, A.I., Creyghton, Y.L.M. et al. (2011). Electrospray crystallization for high-quality submicron-sized crystals. *Chemical Engineering and Technology* 34: 624–630.

40 Guo, X.D., Ouyang, G., Liu, J. et al. (2015). Massive preparation of reduced-sensitivity nano CL-20 and its characterization. *Journal of Energetic Materials* 33: 24–33.

41 Gao, H., Wang, Q.H., Ke, X. et al. (2017). Preparation and characterization of an ultrafine HMX/NQ co-crystal by vacuum freeze drying method. *RSC Advances* 7: 46229–46235.

42 Trask, A.V., Motherwell, W.D.S., and Jones, W. (2005). Pharmaceutical cocrystallization: engineering a remedy for caffeine hydration. *Crystal Growth & Design* 5: 1013–1021.

43 Landenberger, K.B., Bolton, O., and Matzger, A.J. (2015). Energetic-energetic cocrystals of diacetone diperoxide (DADP): dramatic and divergent sensitivity modifications via cocrystallization. *Journal of the American Chemical Society* 137: 5074–5079.

44 Xiao, J.J., Zhu, W.H., Zhu, W. et al. (2013). *Molecular Dynamics of High Energy Materials*. Beijing: Science Press.

45 Zhang, J. and Shreeve, J.M. (2016). Time for pairing: cocrystals as advanced energetic materials. *Crystal Engineering Communications* 18: 6124–6133.

46 Wu, J.T., Zhang, J.G., Li, T. et al. (2015). A novel cocrystal explosive NTO/TZTN with good comprehensive properties. *RSC Advances* 5: 28354–28359.

47 Wang, J.Y., Li, H.Q., An, C.W. et al. (2015). Preparation and characterization of ultrafine CL-20/TNT cocrystal explosive by spray drying method. *Chinese Journal of Energetic Materials* 23: 1103–1106.

48 Xiong, S.L., Chen, S.S., Jin, S.H. et al. (2016). Molecular dynamics simulations on dihydroxylammonium 5,5′-bistetrazole-1,1′-diolate/hexanitrohexaazaisowurtzitane cocrystal. *RSC Advances* 6: 4221.

49 Yuan, S., Gou, B.W., Guo, S.F. et al. (2020). Preparation, characteristic properties of a new CL-20/TKX-50 cocrystal explosive. *Chinese Journal of Explosives and Propellants* 43: 167–173.

50 Xiong, S.L., Chen, S.S., and Jin, S.H. (2017). Molecular dynamic simulations on TKX-50/HMX cocrystal. *RSC Advances* 7: 6795.

51 Song, X.L., Wang, Y., and An, C.W. (2018). Thermochemical properties of nanometer CL-20 and PETN fabricated using a mechanical milling method. *AIP Advances* 8: 065009.

52 Xiao, L., Guo, S.F., Su, H.P. et al. (2019). Preparation and characteristics of a novel PETN/TKX-50 co-crystal by a solvent/non-solvent method. *RSC Advances* 9: 9204.

53 Pang, W.Q., Wang, K., Zhang, W. et al. (2020). CL-20-based cocrystal energetic materials: simulation, preparation and performance. *Molecules* 25: 4311.

54 Zhao, Y., Xie, W.X., Qi, X.F. et al. (2019). Comparison of the interfacial bonding interaction between GAP matrix and ionic/non-ionic explosive: computation simulation and experimental study. *Applied Surface Science* 497: 143813.

55 Tao, J., Wang, X.F., Zhang, K. et al. (2020). Theoretical calculation and experimental study on the interaction mechanism between TKX-50 and AP. *Defence Technology* 16: 825–833.

56 Wang, J.F., Chen, S.S., Yao, Q. et al. (2017). Preparation, characterization, thermal evaluation and sensitivities of TKX-50/GO composite. *Propellants, Explosives, Pyrotechnics* 42: 1104–1110.

57 Mkhoyan, K.A., Contryman, A.W., Silcox, J. et al. (2009). Atomic and electronic structure of graphene oxide. *Nano Letters* 9: 1058–1063.

58 Li, R., Wang, J., Shen, J.P. et al. (2013). Preparation and characterization of insensitive HMX/graphene oxide composites. *Propellants, Explosives, Pyrotechnics* 38 (6): 798–804.

59 Li, H.J., Ren, H., Jiao, Q.J. et al. (2016). Fabrication and properties of insensitive CNT/HMX energetic nanocomposites as ignition ingredients. *Propellants, Explosives, Pyrotechnics* 41: 126–135.

60 Jin, B., Peng, R.F., Chu, S.J. et al. (2008). Study of the desensitizing effect of different [60] fullerene crystals on cyclotetramethylenetetranitramine (HMX). *Propellants, Explosives, Pyrotechnics* 33: 454–458.

61 Yu, L., Ren, H., Guo, X.Y. et al. (2014). A novel ε-HNIW-based insensitive high explosive incorporated with reduced graphene oxide. *Journal of Thermal Analysis and Calorimetry* 117: 1187–1199.

62 Zhang, C.Y. (2010). Understanding the desensitizing mechanism of olefin in explosives versus external mechanical stimuli. *Journal of Physical Chemistry C* 114: 5068–5072.

63 Zhang, C.Y., Cao, X., and Xiang, B. (2012). Understanding the desensitizing mechanism of olefin in explosives: shear slide of mixed HMX-olefin systems. *Journal of Molecular Modeling* 18: 1503–1512.

64 Ma, S., Li, Y.J., Li, Y. et al. (2016). Research on structures, mechanical properties, and mechanical responses of TKX-50 and TKX-50 based PBXs with molecular dynamics. *Journal of Molecular Modeling* 22: 43.

65 Song, X.Y., Xing, X.L., Zhao, S.X. et al. (2020). Molecular dynamics simulation on TKX-50/Fluoropolymer. *Modelling and Simulation in Materials Science and Engineering* 28: 015004.

66 Xiong, S.L., Chen, S.S., and Jin, S.H. (2017). Molecular dynamic simulations on TKX-50/RDX cocrystal. *Journal of Molecular Graphics & Modelling* 74: 171–176.

67 Xie, W.X., Zhao, Y., Zhang, W. et al. (2018). Sensitivity and stability improvements of NEPE propellants by inclusion of FOX-7. *Propellants, Explosives, Pyrotechnics* 43: 308–314.

68 Yu, Y.H., Chen, S.S., Li, X. et al. (2016). Molecular dynamics simulations for 5,5′-bistetrazole-1,1′-diolate (TKX-50) and its PBXs. *RSC Advances* 6: 20034–20041.

69 Li, F.S. and Liu, J. (2018). Advances in micro-nano energetic materials. *Chinese Journal of Energetic Materials* 26: 1061–1073.

70 Cao, X., Shang, Y.P., Meng, K.J. et al. (2019). Fabrication of three-dimensional TKX-50 networklike nanostructures by liquid nitrogen-assisted spray freeze-drying method. *Journal of Energetic Materials* 37: 356–364.

71 Doblas, D., Rosenthal, M., Burghammer, M. et al. (2016). Smart energetic nanosized co-crystals: exploring fast structure formation and decomposition. *Crystal Growth & Design* 16: 432–439.

72 Shen, J.P., Shi, W.M., Wang, J. et al. (2014). Facile fabrication of porous CL-20 for low sensitivity high explosives. *Physical Chemistry Chemical Physics* 16: 23540–23543.

73 Li, T.T., Li, R., Nie, F.D. et al. (2010). Facile preparation of self-sensitized FOX-7 with uniform pores by heat treatment. *Propellants, Explosives, Pyrotechnics* 35: 260–266.

74 Zhang, H.L., Liu, Y., Li, S.C. et al. (2016). Three-dimensional hierarchical 2,2,4,4,6,6-hexanitrostilbene crystalline clusters prepared by controllable supramolecular assembly and deaggregation process. *Crystal Engineering Communications* 18: 7940–7944.

75 Deng, P., Jiao, Q.J., and Ren, H. (2020). Nano dihydroxylammonium 5,5′-bistetrazole-1,1′-diolate (TKX-50) sensitized by the liquid medium evaporation-induced agglomeration self-assembly. *Journal of Energetic Materials* 38 (3): 253–260.

76 Nicolich, S., Amuele, P., Damavarapu, R. et al. (2015). *5,5′-Bis-Tetrazole-1,1′-Diolate (TKX-50) Synthesis and Lab Scale Characterization [R]*. New Jersey: US Army Arments Research, 973-724-3016.

77 Xiong, W.Q., Zhu, W.J., Zheng, G.H. et al. (2018). Performance of TKX-50-based HTPB propellant. *Journal of Solid Rocket Technology* 41 (4): 455–458.

78 Klapötke, T.M. and Suceska, M. (2021). A computational study on the detonation velocity of mixtures of solid explosives with non-explosive liquids. *Propellants, Explosives, Pyrotechnics* 46 (3): 352–354.

79 Klapötke, T.M. (2019). *Chemistry of High Energy Materials*, 5e. Berlin/Boston: Walter de Gruyter.

80 Unger, C., Scharf, R., Klapötke, T.M. (2019). Recent results in toxicity measurements of energetic compounds. *ICT Annual Conference*, Karlsruhe, Germany, pp. 851-859.

81 Sinditskii, V.P., Filatov, S.A., Kolesov, V.I. et al. (2015). Combustion behavior and physico-chemical properties of dihydroxylammonium 5,5′-bistetrazole-1,1′-diolate (TKX-50). *Thermochimica Acta* 614: 85–92.

82 Fitzgerald, R.P. and Brewster, M.Q. (2004). Flame and surface structure of laminate propellants with coarse and fine ammonium perchlorate. *Combustion and Flame* 136: 313–326.

83 Wang, J.F., Yang, Y.F., Zhang, C.Y. et al. (2015). Thermal decomposition reaction kinetics of dihydroxylammonium-5,5′-bistetrazole-1,1′-diolate. *Chinese Journal of Explosives and Propellants* 38 (2): 42–45.

84 van der Heijden, A.E., Bouma, R.H., and van der Steen, A.C. (2004). Physicochemical parameters of nitramines influencing shock sensitivity. *Propellants, Explosives, Pyrotechnics* 29: 304–313.

85 Wang, J.F., Chen, S.S., Jin, S.H. et al. (2019). Pressure characteristics and safety performance of TKX-50 decomposition in confined space. *Journal of Energetic Materials* 37 (1): 1–11.

86 Zhao, C.D., Chi, Y., Yu, Q. et al. (2019). Comprehensive study of the interaction and mechanism between bistetrazole ionic salt and ammonium nitrate explosive in thermal decomposition. *Journal of Physical Chemistry C* 123: 27286–27294.

87 Muravyev, N.V., Monogarov, K.A., Asachenko, A.F. et al. (2017). Pursuing reliable thermal analysis techniques for energetic materials: decomposition kinetics and thermal stability of dihydroxylammonium 5,5′-bistetrazole-1,1′-diolate (TKX-50). *Physical Chemistry Chemical Physics* 19: 436.

88 Walsh, M.E. (2016). Analytical methods for detonation residues of insensitive munitions. *Journal of Energetic Materials* 34: 76.

89 Li, J., Jin, S.H., Lan, G.C. et al. (2020). Reactive molecular dynamics simulations on the thermal decompositions and oxidations of TKX-50 and twinned TKX-50. *Crystal Engineering Communications* 22: 2593.

90 Zhang, J.K., Zhao, F.Q., Yang, Y.J. et al. (2020). Enhanced catalytic performance on the thermal decomposition of TKX-50 by Fe_3O_4 nanoparticles highly dispersed on rGO. *Journal of Thermal Analysis and Calorimetry* 140: 1759–1767.

91 Lu, Z.P., Xiong, Y., Xue, X.G. et al. (2017). Unusual protonation of the hydroxylammonium cation leading to the low thermal stability of hydroxylammonium-based salts. *Journal of Physical Chemistry C* 121: 27874–27885.

92 Zhang, M., Zhao, F.Q., Yang, Y.J. et al. (2020). Synthesis, characterization and catalytic behavior of MFe_2O_4 (M = Ni, Zn and Co) nanoparticles on the thermal decomposition of TKX-50. *Journal of Thermal Analysis and Calorimetry* 141: 1413–1423.

93 Zhu, S.D., Feng, Y.Q., Li, X.X. et al. (2020). Two-dimensional titanium carbide (Ti_3C_2) MXene towards enhancing thermal catalysis decomposition of dihydroxylammonium 5,5′-bistetrazole-1,1′-diolate (TKX-50). *Canadian Journal of Chemistry* 98: 697–700.

94 Yu, C., Yang, L., Chen, H.Y. et al. (2020). Microscale investigations of mechanical responses of TKX-50 based polymer bonded explosives using MD simulations. *Computational Materials Science* 172: 109287.

95 Badgujar, D. and Talawar, M. (2017). Thermal and sensitivity study of dihydroxyl ammonium 5,5′- bistetrazole- 1,1′-diolate (TKX-50)-based melt cast explosive formulations. *Propellants, Explosives, Pyrotechnics* 42: 883–888.

96 Zhou, M.N., Chen, S.S., Wang, D.X. et al. (2019). A comparative study of performance between TKX-50-based composite explosives and other composite explosives. *Journal of Energetic Materials* 37 (2): 162–173.

97 Yu, Y.H., Chen, S.S., Li, T.J. et al. (2017). Study on a novel high energetic and insensitive munitions formulation: TKX-50 based melt cast high explosive. *RSC Advances* 7: 31485.

Part III

Metal-based Pyrotechnic Nanocomposites

7

Recent Advances in Preparation and Reactivity of Metastable Intermixed Composites

Hongqi Nie[1], Wei He[2], and Qilong Yan[1]

[1] *Northwestern Polytechnical University, Science and Technology on Combustion, Thermo-Structure and Internal Flow Laboratory, Xi'an 710072, China*
[2] *Sichuan University, School of Aeronautics and Astronautics, Department of Aerospace Engineering, 24 South Huan Road, Chengdu 610065, China*

7.1 Introduction

Solid propellant is a class of energetic composite materials featured with high volumetric energy density, high combustion calorific value, and easy storage, which is utilized as the main charge in a variety of solid rocket motors [1]. Since the solid propellants serve as the major power source for the rocket and missile systems, to consistently improving their energy densities is an eternal objective for the development of solid propellants [2–4]. Aluminum is commonly used as the metal fuel additives aiming to effectively improve the energy density of the solid propellant because of its high combustion enthalpy and low oxygen consumption rate. Addition of aluminum powder to the solid propellant can increase the flame temperature that results in an improvement of the specific impulse of solid propellants by 15% [5, 6].

However, an inert oxide layer formed on the surface of aluminum during the combustion process prevents the active aluminum content from reacting with the oxidizing atmosphere, which significantly reduces the combustion efficiency of aluminum. In the combustion process of solid propellants containing aluminum, the molten particles of aluminum are observed to be remained on the combustion surface of propellant due to their high ignition temperatures (>1600 K). Those molten particles are likely to agglomerate into larger particles causing significant two-phase flow losses in rocket motor. Additionally, the dependence of combustion rate of propellants on pressure is greatly increased as addition of aluminum, which would lead to an undesirable combustion performance of rocket motor.

The reduction of ignition temperature of aluminum and the degree of agglomeration of the condense phase combustion products could be a strategy for effectively improving the combustion performance of aluminum in solid rocket motors. Reducing particle size (from micron scale to nanoscale) of aluminum is found to markedly decrease its ignition temperature, which is considered as a most

Nano and Micro-Scale Energetic Materials: Propellants and Explosives, First Edition.
Edited by Weiqiang Pang and Luigi T. DeLuca.

simple and straightforward method for improving aluminum combustion [7, 8]. The nanoscale aluminum powder possesses a high specific surface area that greatly enhances the probability of reactions between aluminum and surrounding oxidizers. It enables the aluminum to burn with a higher combustion rate at/near the combustion surface of propellant, which is in favor of the heat feedback to the combustion surface of propellant. However, the nanoscale aluminum powders cannot be practically utilized in the solid propellant formulations, mainly attributed to the high mass ratio of the inert oxide layer with respect to aluminum and the rapidly increased viscosity of solid propellant slurries due to addition of aluminum in nanoscale. Therefore, the focus of research returns to the micron aluminum and their property modifications, which involve surface activation methods, design and preparation of aluminum-based composite fuels.

The highly reactive metastable intermixed composites (commonly named as MICs) are a kind of aluminum based composite fuels widely used in energetic formulations of the applications such as pyrotechnics, solid propellants, gas generators, and micro-energetic devices (e.g. micro-initiation, micro-detonation, and micro-propulsion) [9]. MICs are usually composed of a metal fuel and an oxidizer, including metal oxides and fluoropolymers. The newly developed nanocomposites based on metal/metalloid and metal/metal systems can also be considered as MICs. In both cases, at least one component in the nanocomposite should be on the nanometer scale. Upon the external stimulations, the components in MICs can rapidly react with each other and release a large amount of heat by oxidation reaction. Aluminum is a mostly used metal fuel in MICs forming Al-based MICs ascribed to its high volumetric energy density and combustion heat generated. The choice of oxidizers includes metal oxides (e.g. CuO, Fe_2O_3, and Bi_2O_3), fluoropolymers (e.g. PTFE and PVDF), and energetic salts (e.g. iodate and ammonium perchlorate) [10]. Recently, the energetic metal–organic frameworks (EMOFs) materials are reported as the precursors of metal oxides for the purpose of increasing the energy density and ability of gas generation for MICs. The high porosity and large specific surface area make EMOFs to be a promising candidate of oxidizers with great efficiency [11].

Various preparation methods, such as physical mixing, vapor deposition, and self-assembly have been used to fabricate MICs with an increased interfacial contact of components. By employing these strategies, the MICs have been prepared with distinct structures, including randomly distributed structure, 3D ordered multilayered structures, and core–shell structure. Those microstructures are observed to strongly influence the extent of mixing of metal fuels and oxidizers, which determines the reactivity of composite materials [12–14]. For instance, the microstructures of the Al-based MICs prepared by mechanical milling appear a poor uniformity where the reactants are randomly distributed (as shown in Figure 7.1 on left). It affects the ignition and combustion of MICs negatively due to an ineffective contact between reactants, eventually causing incomplete reactions and low combustion efficiency [5, 6]. However, the multilayer structured MICs have a typical "layer–layer" structure, in which alternating layers of two reactants are stacked with a single periodicity (as shown in Figure 7.1 in middle). This Al-based

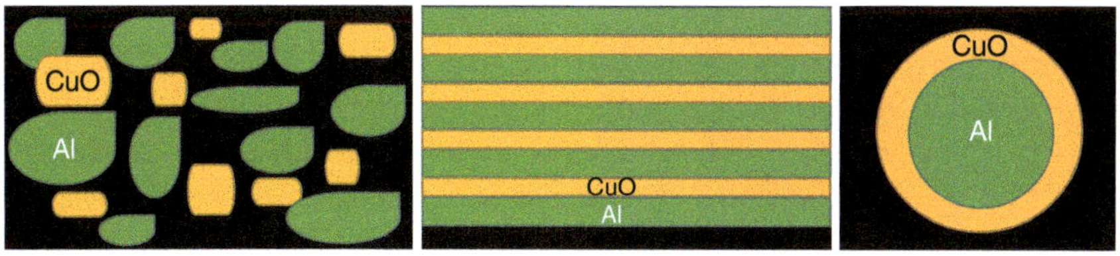

Figure 7.1 Schematics of typical microstructures of aluminum-based MICs: Left, Random distributed structure; Middle, 3D ordered multilayer structure; Right, Core–shell structure. Source: Hongqi Nie.

MICs show significant advantages in fast propagation reaction rate and high energy density. In addition, the reactivity and combustion properties of the MICs with a layered structure can be precisely controlled by adjusting the configurations of microstructure and regulating the physical properties of targeted reactants [15–17].

Over the past years, researchers paid much attention to the MICs having the core–shell structures with the reasons summarized as follow [18–20]. Firstly, the reactants are intimately mixed in a unique way (as shown in Figure 7.1 on right) that enables the sufficient contact between the metal fuel (e.g. Al) and oxidizer, thereby enhancing the reactivity of MICs system. Secondly, a dense oxidizer shell may also prevent the low-temperature oxidation of Al during the preparation and storage of core–shell structured MICs. Last but not the least, the ignition and combustion behaviors of the core–shell structured MICs can be well controlled through the tunable reactive interfacial layers and physiochemical properties.

Therefore, it can be clearly seen that the reactivity of Al-based MICs is heavily dependent on the microstructures fabricated. Nowadays, the studies on the development of novel Al-based MICs with better performances are still in progress, which includes the work relating mainly to the attempts of finding a series of new oxidizers and preparation of Al-based MICs with various microstructures assisted by the advanced material preparation methods and processes [21–23]. Herein, our present review is aimed to provide a comprehensive comparison and in-depth discussions on the recent advancements in preparation and reactivity control of Al-based MICs, with the emphasis on the discussions regarding the effect of microstructures on the reaction kinetics and mechanisms of Al-based MICs. We hope this review would facilitate the development of novel Al-based reactive nanocomposites with properties on demand and be served as an important theoretical basis guiding the reaction model to be established, which can be thereafter used to accurately predict the ignition and combustion behaviors of Al-based MICs.

7.2 The Preparation and Reactivity Control of MICs

7.2.1 Al-Based MICs with Random Distributed Structures

7.2.1.1 Preparation Methods

The Al-based MICs with randomly distributed structures are usually prepared using ultrasonic mixing, mechanical milling, and sol–gel technologies, by which the fuel and oxidizer particles are randomly distributed in the mixtures or in the composites obtained. Among them, ultrasonic mixing is a simple and efficient method. Hence, plenty of studies have been conducted using the ultrasonic mixing method. With the help of ultrasonication, fuel and oxidizer are able to be uniformly distributed in a variety of organic solvents. Nevertheless, such mixture is not capable of obtaining a perfect dispersion: it is more likely to represent a random allocation of the particles for the fuel and oxidizer on nanoscale. For instance, the preparation of aluminum-based MICs involving metal oxides MoO_3, WO_3, CuO, and Bi_2O_3 reported by Sanders's work [24] was based on the ultrasonic mixing method. The agglomeration of MICs products obtained by ultrasonic mixing was observed, which would be harmful to the redox reaction between reactive components as a result of the inhomogeneity of mixture and extended diffusion pathway.

Mechanical ball milling is known as a "top–down" technique for the design and fabrication of MICs. The reactants are mainly composed of metal and metal oxide or fluoropolymer, such combinations can release significant amount of heat energy by exothermic reactions. The high-energy ball milling method used for synthesizing the MICs usually requires one to pay much attention to the self-sustained reaction to be initiated during the preparation. The high-energy ball milling is likely to produce the micron-sized MICs with layered internal morphology at nanoscale, which makes the composite materials to be fully dense and having a better homogeneity. Even though this method is useful for preparation of MICs materials, especially in scale-up production, some limitations still exist.

The sol–gel technique is another means to effectively enhance the uniformity of the resultant composite materials and the safe manufacturing requirements are satisfied. It requires a lower processing temperature to obtain the highly homogeneous composite materials in comparison with that of physically mixed ones. The sol–gel methods were employed firstly by the researchers at Lawrence Livermore National Laboratory for the preparation of MICs. The preparation process of MICs using sol–gel method relates to the generation of dispersed nanoparticles through the reactions taking place in solution, normally defined as "sol." The formation of "gel" is attributed to the condensation process of "sol," due to the linkage of dispersed nanoparticles. The nanoparticles formed with a porous 3D structure can be achieved by well controlling the experimental conditions such as reaction temperature, or by introducing the gelation agents and catalysts. The last step in sol–gel method is aimed to remove the solvent from the solution, which can be done differently: either through the slow-evaporation procedure that results in a xerogel or by utilizing the CO_2 supercritical extraction process to obtain an aerogel. For application in production of MICs composites using sol–gel technique, the aluminum suspension needs to be introduced to the metal oxide sol before the occurrence of gelation.

7.2.1.2 Characterization

Metal Oxide-Based MICs The nano-sized metal oxides are commonly used as the oxidizer in MICs, which can readily react with aluminum by oxidation reaction accompanied by a significant heat release. It was found that the reactivity of

Al-based MICs strongly depends on the types of metal oxides selected. For instance, the Al/CuO system that serves as the high-energy fuel additive in propellants can release the reaction heat as high as 4000 J g^{-1}. Burning rates of Al/CuO MICs are reported to be tunable in a wide range of 1500–2400 m s^{-1}, which are directly correlated with the reactivity of the reactants. Reactants involved in Al/CuO system are in nanoscale with the particle sizes ranging from 50 to 200 nm. As compared, the reaction heat of Al/Fe_2O_3 system is about 2100 J g^{-1}, which is nearly equivalent to half of the value shown for Al/CuO. The heat release measured experimentally in Al/Fe_2O_3 system can be increased close to its theoretical maximum value by improving the interfacial contact conditions for constituting reactants. As an example, the Fe_2O_3 with a porous morphology can be synthesized by sol–gel method, which is then combined with nanoscale Al to form MICs [25]. The burning rate of Al/Fe_2O_3 MICs with a 3D porous structure is enhanced by 30% compared to that of Al/Fe_2O_3 with random distributed structure, which is resulted mainly from the intimate contact between components in the system leading to more reacting surfaces available for the reaction.

The Al-based MICs are the mixed systems in condense phase, reactions of this mixed system leading to the ignition are likely to occur in condensed phase primarily due to the high melting and boiling points of metal oxides, the subsequent combustion would be taken place in both gas-condense phase heterogeneously and in gas phase as well [26]. Piekiel et al. [27] studied the ignition reaction of Al/Bi_2O_3 and found that the transport capability of oxygen atoms in metal oxides determines the oxidation reaction in Al/Bi_2O_3 MICs. It reveals that the ignition of Al/Bi_2O_3 is governed by a condense-phase reaction mechanism. Egan et al. and Jacob et al. observed the reactive sintering taking place prior to the ignition of Al/CuO suggesting that the oxidation reaction and the process of heat release for Al-based MICs occur in condense phase [28, 29]. They also found that the decomposition temperatures and ability of releasing oxygen of metal oxides strongly influence the ignition of aluminum, thus affecting the reactivity of Al-based MICs. In addition, Sullivan et al. conducted the experiments by using a custom burning tube, the results confirm the fact that the ignition of Al/CuO happens in condense phase and the flame propagation speed reaches about 165 m s^{-1} at the initial reaction stage (<0.05 ms).

Umbrajkar et al. [30] characterized the exothermic reaction of Al/CuO MICs and analyzed the corresponded reaction products by using thermal analysis and X-ray diffraction technique, respectively. A multistep heterogeneous reaction mechanism of Al/CuO was proposed in which the reaction undergoes in three steps associated with two exothermic peaks. It shows that CuO decomposes firstly to Cu_2O and O_2 at 450 °C where Al partially reacts with O_2 producing the first exothermic peak. The remaining Al continues to react with the decomposition products of Cu_2O at around 720 °C that forms Al_2O_3 and Cu and generates the second exothermic peak. The acquired thermo-analytical data was then processed by using Kissinger method to obtain the reaction kinetics for each single reaction stage. The resultant activation energies (E_a) are 80, 100, and 265 kJ mol^{-1}, respectively, and the uniformity of reactants in MICs system is observed to affect the E_a delicately. Ermoline et al. [31] established a kinetic model used to describe the oxidation reaction of

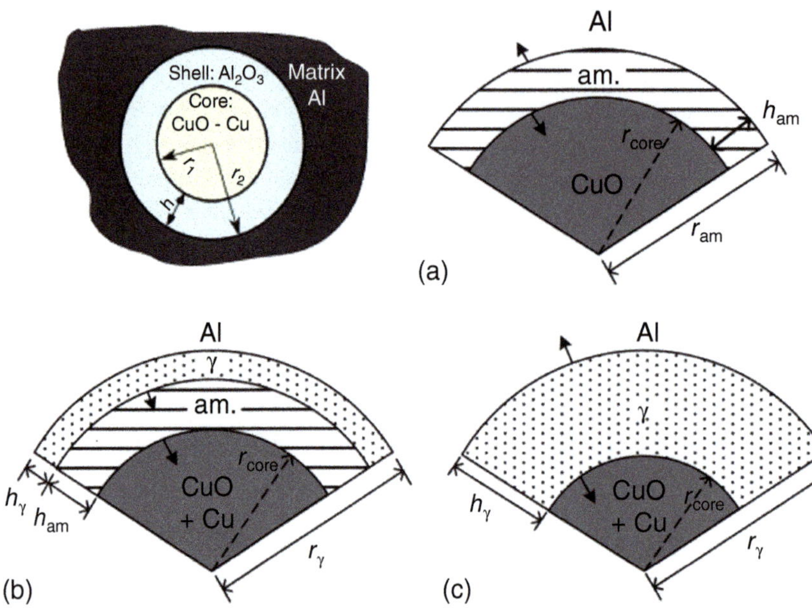

Figure 7.2 Schematic diagram of the core–shell geometry considered in the reaction model and the evolution of reactants and products. Source: Ermoline et al. [31]. Reproduced with permission of Elsevier.

Al/CuO MICs through combing the Cabrera–Mott low-temperature oxidation model and polymorph transformation model of Al_2O_3. The modified reaction model is able to accurately reproduce the ignition behaviors of Al/CuO at elevated heating rates. The model assumes the spherical CuO with a diameter of 100 nm is embedded into the Al matrix where the initial thickness of amorphous Al_2O_3 shell is defined to be 0.9 nm (structure as shown in Figure 7.2 (left)). As the reaction proceeds, amorphous Al_2O_3 grows and starts to transform to a dense and more stable γ-Al_2O_3 at a specific thickness/temperature until the process of polymorph transformation is complete (process as shown in Figure 7.2a–c). In this reaction model, the Cabrera–Mott oxidation mechanism is used to describe the growth of amorphous Al_2O_3, the decrease in the reaction rate of Al/CuO with increasing the oxide thickness is identified. The growth rates of amorphous Al_2O_3 and γ-Al_2O_3 are controlled by the Arrhenius reaction kinetics when the oxide thickness of Al and reaction temperatures reach to a particular range.

Fluoropolymer-Based MICs Besides metal oxides, fluoropolymers are a promising candidate for oxidizers used in MICs that have been successfully applied for various Al-based MICs. On the one hand, the element of fluorine is a strong oxidizer that can react with Al and Al_2O_3 to form AlF_3 producing significant amounts of heat. The volumetric heat output and reaction temperature of aluminum/fluoropolymer system can be 21 kJ cm^{-3} and 3600 K, respectively. On the other hand, the combustion of Al/fluoropolymer system is often associated with the gaseous intermediates produced as a result of the low boiling point of AlF_3. It plays a role in significant reduction of the agglomeration in combustion products, leading to a promoted combustion characteristic. In addition, the naturally formed inert oxide film of Al can be activated by fluoropolymers, which results in an enhanced reactivity and lowered ignition threshold of Al. Studies [32] have suggested that the fluoropolymer is likely to react with the oxide film of Al on nanoscale at a relatively low temperature leading to a so-called pre-ignition reaction (PIR), which is observed to effectively make the inert oxide become the reactive component capable of releasing heat.

Osborne et al. [33] investigated the thermal decomposition of Al/PTFE MICs obtained by solvent mixing method using thermal analysis technique. The results indicate that the oxidation of Al/PTFE consists of two exothermic reactions, in which the first peak appears at 400 °C resulting from the PIR. The intensity of PIR is mainly dependent on the particle size scale of Al (as shown in Figure 7.3b), which affects the reacting interfaces of Al available in reaction significantly. It can be clearly seen that the PIR improves the reactivity of Al to some extent since the peak temperature of PIR is well below that of the oxidation of Al alone (between 500 and 600 °C). The lowered reaction temperature of Al/PTFE due to PIR taking place is in favor of the reduction of ignition temperature of Al. In general, PIR is considered as a reaction that the decomposition product of PTFE ($-(CF_2)_n-$) reacts with alumina assisted by the hydroxyl groups on Al surface. The produced AlF_3 is found to promote the decomposition process of PTFE. Moreover, DeLisio et al. [34] explored the mechanism for the reactions of Al/PVDF MICs by conducting the T-jump ignition measurements in different gaseous environments. The TGA/DSC thermos-analysis technique was employed to identify the PIR reaction, and the T-Jump TOFMS was utilized to detect the gas phase reaction species produced by PIR reaction and Al fluorination reaction. Results indicate that PIR is the main cause of decreasing the initial decomposition temperature of PVDF and the PIR is a condensed-phase reaction due to the negligible mass change in PVDF observed before PIR takes place. It also shows that the fluorination of Al is attributed to the gas-phase product of HF reacting with Al.

To further assess the influence of preparation methods on the microstructures and reaction progresses for fluoropolymer contained MICs, Sippel et al. [35] investigated the ignition and combustion features of Al/PTFE MICs in solid composite propellants. The Al/PTFE composites were formulated in fuel-rich composition with 90 or 70 wt% of Al and 10 or 30 wt% of PTFE by mechanical activation. Using microscopic imaging (as shown in Figure 7.4a), it was found that the modified composite particles are likely to ignite readily at the burning surface of propellants, and the large agglomerates can be broken up producing more particles with fine sizes, which enables to significantly improve the feedback of heat to the burning surface of propellant. The breakage of large agglomerates (~200 μm diameter) into smaller fragments with diameters of 20–100 μm are likely ascribed to AlF_3 sublimation caused by heating (the breakup process of Al/PTFE is illustrated in Figure 7.4b2). This finding is suggestive of a more complete combustion of Al achieved by Al/PTFE system that contrasts with the observation for the propellant comprising the flake Al with comparable particle size, in which the large agglomerates are still retained without breakup to occur. Furthermore, through analyzing the morphology and

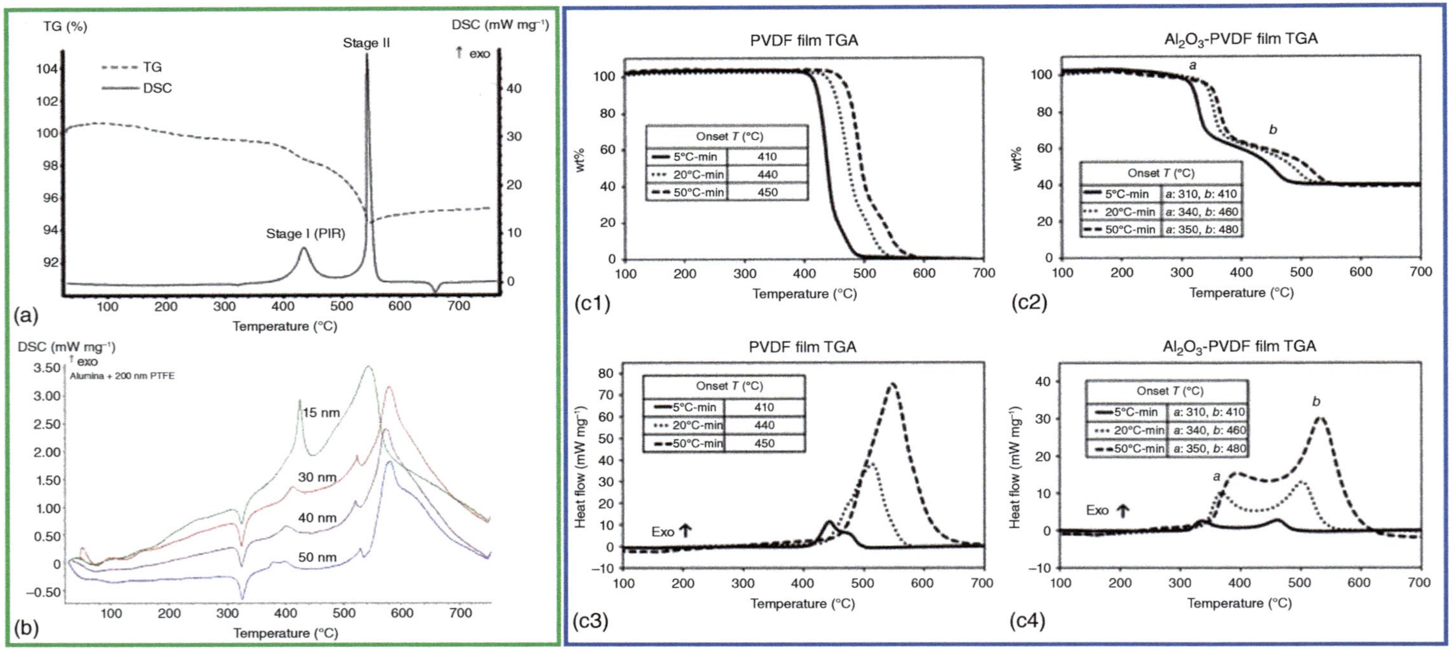

Figure 7.3 (a) DSC/TG traces of Al/PTFE MICs in argon; (b) the dependence of particle size of Al on intensities of pre-ignition reactions; Source: Osborne et al. [33]. Reproduced with permission of Taylor & Francis. (c) DSC/TG traces of PVDF fluoropolymer and Al/PVDF composites in oxygen. Source: DeLisio et al. [34]. Reproduced with permission of American Chemical Society.

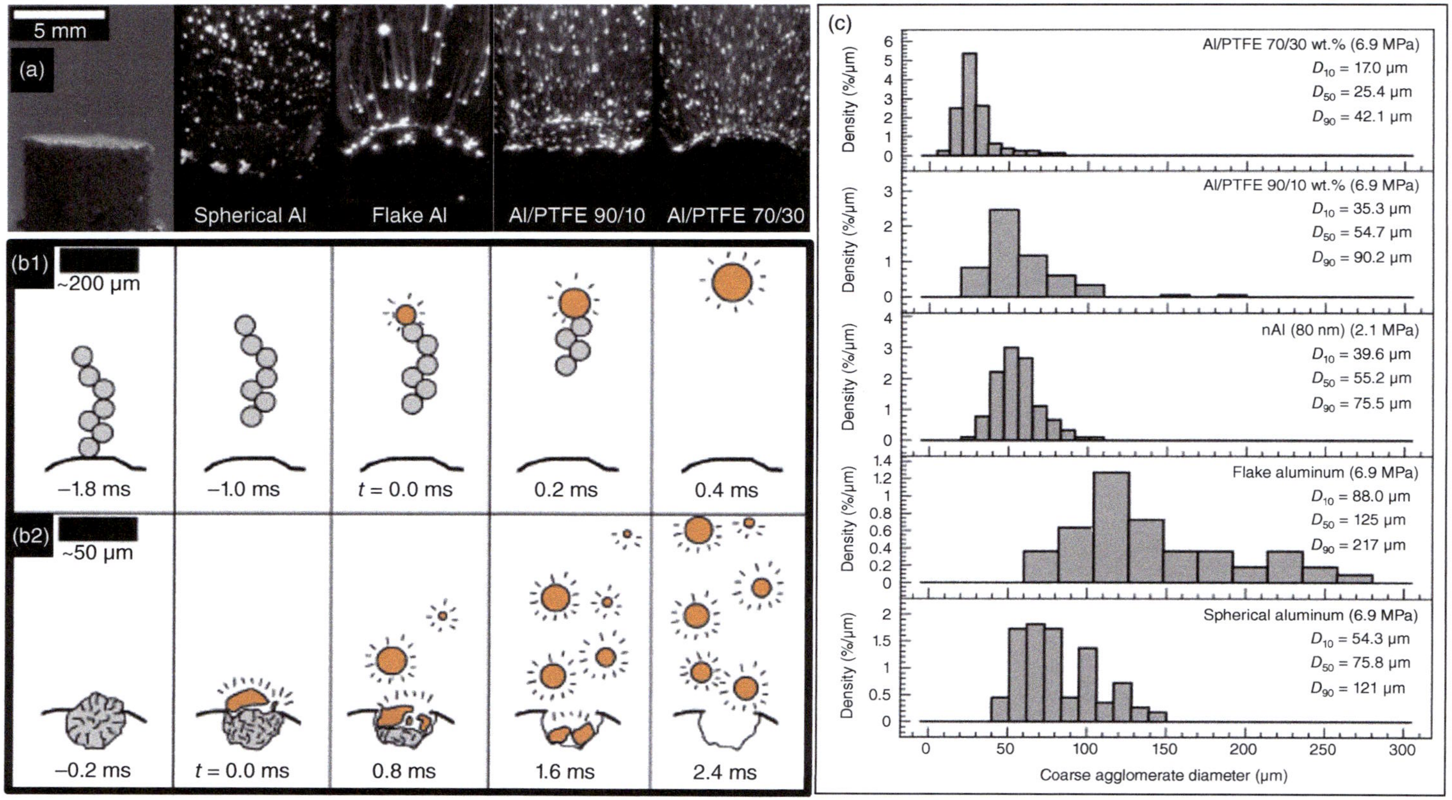

Figure 7.4 (a) propellant sample (on left), the burning surfaces of samples involving Al with various morphologies and Al/PTFE composites (on right); Schematic of ignition mechanism for (b1) spherical aluminum and (b2) Al/PTFE composites (70/30 in wt%); (c) Particle size distribution of agglomerates collected from experiments of combustion of propellant containing Al with various morphologies and Al/PTFE composites. Source: Sippel et al. [35]. Reproduced with permission of Elsevier.

size of collected combustion products, it is indicative of a reduction in agglomerate diameter by 66% or a decrease in volume by 96% for the propellant containing Al/PTFE compared to that formed from the propellant containing reference spherical Al. The condensed-phase reaction products with smaller particle sizes and more gas-phase species generated would mitigate the problems of two-phase flow loss commonly observed in solid rocket motor. Similarly, Nie et al. [36] investigated the combustion of Al/fluoropolymer (Viton) system comprehensively by using filtered photomultiplier tubes and optical spectrometry in different oxidizing environments. Results suggest that measured burn times of Al/Viton composite particles likely represent that of fragments produced during burning, which agree with the observation by Sippel et al. In addition, the combustion temperatures of Al/Viton composites acquired by using two optical approaches are consistently higher than that of reference Al starting powder.

7.2.1.3 Reactivity Control

The comparison in reactivity of MICs with different oxidants is challenging as it is highly affected by preparation method and experimental condition. Some efforts are still undergone to compare the reactivity of MICs with various oxidants by controlling experimental variables. Jian studied the gas-phase oxygen release conditions for different oxidants and compared the effect of oxidants on the ignition performance of MICs [37]. By evaluating the temperature at which oxidants release gaseous oxygen and the ignition temperature of the MICs, the correlation between oxygen release of oxidants and ignition of MICs taking place is obtained. This suggests that the reactivity of MICs can be tuned by using different oxidants associated with different oxygen release temperatures. Sanders compared the pressure output and burning rate of Al-based MICs involving MoO_3, WO_3, CuO, and Bi_2O_3 [24]. Results show that the burning rate of MICs is closely related to formation of gaseous products and the thermodynamic state of the system. For example, liquid Mo has a positive effect on the burning rate, while Al/CuO MICs with less gaseous products to be formed lead to a lower burning rate. Weismiller compared the flame temperature of Al/CuO, Al/MoO_3, and Al/Fe_2O_3 MICs systems [38]. The averaged combustion temperature of Al/CuO, Al/MoO_3, and Al/Fe_2O_3 was determined to be 2390 ± 150 K, 2150 ± 100 K, and 1735 ± 50 K, respectively. The author concludes that flame temperature is also affected by gaseous products, which are produced mainly by the decomposition or evaporation of metal oxides.

In addition, the reactivity of random distributed structured MICs can be tuned by changing the microscopic morphology of reactants. For instance, the detonation rate of the Al/WO_3 can reach 2.5 km s^{-1} under loose accumulation when the WO_3 foam is used as the oxidant, which is superior to that of traditional Al/WO_3 MICs [39]. It could be attributed to the improved surface contact properties that have a positive effect on increasing the burning rate and a large number of holes in the system that increase the mass-transfer efficiency leading to a higher burning rate. The same conclusion has also been confirmed in the three-dimensional porous MICs. The metal oxide is prepared into a three-dimensional porous shape and mixed with aluminum to form a three-dimensional porous MIC. The results suggest that the combustion characteristic of the MICs is highly dependent on the porosity of the oxidant, which confirms that the oxidant structure has an important influence on the performance of MICs. Oxidants with other shapes, such as one-dimensional nanowires, two-dimensional nanorods, nanorings, three-dimensional nanoshells, and nanocapsules have all been applied to alter the reactivity of MICs.

To sum up, hereinabove, the physiochemical properties of oxidizers directly influence the ignition and combustion performances of Al-based MICs with random distributed structures. The reactivity and heat release efficiency of MICs are strongly impacted by their microstructures that relate closely to the interfacial condition between oxidizer and Al. As a result of the microstructure obtained for Al-based MICs as mentioned above, appearing to be randomly distributed, reactants in the system are insufficiently uniform and poorly mixed, and the ignition and combustion behaviors are difficult to be precisely controlled for this type of Al-based MICs. Thus, properly controlling the contact mode of reactants by the on-demand design of the microstructures realized by the advanced preparation methods should be a feasible and effective pathway to obtain the Al-based MICs with tunable ignition and combustion characteristics.

7.2.2 Al-Based MICs with Multilayered Structures

7.2.2.1 Preparation Methods

The multilayered structured Al-based MICs show advantages in high packing density, strengthened reactivity, and great mechanical property. More importantly, the multilayered structures benefit the ability in controlling the reactivity of Al-based MICs compared to randomly distributed structures, resulting from the parameters of layered structure and composition of reactants can be well defined. The physical deposition techniques, such as sputter deposition and electron-beam evaporation, have been broadly utilized to fabricate MICs with an ordered layered structure; two reactants are required at least to be constructed layer by layer alternatively. The obtained laminate materials are usually with the thickness ranging from 0.1 to 300 μm (deposition process is illustrated in Figure 7.5a) [40]. Figure 7.5b shows that the layered structure can be clearly identified where the reactants are in close contact and uniformly stacked. As a result, a decreased ignition temperature, strong self-sustained reaction, and fast flame propagation are achieved for the Al-based MICs with multilayered structures. Except for the preparation of traditional layered MICs composed of metal fuel and metal oxide, the fabrication of multilayered alloy composites with more than two reactive components can also be achieved by this method. With addition of excess components to the system via this method, the exothermicity in reactions and production rate of metal vapors would be strongly affected. For example, the ternary system of multilayered MICs, such as Al/CuO/PVDF, Al/CuO/NC, and Al/CuO/AP composites, have demonstrated a higher reactivity and more complete combustion reactions.

The combination of magnetron sputtering and surface modification methods allows one to prepare MICs with unique features. For instance, superhydrophobic Al/CuO/PVDF MICs prepared by combining magnetron sputtering with atomic

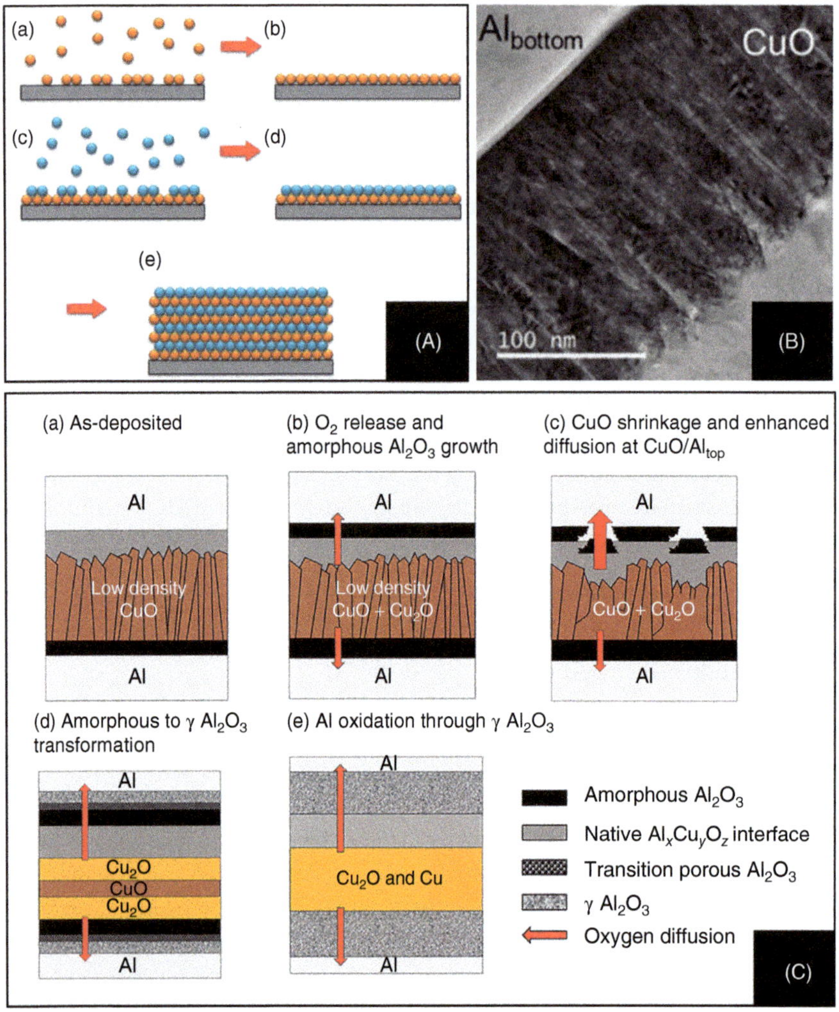

Figure 7.5 (A) Schematic of preparation of energetic multilayer thin film by atomic layer deposition (ALD); Source: Azadmanjiri et al. [40]. Reproduced with permission of Royal Society of Chemistry. (B) TEM image of cross-sectioned multilayer thin film of Al/CuO; (C) Schematic diagram of condense-phase reaction mechanism of Al/CuO binary system. Source: Abdallah et al. [41]. Reproduced with permission of American Chemical Society.

layer deposition (ALD) have a superhydrophobic coating layer on the surface, which prevents the reactants from water permeation, and permits thermite reaction to occur in water or under the condition with high moisture. It suggests that the reactivity of MICs can be well-tailored by precise construction of involved components into multilayered nanostructure.

7.2.2.2 Characterization

Studies on characterization of Al-based MICs with layered structures indicate that the influence of microstructure on reaction pathway and mechanism of MICs

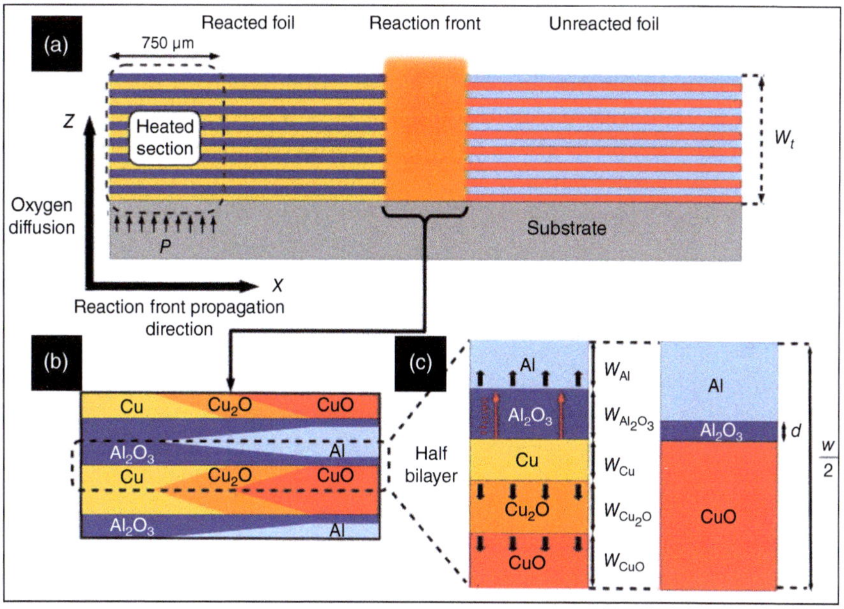

Figure 7.6 Schematic diagrams shown for the reaction model of Al/CuO multilayered system; (a) The reaction front direction propagating from the left to the side of the film; (b) The magnification of the reaction front illustrating different reactants diffusing through interfacial layers and (c) The details in the composition of single layer. Source: Lahiner et al. [17]. Reproduced with permission of AIP Publishing.

system is insignificant. For instance, Abdalah et al. [41] demonstrated that the reaction mechanism of Al/CuO multilayer thin film is similar to that of Al/CuO MICs prepared by mechanical milling having a random distributed structure (reaction mechanism is shown in Figure 7.5c). Lahiner et al. [17] developed a 2D diffusion model for Al/CuO multilayer thin film that is used to simulate the self-propagating reaction occurring in condense phase and predict its ignition threshold (as illustrated in Figure 7.6). Two reaction steps are considered in the model: in the first step, CuO decomposes partly into Cu_2O and O atoms at relatively low temperatures, O atoms then react with Al through diffusion producing Al_2O_3. In the second step, the remaining CuO and Cu_2O proceed to decompose into Cu and O atoms as the temperature of thin film increases. The diffusion of O atoms is intensified under a high-temperature condition, O atoms are able to react with Al by traveling through the interfacial layers of Cu and Al_2O_3 formed during the reaction.

The reactivity of Al-based MICs is demonstrated to be heavily influenced by the microstructures, despite the reaction mechanisms of Al-based MICs are less affected. Egan et al. [42] characterized the reactivity of Al/CuO reactive nanolaminates fabricated by magnetron sputtering using T-Jump/TOFMS technique. Results show that the temperature of ignition is decreased by 46% from 1250 to 670 K and the reactivity of nanolaminates enhances as their bilayers increase from 2 to 6. When the bilayers are greater than 6, the temperature of ignition drops slightly and insignificant effect was observed on the reactivity of nanolaminates. It suggests that increasing the number of bilayers reduces the distances of diffusion effectively and increases the reacting surfaces that promotes the reaction to take place. Lee et al. studied the effect of number of alternating layers and their thickness on the reaction kinetics of energetic Al/CuO multilayers using thermal analysis method. The activation energies of multilayered Al/CuO nanolaminates with 2 and 6 bilayers are respectively 91 and 89 kJ mol^{-1} obtained by processing the resultant DSC traces using Kissinger method. It implies that the number of bilayers in this system affecting the reaction barriers is negligible. However, the calculated pre-exponents that are 3280 and 14500 s^{-1} are indicative of a higher collision frequency among molecules or atoms exhibited in Al/CuO multilayer films with six bilayers. The formation of activated molecules can be facilitated under thermal stimulation that results in the initial reaction temperature for this system lower than the temperature at which Al melts. Hence, the reactivity and ignition property of multilayered Al/CuO can be greatly improved by reducing the thickness of bilayers that leads to the decrease in the distance of mass/heat transport and the increase in reaction interfaces for reactants. DeLisio et al. [43] investigated the influence of the thickness of alternating layers on the condense-phase reaction mechanism of Al/CuO multilayer thin films by using nano-calorimetry system coupled with TOFMS technique. They found that a single exothermic reaction occurs in solid–solid state when the thickness of alternating layer of Al/CuO is 33 nm. Once the thickness of alternating layer is equal or greater than 66 nm, the reaction mechanism alters where the temperatures corresponding to two exothermic peaks are located before and after the temperature of 660 °C where the Al starts to melt respectively, suggesting the redox exothermic reaction of Al/CuO multilayers undergoes a transition process of solid–solid state to solid–liquid state.

7.2.2.3 Reactivity Control

The reactivity control of multilayer structured MICs can be realized by adjusting the thickness of reactant or adding dilution to the interfacial reaction layer. Multilayer structured Al/NiO demonstrated that its reactivity can be regulated by adding different content of dilute Cu [44]. Adding 40% Cu with a mass fraction of Al/NiO has a nearly two orders of magnitude decrease in the combustion rate (Figure 7.7a). The reason is that the added Cu layer hinders the direct contact between the reducing agent and oxidant, and increases the energy barrier of the reaction. In addition, the Cu layer has a negative effect on the mass/heat transfer of the reaction system, leading to reduced reactivity. Another effort to control the reactivity of multilayer structured MICs includes adjustment of the thickness of this metal layer [45]. For instance, the burning rate of multilayer structured Al/CuO can be decreased from 2.0 to 0.4 m s^{-1} by increasing the layer depth of Cu ranging from 0 to 100 nm (Figure 7.7b). Also, the combustion temperature is also reduced by 417 °C when increasing the layer depth of Cu.

In addition to the binary system of Al-based MICs, the preparation methods as mentioned are capable of fabricating the Al-based MICs multilayers involving

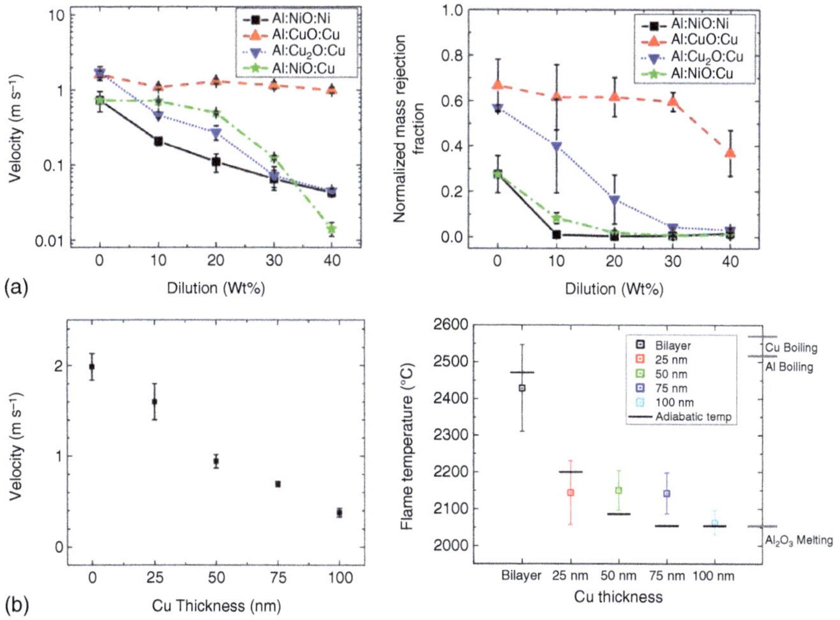

Figure 7.7 The reactivity control of multilayer structured MICs by adding dilution to the interfacial reaction layer (a) and adjusting the thickness of reactant (b). Source: Kinsey et al. [44]. Reproduced with permission of Springer Nature.

three constituents. The outstanding combustion behaviors are demonstrated by ternary system of Al-based MICs due to the great controllability in design of microstructures. Relevant studies report that the reactivity of multilayered MICs becomes more tunable by incorporation of an intermediate layer at the interfaces of metal fuel and oxidizer forming a ternary system. Marin et al. [46] deposited a 5 nm thin layer of Cu between bilayers of Al and CuO in nanolaminates of Al/CuO that considerably promotes the nanolaminates' reactivity because of the diffusion of Cu into Al forming Cu/Al alloy (as shown in Figure 7.8b). The formation of Cu/Al alloy, on the one hand, hinders the growth of inert Al_2O_3 layer; on the other hand, oxygen atoms from CuO can react with the Al in Cu/Al alloy as a result of the melting temperature of Cu/Al alloy is lower than that of Al, thereby increasing the propagation velocity (from 44 to 72 m s^{-1}) of nanolaminates of Al/CuO. Marin et al. also demonstrated that a thin ZnO layer deposited between the bilayers of CuO and Al assisted by ALD deposition would substantially increase the reaction efficiency of Al/CuO nanolaminates. The measured enthalpy of prepared ternary nanolaminates of CuO/ZnO/Al can approach their theoretical values by nearly 98%, with a one-step reaction at 900 °C; however, the traditional CuO/Al nanolaminates without ZnO generate only 78% of their theoretical value, contributed by two discrete reactions happened at 550 and 850 °C. Therefore, it can be clearly seen that the reactivity and combustion performances of Al-based MICs can be well controlled by rational design of the 3D layered structure and selection of components to be involved.

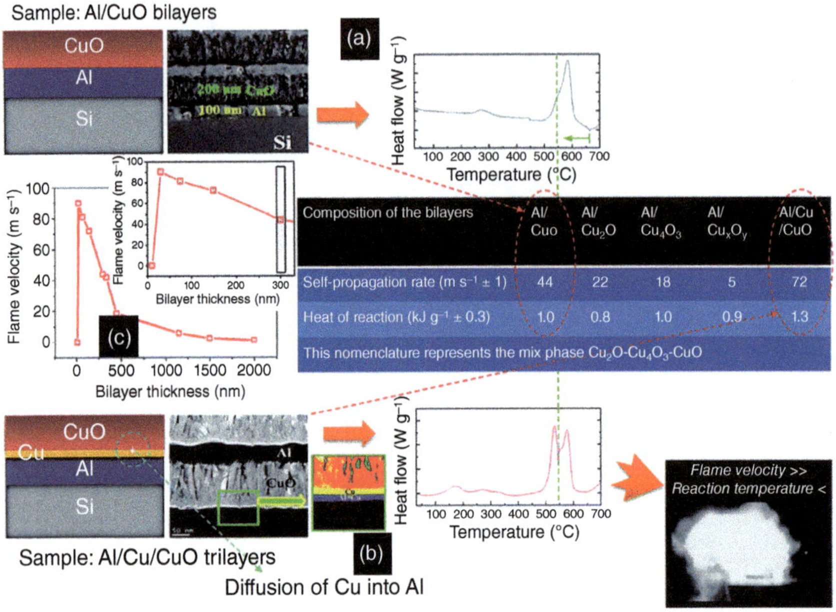

Figure 7.8 Schematic of the self-propagation reaction and the resulting DSC traces of (a) CuO/Al and (b) CuO/Cu/Al nanolaminates; (c) The trends showing the flame velocity evolves as increase of bilayer thickness for two various nanolaminates. Source: Marín et al. [46]. Reproduced with permission of American Chemical Society.

7.2.3 Al-Based MICs with Core–Shell Structures

7.2.3.1 Preparation Methods

The core–shell structure endows the reactants with a unique interfacial configuration compared to the microstructures possessed by the MICs system discussed above. This structure is capable of substantially increasing the reacting surface area and simultaneously the coating materials acting as a protective shell prevent the active Al core from being oxidized by air or solvents during the storage. As a result of the advantages described, researchers have focused on the development of core–shell structured Al-based MICs. The characteristic of prepared MICs with a core–shell structure is demonstrated to be tunable by adjusting the types of coating materials with the properties required.

Several material synthesis techniques, involving layers deposition, self-assembly, and polymerization in situ have been utilized to fabricate the MICs composites with core–shell structure. Metals are one of candidates used to uniformly distribute on Al particles' surfaces producing the core–shell structure with in situ synthesis followed crystal growth method. The obtained Al-based composites have large reaction interfaces and great stability in storage. Lee et al. characterized the thermal reactivity of the core–shell structured Al@Ni composites prepared by wet chemistry method growing the nanoscale nickel particles on Al. Experimental results indicate that a higher reactivity is appeared by the core–shell structured Al/Ni composites, by which the initial reaction temperature is reduced by 100 °C as compared to that of composites prepared by mechanical milling. Rong et al. prepared Al@Fe composites based on the decomposition characteristic of the precursor of iron that results in an even and dense Fe coating on Al surface. Characterization results show that the core–shell structured Al@Fe exhibit an enhanced capability to release heat; specifically, the reaction heat increases from 2680 $J g^{-1}$ for pure Al to 7528 $J g^{-1}$.

Metal oxides and fluoropolymers are studied more extensively as the coating materials in Al-based MICs with core–shell structures over metals. Liu et al. prepared Al@Fe_2O_3 composites by homogeneous precipitation method with a thick and dense iron oxide layer covering the Al surface [47]. The synthesized composites show good chemical and thermal stabilities, and changes in morphology and reactivity of Al@Fe_2O_3 composites can be realized by adjusting the compositions of core/shell materials. Wang et al. [48] utilized a typical fluoropolymer, polytetrafluoroethylene (PTFE), as the carrier of fluorine and the protective layer to create the MICs of Al@PTFE with core–shell structure. The nanostructured Al@PTFE MICs were fabricated using the combined method integrating the in situ deposition with the electrical explosion of wire (EEW) as depicted in Figure 7.9a. Results suggest that the microstructure of the PTFE shell and the reactivity of Al@PTFE composites can be easily tuned by regulating the composition of the precursors used in deposition process. The resultant composites exhibit a substantially promoted reaction heat as compared to that of the one prepared with physical mixing method (the pressurization rate increases from 0.0089 to 0.126 $MPa s^{-1}$ and the burning velocity increases from 0.98 to 3.85 $m s^{-1}$ as shown in Figure 7.9b). This synthetic approach provides a universal methodology to formulate different high-performance composite metal-based fuels with core–shell structures.

Recently, He and Yan et al. [49] proposed a novel strategy for design of a series of Al-based MICs with core–shell structures and explored the methods used for precisely controlling and tuning their combustion characteristics. In their work, the biomimetic materials such as polydopamine (PDA) serve as the interfacial reaction layer by taking advantage of the outstanding anti-oxidation behavior and ability to be chemically grafted for in situ synthesis of metal oxides. Firstly, the experimental methods controlling the thickness of PDA interfacial layer coating on the surface of Al are investigated. The resistance of Al@PDA particles to natural oxidation is found to increase exponentially as increasing the PDA layer's thickness, and the optimized PDA thickness with excellent adhesive properties is in the range of 3–6 nm (Figure 7.10a). Subsequently, multiple oxidizers are successfully constructed on the surface of Al@PDA particle, obtaining respectively the fluoropolymers coated MICs (e.g. Al@PDA@PTFE and Al@PDA@PVDF), metal oxides and energetic metal-organic framework materials coated MICs (Al@PDA@CuO and Al@PDA@EMOF), and binary oxidizer MICs system (Al@PVDF@PDA@EMOF). Taking Al@PDA@EMOF as an example,

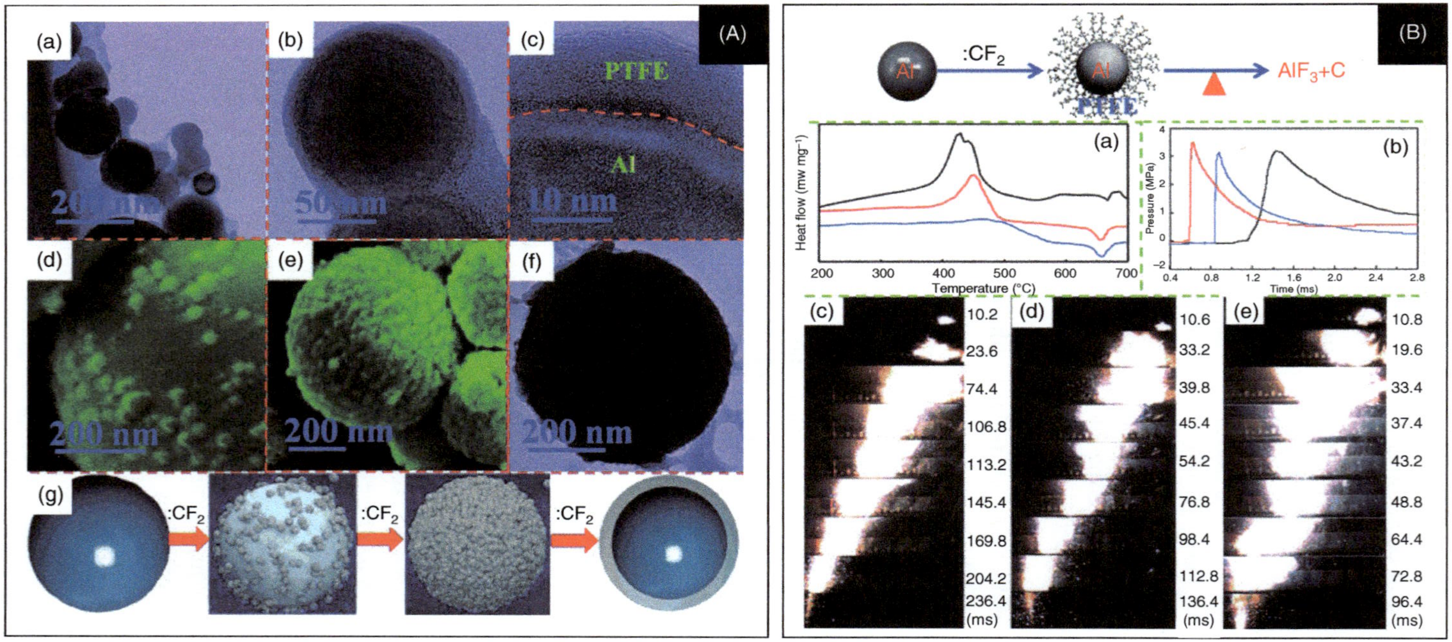

Figure 7.9 (A) TEM images taken on the morphology of core–shell structured Al@PTFE particles and schematic illustrating the formation of core–shell structure; (B) Comparisons in the thermophysical and combustion behaviors of Al@PTFE MICs with core–shell structure and that of physically mixed counterparts. Source: Wang et al. [48]. Reproduced with permission of John Wiley and Sons.

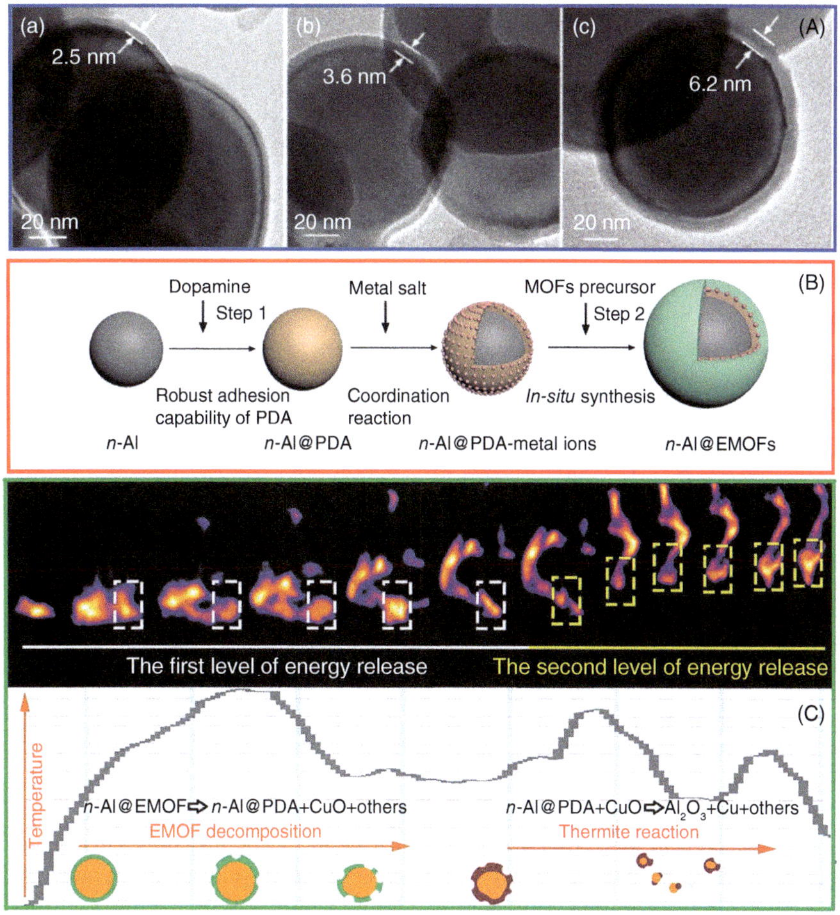

Figure 7.10 (A) TEM images of PDA interfacial layers with various thicknesses; Source: He et al. [10]. Reproduced with permission of American Chemical Society. (B) Schematic of synthesis process of core–shell structured Al@PDA@EMOF MICs; (C) Image of evolutions of flame structure and combustion temperature of Al@PDA@EMOF, and schematic of multilevel energy release mode. Source: He et al. [49]. Reproduced with permission of Elsevier.

the growth process of EMOF on Al surface modified by PDA is illustrated in Figure 7.10b.

7.2.3.2 Characterization

MICs possessing fine core–shell structures usually show intimately mixed components and improved stability. The intimately mixed components lead to a significant reduction in the distance of diffusion and enhancement in reacting interfaces between the involved reactants. For example, the core–shell structured MICs of Al@Fe_2O_3 fabricated by ALD method exhibit the reaction velocity several times higher than that of traditional MICs of Al/Fe_2O_3 prepared by physical mixing. The reactivity of Al@Fe_2O_3 MICs is capable to be further enhanced by the inclusion of carbon nanomaterials such as reduced graphene oxide (rGO) [50]. It has been shown that the core–shell structured Al@Fe_2O_3 MICs synthesized by ALD with 4.8 wt% of rGO addition is found to improve their energy release by 130% as compared to that of the mechanically mixed ones. Moreover, their electrostatic discharge sensitivities can be substantially decreased as the rGO content increases (e.g. up to 9.6 wt%).

Another work employs a different method, known as the opening/filling approach operated in two steps, which constructs the MnO_x@CNF-Al composites materials with core–shell structures by introducing MnO_x into herringbone CNF matrix [51]. This method has demonstrated to be an effective way to reduce the sensitivity of MICs by filling the carbon nanotubes (CNTs) with reactive components. It was observed that their friction sensitivity (threshold > 360 N) and electrostatic discharge sensitivity (from 1.0 to 5300 mJ) were considerably decreased, as compared to that of Al/MnO_2 MICs without addition of CNTs. The combustion velocity and pressurization rate of RDX@Fe_2O_3–Al MICs with a quasi-core–shell structure produced with a surface treatment method assisted by ultrasonication were greatly increased. Additionally, the sensitivity to friction, impact, and electric spark for RDX@Fe_2O_3/Al composite shows to significantly reduced as compared to that of traditional counterparts.

7.2.3.3 Reactivity Control

Core–shell structured Al-based MICs usually are structurally stable and their reactivity can be well controlled by changing the thickness of interfacial layers, using different oxides, or changing the reaction pathways. For instance, the initial reaction temperature, reaction kinetics, and energy output of interfacial layer modified MICs of Al@PDA@PTFE can be adjusted in the range of 352–462 °C, 164–377 kJ mol^{-1}, and 1660–3123 J g^{-1}, respectively, by varying the PDA layer thickness. Regarding the Al@PDA@CuO possessing high energy release characteristic, the reaction heat can be tuned in the range of 1734–2934 J g^{-1} through effectively controlling the thickness of CuO layer. The temperature at which the reaction of Al@PDA@CuO to be initiated is reduced by 140 °C in comparison with that of Al/CuO prepared by physical mixing. Al@PDA@EMOF exhibits a distinct multilevel energy release characteristic as depicted in Figure 7.9c with the generated heat reaching to 4100 J g^{-1}. The first energy release step is likely associated with the decomposition of EMOF to CuO. For binary oxidizer system, Al@PVDF@PDA@EMOF shows the highest reactivity due to the incorporation of PVDF layer that activates Al_2O_3 accompanied by a large amount of heat release [49].

The Al@PTFE MICs with core–shell structure created with an in situ, continuous preparation method present a tunable reactivity and improved energy release [47]. The exothermicity in the reaction of the mechanically mixed Al/PTFE composites and the ones with core–shell structures were systematically compared, and the mechanically blended Al/PTFE has a broad exothermic peak over 450 and 550 °C on the trace of DSC, accompanied with a small exotherm of 470 J g^{-1} heat release.

Table 7.1 Summary of progress in studies on preparation, reaction mechanism and kinetics, and reaction model of Al-based MICs with different microstructures.

Al-based MICs	Preparation methods	Micro structures and reactivity	Reaction mechanisms	Reaction kinetics	Combustion model	References
Mechanically mixed	Solvent mixing, mechanical milling, and sol–gel methods	Randomly distributed Incomplete reaction	Multistep reaction: $4CuO \rightarrow 2Cu_2O + O_2$; $4Al + 3O_2 \rightarrow 2Al_2O_3$; $2Al + 3Cu_2O \rightarrow Al_2O_3 + 6Cu$	E_a values change from 70 to 280 kJ mol^{-1} obtained for different reaction stages	Heterogeneous reaction model	[23–29]
Multilayered	Vapor deposition, atomic layer deposition (ALD), electrospray	3D ordered layered Controllable reactivity	Reaction mechanisms are comparable to that of mechanically mixed MICs	E_a values change from 50 to 100 kJ mol^{-1}	2D diffusion, unstable reaction model	[9–12, 45]
Core-shell structured	In situ synthesis, polymerization, vapor deposition, electrospray, self-Assembly	Multilayer core–shell structure Controllable reactivity	Condense-phase reaction mechanism is not well understood	E_a is only determined for Al@PDA@PTFE system (164–377 kJ mol^{-1})	None	[2, 3]

Source: Hongqi Nie.

For the Al/PTFE MICs with core–shell structures where the thickness of PTFE layer around 10 nm, the temperature of which the reaction to be initiated is found to be 410 °C associated with a stronger heat release of $1810\,J\,g^{-1}$. By adjusting the thickness of PTFE shell, the initial reaction temperatures and energy releases of MICs composites can be further altered.

7.3 Conclusion and Suggestions

The reactivity involving the ignition and combustion behaviors of Al-based MICs is able to be considerably improved and well controlled by the delicate design of microstructures fabricated with advanced preparation methods. Studies on the mechanically mixed and multilayer thin films of Al-based MICs are carried out extensively in decade, the reaction mechanism and corresponding combustion models are established for Al-based MICs, especially with randomly distributed structures and 3D ordered layered structures. Regarding the core–shell structured Al-based MICs developed recently, the relevant works are mainly correlated to the description on experimental phenomena and exploring the scientific rules. Due to multiple reaction interfaces existing in the core–shell structured MICs that directly cause the interactions between reactants to be complicated, the reaction mechanism and kinetics of core–shell structured Al-based MICs with interfacial layers involved are not well known (as summarized in Table 7.1). Considering the coupling effect in the multilevel thermochemical reaction occurring exclusively for the core–shell structured Al-based MICs has not been fully understood, the reaction interface locations and their interactions, the condensed-phase reaction kinetics, and corresponding combustion mechanisms of core–shell structured Al-based MICs need to be systematically investigated. In particular, the effect of reaction interface on the condensed-phase reaction progress is required to be clarified by quantitatively analyzing the physiochemical parameters of reaction interfaces. In addition, the mechanism controlling and promoting the ignition and combustion performances for the Al-based MICs with core–shell structures should be deeply understood, which will be an important theoretical guidance for the design and preparation of the novel Al-based MICs showing superior properties with great potential in engineering applications.

References

1 Sutton, G.P. and Biblarz, O. (2001). *Rocket Propulsion Elements*. Hoboken, NJ: Wiley.
2 Price, E.W. (1984). Fundamentals of solid propellant combustion. In: *Progress in Astronautics and Aeronautics*, vol. 185 (ed. M. Summerfield), 479–513. AIAA.
3 Price, E.W. and Sigman, R.K. (2000). Solid propellant chemistry. In: *Combustion, and Motor Interior Ballistics* (ed. P. Zarchan), 663–687. AIAA.

4 Maggi, F., Bandera, R., Galfetti, R. et al. (2010). Efficient solid rocket propulsion for access to space. *Acta Astronautica* 66 (11-12): 1563–1573.

5 Dreizin, E.L. (2009). Metal-based reactive nanomaterials. *Progress in Energy and Combustion Science* 35 (2): 141–167.

6 Sundaram, D., Yang, V., and Yetter, R.A. (2017). Metal-based nanoenergetic materials: synthesis, properties, and applications. *Progress in Energy and Combustion Science* 61: 293–365.

7 Mench, M.M., Kuo, K.K., Yeh, C.L. et al. (1998). Comparison of thermal behavior of regular and ultra-fine aluminum powders(Alex) made from plasma explosion process. *Combustion Science and Technology* 135 (1): 269–292.

8 Brousseau, P. and Anderson, C.J. (2002). Nanometric aluminum in explosives. *Propellants, Explosives, Pyrotechnics* 27 (5): 300–306.

9 He, W., Liu, P.J., He, G.Q. et al. (2018). Highly reactive metastable intermixed composites (MICs): preparation and characterization. *Advanced Materials* 30 (41): e1706293.

10 He, W., Liu, P.J., Gong, F.Y. et al. (2018). Tuning the reactivity of metastable intermixed composite n-Al/PTFE by polydopamine interfacial control. *ACS Applied Materials & Interfaces* 10 (38): 32849–32858.

11 He, W., Li, Z.H., Chen, S. et al. (2020). Energetic metastable n-Al@PVDF/EMOF composite nanofibers with improved combustion performances. *Chemical Engineering Journal* 383: 123146.

12 Li, X., Guerieri, P., Zhou, W. et al. (2015). Direct deposit laminate nanocomposites with enhanced propellent properties. *ACS Applied Materials & Interfaces* 7 (17): 9103–9109.

13 Zhou, X., Torabi, M., Lu, J. et al. (2014). Nanostructured energetic composites: synthesis, ignition/combustion modeling, and applications. *ACS Applied Materials & Interfaces* 6 (5): 3058–3074.

14 Blobaum, K.J., Wagner, A.J., Plitzko, J.M. et al. (2003). Investigating the reaction path and growth kinetics in CuO_x/Al multilayer foils. *Journal of Applied Physics* 94 (5): 2923.

15 Kwon, J., Ducere, J.M., Alphonse, P. et al. (2013). Interfacial chemistry in Al/CuO reactive nanomaterial and its role in exothermic reaction. *ACS Applied Materials & Interfaces* 5: 605–613.

16 Nicollet, A., Lahiner, G., Belisario, A. et al. (2017). Investigation of Al/CuO multilayered thermite ignition. *Journal of Applied Physics* 121 (3): 034503.

17 Lahiner, G., Nicollet, A., Zapata, J. et al. (2017). A diffusion-reaction scheme for modeling ignition and self-propagating reactions in Al/CuO multilayered thin films. *Journal of Applied Physics* 122 (15): 155105.

18 Crouse, C.A., Christian, J.P., and Jonathan, E.S. (2013). Synthesis and reactivity of aluminized fluorinated acrylic (AlFA) nanocomposites. *Scientific Reports* 3 (12): 3199–3207.

19 He, W., Tao, B., Yang, Z. et al. (2019). Mussel-inspired polydopamine-directed crystal growth of core-shell n-Al@PDA@CuO metastable intermixed composites. *Chemical Engineering Journal* 369: 1093–1101.

20 Tang, D., Chen, S., Liu, X. et al. (2019). Controlled reactivity of metastable n-Al@Bi $(IO_3)_3$ by employment of tea polyphenols as an interfacial layer. *Chemical Engineering Journal* 381: 122747.
21 Patel, V.K. and Bhattacharya, S. (2013). High-performance nanothermite composites based on aloe-vera-directed CuO nanorods. *ACS Applied Materials & Interfaces* 5 (24): 13364–13374.
22 Yan, Q.L., Gozin, M., Zhao, F.Q. et al. (2016). High energetic compositions based on functionalized carbon nanomaterials. *Nanoscale* 8: 4799–4851.
23 Zhang, Y., Yan, Y., Wang, Y. et al. (2018). Enhanced energetic performances based on integration with the Al/PTFE nanolaminates. *Nanoscale Research Letters* 13: 206.
24 Sanders, V.E., Asay, B., Foley, T.J. et al. (2007). Reaction propagation of four nanoscale energetic composites (Al/MoO_3, Al/WO_3, Al/CuO, and Bi_2O_3). *Journal of Propulsion and Power* 23 (4): 707–714.
25 Shen, L., Li, G., Luo, Y. et al. (2014). Preparation and characterization of Al/B/Fe_2O_3 nanothermites. *Science China Chemistry* 57 (006): 797–802.
26 Mursalat, M., Huang, C., Julien, B. et al. (2021). Low-temperature exothermic reactions in Al/CuO nanothermites producing copper nanodots and accelerating combustion. *ACS Applied Nano Materials* 4 (4): 3811–3820.
27 Piekiel, N.W., Zhou, L., Sullivan, K.T. et al. (2012). Initiation and reaction in Al/Bi_2O_3 nanothermites: evidence for the predominance of condensed phase chemistry. *Combustion Science and Technology* 186 (9): 1209–1224.
28 Egan, G.C., Sullivan, K.T., Olson, T.Y. et al. (2016). Ignition and combustion characteristics of nanoaluminum with copper oxide nanoparticles of differing oxidation state. *Journal of Physical Chemistry C* 120 (5): 29023–29029.
29 Jacob, R.J., Jian, G., Guerieri, P.M. et al. (2015). Energy release pathways in nanothermites follow through the condensed state. *Combustion and Flame* 162 (1): 258–264.
30 Umbrajkar, S.M., Schoenitz, M., and Dreizin, E.L. (2006). Exothermic reactions in Al-CuO nanocomposites. *Thermochimica Acta* 451: 34–43.
31 Ermoline, A., Schoenitz, M., and Dreizin, E.L. (2012). Low-temperature exothermic reactions in fully dense Al-CuO nanocomposite powders. *Thermochimica Acta* 527: 52–58.
32 Mccollum, J., Pantoya, M.L., and Iacono, S.T. (2015). Activating aluminum reactivity with fluoropolymer coatings for improved energetic composite combustion. *ACS Applied Materials & Interfaces* 7 (33): 18742–18749.
33 Osborne, D.T. and Pantoya, M.L. (2007). Effects of Al particle size on the thermal degradation of Al/Teflon mixtures. *Combustion Science and Technology* 179 (8): 1467–1480.
34 DeLisio, J.B., Hu, X., Wu, T. et al. (2016). Probing the reaction mechanism of aluminum/fluoropolymer composites. *Journal of Physical Chemistry B* 120 (24): 5534–5542.
35 Sippel, T.R., Son, S.F., Groven, L.J. et al. (2014). Aluminum agglomeration reduction in a composite propellant using tailored Al/PTFE particles. *Combustion amd Flame* 161 (1): 311–321.

36 Nie, H., Pisharath, S., and Hng, H.H. (2020). Combustion of fluoropolymer coated Al and Al-Mg alloy powders. *Combustion and Flame* 220: 394–406.

37 Jian, G., Chowdhury, S., Sullivan, K., and Zachariah, M.R. (2013). Nanothermite reactions: is gas phase oxygen generation from the oxygen carrier an essential prerequisite to ignition. *Combustion and Flame* 160 (2): 432–437.

38 Weismiller, M.R., Lee, J.G., and Yetter, R.A. (2011). Temperature measurements of Al containing nano-thermite reactions using multi-wavelength pyrometry. *Proceedings of the Combustion Institute* 33 (2): 1933–1940.

39 Ke, X., Guo, S., Gou, B. et al. (2019). Superhydrophobic fluorine-containing protective coating to endow Al nanoparticles with long-term storage stability and self-activation reaction capability. *Advanced Materials Interfaces* 6 (19): 1901025.

40 Azadmanjiri, J., Berndt, C.C., Wang, J. et al. (2016). Nanolaminated composite materials: structure, interface role and applications. *RSC Advances* 6 (111): 109361.

41 Abdallah, I., Zapata, J., Lahiner, G. et al. (2018). Structure and chemical characterization at the atomic level of reactions in Al/CuO multilayers. *ACS Applied Energy & Interfaces* 1 (4): 1762–1770.

42 Egan, G.C., Mily, E.J., Maria, J.P. et al. (2015). Probing the reaction dynamics of thermite nanolaminates. *The Journal of Physical Chemistry C* 119 (35): 20401–20408.

43 DeLisio, J.B., Yi, F., LaVan, D.A. et al. (2017). High heating rate reaction dynamics of Al/CuO nanolaminates by nanocalorimetry-coupled time-of-flight mass spectrometry. *The Journal of Physical Chemistry C* 121 (5): 2771–2777.

44 Kinsey, A.H., Slusarski, K., Woll, K. et al. (2016). Effect of dilution on reaction properties and bonds formed using mechanically processed dilute thermite foils. *Journal of Materials Science* 51 (12): 1–12.

45 Blobaum, K.J., Reiss, M.E., Plitzko, J.M., and Weihs, T. (2003). Deposition and characterization of a self-propagating CuO_x/Al thermite reaction in a multilayer foil geometry. *Journal of Applied Physics* 94 (5): 2915–2922.

46 Marín, L., Nanayakkara, C.E., Veyan, J.F. et al. (2015). Enhancing the reactivity of Al/CuO nanolamninates by Cu incorporation at the interfaces. *ACS Applied Materials & Interfaces* 7 (22): 11713–11718.

47 Liu, H.C., Zhang, J.D., Gou, J.Y. et al. (2017). Preparation of Fe_2O_3/Al composite powders by homogeneous precipitation method. *Advanced Powder Technology* 28 (12): 3241–3246.

48 Wang, J., Qiao, Z., Yang, Y. et al. (2016). Core-shell Al-polytetrafluoroethylene (PTFE) configurations to enhance reaction kinetics and energy performance for nanoenergetic materials. *Chemistry* (Weinheim an der Bergstrasse, Germany) https://doi.org/10.1002/chem.201503850.

49 He, W., Ao, W., Yang, G. et al. (2020). Metastable energetic nanocomposites of MOF-activated aluminum featured with multi-level energy releases. *Chemical Engineering Journal* https://doi.org/10.1016/j.cej.2019.122623.

50 Yan, N., Zhou, X., Li, Y. et al. (2013). Fe_2O_3 nanoparticles wrapped in multi-walled carbon nanotubes with enhanced lithium storage capability. *Scientific Reports* 3 (12): 3392.

51 Siegert, B., Comet, M., Muller, O. et al. (2010). Reduced-sensitivity nanothermites containing manganese oxide filled carbon nanofibers. *The Journal of Physical Chemistry C* 114 (46): 19562–19568.

8

Nanothermites: Developments and Future Perspectives

Ahmed Fahd[1,2], Charles Dubois[1], Jamal Chaouki[1], and John Z. Wen[3]

[1] *Polytechnique de Montréal, Chemical Engineering Department, Montréal, H3C 3A7, Canada*
[2] *Technical Research Center, Rockets and Energetic Materials Department, Kobry Elkobbah, Cairo, 11766, Egypt*
[3] *University of Waterloo, Mechanical and Mechatronics Engineering Department, Ontario, N2L 3G1, Canada*

8.1 Introduction

Demand for improved energetic materials (EMs) is increasing as propulsion technologies strive to be smaller, more efficient, and safer to use. Fields of interest in novel EMs include civilian applications such as lab-on-a-chip devices, material synthesis, micro-, and nano-welding as well as defense applications such as micro-thrusters, smart projectiles, modified double base propellant (MDP) [1–4] and pyrotechnical micro electro mechanical systems (PyroMEMS) like pyroswitches or circuit breakers in electrical systems, actuators, and safe-and-arm devices [5, 6]. Consequently, nanothermites are key materials in the development of new Ems, to meet the increasing needs of microscale energetic devices and miniature high energy density propulsion systems [7].

In broad terms, EMs can be defined as materials that store chemical energy and are classified into propellants, explosives, and pyrotechnics. The primary difference between propellants, pyrotechnics, and explosives is the rate of energy release. While propellants and pyrotechnics are characterized by a relatively slow combustion process (several seconds), explosives release their energy through a fast detonation process (microsecond timescale) [8]. Another way of classifying EMs is through their production process, such as monomolecular energetic materials (MEMs), composite and metastable intermolecular composite (MIC) materials. On the one hand, monomolecular EMs have the fuel and oxidizer components in the same molecular species (for example, nitrocellulose, nitroglycerine, and trinitrotoluene), while on the other hand, MICs are produced from mixing fuel (aluminum, sulfur, carbon, etc.) and oxidizers (iron oxide, copper oxide, potassium perchlorate, etc.), like in the case of Al/Fe_2O_3 or Al/CuO mixtures [9].

Thermites are classified as a type of MICs. Compared to MEMs, thermites offer high energy density owing to the higher enthalpies of their metallic fuels [10]. Nanothermites are considered a type of novel MICs that intentionally use nanoscale

Nano and Micro-Scale Energetic Materials: Propellants and Explosives, First Edition.
Edited by Weiqiang Pang and Luigi T. DeLuca.

metallic fuels and oxidizers. They are mixtures of metals and metallic oxides in which at least one-nanometer scale component. Aluminum nanopowder (n-Al) is a key fuel material in nanothermites due to its high heat of formation, high density, low melting temperature, low toxicity, and compatibility with other reactants. More novel nanothermite formulations use combinations where atoms with high electronegativity (O, S, F, etc.) are exchanged between an oxidizer and a fuel, such as in the case of $Al/KClO_4$ or Al/KIO_4 [11–15].

To develop new nanothermite compositions, it is essential to have an in-depth understanding of nanothermite materials, for anticipating the possible effect of the new components on the thermite reaction. It is also required to have a grasp of the reaction mechanism and how those new components can positively enhance the properties of nanothermites to better suit novel applications. A summary of recent findings for newly introduced nanothermite mixtures is presented and discussed in this chapter. In addition, the impact of carbon nanomaterials (CNMs) on the combustion behavior of ternary nanothermites is investigated and thrust performance in millimeter-scale test motors (STMs) is also assessed. Finally, the challenges that still persist in developing the nanothermite composites from both fundamental and application points of view are listed and discussed.

8.2 Nanothermites Versus Microthermites

There are merits and demerits associated with both micron- and nanothermites. The slow rate of energy release and relatively long ignition delays are the main drawbacks of using thermites [16, 17]. Moreover, the high electrostatic sensitivity, in addition to the relatively elevated ignition temperature (~800 °C), are so far considered problematic issues associated with thermites [18–20]. Over and above, less intense maximum pressure and depleted metal content due to early oxidation of Aluminium prior to combustion represent potential problems linked with thermites [21, 22]. Accordingly, the available trust from microthermites is low because of the absence of CHON atoms in the thermite formulation and the oxidation of the Al metal [23, 24]. In another direction, the practical importance of greener thermites (i.e. thermites with low solid residuals and high gaseous products) increases due to the inherent drawbacks associated with traditional thermite compositions. Environmental pollution issues such as very high toxicity and hydrolytic break down are often associated with thermites based on sodium azide and ammonium perchlorate [19, 25]. As a direct result of those dangerous materials on the environment and human health, their use in a wide breath of applications should be limited or controlled [26, 27]. In comparison with their micrometer size counterparts, nanostructured thermites exhibit faster rate of reaction, more intense release of heat, and better trust, owing to characteristics such as low energy of activation, a larger specific area, and a more favorable energy to volume ratio [28, 29]. In addition, the ignition of classical EMs, as opposed to nanothermites, is often slower and their reaction does not spread easily in small channels and tubes [30].

Because of the high cost of Al nanoparticles, which is 30–50 times more expensive than micron-scale particles and possess safety and environmental issues, there are recent reports that suggest that some pretreatment process can improve the reactivity and total performance of microthermite [11, 31–35]. Melt-dispersion mechanism (MDM) is considered the most suitable mechanism to illustrate the combustion behavior of nanothermite at faster heating rates than the traditional diffusion mechanism [32, 36–38]. MDM attributed the high reactivity of nanothermites to the high pressure generated in the molten Al core (1 to 3 GPa), ultimate strength, and tensile hoop stress in the Al oxide shell (>10 GPa) due to the higher volumetric expansion strain accompanied the n-Al melting. Thus, state of art in enhancing the reactivity of microthermite is thermal pretreating of micro-Al (m-Al) to produce desired compressive stress in the alumina shell.

Levitas et al. examined the reactivity of microthermite in terms of flame propagation rate through preheating m-Al fuel (3.5–4 μm) to temperature in the range of 383–473 K. Results showed that 473 K is the best preheating temperature, which yields 36% increase in the flame propagation rate compared to microthermite without heat-treatment and reaches to 68% from the flame propagation rate recorded by corresponding nanothermite [11].

In addition to some qualitative confirmations [11, 31–35], one of the main quantitative studies of the improved performance of preheating of m-Al is related to the results published in [39]. Here, the authors presented a proof of concept for the design of prestressed Al nano- and micron-size particles for energetic applications and their results confirmed the usefulness of MDM as well. The idea of producing prestressed particles to boost particle reactivity and flame speed was the emphasis of their article. In a semi-confined tube, they investigated the flame speed of preheated micro and n-Al with particle sizes of 3.5–4 μm and 80 nm, respectively, mixed with MoO_3 thermites. Prestressing was achieved by heating particles to various high temperatures (25–170 °C), holding them there for 10 minutes to relieve thermal stresses, and then cooling them at various speeds (0.06–0.13 °C s^{-1}) to room temperature. Flame propagation speed increased by 31% for nanoparticles and 41% for micron particles under ideal thermal treatment conditions (heating to 105 °C and cooling at 0.13 °C s^{-1}). After heating to 105 °C, cooling at 0.06 and 0.13 °C s^{-1} did not modify the flame speed or raised it much less.

However, Dikici et al. in [39] demonstrated that for both n-Al and m-Al, combustion performance was dependent on annealing temperature (105 and 170 °C) and quench rate (0.06 and 0.13 °C s^{-1}), there were no established and well-understood direct and reliable relationships between particle size, annealing temperature, and cooling rate. As a result, McCollum et al. subjected Al to a variety of thermal treatments (100, 200, and 300 °C) under the same quench rate to gain a better knowledge of the impacts of annealing Al particles separately from annealing composites [40]. After thermal treatment, the reactivity of Al was determined by mixing it with CuO. Although results showed no change in energy propagation rate for preheated samples at 100 and 200 °C compared to untreated ones, the propagation rate increased from 95.2 to 118 m s^{-1} at 300 °C, which indicated an improvement of over 24% in comparison to the reference sample. Furthermore, synchrotron X-ray diffraction (XRD) validated the earlier findings by evaluating the strain response of annealed m-Al, which was unchanged at 100 and 200 °C but increased by 660% at 300 °C as compared to untreated particles.

Others investigated the mechanical characteristics of Al particles after annealing and quenching [30, 41]. They agreed in [30, 41] that any changes occurring on the prestressed unit particle into a consolidating bulk material affect the mechanical characteristics of the resulting bulk material. The optimal preheating temperature that yields mechanical change is 200 °C for both studies, and this temperature is comparatively lower than the temperature (300 °C) concluded by previously mentioned studies and can be correlated to the high purity of the aluminum particles used in [30, 41].

In terms of combustion performance, all of the previous findings show that the reactivity of m-Al particles increased as the annealing temperature increased. Because prior quenching investigations were conducted on composites in which reactive sintering may have obscured other behaviors, the relevance of quench rate in m-Al reactivity remains unknown. Thus, McCollum and his colleagues explored the effect of quenching rate on increasing or decreasing reactivity of m-Al and intra-particle dilatational strain, as assessed by flame speed and synchrotron XRD analysis [42]. They preheated m-Al with an average diameter of 5 μm to different temperatures (100, 200, and 300 °C) at a heating rate of $10\,K\,min^{-1}$ and held for 15 minutes. The postheating m-Al was conducted to slow and fast cooling regimes with cooling rates of $0.13–0.38\,K\,s^{-1}$ and 0.007 –0.014 for fast and slow regimes. No remarkable change in flame propagation occurred until annealing temperature reached 300 °C, which came with an agreement with the previous results. These findings reveal that annealing temperature has a considerable influence on the mechanical characteristics and combustion performance of Al particles, but the cooling rate has little effect. Variations in strain from 0.15×10^{-5} for untreated Al powder to $9.\ 23 \times 10^{-5}$ for samples annealed at 300 °C have a different effect on flame speeds: when annealed to 300 °C, the flame speed increased by 22%. The governing mechanism is yet unknown, but the findings match previous research that there is a link between increased dilatational strain, increased grain size, decreased hardness, and higher flame speed in Al particles [42].

Kevin J. Hill investigated the influence of faster quenching rates on the impact ignition sensitivity and combustion properties of prestressing m-Al. A Q800 Dynamic Mechanical Analyzer (DMA) was used to anneal the prestressed m-Al powder (3–4.5 μm) in a controlled temperature environment. Heating is done in an air environment at a rate of $10\,K\,min^{-1}$ to 573 K (300 °C), held for 15 minutes then quenched at different rates (200 and $900\,K\,min^{-1}$). To investigate the impact sensitivity and combustion properties of the annealed Al particles, prestressed -Al was combined with bismuth oxide (Bi_2O_3) and tested in low-velocity, drop-weight impact circumstances, and a pressure cell. Impact ignition sensitivity was increased at both quenching rates (i.e. between 83% and 89% decrease in ignition energy). However, when compared to untreated particles, the slower quenching rate showed a 100% rise in pressurization rate, while the faster quenching rate showed a

97% increase in peak pressure, demonstrating that the two quenching rates have different effects on Al particles [43].

Rohit J. Jacob recently employed a similar prestressing method to improve the reactivity of n-Al. They investigated the combustion properties of prestressed n-Al particles that had been annealed at temperatures ranging from 200 to 400 °C and quenched at both slow (exponential) and fast (linear) cooling rates. Prestressing n-Al particles at 300 °C raise strain by an order of magnitude, according to XRD observations. When n-Al coupled with n-CuO was tested in a constant volume combustion cell, prestressed n-Al/CuO composites had greater peak pressures and pressurization rates than their untreated counterparts. These findings show that manipulating fuel particles mechanically could provide fascinating potential for tailoring energy release characteristics in nanothermites [44].

Despite the advantages of nanothermites as compared to microthermites, some of the latter remain afflicted by undesirable characteristics, namely: variable reaction rate due to particles agglomeration, significant ignition deadtimes, and for some instances, unfavorable dependence of the burning rate on temperature, and partial or interrupted combustion [11, 16, 29, 45, 46]. Hence, improvements in nanothermites are required to meet the energetic systems demands as this will be discussed later.

8.3 Nanothermite-friendly Oxidizers

Metallic oxidizer and, conventionally n-Al fuel are the main components of nanothermites. Earlier research on nanothermite field focused on metallic oxides such as iron oxide (Fe_2O_3, FeO), copper oxide (CuO), molybdenum trioxide (MoO_3), tungsten oxide (WO_3), silver oxide (Ag_2O), and bismuth oxide (Bi_2O_3). Recently, new nanothermite compositions aimed at increasing the evolution of gas to simulate the behavior of convention CHON-based EMs. Oxygenated salts (oxygenated oxidisers) could represent an alternative to metallic oxidizers for preparing nanothermites with enhanced combustion characteristics and improved reactivity [47, 48]. The main oxidation–reduction reaction of nanothermite occurs between the metal fuel and oxygen generated by the decomposition of oxygenated salts or metal oxides, which are reduced to original metal or metal salt. The choice of oxygenated salts over metal oxides comes from their higher atomic oxygen content, which results in superior burn rates and gas generation capabilities [13, 49]. In addition, the lower dissociation energies of nonmetal–oxygen bonds (I–O, Cl–O, etc.) in oxygenated salts offer higher oxygen mobility that can lead to decomposition and oxygen release at lower temperatures when compared to metal–oxygen bonds (Cu–O, Fe–O, Bi–O, etc.) [50, 51]. Oxygenated salts produce intermediate radicals (e.g. hydroxyl radical, $OH^{\cdot}$) upon decomposition, which in turn can produce other subsequent radicals and oxidants by initiating a chain of degradation reactions. As such, oxygenated salts have higher oxidation ability than metallic oxides that undergo an electron capture-based oxidation mechanism [52, 53].

Oxidizing agents for nanothermite preparation are usually oxygen-rich ionic solids, which at elevated temperatures decompose and release their oxygen. There are some considerations in selecting the employed oxidizers such as being in suitable pure form, availability of different particle sizes, and cost. Moreover, they should be stable over a broad range of temperatures, not react with humidity, and easily decompose to liberate their oxygen [54]. Although metal sulfide and fluoride can theoretically be employed as oxidizers, they have not yet been utilized in the form of nanostructured materials in nanothermites preparation.

8.3.1 Metallic Oxidizers

When we consider inorganic chemistry, we find a wide variety of metal oxides. However, can any metal oxide be properly used in the preparation of nanothermite composition? The direct answer is no because highly reducing metals such as alkaline or alkaline earth elements can easily form oxides. These oxides are extremely stable in normal conditions and consequently, they do not apply in the preparation of nanothermites. The electronegativity of the metals classifies those suitable for nanothermite formulations. Therefore, metals with electronegativity between 1.5 and 2.5 on the Pauling scale are the best choices for nanothermite mixtures. Fischer and his group arranged thermites and energetic compositions based on their thermochemical characteristics [55].

As we mentioned before, the most commonly used metallic oxidizers for nanothermite compositions are Fe_2O_3, CuO, Bi_2O_3, MoO_3, and WO_3. Here, we will highlight general properties of some nanothermite mixtures based on these oxides.

Weismiller et al. investigated the influence of particle dimensions of nanothermites constituents (Al/CuO and Al/MoO_3) on their performance. They concluded that nanothermites containing both fuel and oxidizer in the nanostructured form offered high combustion rate. Moreover, the effect of nanostructured metallic oxidant is more significant on the combustion performance than n-Al. In the latter, the benefits of a higher specific area are outweighed by the presence of a fraction of Al_2O_3 covering the n-Al particles surface, yielding incomplete combustion. On the other hand, nanostructured oxidizers lead to the liberation of oxygen easily through sublimation in MoO_3 or decomposition in CuO. In addition to the easy oxygen generation, the diffusion distance of gaseous products will decrease and, hence, enhance the nanothermite reactivity [56].

Puszynski summarized and reviewed the combustion characteristics of nanothermite compositions based on different metallic oxidizers such as Bi_2O_3, CuO, MoO_3, and WO_3 and various n-Al particles. He found that the highest combustion rate of nanothermite compositions were achieved with n-Al particle size in the range of 50–80 nm [57].

Furthermore, Puszynski et al. examined the combustion characteristics of nanothermites prepared from n-Al powders (40 nm) and metallic oxidizers (Bi_2O_3, CuO, and MoO_3) with analogous particle size using pressure cell and differential scanning calorimetry (DSC). Direct results from pressure cell analysis demonstrated that Al/Bi_2O_3 sustains higher dynamic pressure than Al/MoO_3 and Al/CuO respectively

[32]. Recently, Glavier and his research group looked at the maximum pressure and the pressurization rate for the same nanothermite compositions studied previously by Puszynski et al. In their investigation, they used pressure cell analysis and compressed nanothermite pellets at different percentages (10%, 30%, and 50%) of their theoretical maximum density (TMD). Nanothermites based on Al/CuO and Al/Bi_2O_3 compressed at 30% and 50% exhibited the maximum pressure and pressurization rate respectively [58].

Sanders et al. reported that burning rate of metallic oxide was enhanced greatly in closed environments. They found that combustion of loose powder Al/Bi_2O_3 and Al/MoO_3 is more rapid than pellets [59]. Trebs et al. studied the possible effect of humidity on the pressure output of aluminothermic compositions reported by Sanders et al. They concluded that water content changed pressure signals produced by these nanothermites significantly due to their acute sensitivity to humidity [60].

Jian et al. examined the ignition behavior of metallic oxide nanothermite mixtures and classified them into three groups. First, compositions that ignite before the release of oxygen from oxidizers such as Al/Bi_2O_3 and Al/SnO_2. Second, mixtures that react after the liberation of oxygen such as composition based on Al and Co_3O_4. Third, nanothermites that ignite without release of oxygen for example Al/WO_3 and Al/MoO_3. The reason behind the ignition of compositions in the first and the third case can be reasonably related to the direct contact between the fuel and oxidizer [45].

8.3.2 Oxidizing Salts

Goldschmidt, a pioneer in the field of thermites, initially used metal oxides and Al as a fuel [61]. Research in the field of EMs is now directed at oxygenated salts as oxidizers for nanothermite compositions to enhance their performance. EMs containing oxidizing salts are employed for a wide breadth of full scale propulsion applications as well as for microthrusters and similar microsystems. The booster rockets of the ESA Ariane 5 launcher are propelled by a solid propellant containing aluminum particles as fuel and ammonium perchlorate (NH_4ClO_4) as an oxidizer. Though developed over a century ago, ammonal is still employed as an explosive with little sensitivity to priming [62]. There are limited investigations in the field of nanothermites based on oxidizing salts, especially with oxygenated salts, so further examination is warranted.

Oxidizing salts are more useful than metallic oxides due to their significant content of oxygen, ease of decomposition, readily release of oxygen, and low oxidation state, which increases their reactivity [13]. The most used oxygenated oxidizers in nanothermites formulations are chlorates, perchlorates, nitrates, iodates, periodates, bromates, permanganates, peroxides, superoxides, and persulfates [63].

Armstrong was one of the first to study the linear burning rate (LBR) of aluminothermic formulations prepared from ammonium perchlorate and nanosized aluminum using pressure cell analysis. His findings revealed that when the nominal average diameter of aluminum particles is reduced from 200 to 40 nm, the

LBR passed from 25 mm s^{-1} under a 0.14 MPa pressure to a much faster value of 600 mm s^{-1} at 15 MPa [64].

Potassium permanganate ($KMnO_4$) is one of the best oxidizing agents for pyrotechnic compositions. Because of its high oxidizing power and low toxicity, it can be reliably used in the development of nanothermites with superior performance without significant environmental harm. Prakash et al. explored the pressurization rate of n-Al/$KMnO_4$ nanothermite using pressure cell, demonstrating that Al/$KMnO_4$ has a pressurization rate double that of CuO/Al and MoO_3/Al nanothermites [65]. The main drawback of $KMnO_4$ compositions is the instability in contact with humidity, acids, and organic materials. The aging of energetic compositions based on $KMnO_4$ occurs in the presence of humidity due to the rapid transformation of $KMnO_4$ to manganese dioxide (MnO_2). Moreover, self-ignition may naturally occur in these compositions when unintentionally in contact with organic materials, such as glycerine or oxalic acid. Prakash and his research group succeeded to solve the instability of Al/$KMnO_4$ nanothermite by coating $KMnO_4$ with ferric oxide (Fe_2O_3). They concluded the reactivity of the new composition can be adjusted by varying the percentage of Fe_2O_3 [66].

Wu et al. encapsulated potassium perchlorate ($KClO_4$) by Fe_2O_3 and CuO individually using the same method described in [66]. Firstly, Fe_2O_3 successfully coated $KClO_4$ with excellent core–shell structure. CuO, which is supposed to be only in the shell, mixed with some $KClO_4$ that in turn disturbed its crystallization. Finally, Wu and his research group succeeded to coat $KClO_4$ with Fe_2O_3 and the successful product was in the form of hollow particles; the composition of the shell comprised a mixture between $KClO_4$ and Fe_2O_3. Nanothermite based on $KClO_4$/Fe_2O_3 and n-Al (50 nm) has higher performance and energy output than Al/CuO and Al/Fe_2O_3 alone. In general, nanothermite based on perchlorates typically has higher pressurization rate and maximum pressure compared to those with metallic oxides alone. The previous behavior can be justly attributed to lower decomposition temperature for $KClO_4$ than Fe_2O_3 and CuO and hence earlier oxygen release, resulting in enhanced reactivity of the exothermic reaction [11].

Coating of perchlorates with metal oxides decreases its susceptibility to humidity and consequently limits its adverse effect on the surrounding environment. Despite this, (ClO_4^-) anions affect endocrine glands even in very small quantities. The National Academy of Sciences reported that the maximum daily safe absorbed amount of (ClO_4^-) is 0.7 μg kg^{-1}. The potential danger of the (ClO_4^-) comes from its toxicity, which can be easily fixed, on a glycoprotein. The essential role of glycoprotein is being transported to iodine across the basolateral membrane of the thyroid cells and consequently disturbs the production of hormones in the thyroid [67]. Subsequently, many published studies have been directed towards the replacement of perchlorates by periodates as a friendly environment oxidizer [47, 49].

The replacement of perchlorate salts by periodates in nanothermites was studied by Sullivan et al., who compared the performance of nanothermite based on silver iodate ($AgIO_3$) and those based on CuO. The reactivity of periodate compositions was higher than compositions based on metallic oxides. While Al/$AgIO_3$ achieved

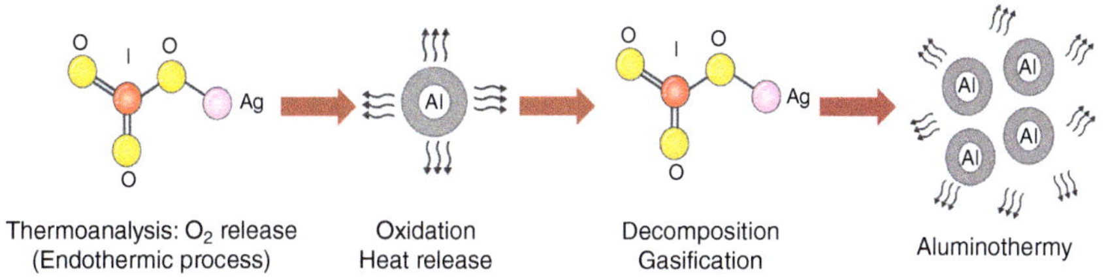

Figure 8.1 Schematic diagram of the reaction mechanism of Al/$AgIO_3$ nanothermite. Source: Adapted from Marc Comet et al. [17].

392 kPa μs^{-1} in a combustion time of 172 μs, Al/CuO attained 61.9 kPa μs^{-1} in 192 μs. More gas release in the case of $AgIO_3$ than CuO is the reason for the elevation of Al/$AgIO_3$ pressure. The reaction mechanism of Al/$AgIO_3$ is illustrated in Figure 8.1 [47].

Also, they reported that heating rate affects the decomposition behavior of $AgIO_3$. On one hand, at a low heating rate (5 K min^{-1}), $AgIO_3$ melts at 692 K, then it decomposes into silver iodide (AgI) and oxygen at 740 K, and finally AgI itself melts at 827 K and evaporates. On the other hand, at a high heating rate (5×10^5 K s^{-1}), the gaseous species were detected at a higher temperature (1150 K) than in the first case (740 K); AgI and Ag are unnoticed in the gaseous phase. Lastly, they concluded that ignition of Al/$AgIO_3$ nanothermite in the case of high heating rate occurs approximately at 1245 K, which is merely above the $AgIO_3$ evaporation temperature. Based on the previous results, researchers hypothesized that ignition of Al is greatly dependent on oxygen partial pressure [47].

Jian et al. studied the combustion behavior of nanothermites based on n-Al, sodium, and potassium periodates ($NaIO_4$ & KIO_4) individually, using pressure cell analysis. Results demonstrated that the rate of pressurization of these compositions greatly increased and reached to 2.4–2.6 MPa μs^{-1} compared with 0.06 MPa μs^{-1} for Al/CuO. Ignition temperature was recorded at 880, 950, and 1040 K for Al/$NaIO_4$, Al/KIO_4, and Al/CuO, respectively. Furthermore, they detected that decomposition of KIO_4 occurs in two stages. Initially, KIO_4 decomposes exothermically to KIO_3 and oxygen. Next, KIO_3 reduced at higher temperatures to potassium iodide (KI) and oxygen. While nanothermites based on KIO_4 react strongly in both air and argon environments, KIO_4 decomposes without reacting with Al in a vacuum chamber at a high heating rate (106 K s^{-1}). Based on these results, it was concluded that oxygen has a crucial role in the ignition and combustion of periodate nanothermites [49].

Recently, Zhou and their research group compared the combustion characteristics of nanothermites prepared from n-Al (50 nm) to the ones obtained from potassium sulfate (K_2SO_4), potassium persulfate ($K_2S_2O_8$), or KIO_4-based MIC. Al/$K_2S_2O_8$ showed the highest maximum pressure and pressurization rate on the pressure cell analysis. Moreover, they illustrated that $K_2S_2O_8$ decomposition occurs in three stages. At the first stage, $K_2S_2O_8$ is converted to pyrosulfate ($K_2S_2O_7$) and released oxygen, followed by the decomposition of $K_2S_2O_7$ into K_2SO_4, oxygen, and sulfur dioxide (SO_2) with increasing temperature. Finally, at higher temperatures, K_2SO_4 decomposes into oxygen and SO_2. In the end, Zhou et al. found that $K_2S_2O_8$ and

K_2SO_4 exhibit different behavior. First, ignition of $Al/K_2S_2O_8$ happens between Al and oxygen released from $K_2S_2O_8$ decomposition. Second, ignition of Al/K_2SO_4 takes place by the condensed-phase reaction between Al and K_2SO_4 like most nanothermites based on metal oxides [13].

In conclusion, nanothermites based on oxidizing salts reveal improved performance over those with metallic oxides. The significant reactivity of these aluminothermic compositions is ascribed to the fast release of high amounts of oxygen at the early stage of the oxidizer decomposition reaction.

8.4 Carbon Nanomaterials and Energetic Compositions

Over the last two decades, a new family of materials based on nanosize carbon particles has been developed. They are often referred to as CNMs. This family includes carbon nanofibers (CNFs), carbon nanotubes (CNTs), derivatives of graphene (namely graphene, graphene oxide (GO), and reduced graphene oxide (RGO)). They have been used as enhancement additives in EMs such as nanothermites, solid propellants, and gas generators. CNMs have been active fields of research in the improvement of nanothermites owing to a favorable specific area and special electrical, thermal, interface, and properties [68–71]. In addition, they can be adequately prepared in various shapes, crystal forms, and sizes. The preference for GO and CNTs results from their significant catalytic properties in addition to simple preparation methods [72]. Here, we will briefly review the potential effect of CNMs, especially GO and CNTs on the combustion properties of EMs.

Compared to other CNMs, GO is considered the worthiest candidate to be integrated in EMs due to its energetic nature. Even without any other additives or co-reagents, GO is a very potent EM. When heated, it can easily transition to an intense violent exothermic decomposition process accompanied by a large heat release. This high energy production is attributed to several oxygenic chemical groups on the main basal plane of its molecular structure as well as carboxylic functions found in the periphery [73]. GO and RGO are typically used to improve the combustion behavior of EMs. For example, Zhang et al. evaluated the effect of doping NC by GO at different percentages (0.1, 0.5, 1, and 2 wt %) on its ignition characteristics. They found the addition of GO increased the burning rate of NC and decreased the laser ignition temperature. GO is not only used in enhancing the burning rate of EMs, but also for decreasing the sensitivity of the energetic crystals as in the case of octahydro-1,3,5,7-tetranitro-1,3,5,7-tetrazocine (HMX) desensitization [74].

Nasir and his colleagues investigated the potential effect of nanocomposite material based on AP and GO on the combustion of composite propellant. They used a recrystallization method to synthesis AP/GO, then integrated it with hydroxyl-terminated polybutadiene (HTPB) composite propellant, and as a result burning rate increased by 15% [75].

Functionalized graphene sheets (FGS) were successfully used to prepare $GO/Al/Bi_2O_3$ nanothermite in a self-assembly method. GO was added with different quantities from 1 wt % to 10 wt %. Self-assembled GO/Al/Bi_2O_3 achieved higher energy output than those of randomly mixed. The addition of GO increased the energy released from 739 to 1421 J g^{-1}. This energy enhancement can be reasonably attributed to the active role of GO as an energetic reactant and the benefits of FGS method [76].

Ning et al. introduced RGO with different percentages on Al/Fe_2O_3 nanothermite using an atomic layer deposition method. They found that 4.8 wt % of RGO improved the heat release of Al/Fe_2O_3 by 130%. This enhancement in the energy can be ascribed to the increase in interfacial contact between the fuel and oxidizer and enhancement of reactant mass diffusion [77].

Ping et al. successfully prepared nanocomposite materials based on different metallic oxides (CoO, CuO, and Fe_2O_3) and graphene. Graphene-based composite materials showed good catalytic activity on the thermal decomposition of $KClO_4$, with the best performance achieved by CoO/graphene. This enhancement is due to the catalytic effect of graphene, good electrical and thermal conductivity, and large specific area. Those properties of graphene improve the adsorption and reaction between products and catalytic active sites and as a result, enhance thermal decomposition of $KClO_4$ [78].

Recently in 2021, Fahd et al. introduced tertiary nanothermites mixtures that can be beneficially used for different energetic applications. In [79], they evaluated the impact of different CNMs (graphene oxide, reduced graphene oxide, CNTs and nanofibers GO, RGO, CNTs, and CNF) on the thermal behavior of nanothermites containing $KClO_4$ (3–5 μm) and n-Al (40 nm). Nanothermite formulations were mixed by employing a sonication probe. They investigated the morphology of nanothermites using both scanning electron microscopy (SEM) and energy dispersive spectroscopy (EDS). The research revealed a homogeneous material where the nanosize particles are well dispersed without the presence of agglomerates, as illustrated in Figure 8.2.

The structure of nanothermite was also characterized by Fourier transform infrared spectroscopy (FTIR), Raman spectroscopy, and XRD. Figure 8.3 shows the Raman spectra of GO, RGO, and nanothermite of 5%GO/Al/$KClO_4$.

As it can be observed, the two principal peaks for GO are displaced by the oxidative process and are now found at 1349 and 1602 cm^{-1} for the D and G bands, respectively. However, with RGO, the same signal is moved back to the expected characteristic peak of graphene (1560 cm^{-1}), thus confirming the elimination of the oxygen chemical functions. The less apparent displacement in RGO signal after the reduction reaction can be attributed to remaining traces of oxygen. It appears that two peaks characterizing GO are also removed from the Raman spectra of 5% GO due to the increase in the ratio between D and G intensities (ID/IG) and consequently it eliminated the defects found in the structure of GO. The homogeneity of the process potassium perchlorate deposition over aluminum substrate is therefore verified by the spectra. It also brings forward the contribution of GO in improving the interfacial contact between the nanothermites constituents and decreasing the contact distance between the fuel and oxidizer, which is expected to improve the performance of

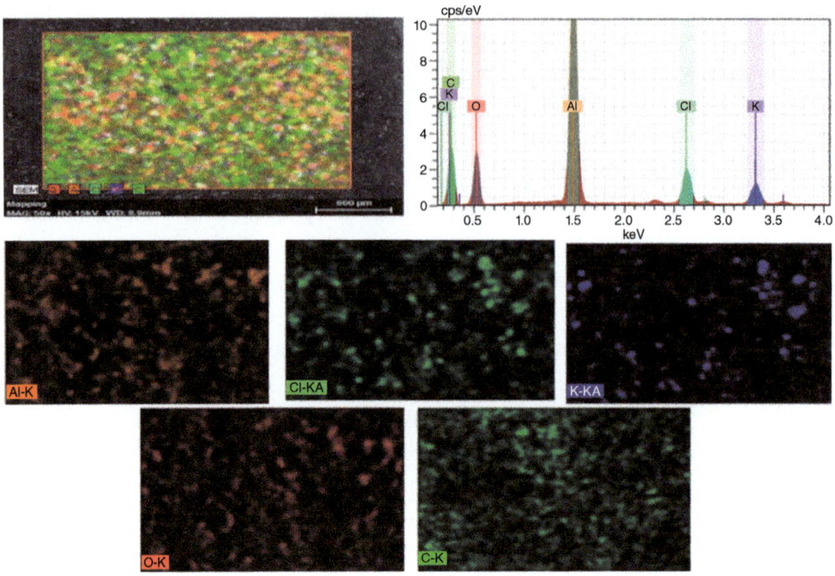

Figure 8.2 EDS characterization of a nanothermite based on 5% GO/Al/$KClO_4$. Source: Ahmed Fahd et al. [79], Reproduced with permission from John Wiley & Sons.

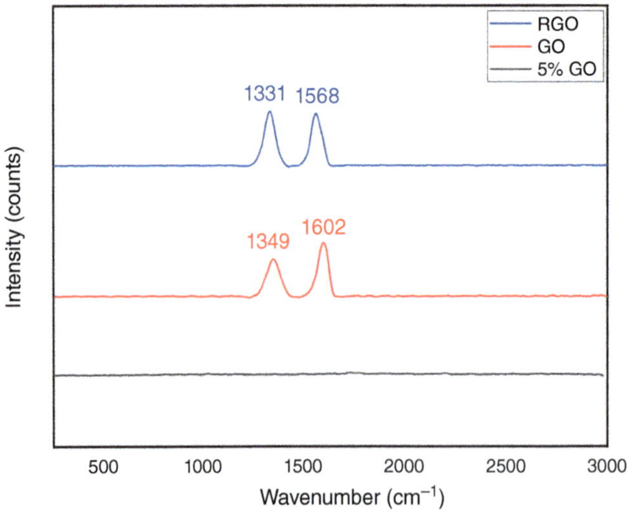

Figure 8.3 Raman spectra of neat GO and RGO powder and 5% GO/Al/$KClO_4$. Source: Ahmed Fahd et al. [79]/John Wiley & Sons.

nanothermites. The X rays diffraction analysis of thermites of interest is presented in Figure 8.4.

The principal diffraction patterns of Al and $KClO_4$ are easily observable in all thermite formulations, confirming that their nanocrystalline structure was preserved during the preparation process [80]. More over, a weak diffraction signal

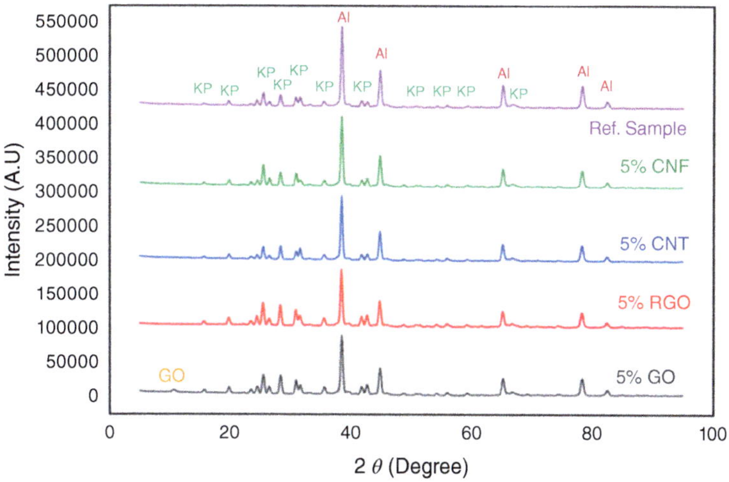

Figure 8.4 XRD patterns of CNMs/thermite samples. Source: [79]/John Wiley & Sons.

indicating the presence of GO was identified. Its small amplitude correlates with the lesser amount of GO in the samples. Characteristic peaks of other CNMs are not found in the XRD patterns, likely owing to the fact that they blend with the background, due to their very small concentration or because of their limited light scattering capability [81].

The effect of CNMs on the combustibility of nanothermite has been studied by thermal gravimetric analysis (TGA), DSC, and bomb calorimetry. Thermograms of TGA analysis of nanothermite mixtures with different percentages of GO are shown in Figure 8.5.

The thermites compositions based on GO decompose in three steps. An initial weight loss (about 5%) is recorded just after 200 °C. It is believed to correspond to the breaking of the oxygenated chemical groups on graphene oxide and the resulting graphite (C). Another significant mass evolution takes place from 400 to 550 °C, revealing the reaction of C-$KClO_4$ (carbon obtained from the decomposition of graphene oxide) and the main thermite reaction in GO/Al/$KClO_4$ [82]. The last weight reduction comes from the melting of metal Al and the melting of unreacted Al and potassium chloride (KCl). The effect of GO concentration on the performance of nanothermite can be observed on a DSC plot as shown in Figure 8.6.

It is clear from the thermograms that the energy released from the reaction of a GO/Al/$KClO_4$ nanothermite increased with the GO content while the ignition temperature followed an inverse trend. The enhancement of the energy released can be ascribed to the compatibilizing effect of GO with respect to fuel-oxidizer mixing and reactants diffusion. As a substrate, GO helps in increasing the dispersion of n-Al and effectively reducing their aggregation during mixing with $KClO_4$. Hence, a homogeneous dispersion of the n-Al metal fuel and $KClO_4$ oxidant takes place. Furthermore, reactions between C formed in the early stages through the decomposition of GO and nanothermite reactants (C–O, C–$KClO_4$, and C–Al) will

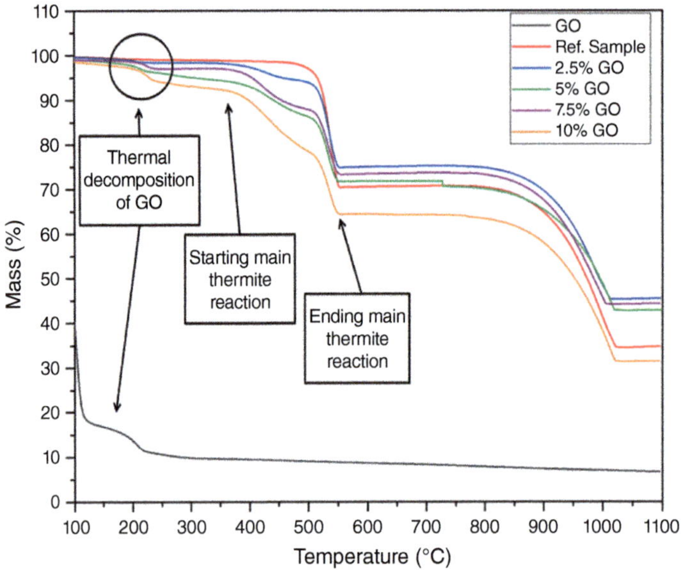

Figure 8.5 TGA thermogram for different percentages of GO over Al/$KClO_4$ nanothermites.

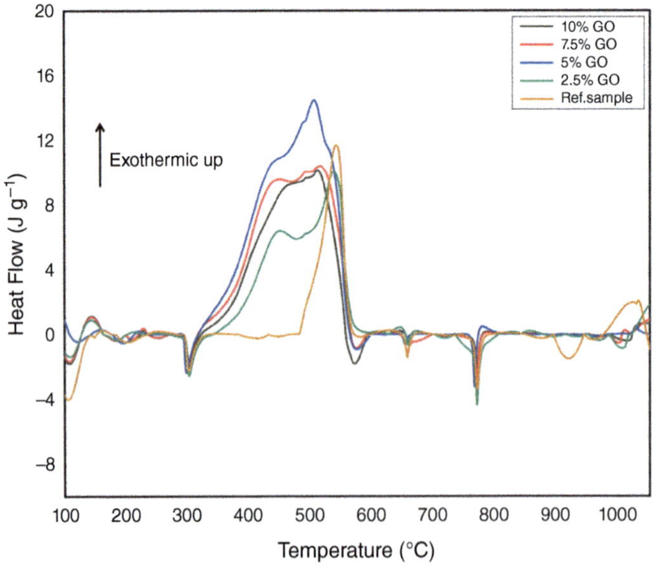

Figure 8.6 DSC thermograms percentages of different % of GO nanothermites.

be effective reaction pathways that help in accelerating the occurrence of main thermite reactions (Al–$KClO_4$ and Al–O) and play the role of preignition reactions. On the other side, the higher content of GO (more than 5%) on nanothermites gives $KClO_4$ the chance to grow directly on GO nano sheets. The separation of the oxidant and the metal fuel will then occur causing the reduction of energy released in the thermite reaction [76, 77, 83]. The highest energy released was achieved by nanothermite composition with 5% GO. The increase in the heat released was approximately 200% greater than the initial value of the reference sample. This high energy release achieved by the new nanothermite composition has many advantages in developing micro-scale propulsion systems.

Fahd et al also measured the average values of the heat of combustion of all CNMs/nanothermite compositions using the oxygen bomb calorimeter. The results confirmed and strongly agreed with the findings of DSC that CNMs have a great influence on improving the performance of nanothermites and, consequently, their role in developing micro energetic applications.

In [84] the authors investigated the thermal behavior and reaction characteristics of tertiary nanothermites based on n-Al, GO, and different types of salt and metallic oxidizers ($KClO_4$, NH_4NO_3, KIO_4, $K_2S_2O_8$, $KMnO_4$, Fe_2O_3, CuO, and WO_3). The findings show that adding GO to nanothermites improves their reactivity with salt and metallic oxidizers by lowering the reaction start temperature, activation energy, and boosting heat release. According to both theoretical and experimental evidence, nanothermites based on oxidizing salts are more reactive than those based on metallic salts. When compared to their reference formulations, the initial decomposition temperature of all nanothermites shifted to a lower temperature. The heat release of the GO/Al/$KClO_4$ nanothermite was the greatest (9614 J g^{-1}), whereas the onset temperature and activation energy of the GO/Al/$K_2S_2O_8$ nanothermite was the lowest observed (380 °C and 105 kJ mol^{-1}). The apparent kinetics parameters were calculated using Kissinger and Ozawa approaches. Although the activation energy measured by the Ozawa technique is slightly larger than that obtained by the Kissinger method, it still has some reference value. The possible ignition mechanism of the prepared ternary nanothermites is illustrated in Figure 8.7.

Because oxygenated salts have a lower thermal decomposition temperature than metallic oxides, the nanothermite components are mixed more uniformly and without visible agglomerations. Also, the exothermic nature of the GO thermal decomposition process might constitute a type of preignition. These findings should serve as guidelines for formula design, safety, storage, and handling of nanothermites, resulting in improved performance in practical applications [84]. Table 8.1 summarizes the practical applications and preparation methods for EMs based on graphene, GO, and RGO.

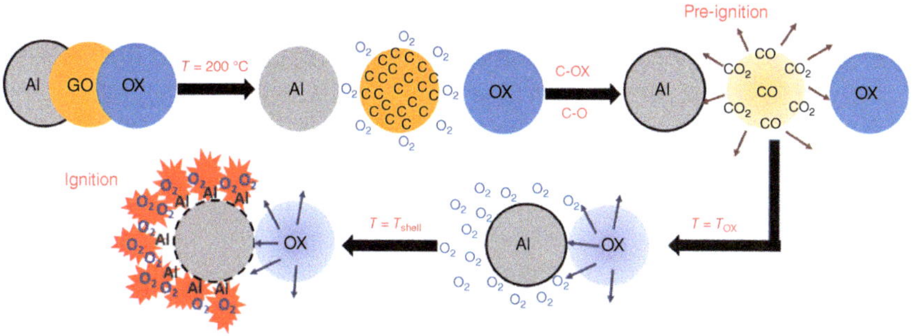

Figure 8.7 Suggested mechanism of tertiary nanothermites reaction.

Table 8.1 Different applications and preparation methods for EMs based on graphene, GO, and RGO.

Compositions	Synthesis method	Applications	Reference
Graphene/Fe_2O_3/NH_4ClO_4	Sol–gel method and supercritical carbon dioxide drying technique	Propellants	[85]
Graphene/NH_4NO_3	Sol–gel and supercritical CO_2 drying method	Propellants Nanothermite	[86]
CL-20/GL/RGO	Emulsion polymerization and mixing methods	Energetic compositions	[87]
Al/Bi_2O_3/GO	Formed in a colloidal suspension phase that ultimately condenses into ultradense macrostructures	Nanothermite	[76]
NH_4NO_3/Graphene	Sol–gel method	Propellants	[85]
HMX/GO	Solvent–nonsolvent method	Propellants and explosives	[74]
GO/NC	Mixing pure NC–acetone solution and various weight ratios of GO water solution	Propellants	[88]
GO/FOX-7	Self-assembly impregnation in N-methyl-2-pyrrolidone at 100–110 °C	Explosives and propellants	[89]
GO/Al/$KClO_4$	Sonication method	Nanothermites Modified propellant	[79]

CNTs integrate with EMs through functionalized polymer binder or combined directly with pyrotechnic compositions and metal fuels. They are used as a combustion catalyst carrier and recently have been functionalized with energetic groups. CNTs can be used as an analog of GO in enhancing the combustion properties of energetic compositions and increase their heat generation. Qian et al. were the pioneer in combining CNTs with pyrotechnic formulations based on $KClO_4$ and KNO_3 in an attempt to increase the burning rate of these compositions. They succeeded in their objective, and CNTs quadrupled the reaction rate of $KClO_4$ [90]. Also, CNTs were used to improve the radiation energy of Zr/$KClO_4$ and it was found that with only 0.5 wt %, the energy release of the pyrotechnic composition increased to 1830 J g^{-1} [91]. Moreover, Sylvain et al. developed flash-ignitable EMs with superior combustion characteristics by integrating CNTs with metal fuels such as (Al, Zn, Fe, Ni, and Cu) [92]. Manjula Sharma and Vimal Sharma studied the beneficial effect of CNTs on the thermite reaction of Al/CuO. They observed a remarkable decrease in ignition temperature and activation energy where both of

Table 8.2 Different applications and preparation methods for EMs based on CNTs.

Materials	Preparation method	Applications	Reference
KNO_3/CNTs	Mechanical mixing and grinding to nano size	Energetic initiator	[93]
Zr/$KClO_4$/CNTs	By grinding a mixture of Zr and $KClO_4$ for 20 minutes	Pyrotechnic reagent	[91]
CNT/BAMO–AMMO	Using CNT-OH as a crosslinking agent, covalently modified BAMO–AMMO	Energetic binder	[94]
Fe_2O_3/MWCNTs	By mild and superior physical absorption method using molten $Fe(NO_3)_3{\cdot}9H_2O$ as the precursor	Thermites and pyrophoric substrates	[95]
GAP/CNTs binder films	Using hydroxylation CNTs as crosslinking agent, GAP as prepolymer, TDI as curing agent	Propellants, PBXs	[96]
Al/CNT composites	Al/CNT mixture was engineered by mechanical pulverization	Thermites and propellants	[97]
CuO/CNTs	Using sol-infusion method under normal pressure and low temperature (at 100 °C)	Catalyst for propellants	[98]
HMX/CNTs	Prepared with an ultrasonic compositing method	Energetic compositions	[99]
Al/Teflon/CNTs	By blending of Al with Teflon and CNTs	Explosives and propellants	[100]

them lowered by 71 °C and 23% respectively. Also, the reaction enthalpy became 13 times higher than reference with addition of 15 wt % CNTs [81]. Table 8.2 summarizes the practical applications and preparation methods of EMs based on CNTs.

8.5 Future Challenges

Nanothermites are now being studied for the purpose of developing new aluminothermic composites and investigating their reactive characteristics. Academically, the future tasks will be to discover new compositions and study their effects. Practically, the endeavor should concentrate on the incorporation of nanothermites and their derivatives into pyrotechnic and explosive systems. Over the coming decade, toxicological concerns are also projected to grow in importance.

Although many extensive works have been performed to develop and enhance the performance of nanothermites, some potential problems like the incomplete combustion of n-Al still exist. The incomplete combustion of n-Al is due to the excessive oxidation of Al before combustion, which in turn decreases the active Al content. In addition to the incomplete combustion of Al, nanothermites have toxicity problems as they release dangerous products from some of their primary components. Recently, researchers have directed much effort to prepare new thermite compositions free of Al to overcome these problems and produce more sustainable products.

To achieve this objective, high-energy metal–organic frameworks (EMOFs) will be an important alternative fuel to Al in preparing new thermite compositions with superior combustion properties. The superiority of EMOFs resulted from their great detonation heat due to organic ligands with high energy density. In addition, EMOFs produce large amount of gas, have a high surface area, good stability, uniform pore size, and customizable structure [101].

The literature about thermites based on EMOFs is limited. Shenghua Li and his research group have reported a new thermite material with ATRZ-1, replacing Al as a fuel and either of perchlorates (NH_4ClO_4 and $KClO_4$) or periodates (KIO_4 and $NaIO_4$) as individual oxidizers. This specific type of thermites exhibited good results in both sensitivity and performance tests, as reported in Table 8.3 [102]. They concluded that thermites based on those EMOFs, and periodate salts show promise as a green secondary gas generator of lesser sensitivity [103].

Fahd et al. introduced new thermites compositions with low ignition temperature, stable propulsive force, and high reactivity. These thermites are based on the $[Cu_4Na(Mtta)_5(CH_3CN)]_n$ (Mtta = 5-Methyl-1H-tetrazole) energetic metal–organic framework (EMOF-1) as a fuel instead of pure metals. They were the pioneers in synthesis of an energetic MOF via the microwave-assisted technique as a more rapid and greener method. The combustion behavior of the novel composites was evaluated by TGA/DSC, bomb calorimetry, and laser ignition. Additionally, the apparent kinetic parameters (activation energy & frequency factor) were calculated by the Kissinger

Table 8.3 Sensitivity and combustion characteristics of ATRZ-1/thermite mixtures [102, 103].

Thermite mixture	Impact sensitivity (J)	Friction sensitivity (N)	Electrostatic sensitivity (J)	Heat of combustion (kJ/g)	Maximum pressure (MPa)	Ignition[a] temperature (°C)
ATRZ-1/NH_4ClO_4	4	118	0.69	3.84[a)](5.41[b)])	6.90	242
ATRZ-1/$KClO_4$	8	110	0.16	1.70[a)](4.45[b)])	5.70	227
ATRZ-1/KIO_4	9	32	0.19	2.05[a)]	1.79	227
ATRZ-1/$NaIO_4$	10	8	0.13	2.63[a)]	1.96	232

a) Heat of combustion and ignition temperature measured by DSC.
b) Heat of combustion measured by calorimetric bomb.

and Ozawa approaches. The results revealed that the new thermite mixtures exhibit superior combustion characteristics of one and a half to two-folds the average heat of combustion compared to aluminum-based ones, at almost half the ignition temperature. In this sense, the combustion reaction proceeds faster, easier (reduced activation energy), the ignition temperatures are noticeably lowered, the heat released has considerably improved, and solid residue significantly decreased. In addition, they exhibited stable force with longer burning time. Among them, EMOF-1/KIO_4 thermite exhibits the highest heat release (4.7 kJ g^{-1}), while EMOF-1/NH4NO_3 thermite shows the lowest onset reaction temperature (224 °C). EMOF-1/$KClO_4$ yields the highest average force (8.4 N), calculated pressure (1365 kPa), pressurization rate (0.32 kPa μs^{-1}), and the longest burning time assigned to EMOF-1/$K_2S_2O_8$ (40 ms). This work showed that thermite mixtures with tailored combustion performance and green combustion products can be developed via EMOF-1 as fuel. The results of this research open the route for more application of EMOFs in the future [104].

In another direction, the practical importance of green primary explosives increases because of the inherent drawbacks associated with traditional initiating explosives. Hyper toxicity and hydrolytic instability are the hazards of using mercury fulminate, lead azide, or lead styphnate. As a direct result of those dangerous materials on the environment and human health, their military and civilian uses are limited for the sake of human and environmental protection [39, 105]. Therefore, mixing nanothermites with explosive nanopowder to create nanostructured thermite explosives (NSTEX) will be better alternatives to primary explosives, but the challenge will be how to overcome the low combustion velocity or pressurization of nanothermites and consequently improve their deflagration to detonation transition (DDT) process.

Marc Comet and his colleagues demonstrated NSTEX by combining nanothermite mixtures (Al/WO_3 & Al/$Bi_2(SO_4)_3$) individually with explosive nanoparticles prepared by spray flash evaporation [106]. NSTEX that has been prepared has superior performance and is far less dangerous than primary explosives. Their explosive composition allows them to regulate their flame propagation velocity from 0.2 to 3.5 km s^{-1}. They were employed to start the detonation of the pentaerythritol tetranitrate high explosive. In semi-confined systems, DDT was achieved using short-length systems (less than 2 cm) and tiny amounts of NSTEX (less than 100 mg). They also have a higher level of sensitivity than primary explosives, making their handling much safe. These findings are critical for industrial applications where the induction of detonation in high explosives without the use of primary explosives has hitherto been a major constraint.

NSTEX based on Al/CuO and hexanitrohexaazaisowurtzitane (CL-20) was introduced in [107] by Ying Zhu. The CL-20 is in situ integrated with CuO/Al in this work using a simple dissolution–recrystallization method making use of the core/shell array structure. This design is fully compatible with MEMS technology, which is advantageous for constructing functioning energetics-on-a-chip devices. The loading amount of CL-20 has been shown to be adjustable, and the total heat of reaction of NSTEX has been improved. When compared to raw CL-20 material, the activation energy for the decomposition of integrated NSTEX is 18.2% lower. Furthermore, for hybrid NSTEX, a two-step reaction is observed with the rate-limiting CuO decomposition followed by reactive species diffusion through alumina shell. The CL-20 incorporated NSTEX also has a desirable burning characteristic, with a vigorous and consistent combustion flame. The simple in situ integration of explosives with nanothermite described in this paper could be extended to other explosives (e.g. PETN, TNT, HMX, RDX) and nanothermites, and thus merits further investigation for possible use as green primary explosives.

Because primary explosives are sensitive to environmental stimuli, they pose a potential safety risk. As a result, their safety features are critical to their employment in both military and civilian applications. Thus, the safety properties of a novel lead-free NSTEX initiatory compound prepared by mixing Al/Fe_2O_3 nanothermite with RDX were examined in detail by Qingping Luo and his colleagues [108]. The sensitivity and safety properties of the prepared NSTEX were found to be a consequence of its specific surface area and constituent quantities. The impact sensitivity of NSTEX decreased as the specific surface area of nano-Fe_2O_3 grew, but the static discharge, flame, and hot bridge wire sensitivities all rose. The sensitivity of the nanocomposite was reduced when both cyclohexane and acetone were utilized as the solvent during mixing, due to the smaller RDX particle size. Impact sensitivity decreased as the Al/Fe_2O_3 nanothermite content increased, whereas flame sensitivity increased. The NSTEX static discharge and hot bridge wire sensitivity changed in an inverted "U" pattern and were governed by both the particle size of the components and the NSTEX resistance. As a result, green NSTEXs are suitable for a variety of initiatory applications and may be made by varying their particular surface area and component amounts.

One of the major challenges limiting the use of EMs in general and thermite specifically in micro-energetic systems is the difficulty of ignition and propagation within narrow tubes and slots. Here, we outline the key researches that have been conducted in this still-new sector, as well as the obstacles that have to be solved.

Certain solid-propellant microthruster applications, such as course correction of high-velocity projectiles or microsatellites, need a fast actuation time. The fuel should be chosen to have the maximum feasible combustion rate for such applications, reducing thrust vector movement. The thrust fuel, on the other hand, should not detonate, as this would result in projectile damage [109]. The pressure and shock wave properties of nanothermite based on CuO nanorods produced using the surfactant-templating technique and self-assembled with n-Al were measured using a shock-tube system, which is extensively described in [110]. Because of the nanoscale mixing, there was a high interfacial contact area between the oxidizer and the fuel. As a result, the reaction of the low-density nanothermite composite produced a propagation of the combustion front, generating shock waves with Mach values of up to 3. The reaction propagation rate of the prepared Al/CuO nanothermites, which is comparable to that of traditional primary explosives but at pressures far lower than those of solid explosives, makes it a viable option for short-impulse microthruster use.

Investigations conducted on Al/CuO nanostructured thermites within a density range of 20–80% TMD in a generic microthruster showed that high-density samples exhibited lower thrust and longer burns (3–5 N, 1.5–3 ms) as compared to higher thrust and shorter burns (75 N, 50 μs) at low packing density. Specific impulse in both cases was fairly constant at around 20–25 s [111]. By using 160 mg of a nanothermite made from porous Cr_2O_3 loaded with RDX, Sourgen et al. shifted in flight a 315 g 40 mm projectile by an angle of 3° as compared to its initial direction, inside a Mach 3 supersonic air flow in a wind tunnel [112] [113].

Fahd et al. in [114] suggest nanothermites with tailored combustion performance can be developed using oxidizing salts instead of metallic oxides and be tuned by addition of small quantity of GO, introducing a new direction for micro-propulsion and micro-energetic applications. They evaluated the thrust-generating properties of nanothermite mixes in a converging–diverging nozzle and an open tube using experimental and computational methods. Mixtures are prepared using n-Al, $KClO_4$, and different types of CNMs additives such as GO, RGO, MWCNTs, and CNF. Analogous to $KClO_4$, oxidizing salts (KIO_4, $K_2S_2O_8$, $KMnO_4$, and NH_4NO_3) are also studied. To study combustion processes and thrust generation, they packed nanothermites at varying densities and ignited them using a 3.5 W continuous wave laser. High-speed imagining and computational fluid dynamics (CFD) modeling are used to analyze the exhaust plume of the reference sample (n-Al and $KClO_4$). All prepared nanothermites ignited successfully and combusted self-sufficiently inside the STMs. With the inclusion of CNMs, thrust production, specific and volumetric impulse (I_{SP} and I_{SV}), and normalized energy all rise significantly. More specifically, I_{FT} and I_{SP} are recorded as 7.95 mN·s and 135.20 s respectively at 20% TMD, which indicates an improvement of over 57% in comparison to the reference sample without GO (5.05 mN·s and 85.88 s). The improved combustion properties of the GO/Al/$KClO_4$ compositions are proposed to be resultant of the catalytic nature of GO in facilitating uniform mixing of n-Al and $KClO_4$, in addition to its effect as an energetic reactant acting as a secondary fuel and adding energy to the system. In combustion of low and high packing densities (20 and 55 TMD, respectively), two different reaction regimes (rapid and slow) are observed, as illustrated in Figure 8.8, and the total impulse (I_{FT}) and I_{SP} achieves a maximum in both regimes when 5% GO/Al/$KClO_4$ is employed.

It is apparent in Figure 8.7 that there are three distinct regimes: a fast reaction regime at low packing density of less than 32.5%TMD (yellow region), a slow reaction regime at high packing density exceeding 55%TMD (gray region), and a transitional regime (green region) in between. It is likely that the low-density samples undergo supersonic combustion: supersonic combustion behavior of low-density nanothermites has been observed in other studies using open-ended combustion tubes [109, 115]. So, fast reaction behavior may be correlated to the rapid discharge of combustion products caused by reflected wave of the combustion front (supersonic combustion performance). Additionally, in open STMs, combustion is controlled by convective heat transfer so, at low packing density, pores within the sample allow hot gas propagation into the fuel grain and thus heat transfer into the sample from the flame front via convection. Removing these pores by compressing thermite into

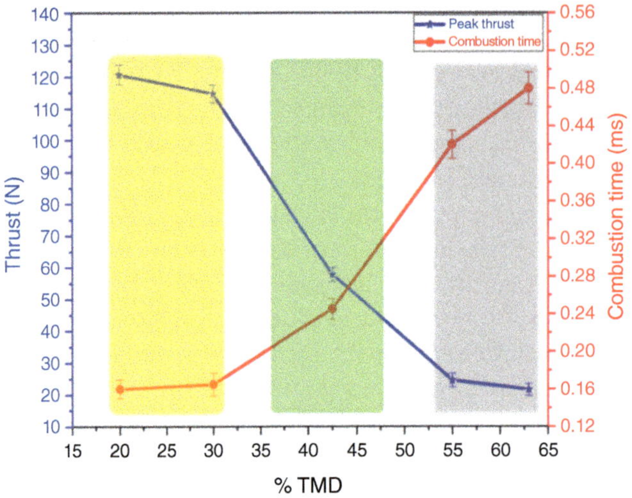

Figure 8.8 Peak thrust and combustion duration of 5% GO/Al/$KClO_4$ with different %TMD showing fast reaction regime (yellow), transitional regime (green), and slow reaction regime (gray).

a pellet (reducing volume but keeping constant mass) reduces the convective pathways, so the role of convective mechanisms on the flame propagation is reduced and conduction becomes dominant.

The introduction of a converging-diverging (CD) nozzle is shown to have two opposing effects on thrust in fast and slow reaction regimes where it reduces thrust output but increased impulse in fast regime samples but increases thrust and reduced impulse in slow regime samples; this is observed numerically and experimentally. Figure 8.9 presents time-resolved thrust-force data for different GO/nanothermite composites for slow and fast reaction regimes (20 and 55%TMD) in CD nozzle and no-nozzle STM configurations.

This work presented by Fahd et al. shows that energetic composites with tailored and controlled thrust performance can be developed via CNMs/n-Al/oxygenated-salt nanothermites and enables development of more focused future propulsion applications.

Because increasing the reactivity of nanothermites, either by manipulating the nature of the oxidizer or by adding energetic gas-generating substances is considered one of the main challenges in the development of nanothermites, Fahd et al. investigated novel quaternary nanothermites based on nitrocellulose (NC), GO, n-Al, and $KClO_4$ for high-speed impulse small-scale propulsion applications in [116]. They used electrospinning as a facile way to prepare quaternary nanothermite composites with well-mixed components. A schematic diagram of the electrospinning process is presented in Figure 8.10.

The composites' combustion behavior was assessed by igniting the samples in open acrylic tubes within a reaction chamber at various packing densities (%TMD) using a 3.5 W continuous wave laser. The flame propagation was recorded using

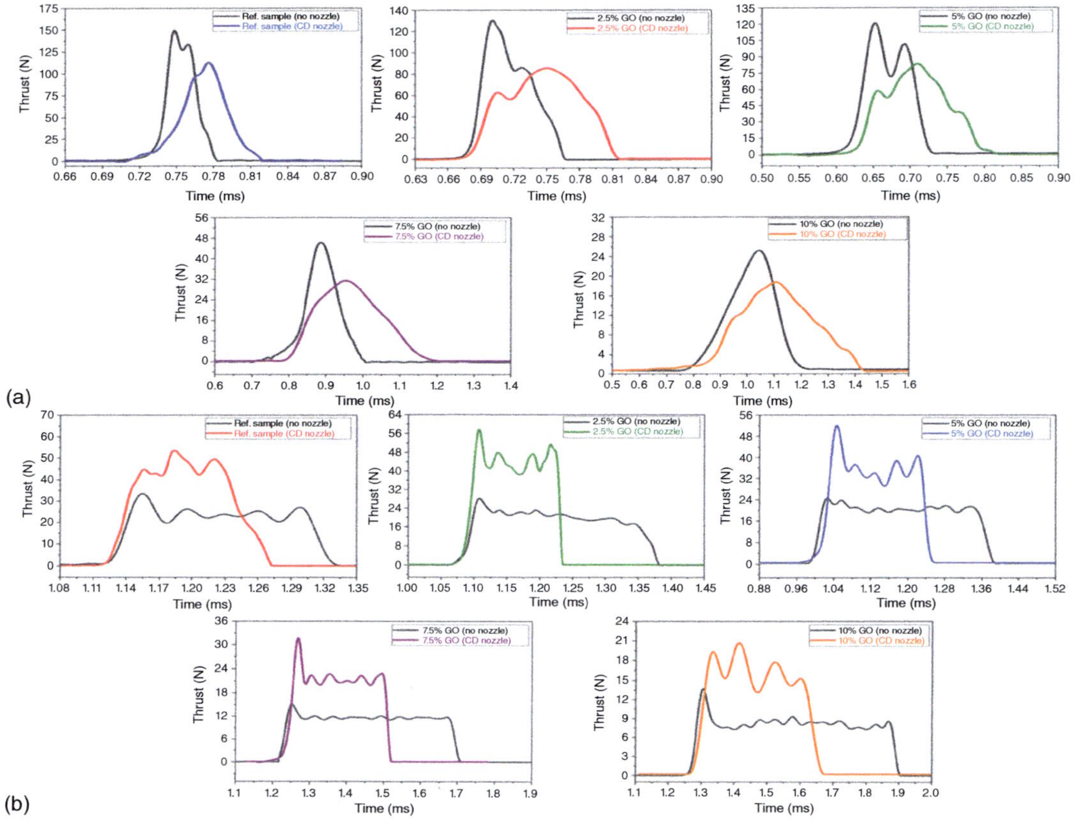

Figure 8.9 Thrust-time curves for different GO/nanothermite samples at (A) 20%TMD (B) 55%TMD with and without CD nozzle.

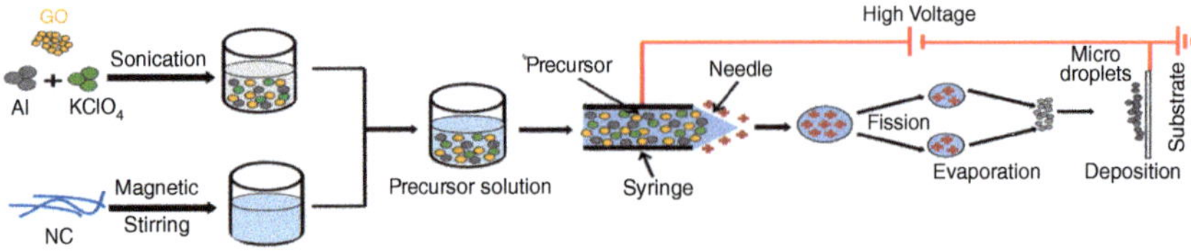

Figure 8.10 Schematic of electrospinning formation of quaternary NC/GO/Al/KClO$_4$ nanothermites.

high-speed imaging, and thermal behavior and energy production were assessed using DSC. A small-scale test motor with a converging/diverging nozzle was utilized to analyze the combustion performance of the produced samples for propulsion applications. With the addition of NC, thrust production, I_{SP} and I_{SV}, and total heat produced all improve significantly. At 50% TMD, the total impulse (I_{FT}) and I_{SP} peaked at 5% NC/GO/Al/KClO$_4$, with I_{FT} of 19.9 mN·s and I_{SP} of 203.2 s, respectively. When compared to the sample without NC, these represent improvements of more than 50%. (13.4 mN·s and 137.4 s). The ignition delay time and power required to ignite the NC-enriched mixture are increased, but the response time is still quick enough for practical applications. According to thermal study, adding NC to the GO/Al/KClO$_4$ reaction added additional steps: a gas–solid phase and liquid–liquid phase diffusion reaction, with the liquid–liquid phase reaction rising with NC percentage. Figure 8.11 illustrates the suggested reaction mechanism of the prepared quaternary nanothermite compositions [116].

Thermite combustion findings reveal that multiphase exhaust flows result in two-phase losses, which might explain the nozzle STM results. It is also possible that fine-tuning the nozzle design will increase performance. The regulated addition of NC to oxygenated salt-based nanothermite mixtures may generally be used to produce adjustable combustion characteristics. To customize performance case specifically, further adjustments to the thermite can be made by altering packing density and nozzle design. This evidence shows that the quaternary nanothermites under consideration might be beneficial in a variety of micro-energetic applications [116].

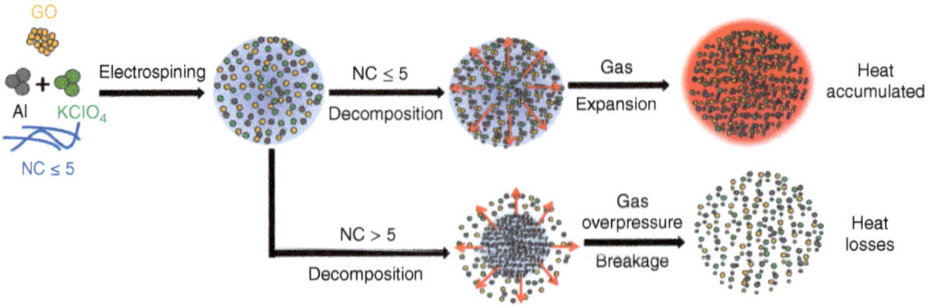

Figure 8.11 Suggested reaction mechanism of quaternary nanothermites.

8.6 Conclusion

Nanothermites are regarded as potential EMs and a critical component in the development of new technologies in the fields of energetics and propulsion systems. A common family of nanothermites is those based on n-Al powders, traditionally mixed with metallic oxides. More novel formulations can be developed by extending nanothermite combinations to include these where atoms with high electronegativity (O, S, F, etc.) are exchanged between an oxidizer and a fuel.

The main challenge thus far has been increasing the reactivity of nanothermites. As a result, significant efforts are required to increase the nanothermite reactivity utilizing either oxygenated and fluorinated oxides or CNMs. Moreover, various nanostructured morphologies, in terms of shape and size, can be exploited for controlling the energy release of the system, as in the case of GO. Another method for increasing the reactivity of nanothermites is to mix them with gas-generating materials such as NC or explosive nanopowders to produce quaternary nanothermites and NSTEX. These quaternary combinations and detonating materials show promise for use in propulsion applications and as a substitute for main explosives containing heavy metals such as lead, cobalt, or nickel.

Understanding the reaction processes of nanothermites and hybrid mixtures is critical for future progress in this field. Intensive research should still be conducted for this aim to get the experimental substrate that will be required for modeling.

Basic and integrative research will drive future advancements. From an academic point of view, the main target will be the discovery of novel nanothermite mixtures with enhanced ballistic effects, notably in the domains of ignition and combustion propagation. From practical point of view, the challenges will revolve around integrating current nanothermites and hybrid compositions into functioning energetic systems to manufacture them on a large scale. Sensitivity and toxicity issues should be addressed as an intermediate step during scaling up production. This review shows energetic composites with tailored and controlled combustion and thrust characteristics can be developed via nanothermites. This enables the development of materials with tailored performance for novel applications.

References

1 Widdis, S.J., Asante, K., Hitt, D.L. et al. (2013). A MEMS-based catalytic microreactor for a H_2O_2 monopropellant micropropulsion system. *IEEE/ASME Transactions on Mechatronics* 18 (10): 1250–1258.

2 Zou, B.L.Q., Zheng, C., Hao, Z. et al. (2015). Novel integrated techniques of drilling–slotting–separation-sealing for enhanced coal bed methane recovery in underground coal mines. *Journal of Natural Gas Science and Engineering* 26: 960–973.

3 Zhou, M.T.X., Lu, J., Shen, R., and Zhang, K. (2014). Nanostructured energetic composites: synthesis, ignition/combustion modeling, and applications. *ACS Applied Materials & Interfaces* 6: 3058–3074.

4 Arkhipova, V.A. and Korotkikh, A.G. (2012). The influence of aluminum powder dispersity on composite solid propellants ignitability by laser radiation. *Combustion and Flame* 159: 409–415.

5 Peng Zhu, G.H., Wang, H., Cong, X. et al. (2018). Design, preparation, and performance of a planar ignitor inserted with pyroMEMS safe and arm device. *Journal of Micro Electro Mechanical Systems* 27: 1186–1192.

6 Andréa Nicollet, L.S., Baijot, V., Estève, A., and Rossi, C. (2018). Fast circuit breaker based on integration of Al/CuO nanothermites. *Sensors and Actuators A* 273 (249–255): 2018.

7 Ji Dai, F.W., Chengbo, R., Jianbing, X. et al. (2018). Ammonium perchlorate as an effective additive for enhancing the combustion and propulsion performance of Al/CuO nanothermites. *Journal of Physical Chemistry C* 122: 10240–10247.

8 Gan, L.Y. and Qiao, L. (2011). Combustion characteristics of fuel droplets with addition of nano and micron-sized aluminum particles. *Combustion and Flame* 158: 354–368.

9 Zhou, L., Piekiel, N., Chowdhury, S., and Zachariah, M.R. (2010). Time-resolved mass spectrometry of the exothermic reaction between nanoaluminum and metal oxides: the role of oxygen release. *The Journal of Physical Chemistry* 114: 14269–14275.

10 Trunov, M.S.M.A., Zhu, X., and Dreizin, E.L. (2005). Effect of polymorphic phase transformations in Al_2O_3 film on oxidation kinetics of aluminum powders. *Combustion and Flame* 140: 310–318.

11 Wu, C., Sullivan, K., Chowdhury, S. et al. (2012). Encapsulation of perchlorate salts within metal oxides for application as nanoenergetic oxidizers. *Advanced Functional Materials* 22: 78–85.

12 Jian, J.F.G., Jacob, R.J., Egan, G.C., and Zachariah, M.R. (2013). Superreactive nanoenergetic gas generators based on periodate salts. *Angewandte Chemie* 52: 9743–9746.

13 Zhou, J.B.D.W., Li, X., Liu, L., and Zachariah, M.R. (2015). Persulfate salt as an oxidizer for biocidal energetic nano-thermites. *Journal of Materials Chemistry* 3: 11838–11846.

14 Zhou, J.B.D.W., Wang, X., and Zachariah, M.R. (2017). Reaction mechanisms of potassium oxysalts based energetic composites. *Combustion and Flame* 177: 1–9.

15 Hu, J.X., Li, X., Zhou, W., and Zachariah, M.R. (2017). Direct deposit of highly reactive $Bi(IO_3)_3$ – polyvinylidene fluoride biocidal energetic composite and its reactive properties: direct deposit of highly reactive $Bi(IO_3)_3$ – polyvinylidene. *Advanced Engineering Materials* 19: 1–9.

16 Sullivan, N.W.P.K.T., Chowdhury, S., Wu, C. et al. (2010). Ignition and combustion characteristics of nanoscale $Al/AgIO_3$: a potential energetic biocidal system. *Combustion Science and Technology* 183: 285–302.

17 Marc Comet, M., Schnell, F., and Spitzer, D. (2019). Nanothermites: a short review. factsheet for experimenters, present and future challenges. *Propellants, Explosives, Pyrotechnics* 44: 18–36.

18 Jian, G., Feng, J., Jacob, R.J. et al. (2013). Super-reactive nanoenergetic gas generators based on periodate salts. *Angewandte Chemie, International Edition* 52: 9743–9746.

19 Luo, Y.R. (2007). *Comprehensive Handbook of Chemical Bond Energies*, 839. Boca Raton, Florida: CRC Press.

20 Stand, N.B. and Darwent, B.D. (1970). *Bond Dissociation Energies in Simple Molecules*, National Standard Reference Data Series, USA, 1–60.

21 Huang, K.-C., Couttenye, R.A., and Hoag, G.E. (2002). Kinetics of heat-assisted persulfate oxidation of methyl tert-butyl ether (MTBE). *Chemosphere* 49: 413–420.

22 Tsitonaki, P.A., Crimi, M., MosbÆk, H. et al. (2010). In situ chemical oxidation of contaminated soil and groundwater using persulfate: a review. *Environmental Science & Technology* 40: 55–91.

23 Balandin, S., Bao, W.Z., Calizo, I. et al. (2008). Superior thermal conductivity of single layer graphene. *Nano Letters* 8: 902–907.

24 Novoselov, K.S., Morozov, S.V., Jiang, D. et al. (2005). Two-dimensional gas of massless diracfermions in graphene. *Nature* 438: 197–200.

25 Stankovich, D.S., Dommett, G., Kohlhaas, K.M. et al. (2006). Graphene based composite materials. *Nature* 442: 282–286.

26 Lee, X., Kysar, J.W., Hone, J. et al. (2008). Measurement of the elastic properties andintrinsic strength of monolayer graphene. *Science* 321: 385–388.

27 Qi-Long Yan, M.G., Zhao, F.-Q., Cohena, A. et al. (2016). Highly energetic compositions based on functionalized carbon nanomaterials. *The Royal Society of Chemistry, Nanoscale* 8: 4799–4851.

28 Ferrari, A.C., Bonaccorso, F., Fal'Ko, V. et al. (2015). Science and technology roadmap for graphene, related two-dimensional crystals, and hybrid systems. *Nanoscale* 7 (11): 4598–4810.

29 Dreizin, E.L. (2009). Metal-based reactive nanomaterials. *Progress in Energy and Combustion Science* 35: 141–167.

30 Pantoya, M.L., Granier, J.J. et al. (2005). Combustion behavior of highly energetic thermites: nano versus micron composites. *Propellants, Explosives, Pyrotechnics* 30: 53–62.

31 Jian, G., Feng, J., Jacob, R.J. et al. (2013). Super-reactive nanoenergetic gas generators based on periodate salts. *Angewandte Chemie* 125: 9925–9928.

32 Jian, G., Chowdhury, S., Sullivan, K., and Zachariah, M.R. (2013). Nanothermite reactions: is gas phase oxygen generation from the oxygen carrier an essential prerequisite to ignition? *Combustion and Flame* 160: 432–437.

33 Kappagantula, K.S., Farley, C., Pantoya, M.L., and Horn, J. (2012). Tuning energetic material reactivity using surface functionalization of aluminum fuels. *The Journal of Physical Chemistry C* 116: 24469–24475.

34 Brege, J.J., Hamilton, C.E., Crouse, C.A., and Barron, A.R. (2009). Ultrasmall copper nanoparticles from a hydrophobically immobilized surfactant template. *Nano Letters* 9: 2239–2242.

35 DiBiaso, H.H., English, B.A., and Allen, M.G. (2004). Solid-phase conductive fuels for chemical microactuators. *Sensors and Actuators A: Physical* 111: 260–266.

36 Zhang, W., Peng, H., Gao, X. et al. (2012). An in situ chemical reaction approach to synthesize zinc picrate energetic thin film upon zinc oxide nanowires array. *Surface and Interface Analysis* 44: 1203–1208.

37 Kim, S.B., Kim, K.J., Cho, M.H. et al. (2016). Micro-and nanoscale energetic materials as effective heat energy sources for enhanced gas generators. *ACS Applied Materials & Interfaces* 8: 9405–9412.

38 Baev, V., Lvov, C., and Shugalei, I. (1998). Influence of the Environment Internal Factors on Warm-blooded Organisms. In: *Proc. Conf. Problems of Military and Emergency Medicine*, 62–63.

39 Talawar, M., Sivabalan, R., Mukundan, T. et al. (2009). Environmentally compatible next generation green energetic materials (GEMs). *Journal of Hazardous Materials* 161: 589–607.

40 Kim, S.H. and Zachariah, M.R. (2004). Enhancing the rate of energy release from nanoenergetic materials by electrostatically enhanced assembly. *Advanced Materials* 16: 1821–1825.

41 Son, S.F., Foley, T.J., Yetter, R.A. et al. (2007). Combustion of nanoscale Al/MoO_3 thermite in microchannels. *Journal of Propulsion and Power* 23: 715–721.

42 Thiruvengadathan, R. (2019). Aluminum-based nano-energetic materials: state of the art and future perspectives. In: *Nano-Energetic Materials* (ed. A.K. Shantanu Bhattacharya, T. Rajagopalan and V.K. Patel), 9–35. Springer Nature Singapore Pvt. Ltd.

43 Valery, J.M. and Levitas, I. (2015). Pre-stressing micron-scale aluminum core-shell particles to improve reactivity. *Scientific Reports* 5: 7879.

44 Levitas, A. and Son, S. (2007). Mechanochemical mechanism for fast reaction of metastable intermolecular composites based on dispersion of liquid metal. *Journal of Applied Physics* 101: 35–43.

45 Levitas, A. (2009). Burn time of aluminum nanoparticles: strong effect of the heating rate and melt-dispersion mechanism. *Combustion and Flame* 156: 543–546.

46 Levitas, V.I. and Pantoya, M.L. (2014). Melt dispersion mechanism for fast reaction of aluminum nano- and micron-scale particles: flame propagation and SEM studies. *Combustion and Flame* 161 (6): 1668–1677.

47 Levitas, V.I., Pantoya, M.L., and Watson, K.W. (2008). Melt-dispersion mechanism for fast reaction of aluminum particles: extension for micron scale particles and fluorination. *Applied Physics Letters* 92 (20): 201917.

48 Levitas, V.I., Pantoya, M.L., and Dikici, B. (2008). Melt dispersion versus diffusive oxidation mechanism for aluminum nanoparticles: critical experiments and controlling parameters. *Applied Physics Letters* 92 (1): 011921.

49 Chowdhury, S., Sullivan, K., Zachariah, M.R. et al. (2010). Diffusive vs explosive reaction at the nanoscale. *Journal of Physical Chemistry C* 114: 9191–9195.

50 Firmansyah, D.A., Sullivan, K., and Lee, K.-S. (2012). Microstructural behavior of the alumina shell and aluminum core before and after melting of aluminum nanoparticles. *Journal of Physical Chemistry C* 116: 404–411.

51 Levitas, V.I., Dikici, B., and Pantoya, M.L. (2011). Toward design of the pre-stressed nano- and microscale aluminum particles covered by oxide shell. *Combustion and Flame* 158 (7): 1413–1417.

52 McCollum, M.L., Tamura, N. et al. (2016). Improving aluminum particle reactivity by annealing and quenchingtreatments: synchrotron X-ray diffraction analysis of strain. *Acta Materialia* 103: 495–501.

53 Bachmaier, A. and Pippan, R. (2011). Effect of oxide particles on the stabilization and final microstructure inaluminium. *Materials Science and Engineering A* 28: 7589–7595.

54 Abdoli, H., Ghanbari, M., and Baghshahi, S. (2011). Thermal stability of nanostructured aluminum powder synthesized byhigh-energy milling. *Materials Science and Engineering A* 528: 6702–6707.

55 McCollum, J., Hill, K.J., Pantoya, M.L. et al. (2016). A slice of an aluminum particle: examining grains, strain and reactivity. *Combustion and Flame* 173: 229–234.

56 Kevin, N.T., Levitas, V.I., Pantoya, M.L. et al. (2018). Impact ignition and combustion of micron-scale aluminum particles pre-stressed with different quenching rates. *Journal of Applied Physics* 124: 115903.

57 Jacob, J., Yang, Y., Zachariah, M.R. et al. (2019). Pre-stressing aluminum nanoparticles as a strategy to enhance reactivity of nanothermite composites. *Combustion and Flame* 205: 33–40.

58 Kappagantula, C.S., Pantoya, M.L., Horn, J. et al. (2012). Tuning energetic material reactivity using surface functionalization of aluminum fuels. *Journal of Physical Chemistry C* 116: 24469–24475.

59 Taylor & Francis Group (2019). Chemistry of pyrotechnics. In: *Component of Energetic Compositions*, 33487–32742. CRC Press, 6000 Broken Sound Parkway NW, Suite 300, Boca Raton, FL.

60 S. Fischer and M. Grubelich (1996). A survey of combustible metals, thermites, and intermetallics for pyrotechnic applications. *Presented at the 32nd AIAA/ASME/SAE/ASEE Joint Propulsion Conference*, Lake Buena Vista, Florida, USA, (July 1996).

61 Weismiller, M.R., Malchi, J.Y., Lee, J.G. et al. (2011). Effects of fuel and oxidizer particle dimensions on the propagation of aluminum containing thermites. *Proceedings of the Combustion Institute* 33 (2): 1989–1996.

62 Puszynski, J.A., Bulian, C.J., and Swiatkiewicz, J.J. (2012). Processing and ignition characteristics of aluminum-bismuth trioxide nanothermite system. *Journal of Propulsion and Power* 23 (4): 698–706.

63 Jan A Puszynski, Chris J Bulian and Jacek J Swiatkiewicz (2006). The effect of nanopowder attributes on reaction mechanism and ignition sensitivity of nanothermites. *Symposium H – Multifunctional Energetic Materials* 896: 0896-H04-01.1-0896-H04-01.12, Boston, USA.

64 Glavier, L., Taton, G., Ducéré, J.M. et al. (2015). Nanoenergetics as pressure generator for nontoxic impact primers: comparison of Al/Bi_2O_3, Al/CuO, Al/MoO_3 nanothermites and Al/PTFE. *Combustion and Flame* 162 (5): 1813–1820.

65 Sanders, V.E., Asay, B.W., Foley, T.J. et al. (2007). Reaction propagation of four nanoscale energetic composites (Al/MoO_3, Al/WO_3, Al/CuO, and Bi_2O_3). *Journal of Propulsion and Power* 23 (4): 707–714.

66 Trebs, A. and Foley, T.J. (2010). Semi-empirical model for reaction progress in nanothermite. *Journal of Propulsion and Power* 26: 772–775.

67 John Alexander Innes, Robin John Batterham, Rod James Dry (1897). Method of producing metals and alloys. US Patent.

68 Fuzellier, H. and Comet, M. (2000). Étude synoptique des explosifs. *Actualite Chimique* 233 (7–8): 4–11.

69 E. Lafontaine and M. Comet (2016). *Nanothermites.* 111 River Street, Hoboken, NJ 07030, USA.

70 Armstrong, B., Booth, D.W., and Samirant, M. (2003). Enhanced propellant combustion with nanoparticles. *Nano Letters* 3: 253–255.

71 Prakash, A. (2005). Synthesis and reactivity of a super-reactive metastable intermolecular composite formulation of Al/$KMnO_4$. *Advanced Materials* 17: 900–903.

72 Prakash, A., McCormick, A.V., and Zachariah, M.R. (2005). Tuning the reactivity of energetic nanoparticles by creation of a core-shell nanostructure. *Nano Letters* 5: 1357–1360.

73 Leung, A.M., Pearce, E.N., and Braverman, L.E. (2010). Perchlorate, iodine and the thyroid. *Best Practice & Research Clinical Endocrinology & Metabolism* 24: 133–141.

74 Li, R., Shen, J.P., Hua, C. et al. (2013). Preparation and characterization of insensitive HMX/graphene oxide composites. *Propellants, Explosives, Pyrotechnics* 38: 798–804.

75 Memon, N.K., AW, M.B., and Son, S.F. (2016). Graphene oxide/ammonium perchlorate composite material for use in solid propellants. *Journal of Propulsion and Power* 32 (3): 682–686.

76 Thiruvengadathan, R., Basuray, S., Balasubramanian, B. et al. (2014). A versatile self-assembly approach toward high performance nanoenergetic composite using functionalized graphene. *Langmuir, American Chemical Society* 9: 158–165.

77 Yan, L., Hao, H., Hui, L. et al. (2017). Iron oxide/aluminum/graphene energetic nanocomposites synthesized by atomic layer deposition: enhanced energy release and reduced electrostatic ignition hazard. *Applied Surface Science* 408: 51–59.

78 Liu, P.A., Wang, M.J., Wang, L. et al. (2019). Effect of nano-metal oxide and nano-metal oxide/graphene composites on thermal decomposition of potassium perchlorate. *Chemical Papers* 73 (6): 1489–1497.

79 Ahmed Fahd, C., Dubois, J.C. et al. (2021). Synthesis and characterization of tertiary nanothermite CNMs/Al/$KClO_4$ with enhanced combustion characteristics. *Propellants, Explosives, Pyrotechnics* 46: 995–1005.

80 Kun Gao, G., Luo, Y., Wang, L. et al. (2014). Preparation and characterization of the AP/Al/Fe_2O_3 ternary nano-thermites. *Journal of Thermal Analysis and Calorimetry* 118: 43–49.

81 Sharma, V. (2017). Effect of carbon nanotube addition on the thermite reaction in the Al/CuO energetic nanocomposite. *Philosophical Magazine* 5: 1921–1938.

82 Peng-Gang Ren, Y., Ji, X., Chen, T. et al. (2011). Temperature dependence of graphene oxide reduced by hydrazine hydrate. *Nanotechnology* 22: 1–8.

83 Piekiel, N., Sullivan, K., Chowdhury, S. et al. (2010). The role of metal oxides in nanothermite reactions: Evidence of condensed phase initiation. DTIC Document.

84 Fahd, A., Dubois, C. et al. (2021). Combustion behavior and reaction kinetics of GO/Al/oxidizing salts ternary nanothermites. *Thermal Analysis and Calorimetry* 8: 1–11.

85 Lan, Y., Jin, M., and Lu, Y. (2015). Preparation and characterization of graphene aerogel/Fe_2O_3/ammonium perchlorate nanostructured energetic composite. *Journal of Sol-Gel Science and Technology* 74: 161–167.

86 Lan, Y. and Luo, Y. (2015). Preparation and characterization of graphene aerogel/ammonium nitrate nano composite energetic materials. *Chinese Journal of Explosives and Propellants* 38: 15–18.

87 Yu, L., Guo, X.Y., Jiang, X.B. et al. (2014). A novel ε-HNIW-based insensitive high explosive incorporated with reduced graphene oxide. *Journal of Thermal Analysis and Calorimetry* 117: 1187–1199.

88 Zhang, W., Zhang, Y., Bhattacharia, S.K. et al. (2013). Direct laser initiation and improved thermal stability of nitrocellulose/graphene oxide nanocomposites. *Applied Physics Letters* 102: 141905.

89 Cai, H., Huang, B., Yang, G. et al. (2013). 1,1-Diamino-2,2-dintroethene FOX-7, nanocrystals embedded in mesoporous carbon FDU-15. *Microporous and Mesoporous Materials* 170: 20–25.

90 Xin-ming, Q.I., Nan, D.E., Si-fan, W., and Zeng-yi, L. (2009). Catalytic effect of carbon nanotubes on pyrotechnics. *Chinese Journal of Energetic Materials* 17: 603–607.

91 Liu, L.M., Yi, Y., Zhang, H.F. et al. (2014). Influence of CNTs on thermal behavior and light radiation properties of Zr/$KClO_4$ pyrotechnics. *Chinese Journal of Energetic Materials* 22: 75–79.

92 Sylvain Desilets, Patrick Brousseau, Nicole Gagnon, Sebastien Cote, Serge Trudel (2004). Flash-ignitable energetic material. Canadian Patent Application CA2434859A1.

93 Guo, R., Shen, R., and Ye, Y. (2014). Electro-explosion performance of KNO_3-filled carbon nanotubes initiator. *Journal of Applied Physics* 115: 174901.

94 Zhang, J.C., Li, Y.J., and Zhang, Z.B. (2015). Preparation and property studies of carbon nanotubes covalent modified BAMO-AMMO energetic binders. *Journal of Energetic Materials* 33: 305–314.

95 Assovskiy, G. and Berlin, A.A. (2009). Metallized carbon nanotubes. *International Journal of Energetic Materials and Chemical Propulsion* 8: 281–289.

96 Zhang, C., Luo, Y., Zhang, X., and Zhai, B. (2013). Preparation and properties of carbon nanotubes modified glycidyl azide polymer binder film. *Polymeric Materials Science and Engineering* 29: 105–108.

97 Jeong, Y., Bae, J.J., Chae, S.H. et al. (2013). Tailoring oxidation of Al particles morphologically controlled by carbon nanotubes. *Energy* 55: 1143–1151.

98 Liu, Y., Liu, J.X., Wang, Y. et al. (2008). Study of catalyzing thermal decomposition and combustion of AP/HTPB propellant with nano Cu/CNTs. *Acta Armamentarii* 29: 1029–1033.

99 Zeng, G.Y., Lin, C.M., Zhou, J.H. et al. (2012). Influences of carbon nanotubes on the thermal decomposition behavior of HMX. *Chinese Journal of Explosives and Propellants* 356: 55–57.

100 Kappagantula, K. and Pantoya, M.L. (2012). Experimentallymeasured thermal transport properties of aluminumpolytetrafluoroethylene nanocomposites with graphene and carbon nanotube additives. *International Journal of Heat and Mass Transfer* 55: 817–824.

101 Zhang, J. and Jean'ne, M.S. (2016). 3D nitrogen-rich metal–organic frameworks: opportuni- ties for safer energetics. *Dalton Transactions* 45: 2363–2368.

102 Hui Su, Z., Du, Y., Zhang, P. et al. (2018). New roles of metal–organic frameworks: Fuels for aluminum-free energetic thermites with low ignition temperatures, high peak pressures and high activity. *Combustion and Flame* 191: 32–38.

103 Hui Su, J., Du, Y., Zhang, P. et al. (2017). New roles for metal–organic frameworks: fuels for environmentally friendly composites. *RSC Advances* 7: 11142–11148.

104 Fahd, A., Dubois, C., Boffito, D.C. et al. (2021). Combustion characteristics of EMOFs/oxygenated salts novel thermite for green energetic applications. *Thermochimica Acta*.

105 Baev, V.I., Lvov, C.N., and Shugalei, I.V. (1998). Influence of the environment internal factors on warm-blooded organisms. In: *Proceedings of the Problems of Military and Emergency Medicine*, St. Petersburg, Russia, 62–63.

106 Comet, M., Klaumunzer, M., Schnell, F. et al. (2015). Energetic nanocomposites for detonation initiation in high explosives without primary explosives. *Applied Physics Letters* 107: 243108.

107 Ying Zhu, X., Xu, J., Ma, X. et al. (2018). In situ preparation of explosive embedded CuO/Al/CL-20 nanoenergetic composite with enhanced reactivity. *Chemical Engineering Journal* 354: 885–895.

108 Luo, X., Nie, F., Liu, G. et al. (2018). The safety properties of a potential kind of novel green primary explosive: Al/Fe_2O_3/RDX nanocomposite. *Materials* 11: 1–10.

109 Steven, A., Apperson, J., Thiruvengadathan, R. et al. (2009). Characterization of nanothermite material for solid-fuel microthruster applications. *Journal of Propulsion and Power* 25: 1–6.

110 Apperson, J., Subramanian, S., Tappmeyer, D. et al. (2007). Generation of fast propagating combustion and shock waves with copper oxide/aluminum nanothermite composites. *Applied Physics Letters* 91: 243109.

111 Apperson, J. (2010). Characterization and MEMS applications of nanothermite materials. Doctoral Faculty of the Graduate School, University of Missouri.

112 Baras C, Spitzer D, Comet M, Ciszek F, Sourgen F, inventors; Institut Franco Allemand de Recherches de Saint Louis ISL, assignee. Piloting device of a missile or of a projectile. United States patent US 8,716,640. 2014 May 6.

113 Comet, M., Pichot, V., Gibot, P., and Spitzer, D. (2008). Preparation of explosive nanoparticles in a porous chromium(III) oxide matrix: a first attempt to control the reactivity of explosives. *Nanotechnology* 19: 285716.

114 Fahd, A., Dubois, C., Chaouki, J. and Wen, J.-Z. (2022). Thrust characteristics of nano-carbon/Al/Oxygenated salt nanothermites for micro-energetic applications. *In preparation.*

115 Bulian, C.J., Smith, S., and Puszynski, J.A. (2006). Experimental and modeling studies of self-sustaining reactions between nanopowders. In: *AIChE Annual Meeting CD-ROM*. New York: American Institute of Chemical Engineers.

116 Fahd, A., Dubois, C., Chaouki, J., and Wen, J.-Z. (2021). Superior performance of quaternary NC/GO/Al/$KClO_4$ nanothermite for high speed impulse small-scale propulsion applications. *Combustion and Flame* 232: 1–13.

9

Engineering Particle Agglomerate and Flame Propagation in 3D-printed Al/CuO Nanocomposites

Haiyang Wang and Michael R. Zachariah

University of California, Department of Chemical and Environmental Engineering, Riverside, CA 92521, USA

9.1 Introduction

Because of high combustion enthalpy (such as with oxygen) and availability, conventional aluminum microparticles (Al MPs) are commonly used in solid rocket propellant and other propulsive systems [1–3]. Although the total energy density of the system is increased after the addition of Al MPs, the burning rate was not found to improve much, resulting in a low energy release rate [4]. Additionally, Al MPs have a high ignition temperature of ~2300 K, delaying the ignition of the system [2, 5, 6]. Luckily, the use of Al NPs lower the ignition temperature to <1000 K, with an enhanced burning rate and a lower ignition delay time [6–12]. Typically, as shown in Figure 9.1, Al MPs-based propellants have a stand-off flame zone (Figure 9.1a), due to the high ignition temperature and delay time of Al MPs, whereas the Al NPs-based propellant surface has a short- stand-off and much smaller burning particles [13]. Therefore, for Al MPs burning, most of the heat release occurs far from the burning surface, which provides weak heat feedback. In contrast, the heat feedback is enhanced with Al NPs as they are burning on or near the propellant surface. Typically, the maximum increase is by a factor of two when replacing Al MPs with NPs in solid propellants.

Another issue in the utilization of nanomaterials is the particle loading limitations in the fabrication of polymer nanocomposites. While high particulate loading can be found in polymer composites these systems involve relatively large super-micron particles [14]. Nanometallic fuels in contrast have been limited by significant processing challenges because the integration of nanoparticles into polymers significantly increases viscosity, making traditional casting methods unfeasible as well as limiting additive manufacturing approaches. As such nanothermites have found limited implementation because particle loadings are not sufficient to obtain high energy densities [5, 15–17]. One approach to resolve the increased viscosity is to assemble these nanoparticles into microspheres while retaining their nanoscale features [15, 16, 18, 19]. Recently, there has been an emerging interest in using additive manufacturing methods to prepare structural energetic materials

Nano and Micro-Scale Energetic Materials: Propellants and Explosives, First Edition.
Edited by Weiqiang Pang and Luigi T. DeLuca.

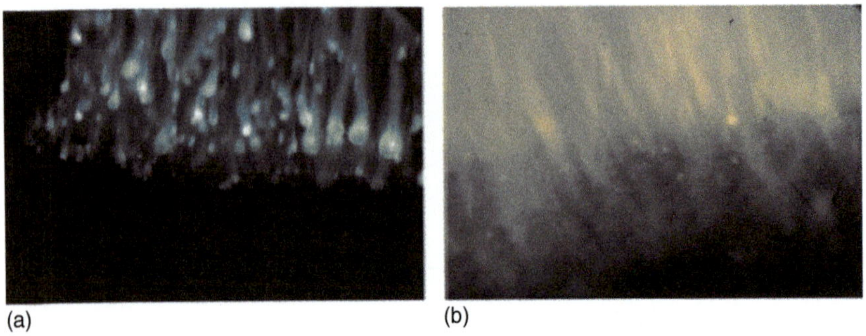

Figure 9.1 Different burning surfaces for solid propellant at 1 MPa with Al MPs (a) and NPs (b). Al MPs are 30 μm spheres in HTPB at 10 bar and Al NPs are 100 nm spheres in polypropylene glycol (PPG) at 10 bar. Source: DeLuca [13], Reproduced with permission from Elsevier.

via templating [20], melting-extruding [21], inkjet printing [22], electrophoretic deposition [23], photopolymerization [14], foaming [24], and more. Among the many methods, direct writing of solvent-based inks is of particularly high interest due to its relative simplicity and convenience [25–28], moreover, the safety of the energetic materials can be dramatically enhanced when processed with a solvent. In a common ink, polymeric binders are used to provide structural integrity to the energetic formulations and afford mechanically stable, relatively insensitive, machinable, and formable energetic architectures. Since binders are generally non-energetic, it is preferred that the binder be added in the minimum quantity necessary to maintain the desired mechanical properties. In the last few decades, hydroxyl-terminated polybutadiene (HTPB) has found common use as the binder in the solid propellant, however, the curing time is too long (days) to be used for the direct-writing approach [29–31]. New binders or binder hybrids are desired to achieve high particle loading and high energy density.

Nanothermites are a class of energetic material with fuel (such as Al) and oxidizer (CuO, Fe_2O_3, MoO_3, Bi_2O_3, etc.) being mixed at the nanoscale, which undergoes a rapid redox reaction upon ignition [32]. Compared to conventional CHNO explosives such as TNT, RDX, and HMX, energetic nanocomposites such as nanothermites are attracting more attention for their high enthalpy of reaction and adjustable reactivity [33–39]. Their high surface area gives them a substantially higher energy release rate in comparison to their micron counterparts, thus making them potential candidates for materials whose reactivity falls in-between primary explosives and conventional pyrotechnics. This unique category, which has both military and civilian applications, such as gas generators [40, 41], nanoscale welding [42, 43], micropropulsion [44, 45], ammunition primers [46], and electric igniters [47, 48], as well as energetic additives in explosives and propellants [46, 49, 50]. However, their implementation is limited due to the inherent complexity of heterogeneous combustion that has yet to be fully understood. For example, theoretically, when reducing the size of the composition constituents to the nanoscale, the energy release rate should be enhanced by orders of magnitudes due to a highly increased interfacial area and reduced diffusion distance between fuel and oxidizer [51–53], however, they have yet to exhibit such impressive enhancements [54–56]. One of the major concerns that may explain this underwhelming enhancement is the loss of nanostructure during the reaction, i.e. agglomeration or sintering.

Agglomeration, or sintering, is an important component in the combustion of Al NPs based energetics, which plays a significant role by rapidly melting and coalescing aggregated Al NPs and increasing the initial size of the reacting composite powders before burning (sintering time $\leq$ reaction time) [56]. Reactive molecular dynamics simulation has found that the sintering behavior may be enhanced by an induced built-in electric field between the metal and native oxide coating resulting in a softening of the alumina coating [57]. Agglomeration of aluminum (Al) particles has been commonly observed in solid propellants, which leads to losses in specific impulse (two-phase loss) and ultimately balances out any potential advantages of Al addition [58–60]. Using capture/quench studies, the agglomerate size of Al has been related to the burn rate of solid propellants and various approaches have been explored to efficiently reduce the effect of agglomerations [61–63]. In addition to ex situ studies, recently developed techniques such as time-resolved X-ray imaging [64, 65], and digital in-line holography [59, 60] have been employed to observe these processes in situ.

The loss of nanostructure due to reactive sintering in Al NPs based energetics such as nanothermite powders dramatically impacts the combustion of the composite and, as a result, burn times of the materials do not significantly shorten as a function of diameter and small particle sizes [53–55]. Other studies have shown that the fractional scaling law observed for the gas phase burning of nanoparticles can be corrected by considering sintering effects prior to burning [66]. Dynamic transmission electron microscopy (TEM) used to observe morphological changes in Al/CuO nanothermite aggregates found the creation of phase-separated adjoining spheroids occurs on the order of ~0.5–5 μs, a range observed to depend on the aggregate size (Figure 9.2) [67]. However, the heating, in this case, was by a laser, and its thermometry was unavailable. Even though the reactive sintering phenomenon was speculated in a motionless heating stage TEM and was simulated by the related models, the direct observation of a dynamic reactive sintering process and subsequent propagation process in a nanothermite is still very challenging.

As we mentioned previously, thermites are a class of energetic materials whose reactivity can be systematically adjusted by changing chemistry (e.g. stoichiometry, reactant choice), particle size, mixing state, and architecture/morphology. Changes in chemical composition can be employed to manipulate the reaction mechanism as a means to modulate performance, while the latter variables are more related to physical effects such as heat transfer and interfacial surface area. However, these factors are inherently entangled since the heat transfer rates can impact reaction rate and vice versa, making it difficult to independently probe these effects. Numerous studies have been conducted with various approaches to directly observe the reaction dynamics of the nanothermites, but none have been able to practically probe a propagating reaction front or reactive sintering on a time or length scale commensurate with the phenomena. Some macroscale studies of thermite reaction dynamics

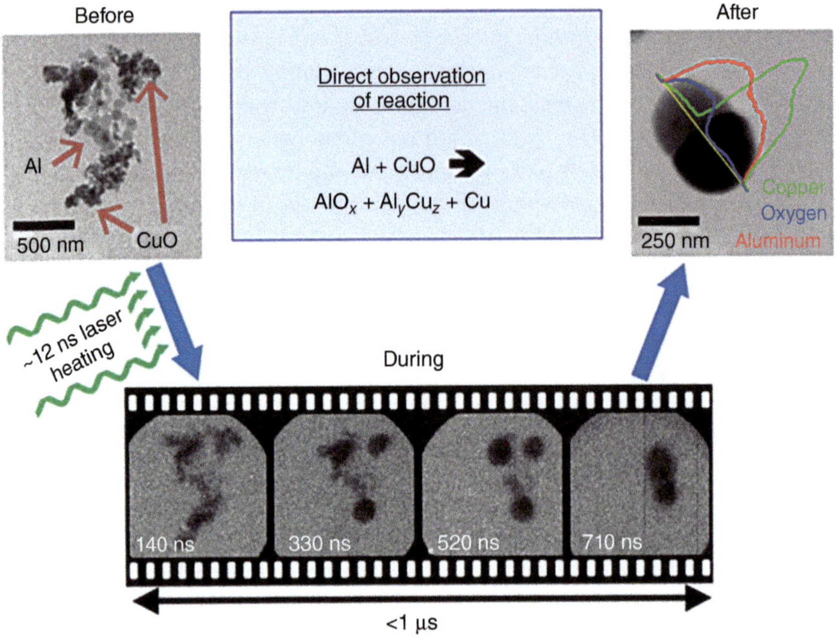

Figure 9.2 Nanothermite materials heated in situ at ~10^{11} K s^{-1} showed significant morphological changes on timescales of 1–5 μs. The reaction between metallic fuel and oxygen carriers produced by the laser heating of aluminum and copper oxide (CuO) nanoparticles (NPs) was investigated (NPs) using movie mode dynamic transmission electron microscopy (MM-DTEM), which enables multi-frame imaging with nanometer spatial and nanosecond temporal resolution. Source: Egan et al. [67], Reproduced with permission from AIP Publishing.

have been captured by T-Jump/time-of-flight mass spectrometer (TOFMS) developed by Zhou et al. [68], which rapidly heats (~10^5 K s^{-1}) material and can provide time resolve ignition and reaction products profiles (Figure 9.3). Another commonly employed device in our lab to quantify reaction dynamics in bulk materials has been a specially modified constant volume combustion cell [69] to simultaneously measure the pressure and optical emission histories of a nanothermite reaction. However, while these measurements provide useful insight on the energy release rate and reaction mechanisms in materials on the appropriate reaction time scales (~μs), they are unable to capture the dynamics and observe the reaction at the particle length scale (μm/nm).

High-speed imaging is quite useful in assessing various combustion regimes as it is the most direct method to interrogate the complexity of such heterogeneous systems during reaction [63–65, 70–72]. Due to the higher flame temperature and burn rate of nanothermites compared to solid propellants, it is difficult to probe the reaction in a high spatial and temporal resolution. To our knowledge, a fast-response and in-operando technique have yet to be employed to observe nanothermite reactions at resolutions high enough to resolve particle-sized phenomena and on the reaction time scale (~μs). In this section, we describe high-speed microscopy and

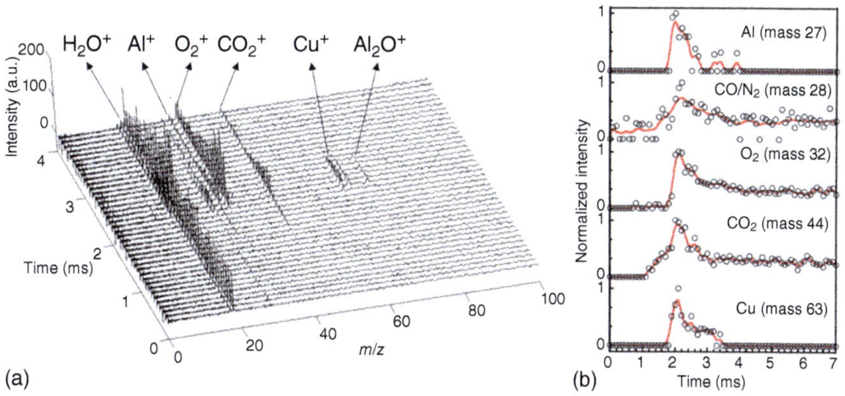

Figure 9.3 Time-resolved mass spectrum (a) and peak intensity (b) of gas species from rapid heating of Al/CuO. Heating rate ~5×10^5 K s^{-1}. Source: Reprinted with permission from Zhou et al. [68]. Wiley Publishing.

pyrometry, which we found to be a useful technique to observe the propagation of printed nanothermites [70–72] at high spatial (μm) and temporal (μs) resolution.

Figure 9.4a shows the major configuration of this study for microscopic characterization, in which a key point is employing a 40 times microscope objective coupling with a high-speed video camera. The microscope objective was focused on the backside of a cover glass slide on which the sample film was printed, thereby allowing the visualization of the flame front without the generated products obscuring the view. With the microscope objective, the pixel/size ratio was ~1 μm/pixel. At this resolution, the flame front (Figure 9.4b,c), as well as a single agglomerating particle (Figure 9.4d,e), could be captured and the corresponding temperature map (Figure 9.4c,) could be obtained. The area was labeled by a thin marker before the ignition, thus allowing us to find the exact area later in a scanning electron microscope (Figure 9.4e). A summary of the kind of information is shown in the panel of images below: Read from right to left are a snapshot taken from a single video frame (Figure 9.4b), the corresponding temperature map (Figure 9.4c) of the corresponding frame, temperature of single agglomeration particle showing a thermal gradient within a particle (Figure 9.4d), and finally the same particle imaged under scanning electron microscopy (SEM) with its energy dispersive spectrometry (EDS) map (Figure 9.4e).

In this chapter, we summarize our recent results about Al/CuO nanothermite composites to systematically show our understanding of how agglomeration affects the propagation rate. Firstly, we developed a universal ink formulation that enables 90 wt% nanoparticle loading with only 10 wt% polymers. Secondly, we extend our understanding of the agglomeration process and its relation to the propagation of Al/CuO nanothermite composites, through an in-operando high spatial (μm) and time (μs) resolution imaging system with pyrometry. Lastly, based on these findings, several approaches to engineer the agglomerations and propagations of Al NPs-based high particle loading nanocomposites are proposed.

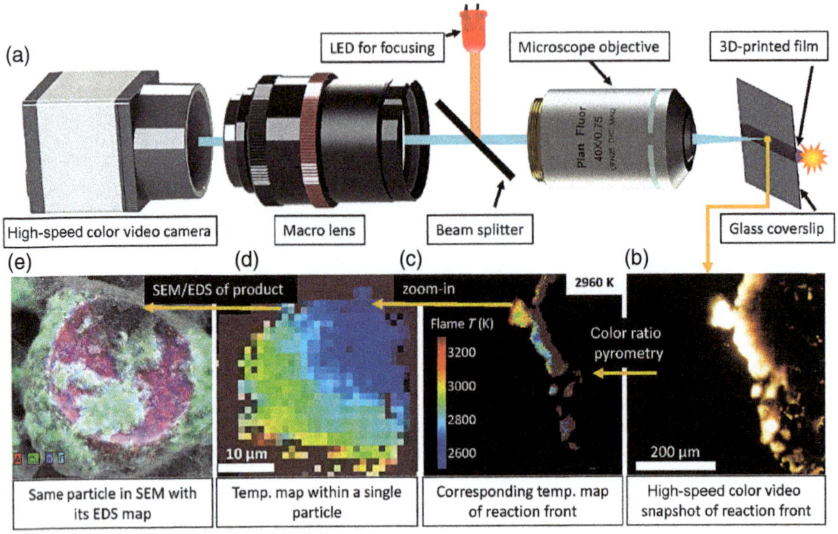

Figure 9.4 (a) Schematic showing high-speed microscope imaging of 3D-printed reactive materials. Read from right to left: (b) High-speed color video snapshot of reaction front; (c) Corresponding temperature map of reaction front; (d) Temperature map within a single particle, and (e) the same particle in SEM with its EDS map. Note: CAD models from Thorlabs (camera and microscope objective) and Ilirjan Leci (macro lens). Source: Wang et al. [73]. Springer Nature, CC BY 4.0.

9.2 Printing High Nanothermite Loading Composite Via a Direct Writing Approach

In this chapter, a simple direct writing approach is employed to print various high Al/CuO loading composite sticks for further characterizations. We develop an ink formulation for 3D printing with ≤10 wt% polymers mixture, which is used to load a total amount of ≥90 wt% nanoparticles. The ability of loading such high percentage of nanoparticles opens up new avenues for the practical applications of 3D printing on energetic materials including propellants, explosives, and pyrotechnics, which have far been unavailable. Adjustment of different energetic materials composition, fuel, and oxidizer equivalence ratio, the measured burn rate, flame temperature, as well as the energy release rate, can be easily tuned significantly. To load nanoparticles in a binder with direct-writing 3D printing, the binder needs to be soluble and its solvent should also have a reasonable boiling point and vapor pressure, to enable its easy removal during the printing, and to ensure safe processing. The ink should also be shear thinning to be extruded easily to make it printable, but beyond this, at such high loadings, unless an extended network is generated, the material will effectively print as a powder with no cohesive strength. Thus the resulting structure should have a mechanical integrity suitable for the application of interest.

We found that to satisfy both the reactive and structural requirements required two binders; a polymer hybrid of hydroxy propyl methyl cellulose (HPMC) and

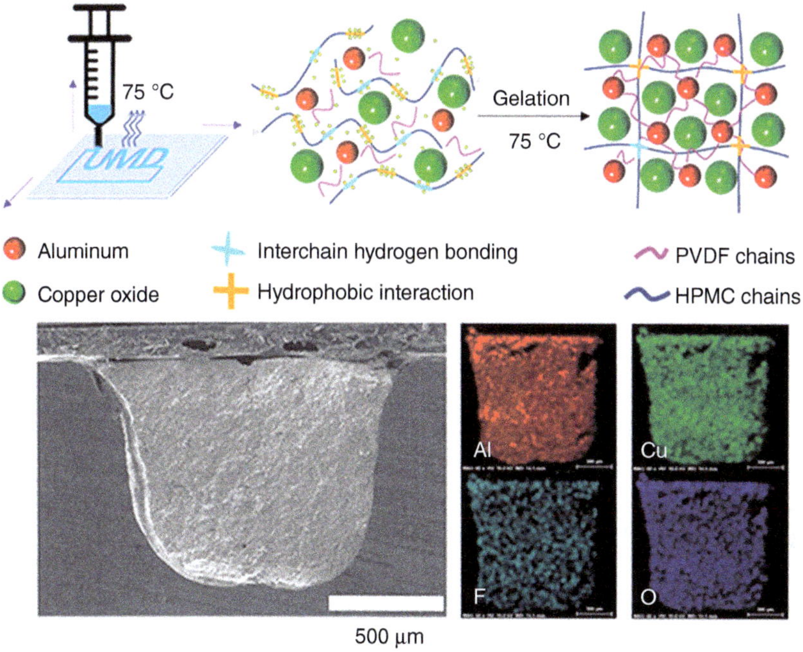

Figure 9.5 Gelation process in 3D printing (upper); Cross-sectional SEM and EDS results (lower) of printed Al/CuO high loading (90 wt%) composite sticks; PVDF: HPMC mass ratio is 2 : 3. Source: Wang et al. [71], Reproduced with permission from John Wiley & Sons.

polyvinylidene fluoride (PVDF) with a mass ratio of 3 : 2. Both polymers are soluble in dimethyl formamide (DMF), and the hybrid mix was found to be very stable with no evidence of separation. PVDF was employed as a polymeric binder since it doubles as an oxidizer and improves the ignitability of the composite *by* promoting the preignition of Al relative to other soluble fluorine-containing polymers such as Viton and THV [74]. HPMC was chosen since it is known to gel via hydrophobic interactions occurring between hydrophobic segments of the polymer chains upon thermal treatment. The very low mass loading of HPMC employed mitigates the fact that it has minimal energy content. This effectively is increasing the degree of crosslinking between polymer chains, forming a continuous 3-dimensional network [75], as schematically shown in Figure 9.5. The apparent viscosity of pure polymer solution is in the scale of 10 Pa s and dramatically increases by two to three orders of magnitude upon addition of solids. However, and most important for this application, the ink viscosity is highly dependent on shear rate. These inks show shear-thinning behavior at a shear rate > 0.05 1/s, with a sharp decline over several orders of magnitude. The ink was directly written on a preheated (~75 °C) glass plate. This thermal treatment is implemented to induce the gelation of HPMC, and evaporation of the solvent (DMF) to form a complete dry layer before a second layer is written [75].

The printing is in predesigned patterns that can be peeled off from the substrate and cut into ~3 cm long sticks for burn rate measurement. The mechanical properties of the sticks are close to that of pure polytetrafluoroethylene (PTFE) [76], confirming the 90% loading nanothermite composite is mechanically strong. For the *fuel-lean* and stoichiometric Al/CuO case, the density was found to be ~2.1 and ~1.8 g cm^{-3}, respectively, which is ~1/3 of the theoretical density. With the increase of Al content, the theoretical and actual density declined gradually owing to the lower density of Al compared to CuO, however, the porosity remains constant at ~66%. While this may seem low, in fact, the theoretical maximum packing density of nanoparticle aggregates is effectively this number, implying that the composite cannot be made denser [77, 78]. A higher packing density could be achieved if the nanoparticles are preprocessed to break the aggregates. The SEM and EDS images (Figure 9.5) of the cross-sectional sticks also confirm the close packing and intimate mixing of Al/CuO in the printed sticks.

For the pure thermite Al/CuO, the adiabatic flame temperature is estimated as ~2840 K and is limited by the boiling point of Cu (~2835 K), one of the major reaction products [79]. The flame propagation snapshots of the burning composite sticks (~3 cm) with normal (Figure 9.6a) and low (Figure 9.6b) exposure (light exposure in high-speed camera) along with the flame temperature maps (Figure 9.6c), and the detailed time-resolved mean/median flame temperature profile (Figure 9.6d) are shown in Figure 9.6. The latter was obtained using a high-speed color camera pyrometry technique detailed in our previous study [74]. All combustion tests were conducted in 1 atm Ar to exclude the effect of additional oxygen [80]. The horizontally propagating flame (Figure 9.6a) indicates very vigorous combustion, generating many hot gas/particles and a bright flame. The flame fronts proceed steadily (Figure 9.6a–c) with time and demonstrate a stable linear burn rate of ~3 cm s^{-1}. The flame temperature remains steady over the length of the burn at 2500–3000 K, and most points are located approximately at ~2800 K. This result is interesting in that it implies we can print a thermite-based material that has a measured flame temperature close to that theoretically expected (~2843 K) and implying the polymer is not impeding the combustion completeness [79]. However, from these snapshots at a resolution near ~100 μm per pixel, it is difficult to observe the Al agglomeration process, and more advanced diagnostic techniques are required for details.

Generally, for a nanothermite, peak reactivity is achieved at or near stoichiometric [81]. Figure 9.7 presents burn rate and temperature measurements as a function of the Al/CuO ratio at 90% loading. The flame temperature peaks at ~22 wt% of Al corresponding to the stoichiometric case ($\Phi = 1$), while the linear burn rate peaks fuel-rich (Figure 9.7), which is probably due to the enhanced gas production and heat convection with more reactive fuel [66, 79, 82]. The relative energy release rate (normalized heat flux Q') in Figure 9.7 peaks between the maximum flame temperature and burn rate implying that if optimization of energy release is desired, a fuel rich formulation ($\Phi = 1.5$–3.5, 30%–50% Al content) is necessary. The calculation of relative energy release rate is as follows:

$$Q' = \frac{Q}{t} = \frac{m \times C_p \times \Delta T}{t} = \frac{\rho \times L \times A \times C_p \times \Delta T}{t} = (A \times C_p) \times \rho \times \left(\frac{L}{t}\right) \times \Delta T \qquad (9.1)$$

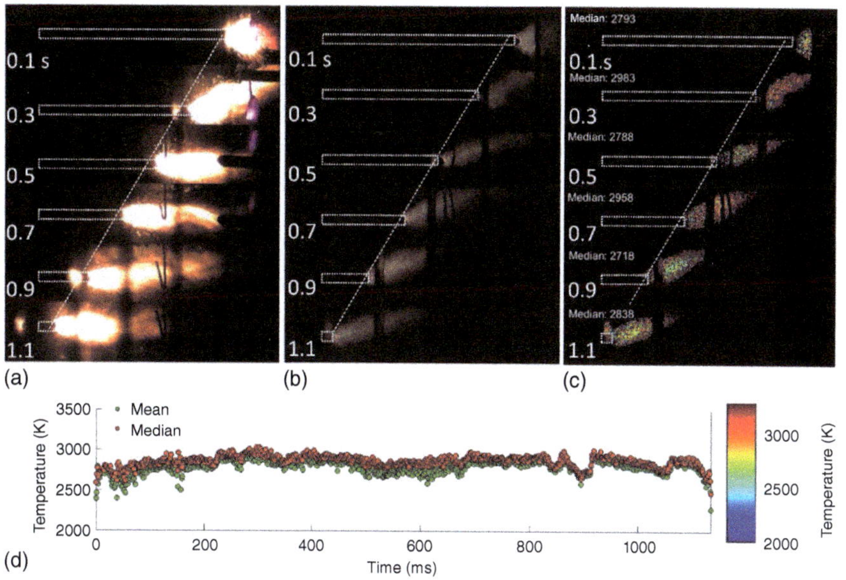

Figure 9.6 Burning snapshots with normal exposure (a), low exposure (b), and the measured temperature map (c) and curve (d) with time for 15-layered stick with 90 wt% Al/CuO nanothermite loading (6 wt% Methocel, 4 wt% PVDF, Al in Al/CuO is 22 wt%). White dotted rectangles represent the burn stick. Source: Wang et al. [71], Reproduced with permission from John Wiley & Sons.

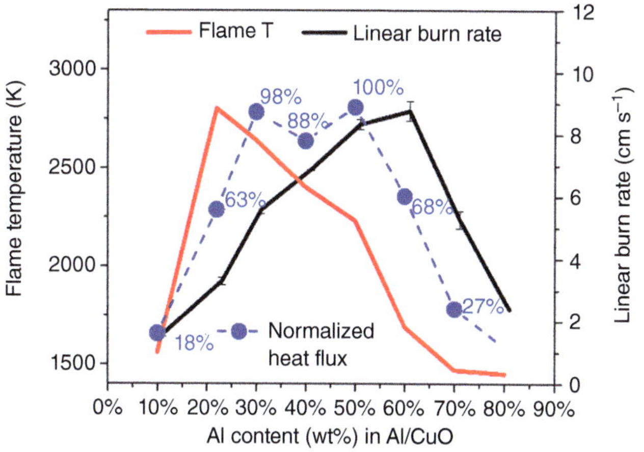

Figure 9.7 Linear burn rate, flame temperature, and normalized heat flux of the burning sticks with 90 wt% nanothermite loading (with different Al content in Al/CuO from 10 to 90 wt%). Source: Reprinted with permission from Wang et al. [71], Wiley Publishing.

where Q is the energy release, C_p is the heat capacity, and T is the flame temperature. ρ is the density, t is the time, A is the cross-sectional area, and L/t is the velocity (ν). If we are assuming the cross-sectional area and heat capacity of different composites are roughly the same, the relative energy release rate (Q') is proportional to $\rho \times \nu \times T$.

In summary for this section, we developed an energetic ink formulation with a particle loading of 90 wt%, which can be used for direct writing of 3D structures. The key additive in the ink is a hybrid polymer of PVDF and HPMC, in which the former serves as an energetic initiator, and the latter is a thickening agent and a binder that can adhere to the particles with a small percentage of polymer. The best polymer ratio (best printing resolution) was found to be 4 wt% PVDF and 6 wt% HPMC enabled particle loadings as high as 90 wt%. The rheology shear thinning properties of the ink was critical to making the formulation at such high loadings printable. The Young's Modulus of the printed stick is found to compare favorably with PTFE, with a particle packing density at the theoretical maximum. The linear burn rate, mass burn rate, flame temperature, and heat flux were found to be easily adjusted by varying the fuel/oxidizer ratio. The average flame temperatures are as high as ~2800 K with near-complete combustion being evident upon examination of the post-combustion products. As we change the equivalence ratio of Al/CuO, the burn rate, flame temperature, and energy release rate vary significantly, as well as the flame structure and Al agglomeration status. It is important to look into these factors in more advanced diagnostic techniques to find out the fundamental mechanism behind this. The Al/CuO high-loading composite sticks ($\Phi = 1$) present in this section are used as the reference (baseline) for the following studies.

9.3 Agglomerating in High Al/CuO Nanothermite Loading Composite

9.3.1 In-Operando Observation of Flame Front

Figure 9.4b show a typical flame front snapshot of the Al/CuO nanothermites (90 wt%) captured in a window of 512 μm × 512 μm with a frame rate of ~18 000 (55.5 μs per frame) and the corresponding flame temperature map (Figure 9.4c) was obtained by a color camera pyrometry through image processing. We can see the flame fronts consist of stochastic bright spots, which discontinuously propagate the reaction. These bright areas are roughly divided into a leading flame front and the following cooling zone, as distinguished by the brightness. The noticeably brighter area spanning ~30 μm was confirmed to be the leading edge from the measured temperature ~ 3000 K, which is close to the previously measured reaction temperature of Al/CuO [80].

Reactive sintering, as discussed above and previously observed [56, 57, 66, 67], is the coalescence of aggregated and/or agglomerated nanoparticles driven by heat released during the reaction and results in the effective loss of nanostructure. As Figure 9.4d shows, after the flame front has passed any given area, the sintered particle with a mean diameter of ~20 μm is observed. It is also notable that the flame front thickness is roughly the same size as the agglomeration size, a reasonable observation considering that these flame fronts were constructed by networks of individually sintering particles propagating the reaction by either advection or heat release.

9.3.2 Mapping Optical to Electron Microscopy of Agglomeration

To investigate the morphology, composition, and size distribution of the post-combustion product, the reaction product-coated slide was examined by SEM and energy EDS. As shown in Figure 9.4e, we were able to find an exact one-for-one correspondence between the *in-operando* microscope imaging and the SEM image. It is also notable that we can achieve a dynamic *in-operando* temperature measurement of a single sintering particle on the resolution of μm. Furthermore, there is a ~1000 K temperature difference across the particle at a distance of ~30 μm, which indicates a temperature gradient of $3 \pm 1 \times 10^7$ K m^{-1} and in the same direction as the flame propagation in this area. From the EDS results, we also concluded that the main composition of the sintered particle is Al_2O_3 (~20 μm) and the coated smaller particles are Cu (<20 nm). Such small Cu nanoparticles indicate that Cu was vaporized during the reaction due to a flame temperature (≥2900 K) above the boiling point of copper (2835 K).

To further explore this difference between micro and macro scales, a group of sintering particles was closely monitored as a bridge between reactive sintering and flame propagation. In Figure 9.8a, flame propagation is from right to left. An initial burning spot (frame 2) spreads to the surrounding area, where another two sintered particles appear. However, before cooling, the adjacent area on the left-top and right-bottom are observed to ignite and then move to the left. The schematic of the above process is demonstrated in Figure 9.8b and the final sintered particles are shown in Figure 9.8c. Even though reactive sintering occurs in ~μs, the propagation of the reaction is relatively slow and limited by the heat conduction and high ignition temperature of Al/CuO (~1000 K) [57] which in the microscope temporal image shows an almost stochastic behavior, despite the overall reaction front moving in a given direction. This suggests that density gradients set up local

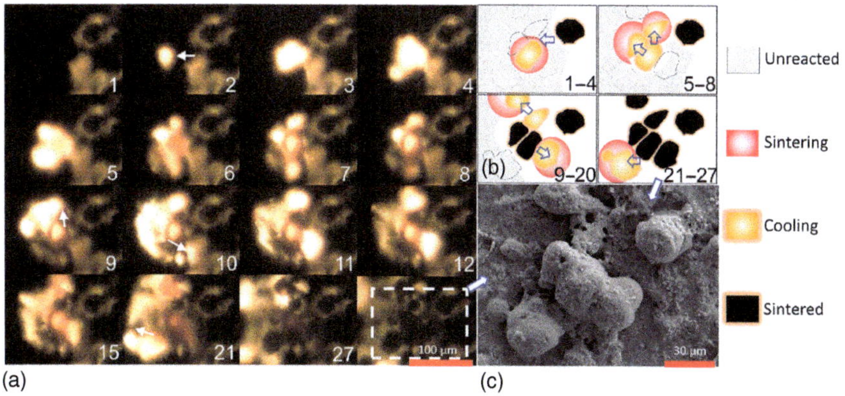

Figure 9.8 (a) Series of reactive sintering and ignition snapshots of a group of particles (labeled is frame sequence, 55.55 μs per frame) and (b) its corresponding schematic cartoon. Note: the marked arrows are propagating direction. (c) The corresponding SEM image of the sintered Al_2O_3 particles coated with Cu nanoparticles, as evident by the EDS results. Source: Wang et al. [73], Springer Nature CC BY 4.0.

sintering regions which comprise heat generation centers. These centers transport energy to neighboring areas that ignite. In this way, propagation is very similar to the general concept of laminar flame theory in which flame velocity ~ (thermal diffusivity × reaction rate)$^{1/2}$. The difference is that the macro scale reaction velocity is limited by the thermal diffusivity between the sintered heat sources.

9.3.3 Agglomeration Affects the Propagation Rate

To evaluate the above argument, consider that during the passing of the reaction front (i) reactive sintering, (ii) cooling, and (iii) a final sintered product is observed. The fact that the flame front consists of these different stages implies an inhomogeneous reaction front with different heat fluxes (Figure 9.9a,). We can estimate heat flux ~ thermal conductivity × temperature gradient = 10^9 W m^{-2} in the reactive sintering stage based on the measured temperature gradient and an estimate thermal conductivity for Al/CuO ~60 W m^{-1} K^{-1}. This large high heat flux supports a rapid front propagating with a velocity as high as ~50 cm s^{-1}. However, when reactive sintering has mostly completed (cooling stage), the heat flux declines by ~3 orders of magnitude to 10^6 W m^{-2}, owing to the much lower thermal conductivity of gas (0.1 W m^{-1} K^{-1}) which separates the reacted from unreacted material. Thus, while we see local rapid reaction these reaction events are slowed down by low conductivity zones resulting in an over propagation rate that is considerably lower (Figure 9.9b). This calculation indicates that the heat flux in the cooling stage in a reactive sintering process is critical to the macroscale flame velocity.

In this section, high-speed microscopy/thermometry enables us to observe them in-operando micro reaction of Al/CuO nanocomposite thermites. We could visualize the agglomerating process (reactive sintering) in nanothermite reaction in high spatial (~1 μm) and temporal (~55 μs per frame) resolutions. The temperature map of the flame fronts and the agglomerating particles were also obtained by a color camera pyrometry of the same images. The flame front thickness was determined

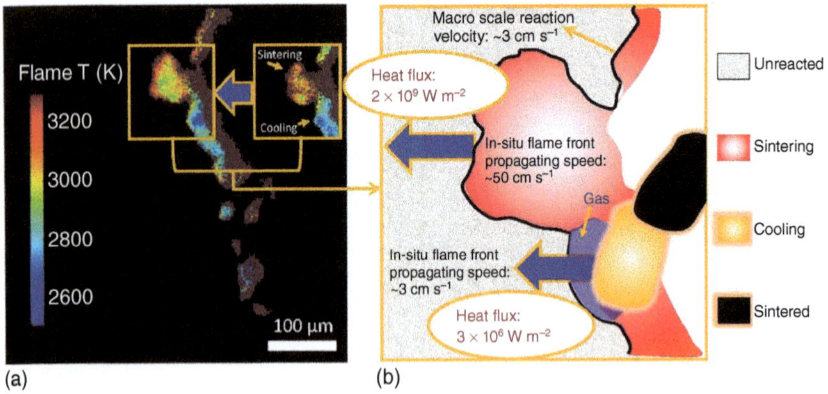

Figure 9.9 (a) Two typical temperature map snapshots of flame front of Al/CuO nanothermite and (b) the schematic showing of heat flux distribution from different stages of reactive sintering particles. Source: Wang et al. [73]. Springer Nature, CC BY 4.0.

from the above results as ~30 μm, which is the same size as the Al agglomeration. Post-reaction analysis of the same area and the corresponding sintered particles were found in the scanning SEM, which enables one-to-one composition mapping. For these studies, we can conclude that the local reaction velocity is an order of magnitude higher than macroscale flame velocity. These results imply that local heat generation with sintering particles limits heat transport to neighboring areas. Heat flux calculations indicate that the macroscale flame velocity was highly dependent on the heat release rate in the cooling stage in a reactive sintering process.

9.4 Engineering Agglomerating and Propagating through Oxidizer Size and Morphology

9.4.1 The Concept of a Pocket Size

The connection between propellant microstructure, agglomerations, and burn rate was first identified in the 1960s by Povinelli and Rosenstein, who compared the effect of oxidizer size on collected agglomerations and burn rate [83]. The heterogeneous nature of agglomerating was also addressed by a "pocket" model. A "pocket" region (Figure 9.10) is where an agglomeration finds a privileged location to take place [84–86]. Cohen [85] established some rules for the definition of final agglomerate size based on local flame temperature and encapsulation criteria, disregarding the random nature of the propellant microstructure. In this section, we are going to change the "pocket" size of Al aggregates by packing them with different sizes and morphology CuO to engineer the Al agglomerations size.

In this section, three different CuO oxidizers were explored in this study with both 100% and 90% thermite loadings: ~5 μm diameter microparticles (CuO MPs),

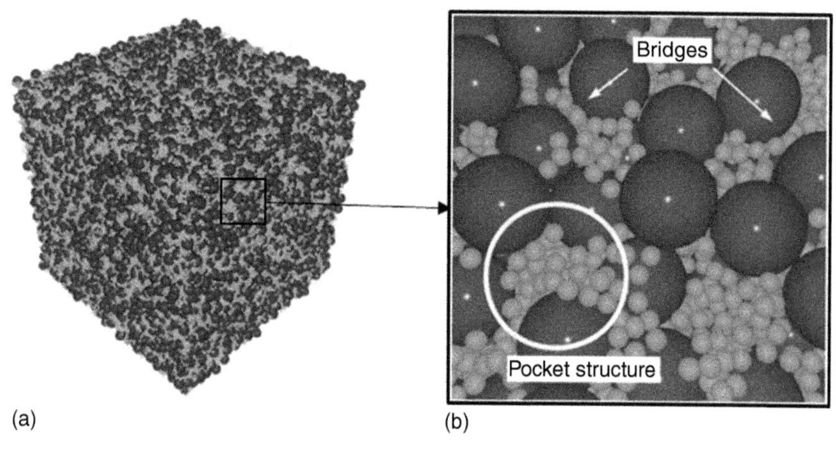

Figure 9.10 Microstructure of a propellant model replicating an Oxidizer/Metal/Binder propellant (61/12/27 vol%, 68/18/14 mass percent). Oxidizer size: 150 μm (dark spheres). Metal size: 30 μm (bright spheres). Source: Reprinted with permission from Ref. [84]. Copyright 2015, by F. Maggi.

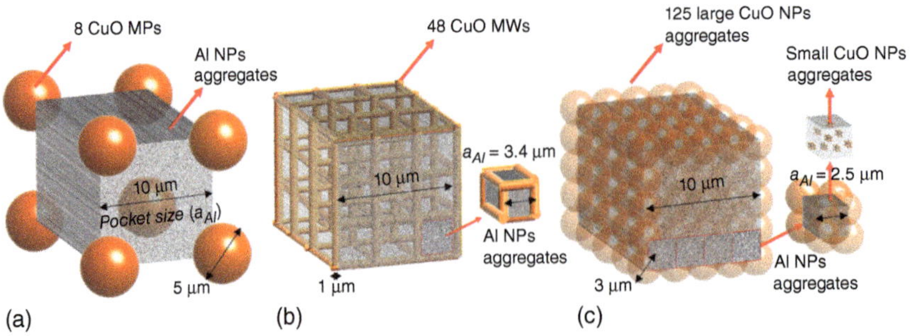

Figure 9.11 Schematic illustrating the "pockets" of Al in Al/CuO composites with different CuO morphologies – MPs (diameter 5 μm), MWs (diameter 1 μm, length 15 μm) and NPs (diameter 3 μm). The resulting pocket size of different composites is labeled as ~10, ~3.4, and ~2.5 μm. Source: Reprinted with permission from Wang et al. [87]. Elsevier Publishing.

~15 μm (length) × 1 μm (diameter) microwires (CuO MWs), ~40 nm diameter nanoparticles (CuO NPs) [87]. In particular, the CuO MWs have a calculated equivalent spherical volume diameter of ~3 μm, so the CuO MWs and CuO MPs are roughly the same volumes. Nevertheless, the calculated specific surface area (per mass) varies widely and can be ranked as CuO NPs (125) ≫ CuO MWs (7) > CuO MPs (normalized to 1).

Herein, we construct a simple model of the pocket size of the three different Al/CuO composites with CuO MPs, MWs, and NPs. The "pocket" of Al NPs is assumed to be a cube whose size is determined by the density and size of the different CuO morphologies employed, which constrains the "pocket." The "pocket-size," (a_{Al}), is the length of the cube (pocket). As shown in Figure 9.11a, to estimate the size of the pocket in the three cases, the volume (Vol.) ratio of Al to CuO was used. To account for the known lower packing density of fractal aggregates [78], the volume of Al and CuO NPs aggregates was increased 3×. For Al/CuO with CuO microparticles and microwires, the Al/CuO volume ratio is 1.95 and 0.65 for Al/CuO with CuO nanoparticles.

The volume-based equivalent diameter (Φ_{CuO}) of CuO MPs and MWs is ~5 and 2.8 μm, respectively. The original size of CuO NPs is ~40 nm and the aggregate size of CuO NPs was estimated as ~3 μm based on SEM observations. Assuming a cubic control volume of 10 μm sides, we will have 8 CuO MPs (8 corners in the cube), 48 CuO MWs (4 × 4 × 3), and 125 CuO NPs (5 × 5 × 5) for this superlattice to construct the structure in Figure 9.11. The size of the smallest confined pocket unit (a_{Al}) of Al NPs in CuO MPs, CuO MWs, and CuO NPs are ~10, ~3.4, and ~ 2.5 μm, respectively (Figure 9.11).

9.4.2 Reducing Agglomeration with CuO Wires

We begin with the microscopic burning images of the CuO MPs-based thermite (90 wt% Al/CuO sticks). The reaction front shows the formation of large, highly emissive, molten droplets (2100–2800 K) formed on the surface (dash lines) of the composite sticks (Figure 9.12a). The temperature of these spheres is significantly higher than the melting point of Al (melting point, MP: 933 K) and CuO (MP: 1600 K) and the formation of spheres confirms they are made up of molten matter. These droplet spheres continually grow in size from the coalescence of smaller droplets driven by surface tension before lifting off from the burning surface (Figure 9.12a–c). The lifting off only occurs just after the Al_2O_3 cap is observed in the droplets and the temperature approaches ~2400 K (Al_2O_3 melting point) [58–63]. The process of coalescence is rapid (<~0.1 ms) and consistent with the timescale we have observed above [73].

The same basic processes appear to be taking place for the CuO MWs and NPs cases (See Figure 9.12b, respectively). However, differences in agglomeration size are obvious in the images and are quantified in the size distribution histograms shown in Figure 9.12a-1~c-1. In both the CuO MW and NP cases, the droplet sizes are much smaller compared to the CuO MPs case (~15 versus ~40 μm). Consistent with the coalescing size in Figure 9.12, the post-combustion residues also confirm that the residue size of CuO MPs case is ~5–10 times larger than that of CuO MWs and NPs.

In summary, we see that the MWs have a behavior closer to that of the NPs over that of the MPs in both microscopic and macroscopic scales. We must now ask why microwires produce such small droplets in comparison to microparticles. There are many complex phenomena involved, but one possible explanation is inspired by the "pocket model" theory where the volume of the agglomerated particles produced is predominately controlled by the effective volume of Al particles that can be aggregated within a surrounding oxidizer matrix (depicted in Figure 9.11) [84–86, 88].

As schematically shown in Figure 9.11, the CuO MWs with a high aspect ratio have ~1/30 smaller pocket volume ($a_{Al}{}^3$) compared to the CuO MPs. The point is that the smaller pocket limits the effective size of the Al aggregate that can be formed during combustion [78]. If we assume that the pocket operates independently due to the gas generation of CuO (which can break apart sintered aggregates), we can estimate the size of the resulting sintered Al particle since we know that the packing density of Al fractal aggregates in the pocket is ~35% [71, 78]. Based on this, the diameter of the average Al sintering size (Φ_{Al}) from the corresponding "pocket" can be calculated as 8.8, 2.9, and 2.2 μm for MPs, MWs, and NPs, respectively ($\Phi_{Al} = 0.874a_{Al}$).

This result indicates that the microwires should yield Al sintered particles that are about the same size as those generated from composites made with CuO NPs, and about a factor of three smaller than those formed during reactions with CuO MPs. This is qualitatively consistent with the experimental observation (see histograms in Figure 9.12). Our calculated particle sizes are smaller than measured, however, this is expected since the assumption of independent pockets is a gross approximation. Nevertheless, the trends are unambiguous.

9.4.3 Promote Propagating through Using CuO Wires

Figure 9.12 also shows that the molten droplets in the cases of the CuO MWs and NPs (~1.2 m s^{-1}) are ejected with higher velocity than the CuO MP case (~0.4 m s^{-1}). The differences in molten droplets velocities suggest higher gas generation rates, and thus faster reaction, and are consistent with the macroscopic flame propagation

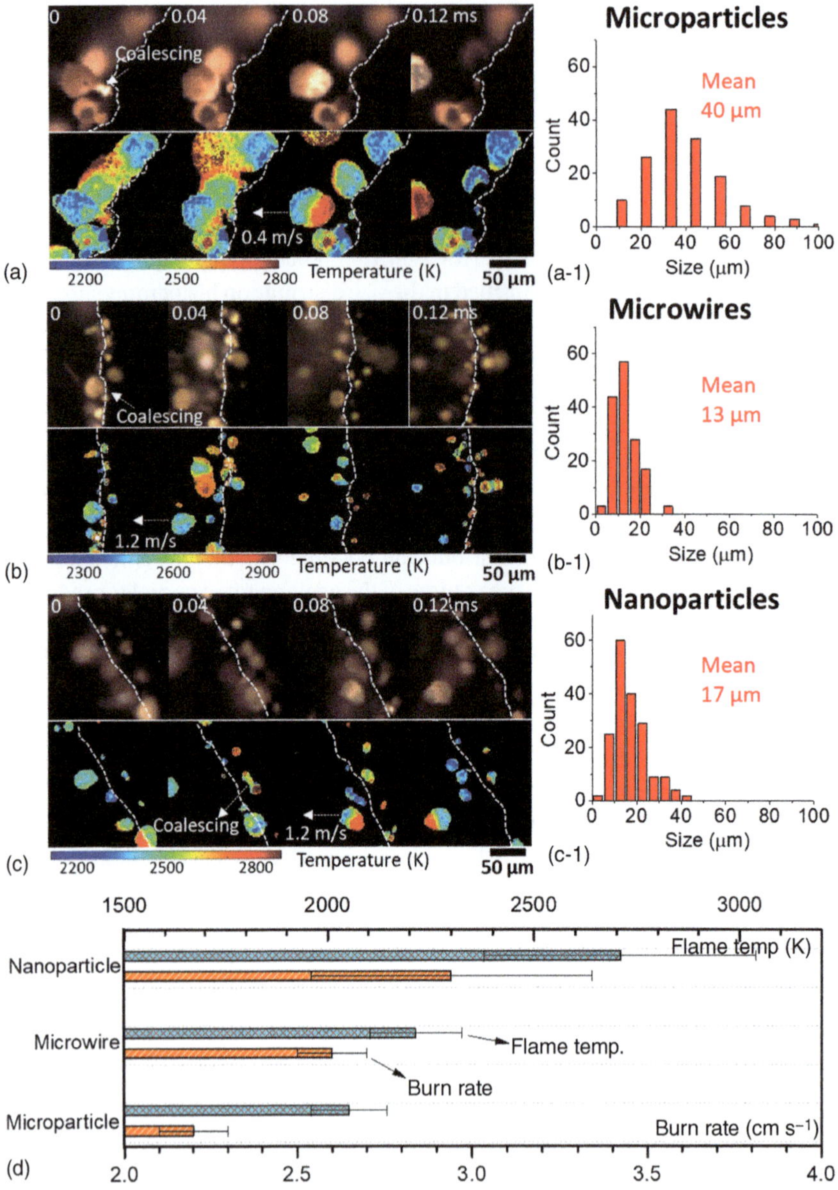

Figure 9.12 The evolution of agglomerations (a–c) with different size distributions (a-1~c-1) was observed in the microscopic burning surface of Al/CuO composite sticks (90 wt%) with CuO microparticles (a), microwires (b), and nanoparticles (c). The sizes of the agglomerations were measured and averaged based on >150 agglomeration droplets in the high-speed videos. Summary of burning rate and flame temperature (d). Δt is ~42 μs between two neighbor frames. Source: Wang et al. [87], Reproduced with permission from ELSEVIER.

results. As shown in Figure 9.12d, the burn rate and flame temperature ranked as CuO NPs > CuO MWs > CuO MPs. The higher temperature for the nanomaterial is not surprising as it suggests more complete combustion.

To more directly demonstrate the effects of CuO size and morphology on thermite propagation, we 3D-printed Al/CuO ink on a glass slide to form a thin thermite film without using any binder. The thermite film is very brittle but could hold its integrity during testing. We estimated their relative density to be 1 (CuO MPs), ~0.83 (CuO MWs), and ~ 1.07 (CuO NPs), based on their thicknesses (weight and width/length is the same). With CuO MWs and NPs, the films burn at an average speed of ~55 and ~33 $m\,s^{-1}$, respectively, which is >15× higher than Al/CuO with CuO MPs (~2 $m\,s^{-1}$). Moreover, the flame temperatures of CuO MW- and NP-based thermites are measured as ~3000–3200 K, respectively, which is ~500 K higher than that of the CuO MPs case (~2550 K).

9.4.4 Polymer Addition Significantly Reduces the Micro-Explosion of the Agglomerations

A notable observation in this study is that only 10 wt% polymer reduced the burn rate of Al/CuO with microspheres by the 2 orders of magnitude (~2 $m\,s^{-1}$ versus ~2 $cm\,s^{-1}$). It is likely that this decrease in propagation rate could be attributed to the polymer acting as a heat sink. A recent study by our group suggests that the agglomeration is only able to transfer a small fraction of the energy required to ignite nearby unreacted regions when considering the polymer and reactants in the burn sticks [72]. However, if the polymer was removed from consideration (as it is in the powder cases), the amount of energy required to ignite unreacted areas decreases, therefore the role of agglomeration as an energy transfer and propagation method becomes more significant. Whereas the agglomeration in the burn stick case provides a limited enhancement in propagation via advection, it is more effective as a heat transfer mechanism and can increase the propagation rate in the thermite powder case.

The polymer used in this section acting as a heat sink also seemingly plays a role in the number of micro-explosions happening in these agglomerations, which further enhance the reactivity. When examining the microscopic combustion videos of the Al/CuO microsphere composite films (100 wt%, no polymer), clear evidence of bubbling and micro-explosions. A complete picture of the overall process which illustrates the coalescing, bubbling, and micro-explosion process, and a schematic cartoon is shown in Figure 9.13. Agglomerations are seen to coalesce and grow into larger droplets within ~0.1 ms, and subsequent bubbling and micro-explosions to start a new cycle of this process. Micro-explosions are known to play a significant role in enhancing reactivity by creating smaller structures with higher surface areas. It is noted that no clear micro-explosion was observed in the low reactivity Al/CuO microsphere burn stick sample (90 wt% reactive material), while the Al/CuO microsphere powder sample (100 wt% reactive material) has numerous clearly-defined micro-explosions in the burning videos (see supporting information).

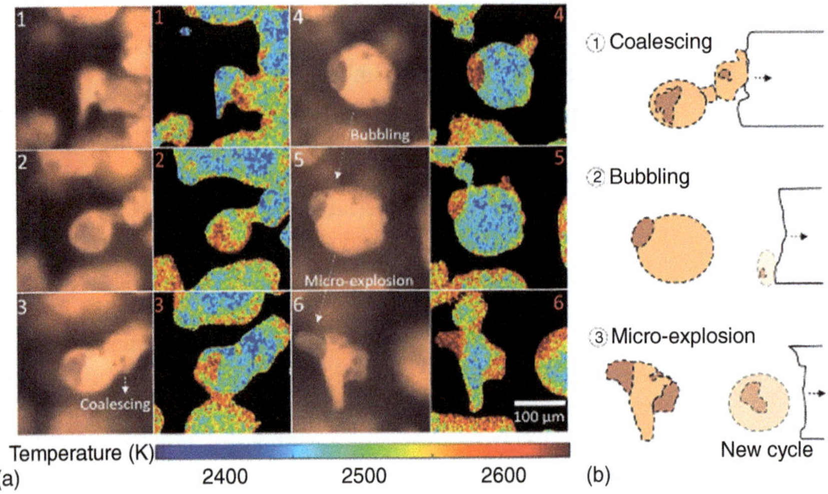

Figure 9.13 Coalescing, bubbling, and micro-explosion processes were observed in the 3D-printed Al/CuO (with CuO microspheres) composites (100 wt%, no polymer) on a thin glass slide. Δt is ~42 μs between two neighbor frames. Source: Haiyang Wang, Michael R. Zachariah.

The absence of micro-explosions in the Al/CuO microsphere sticks (10 wt% polymers) burning could be attributed to a ~ 500 K decrease in flame temperature (mean: ~2050 K) compared to the pure thermite composition (mean: ~2550 K), which makes the vaporization of CuO (boiling point: ~2270 K) much less likely to happen. As the thermite reaction proceeds, the temperature of the molten particles exceeds the boiling point of Al (2743 K) and Cu (2835 K), which causes more micro-explosions to enhance the reactivity.

In summary, the polymer addition to the thermite powder negatively impacts the propagation rate in two ways: (i) the polymer acts as a heat sink which reduces the flame temperature and, in turn, reduces the number of micro-explosions, and (ii) the thermal mass of the polymer makes the energy transferred via advection of micro-explosion byproducts less impactful.

9.4.5 Summary

In this section, high-speed microscopy and pyrometry on a ~μm space and ~μs timescale were used to observe agglomeration from Al/CuO thermite combustion with different oxidizer particle sizes and shapes (microparticles, microwires, and nanoparticles). We found that CuO microwires rather than microparticles make the propagation velocity and the extent of agglomeration behave more like CuO NPs. The agglomeration size in the microscopic burning of the 3D-printed Al/CuO composites was found to be reduced from ~40 to ~13 μm when replacing microparticles with microwires. A simple mechanism based on the "pocket model" was employed to explain why CuO microwires-based thermite produces smaller agglomeration and higher energy release rate. In Al/CuO nanothermite without polymers, replacing

CuO microparticles (5 μm) with similarly sized CuO microwires (equivalent diameter: 3 μm) was shown to dramatically elevate the burn rate by ~27× ($2\,m\,s^{-1}$ versus $55\,m\,s^{-1}$) and increase the flame temperature from 2550 to 3200 K, resulting in ~30 times higher heat flux (energy release rate). Adding 10 wt% polymers into the above three thermite systems slows down the burn rates from ~m/s to ~cm/s, and we proposed the polymer addition might act as a heat sink thus preventing/reducing the micro-explosions in the agglomerations in the loading nanothermite composites.

9.5 Engineering Agglomeration and Propagating through Restraining the Movement of Agglomerations

9.5.1 Adding Carbon Fibers to Promote Energy Release Rate in Energetic Composites

One approach for balancing mechanical integrity, potential energy density, and energy release rate, could be to incorporate additives. For example, replacing conventional binders with energetic ones [28, 74, 89–93] or applying catalyst embedded with ammonium perchlorate [94, 95]. Embedding reactive or metal wires is known to promote combustion [87, 96, 97]. New techniques and formulations have been introduced to increase energetic material content [14, 98] and control combustion behavior via alterations to chemical content [71, 99]. However, since these materials largely rely on nanothermites as their primary energetic component to maximize energy release, it is possible that energy is being inefficiently coupled back into the system to promote propagation given their longer reaction time [100]. Seeing as energy release rates in propellants are determined by how fast reactions occur, and how fast energy can be transferred to unreacted material, one could feasibly use thermally conductive materials as a method to improve heat transfer and energy release rate [101]. Additionally, mechanical integrity may be bolstered by incorporating fibers that have been widely employed to reinforce polymer-based 3D structures [102–105]. Should the thermally conductive additives be fibrous, they could serve the dual purpose of both increasing mechanical integrity and energy release rate [106–108]. In this study, we choose carbon fibers as the additives considering their high melting point (~3700 °C) to hold their shape during the combustion and don't likely have catalytic effects on the other components (polymer decomposition, or Al/CuO reaction), to minimize the effect of possible chemistry between the fibers and the reacting species so that we could better probe the role of carbon fibers on the thermal/physical effects on flame propagation.

In this section, carbon fibers were embedded into ~90 wt% loading Al/CuO nanothermite through a simple direct writing approach. With the addition of only ~2.5 wt% carbon fibers, the propagation rate (burn rate) and heat flux of the sticks were promoted by more than a factor of two. From in-operando microscopic observations, the carbon fibers on the burning surface intercept ejected hot agglomerations, which can provide enhanced heat feedback to the unburnt materials on the flame front. This study provides insight for a new method to potentially accelerate the propagation of 3D-printed energetic composites.

9.5.2 Embedding Carbon Fibers into High Loading Al/CuO Nanothermite Composite

Carbon fibers with a diameter of ~7 μm, and a maximum length of ~3 mm were embedded into high loading (~90 wt%) Al/CuO nanothermite-printed sticks through a direct writing technique. As shown in Figure 9.14a, the carbon fibers are randomly dispersed in the ink and extruded through a blunt stainless-steel needle with an inner diameter of ~1.2 mm (16 gauge). As illustrated in Figure 9.14b, since the carbon fibers are longer than the inner diameter of the needle (>1.2 mm), they are aligned when passing through the needle during the printing process (Figure 9.14b) [71, 100–102, 105]. The morphology of the Al/CuO prints with embedded carbon fibers is depicted in an illustration and an optical microscopy photo in Figure 9.14c. A cross-sectional SEM image (Figure 9.14d) shows that the carbon fibers are generally embedded parallel to the direction of writing. More evidence of the predominantly parallel alignment of the carbon fibers in the Al/CuO

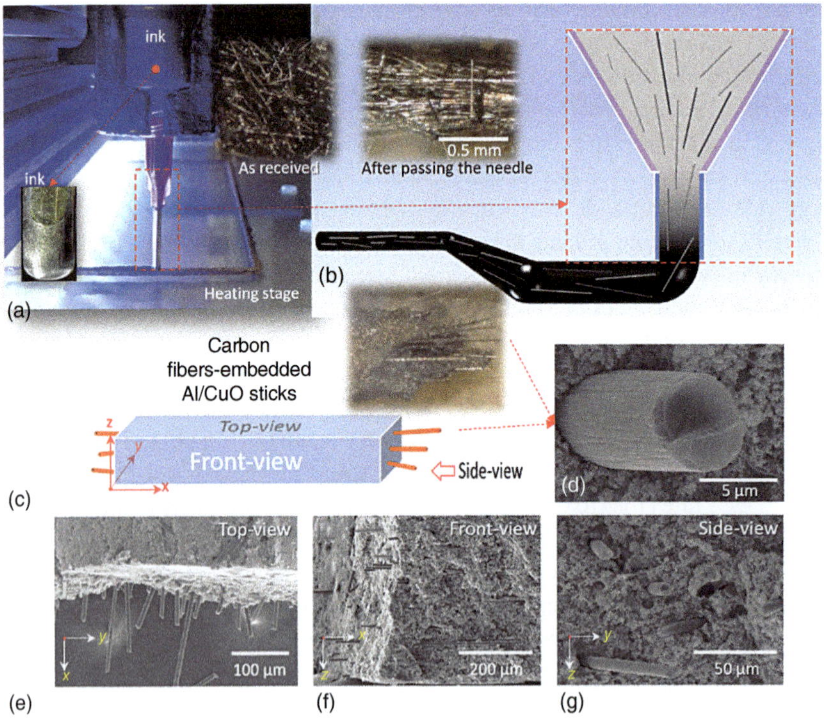

Figure 9.14 Optical image (a) and schematic showing (b) direct writing process of Al/CuO composite sticks containing carbon fibers; Optical images of randomly dispersed (right insert in a) and aligned carbon fibers (insert in b); Prepared ink for direct writing with carbon fiber (left insert in a); Schematic (c) and optical image (insert in c) of Al/CuO (90 wt%) composite sticks with aligned carbon fibers; SEM images of Al/CuO composite with carbon fibers on side view (d and g), top view (e), and front view (f). Note: the carbon fiber content is 2.5 wt%. Source: Wang et al. [109], Reproduced with permission from American Chemical Society.

composites is revealed when the printed sticks are viewed from different angles in the SEM (Figure 9.14e–g).

9.5.3 Enhanced Propagation of Al/CuO Composite with Carbon Fibers

Al/CuO composites with varied carbon fiber content (0, 1, 2.5, 5, and 10 wt%) were printed (cut into 3 cm long sticks, with a height of ~1 mm, a width of ~2 mm) and investigated by measuring propagation rates and flame temperatures in an inert atmosphere (1 atm argon). The macroscopic propagation events were recorded by a high-speed color camera with light exposure settings optimized to obtain the most color information while avoiding overexposure. Figure 9.15a,b show the typical time-resolved snapshots, from which we could obtain the flame propagation rates. The flame temperature of the combustion events was measured using three-color pyrometry [72, 80], and the corresponding temperature maps are shown in Figure 9.15a,b. From these images, one can see a ~350 K variation in flame temperature for materials with different amounts of carbon fiber addition. However, the change is relatively minor considering the errors in the measurement.

The propagation rates (average burn rate v = total stick length/total burn time) and the average flame temperatures (T = an average of all active points in a whole burning event) of the Al/CuO composites with different contents of carbon fibers are summarized and shown in Figure 9.15c. With the increase of carbon fiber content

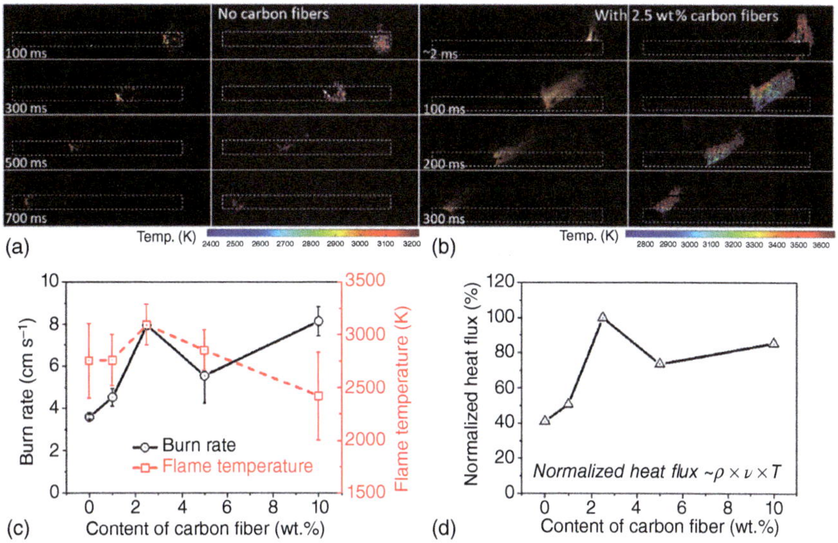

Figure 9.15 Macroscopic imaging snapshots of Al/CuO (Al/CuO is ~82–90 wt%, varying with different carbon fibers) composites without (a) and with 2.5 wt% (b) aligned carbon fibers; Summary of the burn rates and mean flame temperatures (c), the normalized heat flux (d) of Al/CuO composite prints with different contents of carbon fibers; The equivalence ratio of Al/CuO in this figure is 0.8. Note: the marked squares are where the Al/CuO-printed sticks were located. Source: Wang et al. [109], Reprinted with permission from American Chemical Society.

from 0 to 2.5 wt%, the burn rate rapidly increases from ~3.5 to ~8 cm s^{-1} and fluctuates from ~6–8 cm s^{-1} when the carbon fiber content is ≥5 wt%. This result supports the hypothesis of an increase in burn rate with the carbon fiber addition. With the increase of carbon fiber content, flame temperatures peak at ~2.5 wt% and decrease gradually from ~3100 to ~2400 K with the increase in carbon fiber content. Combined with measured densities (ρ) of the prints, we can estimate a normalized heat flux ($\sim\rho \times v \times T$, detailed in supporting information) and summarize the results in Figure 9.15d, which shows that ~2.5 wt% carbon fiber addition enhances the heat flux of the Al/CuO composites by ~250%. Although the snapshots in Figure 9.15a,b indicates that the flames are larger in the composites without carbon fiber additives, the apparent change in heat release suggests that there may be other interactions or modes of heat transfer on smaller length scales that are difficult to resolve with the macroscopic imaging apparatus (~70 μm per pixel).

9.5.4 Enhanced Heat Feedback and Heat Transfer with Carbon Fibers: Restraining the Movement of Agglomerations

To probe reaction dynamics at the flame front, a high-resolution microscopic imaging system (~1.7 μm per pixel) coupled with pyrometry software was employed to closely observe the reaction in a small area of ~1 mm^2. As seen in Figure 9.16a, the reference Al/CuO composites without any carbon fiber addition show growing agglomerations on the burning surface that are up to ~100 μm. As Figure 9.16a shows and Figure 9.16b illustrates, since most of the hot agglomerations are flying away from the burning surface, the heat feedback from these particles to the unreacted material will be relatively low and slow in comparison to those particles that are physically in contact and within the local proximity with the reacting surface. By contrast, the embedded carbon fibers (2.5 wt%, Figure 9.16c), appear to "catch" particles that were being ejected. With the hot particles being closer to and in physical contact with the reacting surface, it is reasonable to expect that the heat feedback increases in those cases with carbon fibers.

As we mentioned previously, a long-existing challenge in Al-based energetics, especially in aluminized solid propellants, is the exhaust Al_xO_y plume that plays a significant role in combustion performance, causing two-phase losses that reduce the specific impulse [57, 59, 110]. Also, as we showed earlier in this chapter, Al nanoparticles start to sinter (agglomerate) when the temperature is near its melting point (933 K), and keep growing (coalescing) into micro-sized spheres on the burning surface [57, 59, 87, 110] until a temperature close to the melting point of Al_2O_3 (2345 K), when the agglomerations start to detach from the burning surface [59, 87]. The enhanced heat feedback from the inclusion of carbon fibers allows the temperature along the reaction front (red dash lines) to reach as high as ~2400–2600 K (Figure 9.16c), ~400 K higher as compared to the reference case (Figure 9.16a). If the temperature of the reaction front is raised high enough (>2400 K), the small droplets leave the surface early without further coalescing into the droplets, while the references keep coalescing and growing due to their relatively lower temperatures. From these results, one could conjecture that these

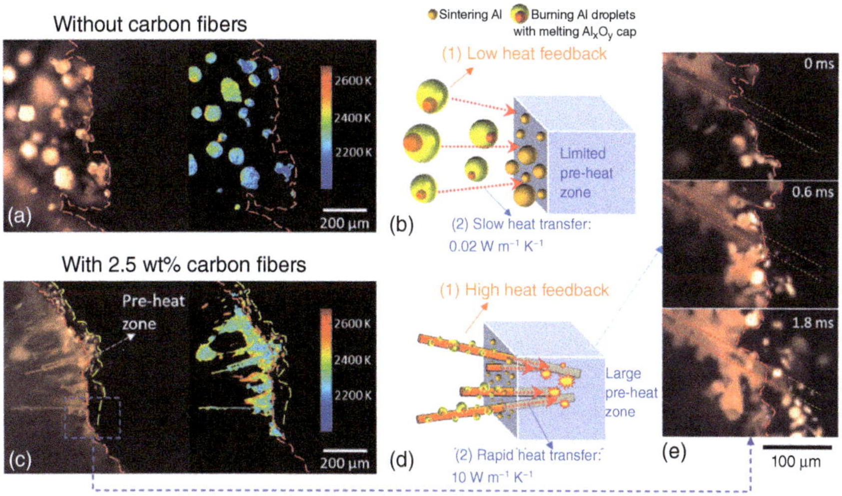

Figure 9.16 Microscopic imaging of Al/CuO composite prints without (a) and with 2.5 wt% (c) carbon fibers; The corresponding cartoons (b and d) illustrate heat feedback to preheat zone; the heated carbon fibers are seen to be inducing the new ignition sites in the preheat zone (e). The propagation is from left to right. Source: Wang et al. [109], Reproduced with permission from American Chemical Society.

embedded carbon fibers can help to reduce the two-phase losses of the aluminized propellants.

9.5.5 Summary

In this section, carbon fibers were embedded into ~90 wt% loading Al/CuO nanothermite sticks that were fabricated via extrusion direct writing. With only ~2.5 wt% carbon fiber addition, the burn rate and heat flux of the sticks were more than doubled. In situ observation on microscopic combustion finds that carbon fibers effectively trap hot agglomerates that could contribute to enhancing heat feedback to the unreacted material. Composite sticks with and without carbon fibers at different equivalence ratios were also investigated, and the results also confirm the enhancement of the carbon fibers on the combustion performance. This study may provide a new means to enhance the propagation and reduce the two-phase flow loss of 3D-printed composite energetics.

9.6 Conclusions and Future Directions

In this chapter, we summarized our recent results on Al/CuO nanothermite composites to show our understanding of how agglomeration affects the energy release rate. Firstly, we developed a universal ink formulation that could bind 90 wt% nanoparticle loading with only 10 wt% polymers for Al/CuO nanothermite composite. The ink rheology and gelation mechanism of ink were studied.

The results show that the shear-thinning and heating-induced-gelation properties make the ink mixture printable. The burn rate, flame temperature, and energy release rate of the printed composites with different equivalence ratios were also investigated. The energy release rate peaks at the equivalence ratio of 1.5–3.5. Secondly, we observed the flame front and agglomeration process of Al/CuO nanothermite composites through an in-operando high spatial (μm) and time (μs) resolution imaging system with pyrometry. We got the correlation between agglomerations and flame front, and the relationship between local agglomerating rate and global propagation rate. Based on the above findings, we demonstrated two typical approaches to engineer the agglomerations and propagations of Al NPs-based high particle loading nanocomposites via the addition of reactive CuO wires and unreactive carbon fibers. In the first approach, by replacing spherical CuO particles with high length-to-diameter CuO wires, we successfully reduced the agglomeration size by three times, evident by our calculations of three times smaller "pocket-size" of Al NPs aggregates reserved by CuO wires. In the other approach, we found 2.5 wt% carbon fiber addition can increase the energy delivery rate by two times. Our further microscopic imaging results reveal that these carbon fibers can effectively trap hot agglomerates, which can enhance heat feedback from the flame front to the unreacted material, thus reducing the agglomerations and promoting the propagations of Al NPs-based composite energetics.

Even though these results and observations offer key insights into agglomeration processes and their relationship to propagation as well as approaches to circumvent some of these limitations. There are still significant gaps in our understanding. For example, previous results about Al agglomeration in Al/CuO nanothermite in late stages include rotating, coalescing, bubbling, and sometimes micro-explosions. However, how the Al agglomerations are formed starting from Al aggregations is unclear. Also, evidence suggests that temperature is one of the key factors that control the micro-explosion of Al agglomerations, but are there other promotion factors that could be applied to enhance the micro-explosions and promote the propagation? Any approaches to turn these factors on and off to alter the energy release rate in situ? Lastly, as our previous studies indicate, coating a gas generator on Al NPs reduces agglomeration in both Al and Al/CuO mesoparticles. But how to make this happen upon adding gas generators into a free-standing energetic composite is an open question. Our current capacity to print high nanothermite particle loading composite and to characterize microscopic combustion in a high spatial and temporal resolution with pyrometry offers a window to study many particle-scale processes occurring in energetic propagation.

Acknowledgments

We acknowledge the following researchers (list in the citing order in the text) for the results we notably cited in this book chapter.

Luigi T. DeLuca, Garth C. Egan, Kyle T. Sullivan, Thomas LaGrange, Bryan W. Reed, Lei Zhou, Nicholas Piekiel, Snehaunshu Chowdhury, Jinpeng Shen,

Dylan J. Kline, Noah Eckman, Niti R. Agrawal, Tao Wu, Peng Wang, Filippo Maggi, Alessio Bandera, Prithwish Biswas, Miles C. Rehwoldt.

We also acknowledge the following publishers for granting us the permission to reuse some texts that describe the figures from reference [71, 73, 87, 109] as mentioned above:

Wiley, Springer, Elsevier, ACS

References

1 Berthe, J.E., Comet, M., Schnell, F. et al. (2016). Propellants reactivity enhancement with nanothermites. *Propellants, Explosives, Pyrotechnics* 41 (6): 994–998.
2 Comet, M., Martin, C., Klaumünzer, M. et al. (2015). Energetic nanocomposites for detonation initiation in high explosives without primary explosives. *Applied Physics Letters* 107 (24): 243108.
3 Dokhan, A., Price, E.W., Seitzman, J.M., and Sigman, R.K. (2002). The effects of bimodal aluminum with ultrafine aluminum on the burning rates of solid propellants. *Proceedings of the Combustion Institute* 29 (2): 2939–2946.
4 Bocanegra, P.E., Chauveau, C., and Gökalp, I. (2007). Experimental studies on the burning of coated and uncoated micro and nano-sized aluminum particles. *Aerospace Science and Technology* 11 (1): 33–38.
5 Dreizin, E.L. (2009). Metal-based reactive nanomaterials. *Progress in Energy and Combustion Science* 35 (2): 141–167.
6 Yetter, R.A., Risha, G.A., and Son, S.F. (2009). Metal particle combustion and nanotechnology. *Proceedings of the Combustion Institute* 32 (2): 1819–1838.
7 Jian, G., Chowdhury, S., Sullivan, K., and Zachariah, M.R. (2013). Nanothermite reactions: is gas phase oxygen generation from the oxygen carrier an essential prerequisite to ignition? *Combustion and Flame* 160 (2): 432–437.
8 Muthiah, R.M., Krishnamurthy, V.N., and Gupta, B.R. (1992). Rheology of HTPB propellant. I. Effect of solid loading, oxidizer particle size, and aluminum content. *Journal of Applied Polymer Science* 44 (11): 2043–2052.
9 Meda, L., Marra, G., Galfetti, L. et al. (2007). Nano-aluminum as energetic material for rocket propellants. *Materials Science and Engineering: C* 27 (5–8): 1393–1396.
10 Puszynski, J.A., Bulian, C.J., and Swiatkiewicz, J.J. (2007). Processing and ignition characteristics of aluminum-bismuth trioxide nanothermite system. *Journal of Propulsion and Power* 23 (4): 698–706.
11 Sullivan, K., Young, G., and Zachariah, M.R. (2009). Enhanced reactivity of nano-B/Al/CuO MIC's. *Combustion and Flame* 156 (2): 302–309.
12 Armstrong, R.W., Baschung, B., Booth, D.W., and Samirant, M. (2003). Enhanced propellant combustion with nanoparticles. *Nano Letters* 3 (2): 253–255.
13 DeLuca, L.T. (2018). Overview of Al-based nanoenergetic ingredients for solid rocket propulsion. *Defence Technology.* 14 (5): 357–365.

14 McClain, M.S., Gunduz, I.E., and Son, S.F. (2019). Additive manufacturing of ammonium perchlorate composite propellant with high solids loadings. *Proceedings of the Combustion Institute* 37 (3): 3135–3142.

15 Wang, H., Jian, G., Egan, G.C., and Zachariah, M.R. (2014). Assembly and reactive properties of Al/CuO based nanothermite microparticles. *Combustion and Flame* 161 (8): 2203–2208.

16 Wang, H., Jacob, R.J., DeLisio, J.B., and Zachariah, M.R. (2017). Assembly and encapsulation of aluminum NP's within AP/NC matrix and their reactive properties. *Combustion and Flame* 180: 175–183.

17 Teipel, U. and Förter-Barth, U. (2001). Rheology of nano-scale aluminum suspensions. *Propellants, Explosives, Pyrotechnics* 26 (6): 268–272.

18 Kim, S.H. and Zachariah, M.R. (2004). Enhancing the rate of energy release from nanoenergetic materials by electrostatically enhanced assembly. *Advanced Materials* 16 (20): 1821–1825.

19 Severac, F., Alphonse, P., Estève, A. et al. (2012). High-energy Al/CuO nanocomposites obtained by DNA-directed assembly. *Advanced Functional Materials* 22 (2): 323–329.

20 Slocik, J.M., McKenzie, R., Dennis, P.B., and Naik, R.R. (2017). Creation of energetic biothermite inks using ferritin liquid protein. *Nature Communications* 8 (1): 1–7.

21 Bencomo, J.A., Iacono, S.T., and McCollum, J. (2018). 3D printing multifunctional fluorinated nanocomposites: tuning electroactivity, rheology and chemical reactivity. *Journal of Materials Chemistry A* 6 (26): 12308–12315.

22 Murray, A.K., Novotny, W.A., Fleck, T.J. et al. (2018). Selectively-deposited energetic materials: a feasibility study of the piezoelectric inkjet printing of nanothermites. *Additive Manufacturing* 22: 69–74.

23 Sullivan, K.T., Zhu, C., Duoss, E.B. et al. (2016). Controlling material reactivity using architecture. *Advanced Materials* 28 (10): 1934–1939.

24 Comet, M., Martin, C., Schnell, F., and Spitzer, D. (2017). Nanothermite foams: from nanopowder to object. *Chemical Engineering Journal* 316: 807–812.

25 Ghosh, S., Parker, S.T., Wang, X. et al. (2008). Direct-write assembly of microperiodic silk fibroin scaffolds for tissue engineering applications. *Advanced Functional Materials* 18 (13): 1883–1889.

26 Chang, C., Tran, V.H., Wang, J. et al. (2010). Direct-write piezoelectric polymeric nanogenerator with high energy conversion efficiency. *Nano Letters* 10 (2): 726–731.

27 Xu, C., An, C., He, Y. et al. (2018). Direct ink writing of DNTF based composite with high performance. *Propellants, Explosives, Pyrotechnics* 43 (8): 754–758.

28 Ruz-Nuglo, F.D. and Groven, L.J. (2018). 3-D printing and development of fluoropolymer based reactive inks. *Advanced Engineering Materials* 20 (2): 1700390.

29 McDonald, B.A., Rice, J.R., and Kirkham, M.W. (2014). Humidity induced burning rate degradation of an iron oxide catalyzed ammonium perchlorate/HTPB composite propellant. *Combustion and Flame* 161 (1): 363–369.

30 Kurva, R., Gupta, G., Dhabbe, K.I. et al. (2017). Evaluation of 4-(dimethylsilyl) butyl ferrocene grafted HTPB as a burning rate modifier in composite propellant formulation using bicurative system. *Propellants, Explosives, Pyrotechnics* 42 (4): 401–409.

31 Chandru, R.A., Balasubramanian, N., Oommen, C., and Raghunandan, B.N. (2018). Additive manufacturing of solid rocket propellant grains. *Journal of Propulsion and Power* 34 (4): 1090–1093.

32 Piercey, D.G. and Klapoetke, T.M. (2010). Nanoscale aluminum-metal oxide (thermite) reactions for application in energetic materials. *Central European Journal of Energetic Materials* 7 (2): 115–129.

33 Hastings, D.L., Schoenitz, M., and Dreizin, E.L. (2018). High density reactive composite powders. *Journal of Alloys and Compounds* 735: 1863–1870.

34 Kim, W.D., Lee, S., and Lee, D.C. (2018). Nanothermite of Al nanoparticles and three-dimensionally ordered macroporous CuO: mechanistic insight into oxidation during thermite reaction. *Combustion and Flame* 189: 87–91.

35 Calais, T., Bancaud, A., Estève, A., and Rossi, C. (2018). Correlation between DNA self-assembly kinetics, microstructure, and thermal properties of tunable highly energetic Al–CuO nanocomposites for micropyrotechnic applications. *ACS Applied Nano Materials* 1 (9): 4716–4725.

36 He, W., Liu, P.J., He, G.Q. et al. (2018). Highly reactive metastable intermixed composites (MICs): preparation and characterization. *Advanced Materials* 30 (41): 1706293.

37 Meeks, K., Smith, D.K., Clark, B., and Pantoya, M.L. (2017). Percolation of a metallic binder in energy generating composites. *Journal of Materials Chemistry A* 5 (15): 7200–7209.

38 Weismiller, M.R., Huba, Z.J., Tuttle, S.G. et al. (2017). Combustion characteristics of high energy Ti–Al–B nanopowders in a decane spray flame. *Combustion and Flame* 176: 361–369.

39 Koenig, J.T., Shaw, A.P., Poret, J.C. et al. (2017). Performance of W/MnO_2 as an environmentally friendly energetic time delay composition. *ACS Sustainable Chemistry & Engineering* 5 (10): 9477–9484.

40 Martirosyan, K.S. (2011). Nanoenergetic gas-generators: principles and applications. *Journal of Materials Chemistry* 21 (26): 9400–9405.

41 Jian, G., Liu, L., and Zachariah, M.R. (2013). Facile aerosol route to hollow CuO spheres and its superior performance as an oxidizer in nanoenergetic gas generators. *Advanced Functional Materials* 23 (10): 1341–1346.

42 Rossi, C., Estève, A., and Vashishta, P. (2010). Nanoscale energetic materials. *Journal of Physics and Chemistry of Solids* 2 (71): 57–58.

43 Shen, Z., Ding, Y., Chen, J. et al. (2019). Interfacial bonding mechanism in Al/coated steel dissimilar refill friction stir spot welds. *Journal of Materials Science and Technology* 35 (6): 1027–1038.

44 Ali, A.N., Son, S.F., Hiskey, M.A., and Naud, D.L. (2004). Novel high nitrogen propellant use in solid fuel micropropulsion. *Journal of Propulsion and Power* 20 (1): 120–126.

45 Rossi, C., Orieux, S., Larangot, B. et al. (2002). Design, fabrication and modeling of solid propellant microrocket-application to micropropulsion. *Sensors and Actuators A: Physical* 99 (1–2): 125–133.

46 Yen, N.H. and Wang, L.Y. (2012). Reactive metals in explosives. *Propellants, Explosives, Pyrotechnics* 37 (2): 143–155.

47 Sanders, V.E., Asay, B.W., Foley, T.J. et al. (2007). Reaction propagation of four nanoscale energetic composites (Al/MoO_3, Al/WO_3, Al/CuO, and $B_{12}O_3$). *Journal of Propulsion and Power* 23 (4): 707–714.

48 Ilyushin, M.A., Tselinsky, I.V., and Shugalei, I.V. (2012). Environmentally friendly energetic materials for initiation devices. *Central European Journal of Energetic Materials* 9 (4): 293–327.

49 Weismiller, M.R., Malchi, J.Y., Yetter, R.A., and Foley, T.J. (2009). Dependence of flame propagation on pressure and pressurizing gas for an Al/CuO nanoscale thermite. *Proceedings of the Combustion Institute* 32 (2): 1895–1903.

50 Yan, S., Jian, G., and Zachariah, M.R. (2012). Electrospun nanofiber-based thermite textiles and their reactive properties. *ACS Applied Materials and Interfaces* 4 (12): 6432–6435.

51 Baijot, V., Mehdi, D.R., Rossi, C., and Esteve, A. (2017). A multi-phase micro-kinetic model for simulating aluminum based thermite reactions. *Combustion and Flame* 180: 10–19.

52 Chiang, Y.C. and Wu, M.H. (2017). Assembly and reaction characterization of a novel thermite consisting aluminum nanoparticles and CuO nanowires. *Proceedings of the Combustion Institute* 36 (3): 4201–4208.

53 Hübner, J., Klaumünzer, M., Comet, M. et al. (2017). Insights into combustion mechanisms of variable aluminum-based iron oxide/−hydroxide nanothermites. *Combustion and Flame* 184: 186–194.

54 Sullivan, K.T., Kuntz, J.D., and Gash, A.E. (2014). The role of fuel particle size on flame propagation velocity in thermites with a nanoscale oxidizer. *Propellants, Explosives, Pyrotechnics* 39 (3): 407–415.

55 Zohari, N., Keshavarz, M.H., and Seyedsadjadi, S.A. (2013). The advantages and shortcomings of using nano-sized energetic materials. *Central European Journal of Energetic Materials* 10 (1): 135–147.

56 Sullivan, K.T., Piekiel, N.W., Wu, C. et al. (2012). Reactive sintering: an important component in the combustion of nanocomposite thermites. *Combustion and Flame* 159 (1): 2–15.

57 Chakraborty, P. and Zachariah, M.R. (2014). Do nanoenergetic particles remain nano-sized during combustion? *Combustion and Flame* 161 (5): 1408–1416.

58 Ao, W., Liu, X., Rezaiguia, H. et al. (2017). Aluminum agglomeration involving the second mergence of agglomerates on the solid propellants burning surface: experiments and modeling. *Acta Astronautica* 136: 219–229.

59 Chen, Y., Guildenbecher, D.R., Hoffmeister, K.N. et al. (2017). Study of aluminum particle combustion in solid propellant plumes using digital in-line holography and imaging pyrometry. *Combustion and Flame* 182: 225–237.

60 Jin, B.N., Wang, Z.X., Xu, G. et al. (2020). Three-dimensional spatial distributions of agglomerated particles on and near the burning surface of aluminized solid propellant using morphological digital in-line holography. *Aerospace Science and Technology* 106: 106066.

61 Tejasvi, K., Venkateshwara Rao, V., Pydi Setty, Y., and Jayaraman, K. (2020). Ultra-fine aluminium characterization and its agglomeration features in solid propellant combustion for various quenched distance and pressure. *Propellants, Explosives, Pyrotechnics* 45 (5): 714–723.

62 Li, L.B., Chen, X., Zhou, C.S. et al. (2020). Experimental and model investigation on agglomeration of aluminized fuel-rich propellant in solid fuel ramjet. *Combustion and Flame* 219: 437–448.

63 Ao, W., Liu, P., Liu, H. et al. (2020). Tuning the agglomeration and combustion characteristics of aluminized propellants via a new functionalized fluoropolymer. *Chemical Engineering Journal* 382: 122987.

64 Wainwright, E.R., Lakshman, S.V., Leong, A.F. et al. (2019). Viewing internal bubbling and microexplosions in combusting metal particles via x-ray phase contrast imaging. *Combustion and Flame* 199: 194–203.

65 Grapes, M.D., Reeves, R.V., Fezzaa, K. et al. (2019). In situ observations of reacting Al/Fe_2O_3 thermite: relating dynamic particle size to macroscopic burn time. *Combustion and Flame* 201: 252–263.

66 Zong, Y., Jacob, R.J., Li, S., and Zachariah, M.R. (2015). Size resolved high temperature oxidation kinetics of nano-sized titanium and zirconium particles. *The Journal of Physical Chemistry A* 119 (24): 6171–6178.

67 Egan, G.C., Sullivan, K.T., LaGrange, T. et al. (2014). In situ imaging of ultra-fast loss of nanostructure in nanoparticle aggregates. *Journal of Applied Physics* 115 (8): 084903.

68 Zhou, L., Piekiel, N., Chowdhury, S., and Zachariah, M.R. (2009). T-Jump/time-of-flight mass spectrometry for time-resolved analysis of energetic materials. *Rapid Communications in Mass Spectrometry* 23 (1): 194–202.

69 Sullivan, K. and Zachariah, M. (2010). Simultaneous pressure and optical measurements of nanoaluminum thermites: investigating the reaction mechanism. *Journal of Propulsion and Power* 26 (3): 467–472.

70 Wang, H., Julien, B., Kline, D.J. et al. (2020). Probing the reaction zone of nanolaminates at ~ μs time and~ μm spatial resolution. *The Journal of Physical Chemistry C.* 124 (25): 13679–13687.

71 Wang, H., Shen, J., Kline, D.J. et al. (2019). Direct writing of a 90 wt% particle loading nanothermite. *Advanced Materials* 31 (23): 1806575.

72 Kline, D.J., Alibay, Z., Rehwoldt, M.C. et al. (2020). Experimental observation of the heat transfer mechanisms that drive propagation in additively manufactured energetic materials. *Combustion and Flame* 215: 417–424.

73 Wang, H., Kline, D.J., and Zachariah, M.R. (2019). In-operando high-speed microscopy and thermometry of reaction propagation and sintering in a nanocomposite. *Nature Communications* 10 (1): 1–8.

74 Wang, H., Rehwoldt, M., Kline, D.J. et al. (2019). Comparison study of the ignition and combustion characteristics of directly-written Al/PVDF, Al/Viton and Al/THV composites. *Combustion and Flame* 201: 181–186.

75 Sadasivuni, K.K., Cabibihan, J.J., Ponnamma, D. et al. (ed.) (2016). *Biopolymer Composites in Electronics.* Elsevier.

76 Rae, P.J. and Dattelbaum, D.M. (2004). The properties of poly (tetrafluoroethylene)(PTFE) in compression. *Polymer* 45 (22): 7615–7625.

77 Scott, G.D. and Kilgour, D.M. (1969). The density of random close packing of spheres. *Journal of Physics D: Applied Physics* 2 (6): 863.

78 Zangmeister, C.D., Radney, J.G., Dockery, L.T. et al. (2014). Packing density of rigid aggregates is independent of scale. *Proceedings of the National Academy of Sciences* 111 (25): 9037–9041.

79 Egan, G.C. and Zachariah, M.R. (2015). Commentary on the heat transfer mechanisms controlling propagation in nanothermites. *Combustion and Flame* 162 (7): 2959–2961.

80 Jacob, R.J., Kline, D.J., and Zachariah, M.R. (2018). High speed 2-dimensional temperature measurements of nanothermite composites: probing thermal vs. gas generation effects. *Journal of Applied Physics* 123 (11): 115902.

81 Dutro, G.M., Yetter, R.A., Risha, G.A., and Son, S.F. (2009). The effect of stoichiometry on the combustion behavior of a nanoscale Al/MoO_3 thermite. *Proceedings of the Combustion Institute* 32 (2): 1921–1928.

82 Henz, B.J., Hawa, T., and Zachariah, M.R. (2010). On the role of built-in electric fields on the ignition of oxide coated nanoaluminum: ion mobility versus Fickian diffusion. *Journal of Applied Physics* 107 (2): 024901.

83 Povinelli, L.A. and Rosenstein, R.A. (1964). Alumina size distributions from high-pressure composite solid-propellant combustion. *AIAA Journal* 2 (10): 1754–1760.

84 Maggi, F., DeLuca, L.T., and Bandera, A. (2015). Pocket model for aluminum agglomeration based on propellant microstructure. *AIAA Journal* 53 (11): 3395–3403.

85 Cohen, N.S. (1983). A pocket model for aluminum agglomeration in composite propellants. *AIAA Journal* 21 (5): 720–725.

86 Maggi, F., Bandera, A., DeLuca, L.T. et al. (2011). Agglomeration in solid rocket propellants: novel experimental and modeling methods. *Progress in Propulsion Physics* 2: 81–98.

87 Wang, H., Kline, D.J., Biswas, P., and Zachariah, M.R. (2021). Connecting agglomeration and burn rate in a thermite reaction: role of oxidizer morphology. *Combustion and Flame* 231: 111492.

88 Sambamurthi, J.K., Price, E.W., and Sigman, R.K. (1984). Aluminum agglomeration in solid-propellant combustion. *AIAA Journal* 22 (8): 1132–1138.

89 Göçmez, A., Erişken, C., Yilmazer, Ü. et al. (1998). Mechanical and burning properties of highly loaded composite propellants. *Journal of Applied Polymer Science* 67 (8): 1457–1464.

90 Huang, C., Jian, G., DeLisio, J.B. et al. (2015). Electrospray deposition of energetic polymer nanocomposites with high mass particle loadings: a prelude to 3D printing of rocket motors. *Advanced Engineering Materials* 17 (1): 95–101.

91 Shen, J., Wang, H., Kline, D.J. et al. (2020). Combustion of 3D printed 90 wt% loading reinforced nanothermite. *Combustion and Flame* 215: 86–92.

92 Wang, Y., Luo, T., Song, X., and Li, F. (2019). Electrospinning preparation of nc/gap/submicron-hns energetic composite fiber and its properties. *ACS Omega* 4 (10): 14261–14271.

93 Cheng, L., Yang, H., Yang, Y. et al. (2020). Preparation of B/nitrocellulose/Fe particles and their effect on the performance of an ammonium perchlorate propellant. *Combustion and Flame* 211: 456–464.

94 Fehlberg, S., Örnek, M., Manship, T.D., and Son, S.F. (2020). Decomposition of ammonium-perchlorate-encapsulated nanoscale and micron-scale catalyst particles. *Journal of Propulsion and Power* 36 (6): 862–868.

95 Pang, W.Q., DeLuca, L.T., Fan, X.Z. et al. (2020). Combustion behavior of AP/HTPB/Al composite propellant containing hydroborate iron compound. *Combustion and Flame* 220: 157–167.

96 Kubota, N., Ichida, M., and Fujisawa, T. (1982). Combustion processes of propellants with embedded metal wires. *AIAA Journal* 20 (1): 116–121.

97 Isert, S., Lane, C.D., Gunduz, I.E., and Son, S.F. (2017). Tailoring burning rates using reactive wires in composite solid rocket propellants. *Proceedings of the Combustion Institute* 36 (2): 2283–2290.

98 Golobic, A.M., Durban, M.D., Fisher, S.E. et al. (2019). Active mixing of reactive materials for 3D printing. *Advanced Engineering Materials* 21 (8): 1900147.

99 Westphal, E.R., Murray, A.K., McConnell, M.P. et al. (2019). The effects of confinement on the fracturing performance of printed nanothermites. *Propellants, Explosives, Pyrotechnics* 44 (1): 47–54.

100 Wang, H., Biswas, P., Kline, D.J., and Zachariah, M.R. (2022). Flame stand-off effects on propagation of 3D printed 94 wt.% nanosized pyrolants loading composites. *Chemical Engineering Journal* 434: 134487.

101 Rehwoldt, M.C., Kline, D.J., and Zachariah, M.R. (2021). Numerically evaluating energetic composite flame propagation with thermally conductive, high aspect ratio fillers. *Chemical Engineering Science* 229: 116087.

102 Lewicki, J.P., Rodriguez, J.N., Zhu, C. et al. (2017). 3D-printing of meso-structurally ordered carbon fiber/polymer composites with unprecedented orthotropic physical properties. *Scientific Reports* 7 (1): 1–4.

103 Li, N., Li, Y., and Liu, S. (2016). Rapid prototyping of continuous carbon fiber reinforced polylactic acid composites by 3D printing. *Journal of Materials Processing Technology* 238: 218–225.

104 Love, L.J., Kunc, V., Rios, O. et al. (2014). The importance of carbon fiber to polymer additive manufacturing. *Journal of Materials Research* 29 (17): 1893–1898.

105 Tekinalp, H.L., Kunc, V., Velez-Garcia, G.M. et al. (2014). Highly oriented carbon fiber–polymer composites via additive manufacturing. *Composites Science and Technology* 105: 144–150.

106 Yi, Z., Cao, Y., Yuan, J. et al. (2020). Functionalized carbon fibers assembly with Al/Bi2O3: a new strategy for high-reliability ignition. *Chemical Engineering Journal* 389: 124254.

107 Siegert, B., Comet, M., Muller, O. et al. (2010). Reduced-sensitivity nanothermites containing manganese oxide filled carbon nanofibers. *The Journal of Physical Chemistry C*. 114 (46): 19562–19568.

108 Elder, B., Neupane, R., Tokita, E. et al. (2020). Nanomaterial patterning in 3D printing. *Advanced Materials* 32 (17): 1907142.

109 Wang, H., Kline, D.J., Rehwoldt, M.C., and Zachariah, M.R. (2021). Carbon fibers enhance the propagation of high loading nanothermites: in situ observation of microscopic combustion. *ACS Applied Materials and Interfaces* 13 (26): 30504–30511.

110 Karasev, V.V., Onischuk, A.A., Glotov, O.G. et al. (2004). Formation of charged aggregates of Al_2O_3 nanoparticles by combustion of aluminum droplets in air. *Combustion and Flame* 138 (1–2): 40–54.

Part IV

Solid Propellants and Fuels for Rocket Propulsion

10

Glycidyl Azide Polymer Combustion and Applications Studies Performed at ISAS/JAXA

Keiichi Hori[1], Yutaka Wada[2], Makihito Nishioka[3], Motoyasu Kimura[4], Iwao Komai[5], Koh Kobayashi[6], and Jyun Ohba[5]

[1] *Institute of Space and Astronautical Science (ISAS)/Japan Aerospace Exploration Agency (JAXA), Department of Space Flight Dynamics, 3-1-1 Yoshinodai, Chuo-ku, Sagamihara, Kanagawa 252-5210, Japan*
[2] *Chiba Institute of Technology, Department of Innovative Mechanical and Electronic Engineering, Faculty of Engineering, 2-17-1, Tsudanuma, Narashino, Chiba 275-0016, Japan*
[3] *University of Tsukuba, Department of Engineering Mechanics and Energy, Faculty of Engineering, Information and Systems, 1-1-1, Tennodai, Ibaraki 305-8573, Japan*
[4] *NOF Corporation, Purchasing Department, Yebisu Garden Place Tower, 20-3, Ebisu 4-Chome, Shibuya-ku, Tokyo 150-6019, Japan*
[5] *NOF Corporation, Aichi Works, Taletoyo-plant, 61-1, Kitakomatsudani, Taketoyo-cho, Chita-gun, Aichi 470-2379, Japan*
[6] *NOF Corporation, Tanegashima Works, 3138-4, Hirayama, Minamitane-cho, Kumage-gun, Kagoshima 891-3702, Japan*

10.1 Introduction

The high-energy Glycidyl Azide Polymer (GAP) has great prospects as a fuel-binder for composite propellants or solid fuel for hybrid rocket; furthermore, numerous studies on its aspects, such as chemistry and combustion, have been carried out [1–40]. This paper summarizes the study on combustion in GAP conducted by the author group and its application to hybrid rocket.

Composite propellants mainly comprise an oxidant, a metallic fuel, and a fuel binder. With the advancements in the polymer industry, the binder that contained liquid rubber as its main component has changed to polysulfide, polyol-based polyurethane, and carboxyl-terminated polybutadiene (CTPB), and currently hydroxyl-terminated Polybutadiene (HTPB) is the most used variant. HTPB can be cured by polyurethane polyaddition reaction utilizing terminal hydroxyl group, and the reaction rate can be easily controlled. Since the glass transition temperature is low, it can achieve high elongation and low Young's modulus.

Meanwhile, many high-density and high-energy substances have been researched and developed as objects for research and development of explosives. Azide compounds were developed by Rocketdyne Inc. under the guidance of the U.S. Air Force Laboratory for use in high-performance solid and gun propellants [41]. In this

Nano and Micro-Scale Energetic Materials: Propellants and Explosives, First Edition.
Edited by Weiqiang Pang and Luigi T. DeLuca.

project, organic azide compounds were considered since they have great potential in building a new generation of solid propellants. A study of the synthesis of hydroxyl-terminated azide prepolymer was carried out in 1976 [41].

GAP was originally attempted to be polymerized after the azidation of epichlorohydrin (ECH); however, the monomer was poorly reactive, which led to failure. Since then, GAP is synthesized through the azidation of polyepichlorohydrin (PECH), and a practical test using a tri-ol GAP was conducted until 1981 [42].

Nippon Oil and Fat Corporation (NOF) has the technology and production facilities for manufacturing glycol-based polymers. It was the first company in Japan to focus on GAP as the next-generation propellant fuel and started synthesizing hydroxyl-terminated PECH prepolymer in 1983.

In addition to GAP, other high-energy azide polymers such as 3-azidomethyl-3-methyloxetane (AMMO) [43] and 3,3-bisazidomethyloxetane (BAMO) [44] have been developed. NOF has been engaged in research and development of GAP that can be applied to large rocket motors, mainly focusing on developing tetra-ol GAP, which can be cured for the fuel of hybrid rockets, and di-ol GAP that can be applied as a fuel-binder for composite propellants.

However, GAP has a higher glass transition temperature (−53 °C) than that of HTPB (−84 °C); therefore, from a perspective of the mechanical properties, there were challenges in applying it to propellants for large rockets and in operation in low temperature areas. To solve this problem, the studies on multiple copolymer GAP have been carried out to lower the glass transition temperature. Among them, the propellant using the tetrahydrofuran (THF) copolymerization typed GAP has succeeded in the large-scale motor combustion test in the low temperature area, and there is a case where it has been put into practical use [45].

In the recent years, its applications to general chemical products have also been studied using its property that includes easy degradation of azide groups by ultraviolet rays to generate nitrogen gas [46].

In 1996, K. Hori and M. Kimura conducted a study of combustion mechanism [47] utilizing di-ol GAP using the mathematical technique of asymptotic analysis based on the large value of the surface degradation activation energy [48]. By adding polypropylene glycol (PPG) as a diluent and measuring the burning rate characteristic and temperature field, the process of azide group degradation in the condensed phase was found to be dominant and determined the activation energy of the process.

In the 2000s, Y. Wada, M. Nishioka and K. Hori conducted a combustion experiment using a tetra-ol GAP, and constructed a three-phase one-dimensional model that used detailed chemical reaction calculation for the gas phase from the results, and found that the combustion residue played a major role, and thus made it possible to accurately simulate the burning rate and temperature field [49, 50]. Y. Wada and K. Hori applied the tetra-ol GAP to a gas hybrid rocket, and predicted its feasibility, although it was a small scale experiment [2, 51–53].

10.2 Combustion Mechanism

10.2.1 Simplified Model by Asymptotic Analysis [47]

The burning rate was measured using NOF di-ol GAP as a sample, and it was found that the pressure index changed at 2.3 MPa and reduced (about 0.2) on the high-pressure side (Figure 10.1). Furthermore, when the temperature field was measured using a fine thermocouple with a wire diameter of 20 μm, it was found that the temperature rose rapidly from the initial temperature, and slowly from a certain temperature (T_s), and finally reached a constant value (T_f). T_s and T_f were measured using as a function of pressure, and both were found to increase slightly with pressure (Figure 10.2). In addition, when 40% of PPG was added to GAP, the burning rate was found to be proportional to the azide group concentration and the pressure index was found to not change (Figure 10.1). From this, it was experimentally found that the burning rate is controlled by the amount of heat generated during the degradation of the azide group.

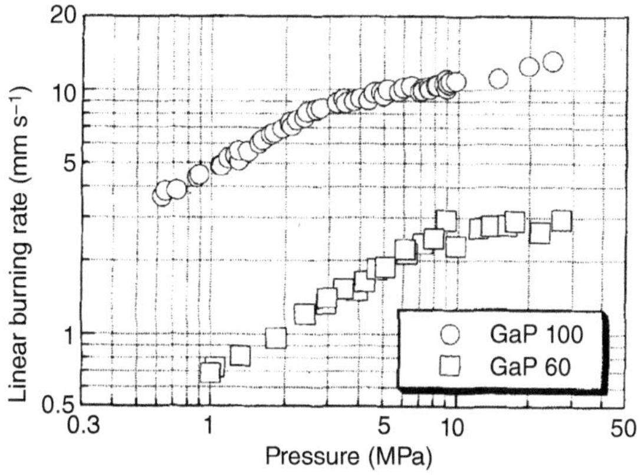

Figure 10.1 Linear burning rates of GAP and GAP60PPG40. Source: Hori and Kimura [47]/John Wiley & Sons.

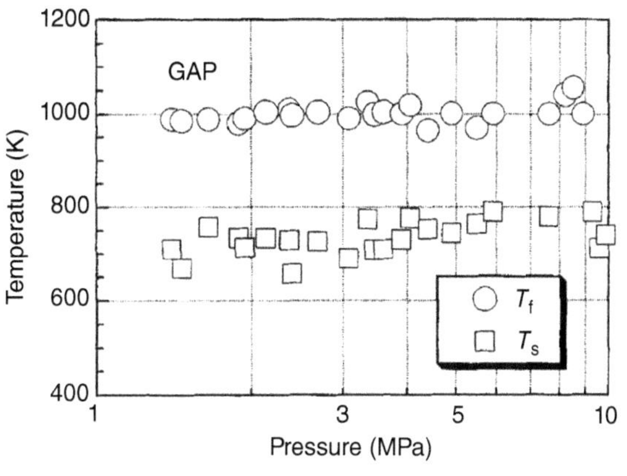

Figure 10.2 T_s and T_f as a function of pressure. Source: Hori and Kimura [47]/John Wiley & Sons.

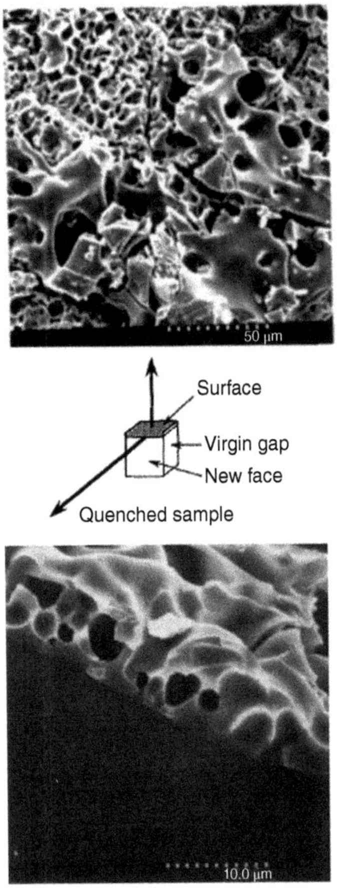

Figure 10.3 SEM images of quenched sample @2.3 MPa. Source: Hori and Kimura [47]/John Wiley & Sons.

When the surface of the sample, which was extinguished by rapidly depressurizing the burning GAP sample, was observed using a scanning electron microscope (SEM), it was found that there was a liquid zone at the burning surface that actively generates bubbles (Figure 10.3). This fact revealed that the azide group degradation occurred in the liquid layer on the burning surface, and the critical point T_s observed by the temperature field measurement was the temperature of the burning surface.

Based on these experimental findings, we constructed a simplified combustion model using activation energy asymptotic analysis. Figures 10.4 and 10.5 show the physical image underlying the model. The chemical reaction model used a simplified chemical formula as shown in Eq. (10.1).

$$\text{GAP} \xrightarrow[Q_1]{N_2\uparrow} \text{Fragments} \xrightarrow[Q_2]{} \text{Final Products} \tag{10.1}$$

The nitrogen release process at the first step of the reaction is the process of controlling the burning rate, and the governing mechanism in the surface liquid

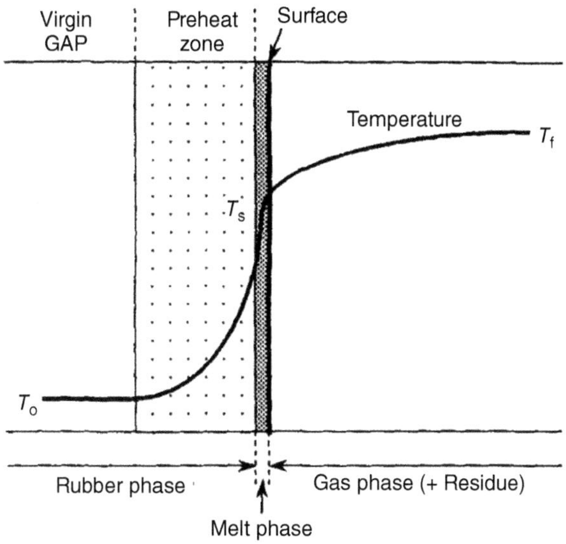

Figure 10.4 Schematic of the structure of GAP combustion. Source: Hori and Kimura [47]/John Wiley & Sons.

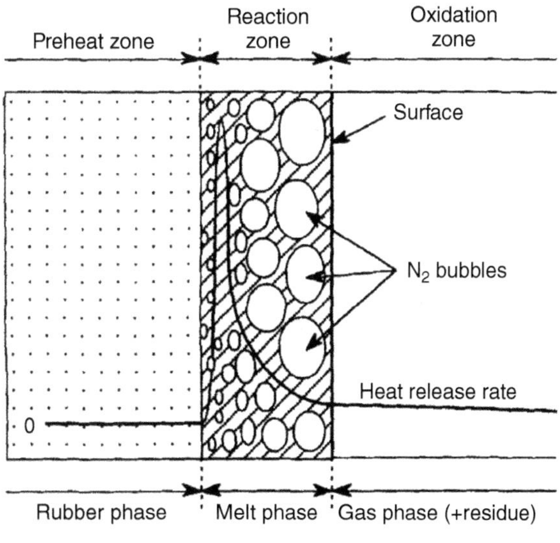

Figure 10.5 Schematic of the microstructure of GAP combustion. Source: Hori and Kimura [47]/John Wiley & Sons.

layer is represented by the following two dimensionless equations (Chemical species conservation formula: Eq. (10.2), Energy conservation formula: Eq. (10.3)).

$$\frac{1}{\mathrm{Le}, m}\frac{\mathrm{d}^2Y}{\mathrm{d}x^2} - \frac{\mathrm{d}Y}{\mathrm{d}x} = \frac{D_1}{M^2}Y\exp\left(-\frac{E_1}{T}\right) \tag{10.2}$$

$$\frac{\mathrm{d}^2T}{\mathrm{d}x^2} - \frac{\mathrm{d}T}{\mathrm{d}x} = -Q_1\frac{D_1}{M^2}Y\exp\left(-\frac{E_1}{T}\right) \tag{10.3}$$

Le,m: Lewis number in the liquid layer, Y: Azide group concentration, D_1: Damköhler Number of the first step, M: mass flux, E_1: Activation energy in the first step, T: Temperature, Q_1: Calorific value of the first step.

To investigate the structure of the liquid layer, the large parameter needed for the asymptotic analysis is selected as $\beta(=\frac{E_1}{T_s})$ and the temperature is expanded as shown in Eq. (10.4)

$$T = T_s + \frac{1}{\beta}T_1 + o\left(\frac{1}{\beta}\right) \tag{10.4}$$

where T_1 is the first perturbation. Examining the liquid layer revealed [47] that the orders of magnitude of Le,m and β are both $O(10)$. Thus, Le,m = $O(\beta)$, and the Shvab-Zel'dovich formulation (Eq. (10.5)) held true even in the liquid layer

$$T + \frac{Y}{\text{Le}, m} = T_s \tag{10.5}$$

In addition to Eqs. (10.2)–(10.5), when stretched spatial valuable is entered, the governing equation in the liquid layer is expressed as Eq. (10.6)

$$\frac{d^2T_1}{d\xi^2} = \frac{\overline{D_1}\text{Le}, m}{M^2}T_1\exp(T_1) \tag{10.6}$$

When the Damköhler number is $O(\beta^2)$, and it can be expanded, as shown in Eq. (10.7)

$$D_1 = \beta^2(\overline{D_1} + \cdots\cdots)\exp\left(\frac{E_s}{T_1}\right) \tag{10.7}$$

Using the boundary condition for Eq. (10.6)

$$\xi \to -\infty : T_1 = \xi(T_s - T_o), \xi \to 0 : \frac{dT_1}{d\xi} = T_s' + o(1) \tag{10.8}$$

the mass flux M is expressed as Eq. (10.9)

$$M \propto \frac{T_s^2}{\sqrt{Q_1}}\exp\left(-\frac{E_1}{2T_s}\right) \tag{10.9}$$

Here, the temperature gradient (gas phase side) T_s' at the burning surface is treated as a constant because the subsequent reaction mechanism in the gas phase is considered to be almost constant regardless of PPG addition and pressure.

From this, it became clear that the burning rate depends on the burning surface temperature and the calorific value (effect of PPG addition). From the measured values of the burning rate and the burning surface temperature, the activation energy E_1 was found to be approximately 150 kJ (mol K)$^{-1}$.

In this report, experimental observations revealed that the burning rate is governed by the denitrogenation step that occurs on the burning surface, and a solution of the burning rate was determined using asymptotic analysis. However, at this step, it was not clear why the final gas phase temperature T_f was approximately 200 K lower than the adiabatic flame temperature, and it was also not clear as to why the burning surface temperature T_s was not theoretically obtained. Therefore, the combustion model could not be refined until three-phase one-dimensional numerical analysis was done, as described in Section of 10.2.2.

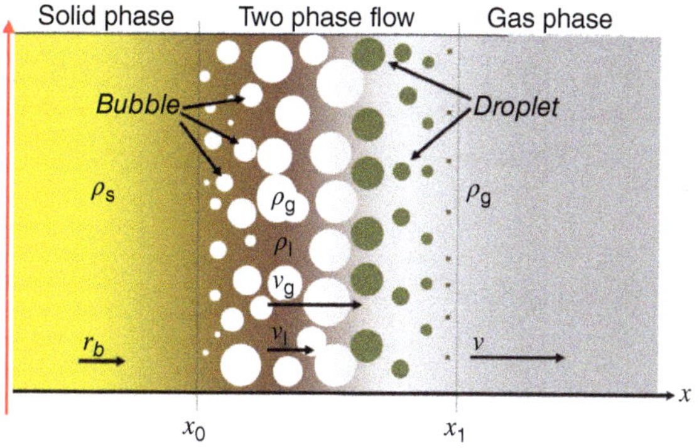

Figure 10.6 Three phase-one dimensional GAP combustion model with respect to an observer fixed at the burning surface x_0. Source: Wada et al. [49]/With permission of Springer Nature.

10.2.2 Three Phase-One Dimensional Full Kinetics Model [49, 50]

In the 1990s, CHEMKIN, a numerical calculation tool for combustion phenomena, was developed, and research on numerical calculation in the study of combustion of high-energy materials was initiated. Beckstead and coworkers used CHEMKIN to conduct a combustion study on tri-ol GAP manufactured in the U.S. and obtained good results [54, 55]. In the combustion study on NOF tetra-ol GAP, we modified CHEMKIN PREMIX, the calculation code for one-dimensional laminar flame of CHEMKIN-II for high-energy materials in the condensed phase. It is called HEM (High Energy Material) PREMIX. Based on the model of Beckstead and coworkers [54, 55], we divided it into three regions, respectively featuring (i) solid phase, (ii) gas-liquid two-phase, and (iii) gas phase. One-dimensional modeling was attempted of the schematic shown in Figure 10.6; the assumed physical changes and chemical reactions are shown in R_0–R_3. Based on experimental observations by video inspection [49, 50], the combustion of GAP is considered to go through the liquifying step (R_0) first followed by the occurrence of the two gas generation reactions R_1 and R_2 and formation of a gas–liquid two-phase flow. The combustion residue formation process is simplified here to R_3.

Liquifying Step

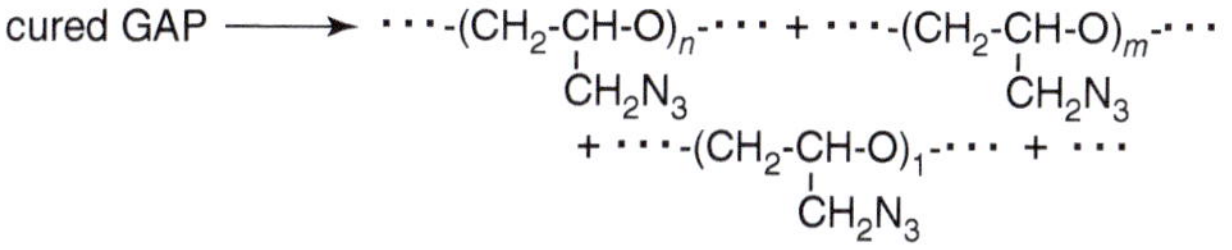

(R_0)

1st Chemical Step

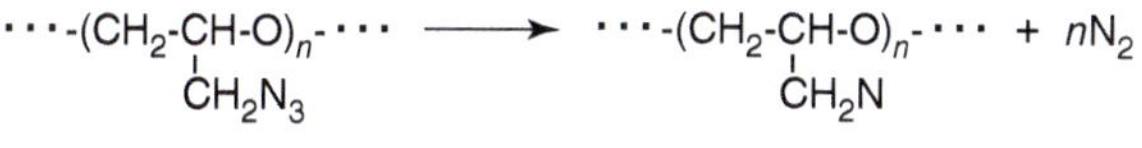

(R_1)

2nd Chemical Step

$$\cdots\text{-(CH}_2\text{-CH-O)}_n\text{-}\cdots \underset{\text{CH}_2\text{N}}{\longrightarrow} \begin{array}{l} \alpha_1 H_2 + \alpha_2 HCN + \alpha_3 NH_3 + \alpha_4 CO + \alpha_5 CH_2 O \\ + \alpha_6 HO_2 + \alpha_7 CH_3 CHO + \alpha_8 C_2 H_2 + \alpha_9 H_2 O \\ + \alpha_{10} C_2 H_4 + \alpha_{11} CH_4 + \alpha_{12} C_{(S)} \end{array}$$

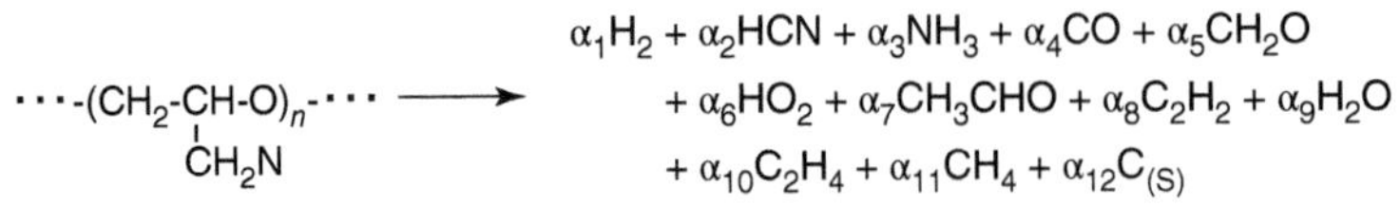

(R$_2$)

3rd Chemical Step

$$2C_2H_2 \longrightarrow CH_4 + 3C(S)$$

(R$_3$)

R$_1$ is an elimination reaction of N$_2$, which is accompanied by large exothermic heat and is widely recognized as the primary pyrolysis reaction in the combustion of GAP [47, 49, 50, 54, 55]. The residual polymer fragments are degraded into gaseous chemical species, which are defined by R$_2$. The coefficients of each chemical species (α_1 to α_{12}) are determined by the chemical species that can be produced from the molecular structure of GAP, in addition to the results from past studies. In this study, R$_3$ which is used to reproduce the pressure dependence of the linear burning rate in the Beckstead model is used as a step to generate the soot composition of the combustion residue.

The chemical reaction calculation of gas phase considered 64 chemical species and 443 elementary reactions, which was constructed from the Research Department eXplosive (RDX) mechanism by Yetter and coworker [56], and Gas Research Institute (GRI)-mech 3.0 [57] except for the reaction that cannot be produced by GAP combustion. Moreover, we also introduced the pyrolysis reaction mechanism of acetaldehyde. In fact, being included in the R$_2$ products but neglected by the two previous schemes [56, 57], acetaldehyde was introduced resorting to the mechanism proposed by Ehime University [58, 59].

In the simulation of this study, it was calculated by dividing into the following three regions based on the three-phase one-dimensional model: (i) solid phase region considering only heat conduction, (ii) gas–liquid two-phase region formed by gas generation (R$_1$) by the GAP degradation reaction occurring in the liquid phase and subsequent reactions (R$_2$, R$_3$), and (iii) gas phase region.

The governing equations in two-phase region are shown in Eqs. (10.10)–(10.14).

Mass conservation formula of gas and liquid;

$$\dot{m} = \varphi \rho_g v_g + (1-\varphi) \rho_l v_l = \varphi \dot{m}_g + (1-\varphi) \dot{m}_l = \text{const} \quad (10.10)$$

Mass conservation formula of liquid phase;

$$\frac{d}{dx}\{(1-\varphi)\rho_l v_l\} = \frac{d}{dx}\{(1-\varphi)\dot{m}_l\} = \sum_{k=1}^{K} W_k \dot{\omega}_{k,l} \quad (10.11)$$

Energy conservation formula;

$$\frac{d}{dx}\{(1-\varphi)\dot{m}_l h_l + \varphi \dot{m}_g h_g\} - \frac{d}{dx}\left(\lambda_{gl}\frac{dT}{dx}\right) = 0$$

where

$$h_l = Y_{G1}\left\{h^0_{G1,l} + c_{p,G1,l}(T - T^0)\right\} + (1 - Y_{G1,l})\left\{h^0_{G2,l} + c_{p,G2.l}(T - T^0)\right\}$$

$$h_g = \sum_{k=1}^{K} Y_{k,g}h_{k,g} = \sum_{k=1}^{K} Y_{k,g}\left(h^0_{k,g} + \int_{T^0}^{T} c_{p,k,g}dT\right)$$

$$\lambda_{gl} = (1 - \varphi)\,\lambda_l + \varphi\,\lambda_g \tag{10.12}$$

Formula for species conservation formula of gas phase;

$$\frac{\mathrm{d}}{\mathrm{d}x}(\varphi\rho_g v_g Y_{k,g}) = \frac{\mathrm{d}}{\mathrm{d}x}(\varphi\dot{m}_g Y_{k,g}) = W_k\{\varphi\dot{\omega}_{k,g} + (1 - \varphi)\dot{\omega}_{k,l}\} \qquad (k = 1, \ldots, K) \tag{10.13}$$

Formula for species conservation of liquid phase;

$$\frac{\mathrm{d}}{\mathrm{d}x}\{(1 - \varphi)\rho_l v_l Y_{G1,l}\} = \frac{\mathrm{d}}{\mathrm{d}x}\{(1 - \varphi)\dot{m}_l Y_{G1,l}\} = (1 - \varphi)W_{G1}\dot{\omega}_{G1,l} \tag{10.14}$$

Here, $\dot{m}$: mass flux (kg m^{2}s^{-1}), φ: void fraction, λ: thermal conductivity (W (m K)$^{-1}$), v: velocity (m s^{-1}), $c_{p,k}$: constant pressure specific heat of species k (J (kg K)$^{-1}$), Y_k: mass fraction of species k, ρ: density (kg m^{-3}), h: enthalpy (J kg^{-1}), $\dot{\omega}_k$: molar formation rate (mol (m^3 s)$^{-1}$), W_k: molecular weight of species k, K: total number of species, g: gas phase, l: liquid phase, gl: two-phase averaged, $G1$: cured GAP, $G2$: GAP polymer backbone after N_2 liberation.

First, to confirm the stability of the HEM PREMIX developed in this study, we simulated the tri-ol GAP calculation by Beckstead et al. and confirmed that the same results were obtained. Therefore, we simulated the combustion of the tetra-ol GAP that we used in this study. Table 10.1 shows the physical properties such as density ρ_s, heat capacity at constant pressure c_{ps} and thermal conductivity λ_s. The simulated burning rate and temperature field significantly exceeded the experimental data; moreover, the imperfection of the combustion had to be taken into consideration.

When the burning surface of GAP was observed by video, it was confirmed that many bubbles were generated in the liquid layer on the surface, and black soot-like residue was generated on the liquid surface, which disappeared to the gas phase at the same rate as the surrounding gas velocity. Furthermore, after burning the sample, along with soot a high-viscosity brown residue was collected and, in the experiments performed at high pressures, a residue of yellow solid fine particles was also detected. Many small spherical residues were found in the SEM observations. Figure 10.7 shows that the spheres were made of a highly viscous substance,

Table 10.1 Tetra-ol GAP physical properties [49].

Density (kg m^{-3})	1290
Heat capacity (kJ (kg K)$^{-1}$)	3.3
Thermal conductivity (J (m s K)$^{-1}$)	2.17×10^{-2}

Source: Wada et al. [49]/With permission of Springer Nature.

Figure 10.7 SEM image of HVR sphere. Source: Wada et al. [49]/With permission of Springer Nature.

and foaming is actively occurring inside. This is considered to be and called here High Viscosity Residue (HVR). Experimental results of the HVR elemental analysis (in molar base) pointed out C:H:N = 3.70 : 6.11 : 1.69. Being the tetra-ol GAP molecular formula $C_{96}H_{164}N_{90}$, the collected data show that while C:H was nearly kept, most of the N_2 elimination reaction was accomplished. Thus, the HVR elemental analysis of showed it to be close to the element ratio of the polymer backbone where nitrogen gas was removed from the GAP, so HVR is considered to be the polymer backbone of the GAP produced by R_1 or its degradation product (still high molecular weight). Furthermore, the degradation and low molecular weight are considered to advance under high pressures, and it becomes oligomers and yellow particulate combustion residues.

In previous combustion models, soot was treated as C(S), R_2 was thought to progress completely, and HVR was not considered. However, the HVR was blown off from the burning surface to the gas phase by the GAP degradation gases. The blown HVR droplets take away thermal energy from the burning surface, and also the chemical enthalpy that contributes to combustion. Figure 10.7 shows this droplet, and it can be observed that the gasification reaction is actively proceeding inside the droplet even being blown off. We consider this phenomenon to be the main cause of incomplete combustibility of GAP and referred to it as the Blow Off Mechanism. Figure 10.8 summarizes the physical sketch of combustion residue formation.

Considering the behavior of the above-mentioned combustion residue in the GAP combustion model, the concept of critical void fraction φ_{cr} is introduced. In the previous models, the point at which the void fraction φ, which is the volume fraction of the gas phase in the gas–liquid two-phase region, reaches unity, is the interface (burning surface) between the gas–liquid two-phase and the gas phase, and only gas ingredients and C(S) were considered in the gas phase. However, herein, the burning surface is defined as the point of $\varphi = \varphi_{cr}$ and it is assumed that HVR exists in the gas phase. In this simulation, the HVR is frozen in the gas phase, does not contribute

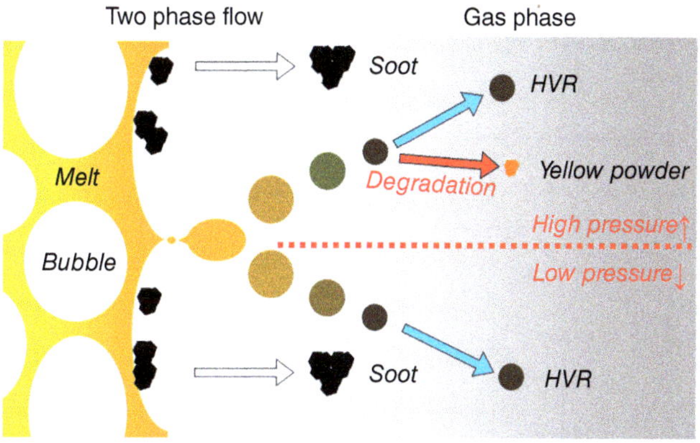

Figure 10.8 Physical sketch of combustion residue formation: HVR + yellow powder at high pressures (7–10 MPa) versus HVR only at low pressures (1–3 MPa). Source: Wada et al. [49]/With permission of Springer Nature.

Table 10.2 Heat release of R_1 and kinetic parameters of R_1 and R_2 [49].

Heat release of R_1		$1507\ kJ\ kg^{-1}$
Pre-exponential factor: A		Activation energy: E
R_1	0.9×10^{10} 1/s	$163\ kJ\ mol^{-1}$
R_2	0.1×10^{90} 1/s	$126\ kJ\ mol^{-1}$

Source: Wada et al. [49]/With permission of Springer Nature.

to the chemical reaction, and it is output as a combustion residue in the calculation results. The soot-like combustion residue is assumed to be the produced from R_3 as C (S).

In this simulation, the heat generated by N_2 elimination reaction, the rection rate parameters of R_1 and R_2, and φ_{cr} are used as parameters to adjust the mass ratios of the combustion residues. These parameters are shown in Table 10.2. The mass fractions of the combustion residues generated in the numerical simulation were well consistent with those of the combustion residues obtained in the experiments by adjusting φ_{cr} as a function of pressure (Table 10.3). Testing was conducted by burning strands in a closed chamber under N_2 atmosphere at the indicated pressures and then measuring the mass of HVR as well as other particles. It can be observed that φ_{cr} rises with pressure and combustion completion is improved. Figure 10.9 shows a comparison of temperature histories at 1 and 5 MPa. For both 1 and 5 MPa, the temperature rise in the condensed phase, burning surface temperature, and final gas temperature in the simulation were all highly consistent with the experimental data. Figure 10.10 shows a comparison of linear burning rates. The calculated results were highly consistent with the experimental data, and the pressure index could be well simulated.

In this study, it was proved that applying the blow off mechanism to the GAP combustion model made it possible to perform advanced simulations of the mass fractions of combustion residues, temperature profile, and linear burning rate.

Table 10.3 Comparison of residue mass fractions [49].

Experimental results				Numerical simulation				
Pressure (MPa)	Gas (%)	HVR (%)	Particle (%)	Pressure (MPa)	Gas (%)	Residue (%)	C(S)	φ_{cr} (%)
1	59.6	31.4	9.0	1	60.2	34.4	5.4	0.66
3	66.4	26.9	6.7	3	64.5	30.4	5.1	0.70
5	75.4	18.4	5.7	5	75.1	20.1	4.8	0.80
7	80.1	14.8	5.1	7	79.3	16.0	4.7	0.84
10	91.1	5.3	3.6	10	90.7	5.0	4.3	0.95

Source: Wada et al. [49]/With permission of Springer Nature.

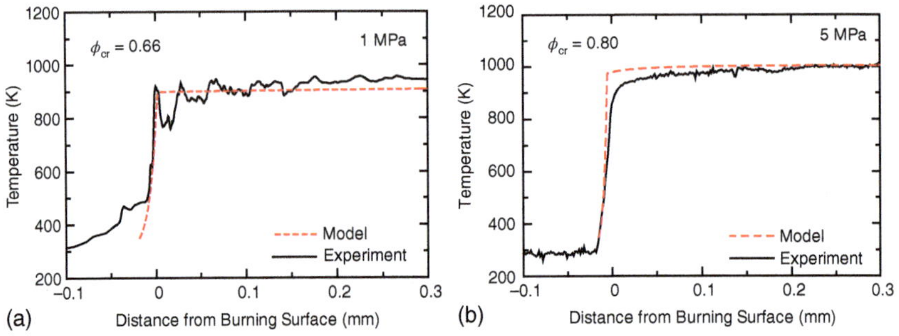

Figure 10.9 Comparison of temperature profile between experiment and simulation. (a) @1 MPa. (b) @5 MPa. Source: Wada et al. [49]/With permission of Springer Nature.

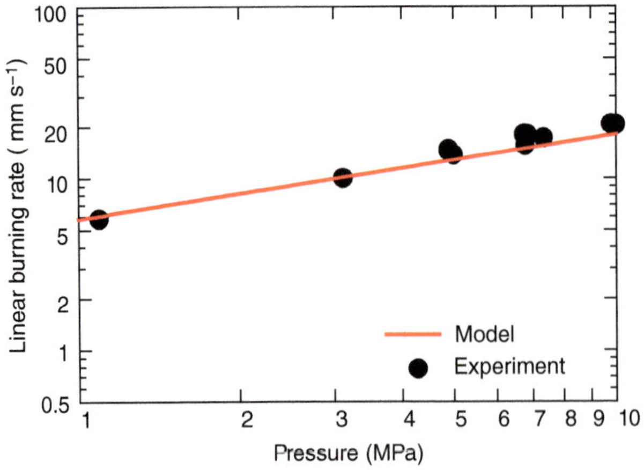

Figure 10.10 Comparison of linear burning rate between experiment and simulation. Source: Wada et al. [49]/With permission of Springer Nature.

We believe that the blow off mechanism accurately captures the essence of GAP combustion and provides the correct guidelines for designing practical rocket motor applications.

10.3 Application of GAP to Gas Hybrid Rocket Motor [2, 51–53]

We experimentally verified the combustion performance and the possibility of thrust control of a gas hybrid rocket using GAP as solid fuel. Figure 10.11 shows a schematic of a gas hybrid rocket. The solid fuel is burned as a gas generator (GG) in the primary combustion chamber, and the liquid oxidizer is injected into the secondary combustion chamber by the combustion pressure. The combustion gas of solid fuel enters the secondary combustion chamber as the fuel gas, mixes with oxidizers, and is completely burned. It is a simple and lightweight system that does not require the heavy and complicated oxidizer injection system required by the traditional hybrid rocket systems. GAP is a high-energy substance with self-combustibility and can achieve high specific impulse. In addition, since it does not contain metallic fuel or chlorine atoms, an ideal propulsion system with a low environmental load can be realized.

First, combustion experiments of two small motors (diameter 60 and 80 mm, respectively) were conducted as GAP GG test. Figure 10.12 shows a schematic diagram of a φ 80 mm motor. GAP grain is of the end-burning type. Two igniters are arranged and tilted 15 ° from the normal direction of the motor case. Combustion gases from both igniters collide at the center of the motor, GAP grains ignite stably in the center, and combustion proceeds smoothly.

Figure 10.13 shows a comparison of the linear burning rate obtained from tests conducted for each motor and the linear burning rate obtained from the strand burner. The two data show a good agreement, which implies that the GAP combustion in the motor is consistent with the strand test. Figure 10.14 shows the C^* efficiency of GAP GG. C^* efficiency is in the range of 70–85% and rises with the combustion pressure. The value of C^* efficiency is almost the same as φ_{cr} (Table 10.3), indicating that the completeness of GAP combustion in the motor is consistent with that at the strand level.

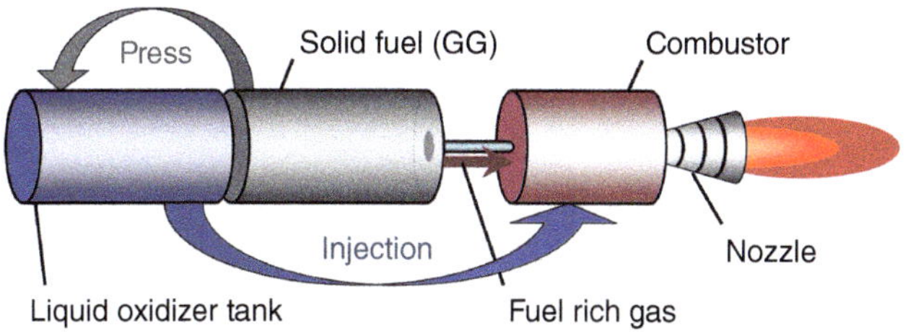

Figure 10.11 Conceptual scheme of gas hybrid rocket. Source: Wada [51]/With permission of Begell House Inc.

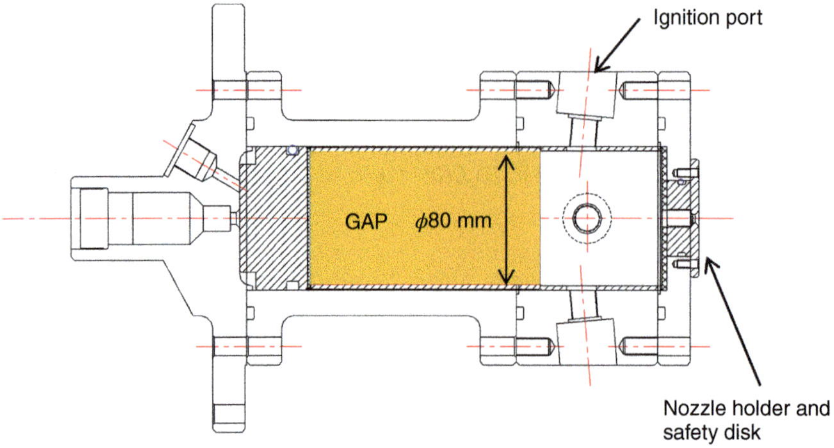

Figure 10.12 ϕ80 mm rocket motor. Source: Wada [51]/With permission of Begell House Inc.

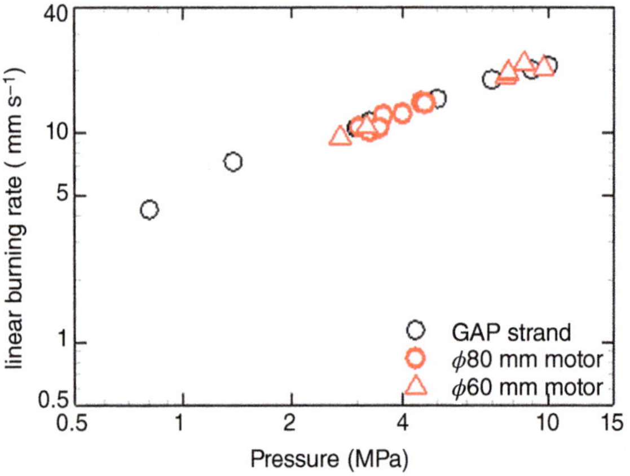

Figure 10.13 Comparison of the burning rate between small motors and strand burner. Source: Wada [51]/With permission of Begell House Inc.

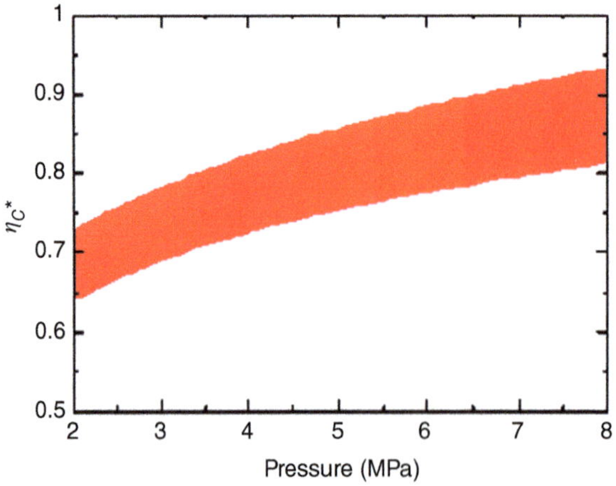

Figure 10.14 C^* efficiency of GAP gas generator. Source: Wada [51]/With permission of Begell House Inc.

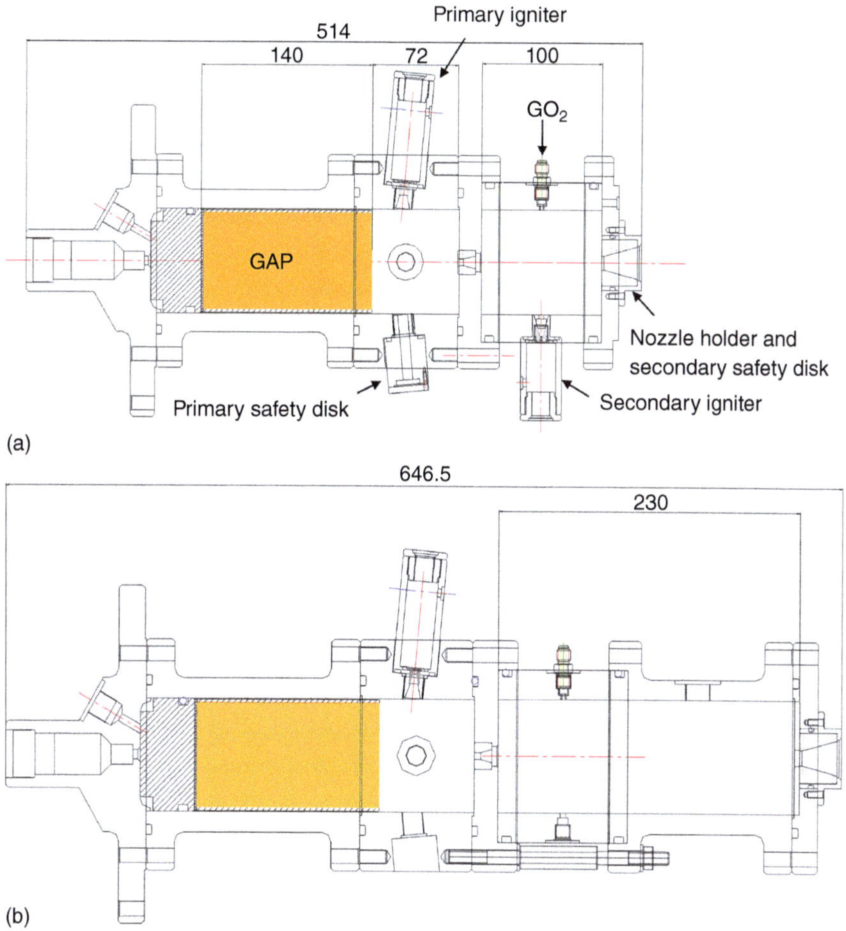

Figure 10.15 Gas hybrid rocket motors with different length of secondary combustor. (a) Basic type: 100 mm of secondary combustor length. (b) Extended type: 230 mm of secondary combustor length. Source: Wada [51]/With permission of Begell House Inc.

Next, a secondary combustor was attached to the φ 80 mm motor and operated as a gas hybrid rocket using gas oxygen. Figure 10.15 shows a schematic of a φ 80 mm gas hybrid rocket with different length of a secondary combustion chamber.

Figure 10.16 shows a photograph of the firing test of a gas hybrid rocket. Figure 10.17 shows the C^* efficiency when changing the length of the secondary combustion chamber. The results of these combustion tests showed that the C^* efficiency reached 98% when the length of the secondary combustor was optimal, and that the GAP gas hybrid rocket could be a practical system.

As for thrust control, (i) the method of controlling the thrust by changing the mass flow rate (equivalent ratio) of oxygen, and (ii) the method of controlling the thrust by changing the linear burning rate of the solid fuel according to the composition were performed. To change the mass flow rate of gas oxygen (GOX), a test was

Figure 10.16 Firing test of GAP gas hybrid rocket. Source: Wada et al. [51], JAXA-Japan Aerospace Exploration Agency.

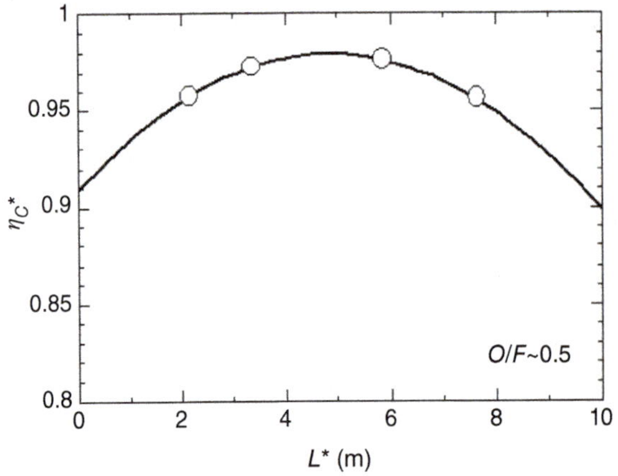

Figure 10.17 C^* efficiency as a function of secondary combustor L^*. Source: Wada [51]/With permission of Begell House Inc.

conducted to control the thrust. Figure 10.18 shows the pressure history, while changing the O/F by controlling the flow rate of gas oxygen during combustion using a solenoid valve. The test included laying two gas oxygen supply lines, with large (main GOX) and small (sub GOX) flow rate. At 0 seconds, ignition of the primary chamber was performed and burning of the GAP solid fuel started. At 0.5 seconds, oxygen flow was started at full through both the main and sub lines for 1 seconds (0.5–1.5 seconds in Figure 10.18) and the secondary combustion chamber was ignited at the same time (triggering the hybrid burn). Afterwards, oxygen was supplied for 1 seconds only to the main line (1.5–2.5 seconds in Figure 10.18) and for another 1 seconds only to the sub line (2.5–3.5 seconds in Figure 10.18). At 3.5 seconds only GAP could burn (no oxygen flow and no hybrid burn).

Although the O/F ratio changed sharply during combustion, the secondary combustion pressure exhibited a very good response, and the combustion was stable without inducing vibration. It was demonstrated that it is possible to control the thrust by changing the mass flow rate of oxygen.

In this study, the linear burning rate of GAP based fuel was changed by adding polyethylene glycol (PEG). PEG was used with the number of repeating units aligned with GAP whose degree of polymerization is same as that of GAP prepolymer. PEG

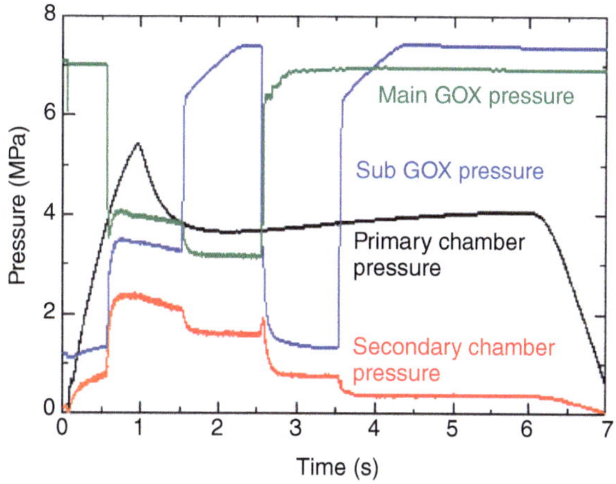

Figure 10.18 Change in secondary combustor pressure when changing *O/F*. Source: Wada [51]/With permission of Begell House Inc.

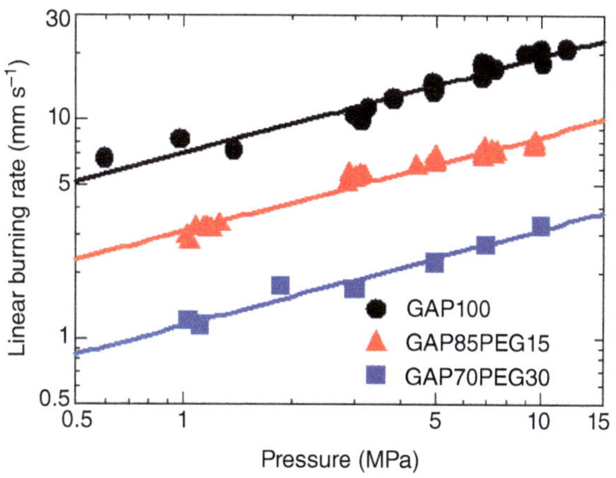

Figure 10.19 Linear burning rates of GAP+PEG mixtures. Source: Wada [51]/With permission of Begell House Inc.

has good compatibility with GAP, and the linear burning rate easily changes with the addition of PEG. Figure 10.19 shows the linear burning rate of GAP fuel with the addition of 15% and 30% PEG based on the mass ratio. The addition of 30% PEG reduces the linear burning rate to about 1/10.

A gas hybrid rocket test was conducted using double layered end-burning type grains by GAP100 and GAP95PEG5 as GG (Figure 10.20). Figure 10.21 shows the results of the combustion test. In the primary combustion chamber, GAP100 was first burned, followed by GAP95PEG5. The double layered fuel burned stably. Based on the pressure of the secondary combustor, the thrust of GAP95PEG5 fuel during combustion is about 78% of the GAP100, and it is demonstrated that about 20% of thrust can be controlled with the addition of only 5% PEG.

The above test results showed that the thrust control of the rocket motor can be realized by both the oxidizer flow rate control and the fuel flow rate control. Combining the two makes it possible to have a very wide range of thrust control. This showed a strong possibility that a gas hybrid rocket using GAP could handle a variety of missions.

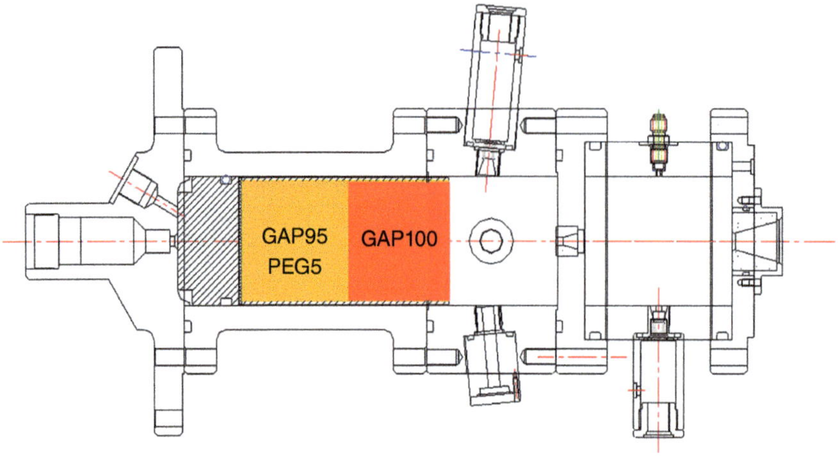

Figure 10.20 Experimental setup for the double layered grain GAP100 and GAP95PEG5. Source: Wada [51]/With permission of Begell House Inc.

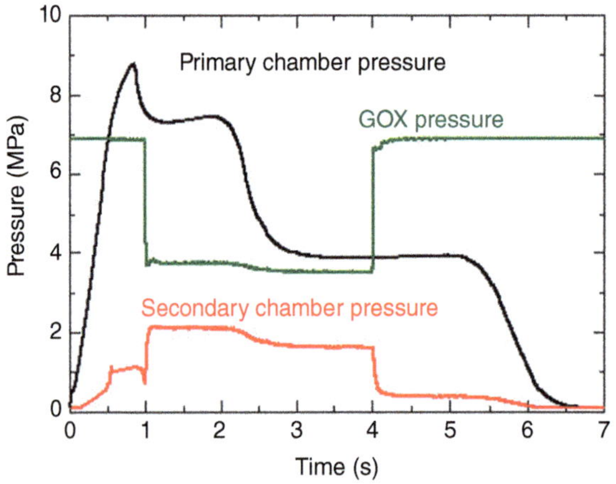

Figure 10.21 Results from gas hybrid rocket test using double layered grain of GAP100 and GAP95PEG5. Source: Wada [51]/With permission of Begell House Inc.

10.4 Summary

The history of authors activities on GAP performed at ISAS/JAXA is summarized. The production of GAP by NOF Corporation is briefly described. Studies on the combustion mechanism are presented in Sections 10.2.1 and 10.2.2. Through these studies, the essence of GAP combustion was well understood and applied to the gas hybrid rocket. GAP gas hybrid rocket system shows high propulsive performance at 98% of C^* efficiency and system design freedom in the thrust modulation. The wide variety of application of GAP gas hybrid rocket, especially to the satellite propulsion in term of high performance and debris-free system is highly recommended.

References

1 Jiao, L., Cai, J., Ma, H. et al. (2015). Application of energetic materials in LASER ablative micropropulsion. *International Journal of Energetic Materials and Chemical Propulsion* 14: 57–69.

2 Chang, P.-J., Wada, Y., Garg, A. et al. (2015). Combustion and performance studies of glycidyl azide polymer and its mixtures as hybrid rocket fuel. *International Journal of Energetic Materials and Chemical Propulsion* 14: 221–239.

3 Wingborg, N., Lindborg, A., Oscarson, C., and Sjöblom, M. (2015). Characterization of solid propellants based on ADN and GAP. *51st AIAA/SAE/ASEE Joint Propulsion Conference* https://doi.org/10.2514/6.2015-4222.

4 Zhang, Z., Wang, G., Wang, Z. et al. (2015). Synthesis and characterization of novel energetic thermoplastic elastomers based on glycidyl azide polymer (GAP) with bonding functions. *Polymer Bulletin* 72: 1835–1847.

5 Gołofit, T. and Zysk, K. (2015). Thermal decomposition properties and compatibility of CL-20 with binders HTPB, PBAN, GAP and poly NIMMO. *Journal of Thermal Analysis and Calorimetry* 119: 1931–1939.

6 Li, G., Liu, M., Zhang, R. et al. (2015). Synthesis and properties of RDX/GAP nano-composite energetic materials. *Colloid and Polymer Science* 293: 2269–2279.

7 Pang, W.Q., DeLuca, L.T., Xu, H.X. et al. (2016). Effects of CL-20 on the properties of glycidyl azide polymer (GAP) solid rocket propellant. *International Journal of Energetic Materials and Chemical Propulsion* 15: 49–64.

8 Li, H., An, C., Du, M. et al. (2016). Preparation and characterization of HMX/GAP-ETPE nanocomposites. *International Journal of Energetic Materials and Chemical Propulsion* 15: 131–140.

9 Li, H., An, C., Ye, B. et al. (2016). RDX/GAP-ETPE nanocomposites for remarkably reduced impact sensitivity. *International Journal of Energetic Materials and Chemical Propulsion* 15: 275–284.

10 Abirami, A., Soman, R.R., Agawane, N.T. et al. (2016). Studies on curing of glycidyl azide polymer using isocyanate, acrylate and processing of GAP-BORON-based, fuel-rich propellants. *International Journal of Energetic Materials and Chemical Propulsion* 15: 215–230.

11 Wada, Y., Kawabata, Y., Kato, R. et al. (2016). Observation of combustion behavior of low melting temperature fuel for a hybrid rocket using double slab motor. *International Journal of Energetic Materials and Chemical Propulsion* 15: 351–369.

12 Lima, R.J.P., Dubois, C., Stowe, R., and Ringuette, S. (2016). Enhanced reactivity of aluminum powders by capping with a modified glycidyl azide polymer. *International Journal of Energetic Materials and Chemical Propulsion* 15: 481–500.

13 Hayashi, K. and Kuwahara, T. (2016). Combustion characteristics of GAP for the liquid monopropellant–effects of the spray droplet diameters. *52nd AIAA/SAE/ASEE Joint Propulsion Conference* https://doi.org/10.2514/6.2016-4664.

14 Ao, W., Liu, P., and Yang, W. (2016). Agglomerates, smoke oxide particles, and carbon inclusions in condensed combustion products of an aluminized GAP-based propellant. *Acta Astronautica* 129: 147–153.

15 Gettwert, V., Franzin, A., Bohn, M.A. et al. (2017). Ammonium dinitramide/glycidyl azide polymer (ADN/GAP) composite propellants with and without metallic fuels. *International Journal of Energetic Materials and Chemical Propulsion* 16: 61–79.

16 Pei, J., Zhao, F., Wang, Y. et al. (2017). Energy and combustion characteristics of propellants based on BAMO-GAP copolymer. *Springer Aerospace Technology* 341–363.

17 Weiser, V., Franzin, A., DeLuca, L.T. et al. (2017). Combustion behavior of aluminum particles in ADN/GAP composite propellants. *Springer Aerospace Technology* 253–270.

18 Wang, S., Liu, C., Guo, X. et al. (2017). Effects of crosslinking degree and carbon nanotubes as filler on composites based on glycidyl azide polymer and propargyl-terminated polyether for potential solid propellant application. *Journal of Applied Polymer Science* 134: 39.

19 Deng, J., Wang, X., Li, G., and Luo, Y. (2017). Effect of bonding agent on the mechanical properties of GAP high-energy propellant. *Propellants, Explosives, Pyrotechnics* 42: 394–400.

20 Singh, H. (2017). Survey of new energetic and eco-friendly materials for propulsion of space vehicles. *Springer Aerospace Technology* https://doi.org/10.1007/978-3-319-27748-6_4.

21 Guo, M., Ma, Z., He, L. et al. (2017). Effect of varied proportion of GAP-ETPE/NC as binder on thermal decomposition behaviors, stability and mechanical properties of nitramine propellants. *Journal of Thermal Analysis and Calorimetry* 130: 909–918.

22 Jensen, T.L., Unneberg, E., and Kristensen, T.E. (2017). Smokeless GAP-RDX composite rocket propellants containing diaminodinitroethylene (FOX-7). *Propellants, Explosives, Pyrotechnics* 42: 381–385.

23 Lu, Y., Shu, Y.J., Liu, N. et al. (2017). Theoretical simulations on the glass transition temperatures and mechanical properties of modified glycidyl azide polymer. *Computational Materials Science* 139: 132–139.

24 Elghany, M.A., Klapötke, T.M., and Elbeih, A. (2018). New green and thermally stable solid propellant formulations based on TNEF. *International Journal of Energetic Materials and Chemical Propulsion* 17: 349–357.

25 Zhang, W., Ren, H., Sun, Y. et al. (2018). Effects of ester-terminated glycidyl azide polymer on the thermal stability and decomposition of GAP by TG-DSC-MS-FTIR and VST. *Journal of Thermal Analysis and Calorimetry* 132: 1883–1892.

26 Qi, J., Zhang, S., Liang, T. et al. (2018). Plasma characteristics of energetic liquid polymer ablated by nanosecond laser pulses. *Frontiers of Optoelectronics* 11: 261–266.

27 Hafner, S., Keicher, T., and Klapötke, T.M. (2018). Copolymers based on GAP and 1,2-epoxyhexane as promising prepolymers for energetic binder systems. *Propellants, Explosives, Pyrotechnics* 43: 126–135.

28 Elghany, M.A., Elbeih, A., and Klapötke, T.M. (2018). Thermo-analytical study of 2,2,2-trinitroethyl-formate as a new oxidizer and its propellant based on a GAP matrix in comparison with ammonium dinitramide. *Journal of Analytical and Applied Pyrolysis* 133: 30–38.

29 Hussein, A.K., Zeman, S., and Elbeih, A. (2018). Thermo-analytical study of glycidyl azide polymer and its effect on different cyclic nitramines. *Thermochimica Acta* 660: 110–123.

30 Tao, J., Jin, B., Peng, R., and Chu, S. (2018). Isothermal curing of the glycidyl azide polymer binder system by microcalorimetry. *Polymer Testing* 71: 231–237.

31 Wada, Y., Hatano, S., Banno, A. et al. (2019). Development of a direct injection gas-hybrid rocket system using glycidyl azide polymer. *International Journal of Energetic Materials and Chemical Propulsion* 18: 157–170.

32 Ma, S., Fan, H., Zhang, N. et al. (2020). Investigation of a low-toxicity energetic binder for a solid propellant: curing, microstructures, and performance. *ACS Omega* 5: 30538–30548.

33 Liu, D., Geng, D., Yang, K. et al. (2020). Decomposition and energy-enhancement mechanism of the energetic binder glycidyl azide polymer at explosive detonation temperatures. *Journal of Physical Chemistry A* 124: 5542–5554.

34 Comtois, E., Favis, B.D., and Dubois, C. (2020). Phase transitions and mechanical properties of nitrocellulose plasticized by glycidyl azide polymer and nitroglycerine. *Polymer Engineering and Science* 60: 2301–2313.

35 Jafari, N., Arab, B., Zekri, N., and Alamdari, R.F. (2020). Experimental and simulation study on glass transition temperatures of GAP with ionic-liquid-based energetic plasticizers. *Propellants, Explosives, Pyrotechnics* 45: 615–620.

36 Liu, H., Ao, W., Liu, P. et al. (2020). Experimental investigation on the condensed combustion products of aluminized GAP-based propellants. *Aerospace Science and Technology* 97: 2.

37 Vo, T.A., Jung, M., Adams, D. et al. (2020). Dynamic modeling and simulation of the combustion of aluminized solid propellant with HMX and GAP using moving boundary approach. *Combustion and Flame* 213: 409–425.

38 Comtois, E., Favis, B.D., and Dubois, C. (2021). Linear burning rate and erosivity properties of nitrocellulose propellant formulations plasticized by glycidyl azide polymer and nitroglycerine. *Propellants, Explosives, Pyrotechnics* 46: 494–504.

39 Xu, S., Pang, A.M., Wu, Z. et al. (2021). Synthesis of polyetheramine based bonding agents and their effect on mechanical properties of an AP/CL-20/GAP formulation. *Propellants, Explosives, Pyrotechnics* 46: 1216–1226.

40 Li, M., Hu, R., Xu, M. et al. (2021). Burning characteristics of high density foamed GAP/CL-20 propellants. *Defense Technology* 8: https://doi.org/10.1016/j.dt.2021.08.006.

41 M. Frankel, L. Grant, J. Flanagan (1989). "Historical Development of GAP", AIAA Paper, 89-2307.

42 Norman J. Vander Hyde, Robert L. Geisler and Charles R. Cooke (1977). "Manufacture of PCDE Pre-polymer", AFRPL-TR-77-45, Hercules Ins.

43 Bazaki, H., Kai, T., and Mitarai, Y. (1992). Combustion performance of AP/AMMO composite propellants. *Science and Technology of Energetic Materials* 53 (1): 25–30.

44 Bando, K., Sato, K., Inokami, K. et al. (1990). 3,3-Bis azidomethyl oxetane (BAMO)-based propellants (1) – synthesis and characterization of BAMO-based polymers. *Science and Technology of Energetic Materials* 51 (4): 228–239.
45 K. Kato, K. Kobayashi, Y. Seike, K. Sakai, Y. Matuzawa (1996). "Mechanical Properties of GAP Reduced Smoke Propellants", AIAA Paper, 96-3253.
46 T. Ogura, K. Fujiwara, M. Naijyo: JP 2020-186294A.
47 Hori, K. and Kimura, M. (1996). Combustion mechanism of glycidyl azide polymer. *Propellants, Explosives, Pyrotechnics* 21 (3): 160–165.
48 https://en.wikipedia.org/wiki/Activation_energy_asymptotics
49 Wada, Y., Seike, Y., Tsuboi, N. et al. (2008). Combustion mechanism of tetra-ol glycidyal azide polymer. *Science and Technology of Energetic Materials* 69 (5): 143–148.
50 Wada, Y., Seike, Y., Tsuboi, N. et al. (2009). Combustion model of tetra-ol glycidyl azide polymer. *Proceedings of the Combustion Institute* 32: 2005–2012.
51 Wada, Y. (2009). Combustion study of tetra-ol glycidyl azide polymer – combustion mechanism and its application to hybrid rocket. PhD Thesis. The Graduate University for Advanced Studies, SOKENDAI, Ko-1229.
52 Wada, Y., Seike, Y., Nishioka, M. et al. (2009). Combustion mechanism of tetra-ol glycidyl azide polymer and its application to hybrid rocket. *International Journal of Energetic Materials and Chemical Propulsion* 8: 555–570.
53 Hori, K., Nomura, Y., Fujisato, K. et al. (2012). Glycidyl azide polymer and polyethylene glycol mixtures as hybrid rocket fuels. *International Journal of Energetic Materials and Chemical Propulsion* 11: 97–106.
54 K. V. Puduppakkam, M. W. Beckstead, 37th JANNAF combustion meeting, CPIA, 2000, 701, 639.
55 Puduppakkam, K.V. and Beckstead, M.W. (2005). Combustion modeling of glycidyl azide polymer with detailed kinetics. *Combustion Science and Technology* 177: 1661–1667.
56 Prasad, K., Yetter, R.A., and Smooke, M.D. (1997). An Eigenvalue method for computing the burning rates of RDX propellants. *Combustion Science and Technology* 124: 35–82.
57 G. P. Smith, D. M. Golden, M. Frenklach, N. W. Moriarty, B. Eiteneer, M. Goldenberg, C. T. Bowman, R. K. Hanson, S. Song, W.C. Gardiner, V. V. Lissianski, Z. Qin, GRI-Mech 3.0, http://www.me.berkeley.edu/gri_mech.
58 San Diego Mechanism 2005/12/01, http://maeweb.ucsd.edu/~combustion/cermech.
59 Hidaka Laboratory, Department of Chemistry, Ehime University, Dimethyl ether reaction mechanism, http://chem.sci.ehime-u.ac.jp/%7Ephychem1/hidaka/work.htm.

11

Effect of Different Binders and Metal Hydrides on the Performance and Hydrochloric Acid Exhaust Products Scavenging of AP-Based Composite Solid Propellants: A Theoretical Analysis

Fateh Chalghoum[1,2], Djalal Trache[1], Ahmed M'cili[1], Mokhtar Benziane[2], Ahmed F. Tarchoun[1], and Amir Abdelaziz[1]

[1] *Ecole Militaire Polytechnique, Energetic Materials Laboratory, Teaching and Research unit of Energetic Processes, Zip code 17, Bordj El-Bahri, 16046 Algiers, Algeria*
[2] *Laboratory of Research in Pyrotechnics and Propulsion, Scientific and Technological Research Direction, Superior School of Material, Zip Code 188, Beau-Lieu, Algiers, Algeria*

Nomenclature

AP	ammonium perchlorate
CCPs	condensed combustion products
CEA	Chemical Equilibrium with Application
CL-20	hexanitrohexaazaisowurtzitane
CSPs	composite solid propellants
GAP	glycidyl azide polymer
HCl	hydrochloric acid
HMX	cyclotetramethylene tetranitramine
HPHS	high-performance halide scavenger parameter
HTPB	hydroxyl-terminated polybutadiene
$I_{sp,v}$	volumetric specific impulse
I_{sp}	gravimetric specific impulse
M_w	average molecular weight of the combustion products
OB	oxygen balance
PAMMO	poly(3-azidomethyl-3-methyloxetane)
PBAMO	poly(3,3-bis-azidomethyloxetane)
PGN	poly(glycidyl nitrate)
PNMMO	poly(3-nitratomethyl-3-methyloxetane)
RDX	cyclotrimethylene trinitramine
T_c	adiabatic flame temperature
$\Delta H(T)_r$	enthalpy of reaction
$\Delta H_f(T)_p$	formation enthalpy of combustion products
$\Delta H_f(T)_r$	formation enthalpy of reactants
ρ_p	propellant density

Nano and Micro-Scale Energetic Materials: Propellants and Explosives, First Edition.
Edited by Weiqiang Pang and Luigi T. DeLuca.

11.1 Introduction

One of the main expected objectives in the field of solid rocket propulsion is to increase the operating range of rockets by using high-performance propellant formulations. The addition of highly energetic particles as supplementary fuels is one of the effective tools for improving propellant performance. Broadly, metallic powders such as aluminum and boron are used in propellant formulations to increase their heat of combustion, density and combustion temperature, and thus their specific impulse [1–3]. However, these metallic fuels induce several negative consequences such as the formation of condensed combustion products (CCPs), agglomeration phenomena, and nozzle erosion [4, 5]. One of the interesting alternatives, widely proposed in the literature to overcome such drawbacks, is the incorporation of these energetic metals in the form of simple or complex hydrides [4, 6, 7]. In fact, it was reported that metal hydrides dehydrogenation that may occur at relatively low temperatures can enhance the heat release and increase the specific impulse, and subsequently the resulting base metals will react at a higher temperature, producing additional energy [7–9]. Particularly, alkali metal hydrides are widely considered as a potential fuel substitute to increase the delivered specific impulse. Indeed, these hydrogenated additives can significantly reduce CCPs and two-phase losses due to the lightweight nature of their combustion products and the low volatilization temperatures of their oxides and chlorides [6–8, 10]. On the other hand, several negative effects are attributed to the employment of ammonium perchlorate (AP), as the universal oxidizer for modern solid rocket propellants. This strong oxidizer has been recently classified as a persistent environmental contaminant [11, 12] due to its extensive military and civilian uses. It has been reported that exposure to AP can cause hypothyroidism. The latter is a common condition that refers to the inability of the thyroid gland to produce enough thyroid hormone and release it into the bloodstream. These particular hormones have an important role in the metabolic regulation and oxygen consumption in our bodies. Recently, some research has demonstrated that chlorate and chlorite produced during the perchlorate reduction process can also cause hemolytic anemia in laboratory animals, methemoglobin development in mammals, as well as toxicity to microorganisms and plants [13–15]. Another major disadvantage, related particularly to the combustion process of AP-based solid propellants is the hydrochloric acid (HCl) emission, which provokes several environmental issues, including ozone layer depletion and acid-rain phenomena enhancement [10, 16]. Several investigations have focused on the employment of metal hydride compounds to increase energy propellants and reduce their HCl-gaseous products emission [6, 8, 10]. One of the main approaches applied to reduce HCl from combustion products of AP-based solid propellants involves the use of scavenger compounds that react with chlorine ions to produce chloride species. In this respect, it should be noted that chloride salts, at high concentrations in soil and water, present some challenges. Indeed, it has been reported that chlorides, which are highly soluble and mobile, tend to decrease the biodiversity of aquatic plants and animals. Furthermore, Cl reduces aquatic self-purification processes by decreasing nutrient accumulation in macrophytes and reducing the degradation of organic matter [17, 18].

One of the most attractive candidates for propellant fuel replacement is lithium-based hydrides. Indeed, due to their halophilic characteristics, they can reduce the negative environmental impacts of AP-based propellants by removing almost all the hydrochloric acid produced during propellant combustion [8]. It was found, when analyzing the effect of numerous complex metal hydrides on the theoretical performance of composite solid propellants (CSPs) based on AP and hydroxyl-terminated polybutadiene (HTPB), that some ternary and quaternary metal hydrides demonstrate a synergistic effect of promising performance improvement and high chlorine scavenging capacity [6]. These hydrogenated compounds have highlighted the important contribution that oxophilic base metals such as Al, Mg, and B can make to Li-based complex metal hydrides to improve their chlorine scavenging ability and increase the specific impulse of the resulting propellant formulations [6]. On the other hand, HTPB is an inert binder, immiscible with most nitrate ester-based energetic plasticizers, which limits its contribution to improving the energetic performances of CSPs [19]. Accordingly, several investigations have been focused on the development of advanced propellant compositions formulated with new active binders such as glycidyl azide polymer (GAP), poly(3-nitratomethyl-3-methyloxetane) (PNMMO), and poly(3,3-bis-azidomethyloxetane) (PBAMO) [19, 20]. It is therefore interesting to examine the theoretical performance of these advanced composite propellants elaborated with energetic binders and enriched with different metal hydrides. The present chapter follows this approach through an extended theoretical analysis of the energetic performances of 29 AP-based propellant formulations in addition to the AP/HTPB aluminized propellant, which is used as a reference propellant to compare the theoretical performances of the different investigated metal hydrides-based propellants. It should be noted that the investigated metal hydrides are selected based on their performance (high specific impulse and high HCl removal capacity), as computed in our previous work [6]. The computations performed in this study were carried out using NASA Lewis Code, Chemical Equilibrium with Application (CEA). Gravimetric specific impulse (I_{sp}), volumetric specific impulse ($I_{sp,v}$), adiabatic flame temperature (T_c), the average molecular weight of the combustion products (M_w), the concentration of the CCPs as well as the high-performance halide scavenger parameter (HPHS) are the computed parameters used to evaluate the performance of the studied composite propellants.

11.2 Theoretical Background and Computation Procedure

11.2.1 Performance of Composite Solid Propellants

Gravimetric specific impulse (I_{sp}), is one of the main parameters commonly used to evaluate the propellant performance. It is well known that under frozen chemical expansion, the mathematical expression of this parameter is directly proportional to the adiabatic flame temperature (T_c) and inversely proportional to the molar mass of the combustion products (M_w) as shown by the following formula (Eq. (11.1)):

$$I_{sp} \propto \left(\sqrt{\frac{T_c}{M_w}}\right) \tag{11.1}$$

Accordingly, one approach to improving I_{sp} is to increase the ratio of the adiabatic flame temperature to the average molecular weight of the combustion products by using energetic additives such as metal hydrides based on light high-energy metals. In fact, the combustion of these energetic particles promotes the formation of light gaseous species that enhance the flow rate of the expanded gases, thus improving the overall performance of the propellant. On the other hand, the heat of combustion depends on the formation enthalpy of the different propellant ingredients as given in Eq. (11.2).

$$\Delta H_r(T) = \Delta H_f(T)_p - \Delta H_f(T)_r \tag{11.2}$$

where, $\Delta H_f(T)_p$ and $\Delta H_f(T)_r$ denote the formation enthalpies of combustion products and reactants, respectively.

As the formation enthalpy of reactants increases, it will contribute to increasing the propellant heat combustion. The value of this important thermodynamic parameter is directly dependent on the type of chemical bonds between the atoms in the propellant ingredient molecules, which are typically C, H, O, and N in the binder network. Among the most used inert binders, the one that has the lowest negative enthalpy of formation is the more susceptible to improving the combustion heat of the composite propellant. However, energetic binders, which exhibit the highest positive enthalpy of formation, are more expected to increase the combustion flame temperature. One of the effective tools to improve the energy of the propellant binder is the inclusion of energetic functional groups, such as nitro, nitrato, and azido moieties in the hydrocarbon backbone of the base polymer of the propellant binder [21, 22].

Another important propellant performance criterion is the volumetric specific impulse ($I_{sp,v}$) defined as:

$$I_{sp,v} = I_{sp} * \rho_p \tag{11.3}$$

As shown in Eq. (11.3), the improvement of $I_{sp,v}$ can be obtained by increasing the propellant density (ρ_p) through the employment of dense ingredients. The density of many energetic binders is higher than that of HTPB, which is the most common binder for composite solid propellants, making them suitable for improving the volumetric specific impulse.

11.2.2 Propellant Energetic Ingredients

Recent advances in chemical propulsion systems are marked by the emergence of new propellant ingredients, such as the employment of energetic binders instead of conventional organic polymers. They can provide additional energy to the composite propellant through highly exothermic reactions of its energetic functions such as azido and nitro groups, which improve the propellant's overall performance. Azido-polymers are macromolecules containing in their hydrocarbon backbone the $^{-}N{=}N{=}N^{+}$ bonds. These active organic materials can be burned without an external oxidizer producing nitrogen and a relatively high content of fuel fragments such as H_2, CO, and C(s); the subsequent combustion of these products provides more heat released by the propellant. The bond breaking of $-N_3$ moieties is the initial step for azido-polymer decomposition, followed by other reaction processes such as melting and gasification [23–25]. Active azido-polymers including GAP, PBAMO, poly(3-azidomethyl-3-methyloxetane) (PAMMO) are widely cited in the literature as potential candidates for the future propulsion applications. GAP as an active binder has attracted considerable attention due to its interesting properties such as high burn rate, high enthalpy of formation, relatively low impact sensitivity, and high density. PBAMO showed a high content of azido groups and was found to be compatible with some energetic oxidizers such as hexanitrohexaazaisowurtzitane (CL-20), cyclotrimethylene trinitramine (RDX), cyclotetramethylene tetranitramine (HMX). The mechanical features of PAMMO are very attractive due to the flexibility of its carbon backbone [26–28]. Nitro polymers were also broadly reported as promising energetic binders. Indeed, grafting nitro groups into the hydrocarbon backbone of the binder can contribute to the global oxygen balance of the composite propellant, thus providing the possibility to reduce the oxidizer content in the propellant formulation and substitute it with an additional amount of energetic fuel [28]. Poly(glycidyl nitrate) (PGN) is one of the most important nitro polymers extensively researched due to its promising characteristics, encompassing high density, high oxygen balance, high burn rate, in addition to the high energy and low vulnerability imparted to the resulting propellant formulations [28, 29].

Simple and complex metal hydrides, typically used as H_2 storage materials, are widely proposed as fuel additives in propellant and pyrotechnic formulations due to their promising combustible contents, including hydrogen and energetic metals substrate. It was reported that the addition of hydrogen in the propellant formulation can decrease the propellant ignition delay and significantly enhance the flame combustion characteristics. Based on the Eq. (11.1), it can clearly be seen that one way to improve specific impulses is to increase T_c/M_w ratio. This can be achieved by reducing the average molecular weight of gaseous combustion products and/or increasing the adiabatic flame temperature. Correspondingly, it is anticipated that the incorporation of the appropriate metal hydride into propellant formulations can improve its energetic performance by providing high hydrogen and energetic light metal content. Complex metal hydrides, generally, have better hydrogen storage capacity and long-term stability than simple metal hydrides. Moreover, complex alkali-metal hydrides have shown a potential effect on decreasing the propellant condensed-combustion products and increasing its hydrochloric-acid removal capacity. The promising class of Li-based complex metal hydrides is that contains oxophilic metals such as aluminum, magnesium, and boron, due to their significant contribution to the stability of the metal hydride and the chlorine scavenging ability of the resulting composite propellants [6, 10, 30, 31].

Table 11.1 Some properties of the investigated propellants ingredients.

Ingredient	Formula	ρ (g cm^{-3})	ΔH_f (kJ mol^{-1})	OB (%)	References
Ammonium perchlorate	NH_4ClO_4	1.95	−295.8	+34	[10]
Aluminum	Al	2.7	/	/	[10]
Aluminum hydride	AlH_3	1.48	−11.4	/	[10]
Lithium aluminum hydride	$LiAlH_4$	0.92	−113.4	/	[10]
Lithium magnesium hexa-alanate	$LiMgAlH_6$	/	−186.4	/	[6]
Magnesium hydride	MgH_2	1.45	−74.5	/	[10]
Magnesium iron hexahydride	Mg_2FeH_6	2.74	−232.2	/	[6]
HTPB	$C_{7.075}H_{10.65}O_{0.223}N_{0.063}$	0.92	−58	−65	[10, 32]
GAP	$C_{3.028}H_{5.046}O_{1.009}N_{3.028}$	1.3	+117	−45	[32–34]
PBAMO	$C_{5.08}H_{8.45}O_{1.41}N_{5.44}$	1.3	+413	−45	[32, 34, 35]
PAMMO	$C_{5.07}H_{9.37}O_{1.38}N_{2.48}$	1.06	+179	−35	[32, 34, 35]
PGN	$C_{2.71}H_{4.73}O_{3.34}N_{0.67}$	1.42	−285	−35	[32, 34, 36]

11.2.3 Computation Procedure of CSPs Performance

Based on the CEA code, some important theoretical performance of different propellant formulations has been calculated. Some physicochemical properties of the propellant ingredients used in the current study are shown in Table 11.1. It is worth noting that the empirical formula ($C_{x1}H_{x2}N_{x3}O_{x4}$) of PBAMO and PAMMO was calculated based on their experimental carbon, hydrogen, nitrogen and oxygen (CHNO) contents measured by the mid-infrared technique [35], using the following Formula:

$$x_i = \frac{[M_{(\text{monomer})} + \%(i\text{th atom})]}{M_{(i\text{th atom})}} \tag{11.4}$$

In Eq. (11.4), x_i is the number of the ith atom in the empirical polymer formula, $M_{(\text{monomer})}$ stands for the molecular weight of the polymer unit, %(ith atom) corresponds to the experimental content of the ith atom in the polymer, while $M_{(i\text{th atom})}$ denotes its molar mass. The computation procedure assumes that the propellant is completely burned in an ideal-rocket motor chamber and the propellant combustion products are expanded, as an isentropic and uniaxial flow, as well as the produced species, are in chemical equilibrium during the combustion process and gas expansion [35, 36]. Moreover, for all computational tests, the combustion chamber pressure, the exit pressure and the expansion ratio were maintained at 7 MPa, 0.1 MPa, and 10, respectively. The calculation approach targets the ternary mixture that gives the optimal specific impulse, which corresponds to the highest I_{sp} value obtained by imposing a constraint on the binder ratio that must be at least 10% and without exceeding 20%. It is important to note that this required binder fraction is only approximate and depends on several parameters such as the density and particles size of the propellant ingredients, as well as the compatibility of the mixture. The output data obtained for this optimal ternary composition was then used to calculate the other energetic performance of the tested propellant formulations. An analysis of the environmental impact of these AP-based propellant formulations has been performed as well using Eq. (11.5), which provides an accurate estimation of the HCl gas fraction scavenged by the studied propellants.

$$\mathrm{HPHS} = \left(\frac{I_{sp}}{I_{sp,max}}\right) [100 - (\%\mathrm{Cl} \rightarrow \mathrm{HCl})] \tag{11.5}$$

where I_{sp} is the specific impulse of the optimal (AP/binder/additive) ternary mixture, $I_{sp,max}$ is the highest specific impulse that is achieved when varying the ingredient fractions of this ternary composition, while the symbol [%Cl → HCl] stands for the mole fraction of the available chlorine, which is transformed to hydrochloric acid (HCl).

According to Eq. (11.5), as the chlorine scavenging effect of the explored propellant composition is improved, the HCl emission concentration decreases, and thus the HPHS value increases. Theoretically, the maximum value of this parameter is 100%, which means that no hydrochloric acid species are released into the exhaust products.

11.3 Results and Discussion

The obtained theoretical values of the gravimetric specific impulse of the investigated CSPs were plotted in Gibbs-triangle diagrams as shown in Figures 11.1–11.5. As can be seen from these ternary graphs, by changing the mass fractions of the AP/binder/fuel additive ternary mixture, different zones of specific impulse values are obtained. A noticeable improvement in I_{sp} was observed when energetic binders replaced HTPB, except for PGN polymer. Based on the highest I_{sp} values (red area in the ternary graphs), the energetic performance of the used binders can be classified in the following trend: [PAMMO] > [PBAMO] > [GAP] > [HTPB] > [PGN], where the greatest I_{sp} values are achieved with the ternary mixture of AP/PAMMO/AlH_3 ($273.9\,s < I_{sp} < 289\,s$). The excellent performance of azido polymers compared to nitro one may be due to the high-energy content of the chemical bonds within the azido groups, which gives them a high enthalpy of formation. Another interesting finding, which can be deduced from this thermochemical calculation, is that PAMMO-based formulation displays a higher performance compared to that of PBAMO-based one despite the low enthalpy of formation of PAMMO. This result can be attributed to the higher oxygen balance of PAMMO, which may further improve the combustion process and the overall performance.

Regarding the performance of fuel additives as compared to aluminum, it was found that all the investigated metal hydrides showed enhancement in specific

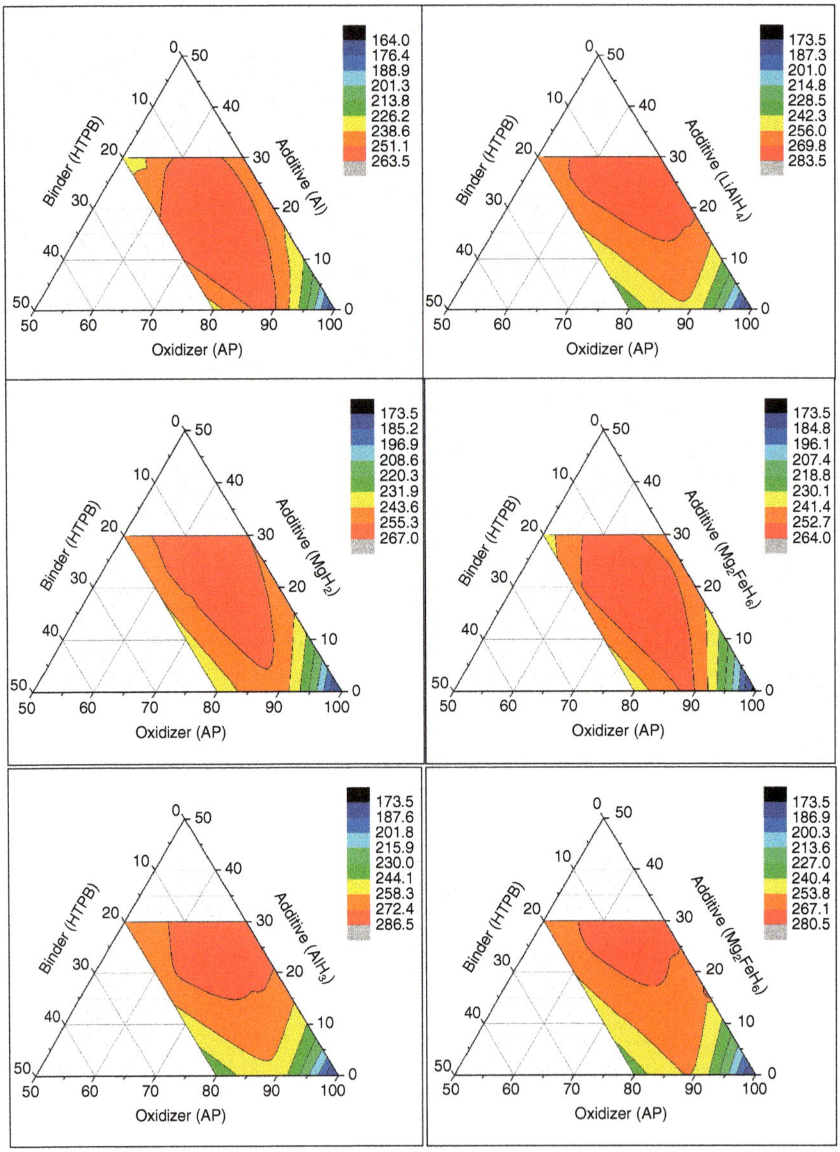

Figure 11.1 Ideal specific impulse evolution versus ingredients mass fractions of HTPB based-propellant formulations.

impulse, where AlH_3 and $LiAlH_4$ exhibited the highest improvement. As mentioned in Section 11.2.3, the optimal specific impulse of the studied CSPs was calculated with the assumption that the binder content was between 10% and 20%, and the maximal specific impulse was considered as the highest I_{sp} obtained by changing the ingredients fractions of the ternary mixture. Tables 11.2–11.6 show the optimal and maximal specific impulse, optimal propellant composition as well as the ternary mixture giving the highest specific impulse for each tested propellant. The

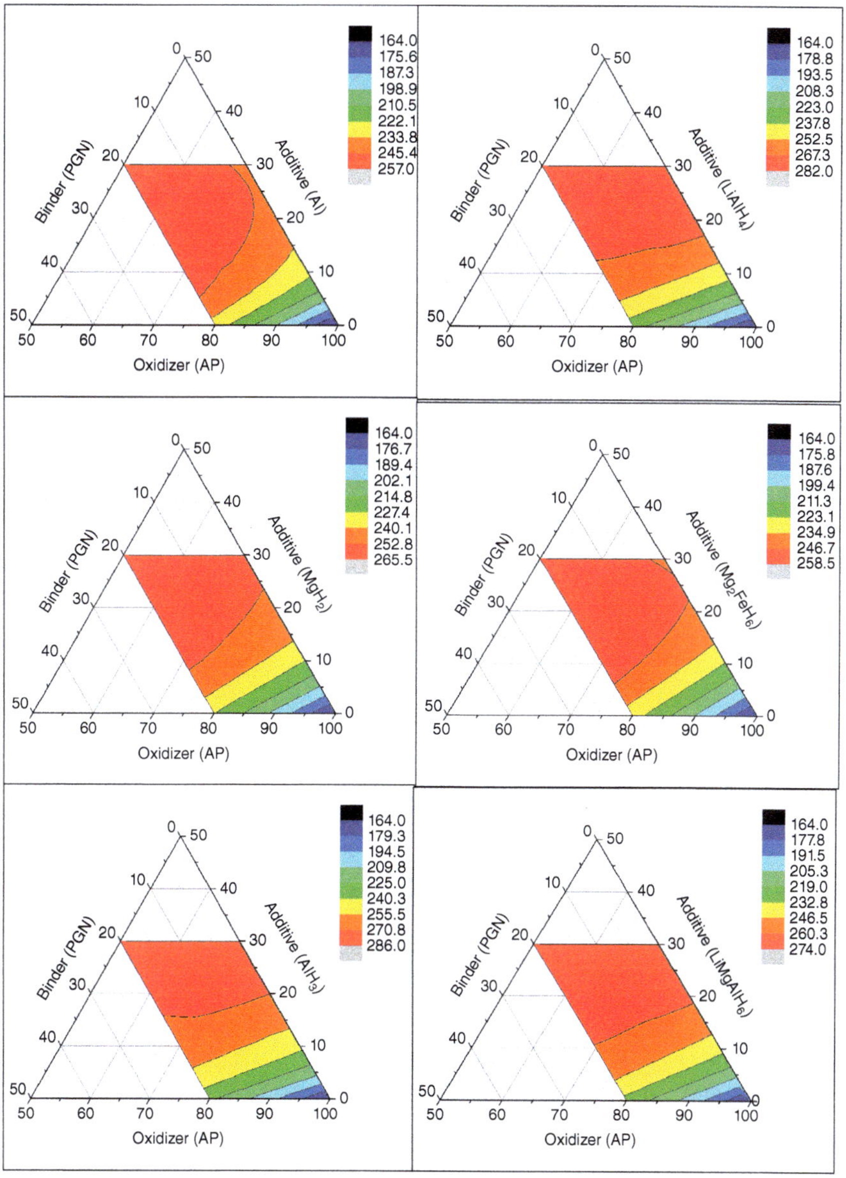

Figure 11.2 Ideal specific impulse evolution versus ingredients mass fractions of PGN based-propellant formulations.

first casual remark from the data of these tables is that the amount of AP used in the optimal formulations based on energetic binders is less than that used in the HTPB propellants. This can be explained by the fact that energetic binders can provide additional oxidizing species that contribute to the total oxidation of the available fuel species. Indeed, as can be seen from Table 11.1, the oxygen balance values of all the used energetic binders are higher than that of HTPB. Concerning

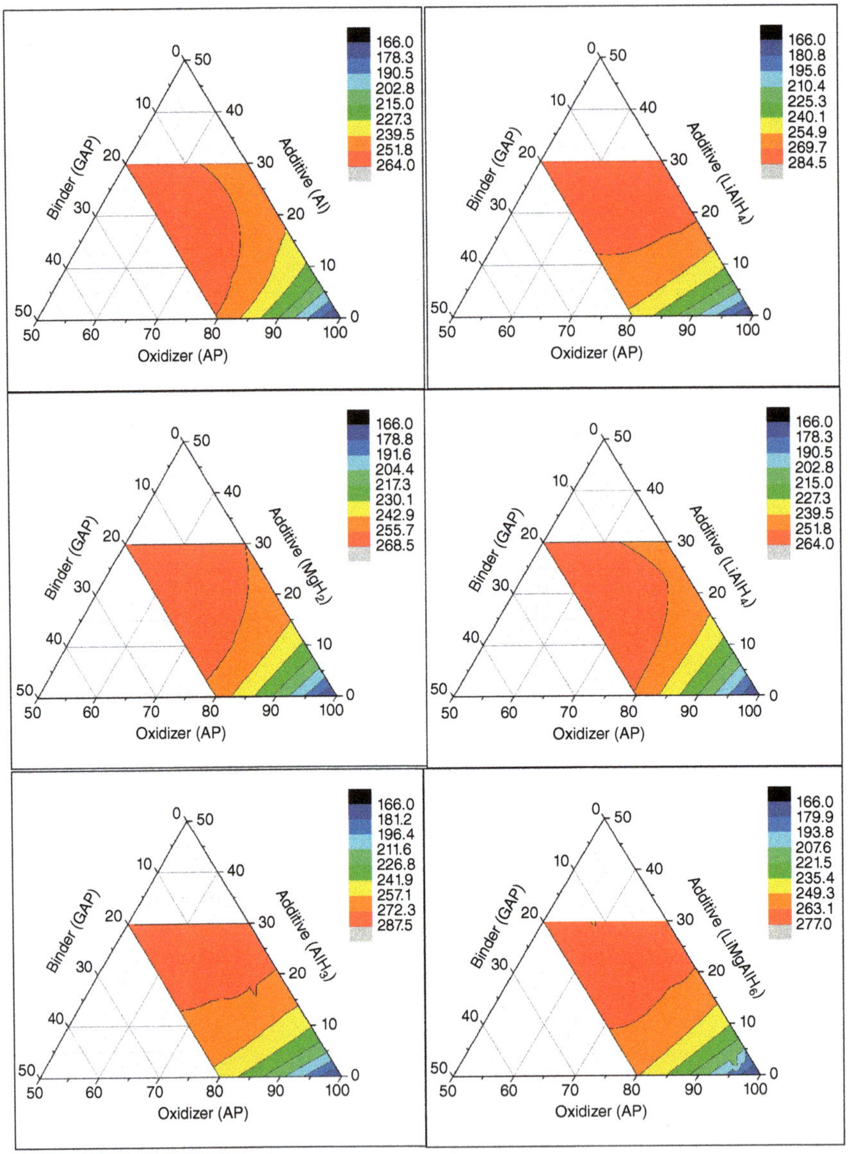

Figure 11.3 Ideal specific impulse evolution versus ingredients mass fractions of GAP based-propellant formulations.

the optimal unconstrained compositions, it is clear that some HTPB-based formulations are practically unfeasible due to the low binder content (HTPB content <10%). This limitation was not encountered with the energetic binders, which offer the potential to significantly improve I_{sp} without depleting the propellant binder content (ensuring propellant slurry consistency). Hence, these energetic binders are more prone to contain a higher content of energetic fuels such as AlH_3 and MgH_2.

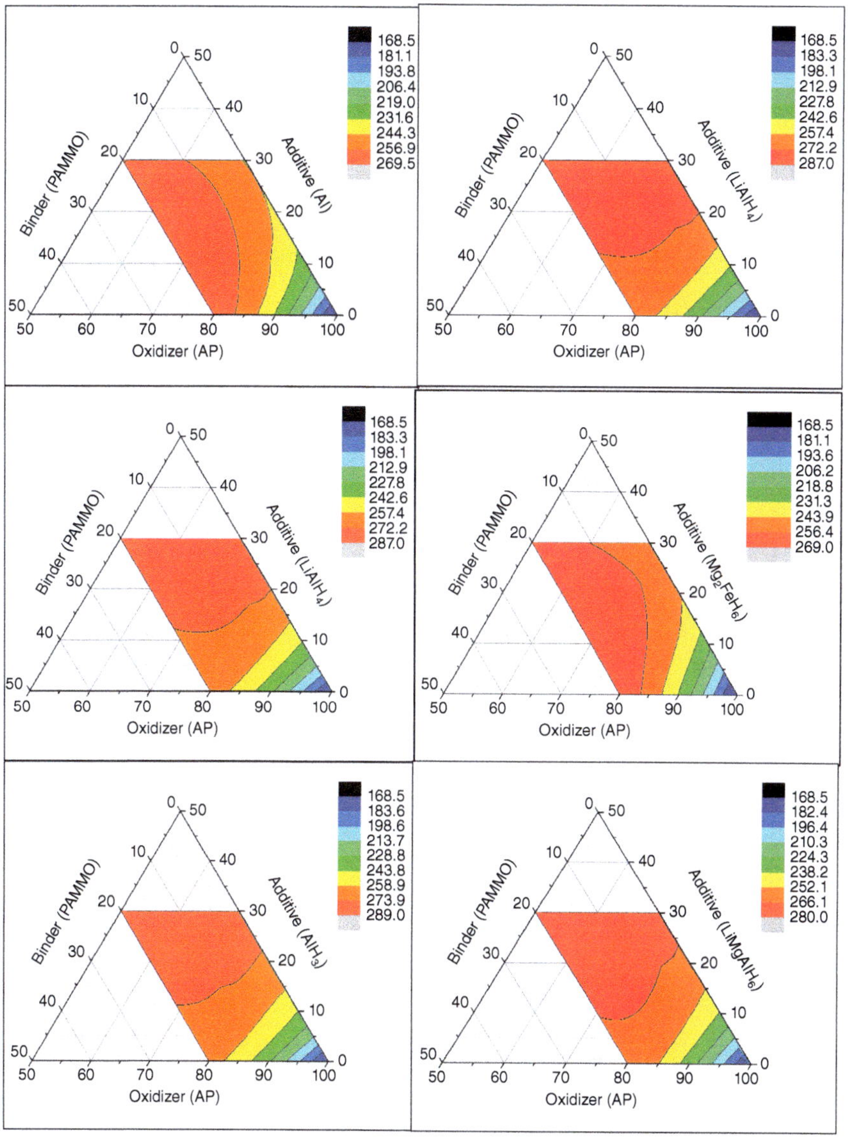

Figure 11.4 Ideal specific impulse evolution versus ingredients mass fractions of PAMMO based-propellant formulations.

Concerning the adiabatic flame temperatures of the studied propellants, it was found that aluminum-based formulations produce higher combustion flame temperatures compared to the metal hydrides (Figures 11.6 and Tables 11.7–11.11). This may be due to the dehydrogenation reaction, which absorbs some portion of the heat, causing a significant drop in the gaseous species' temperature. Furthermore, the use of energetic polymers in the optimal formulations instead of the HTPB binder increased the adiabatic flame temperature, with the highest value obtained

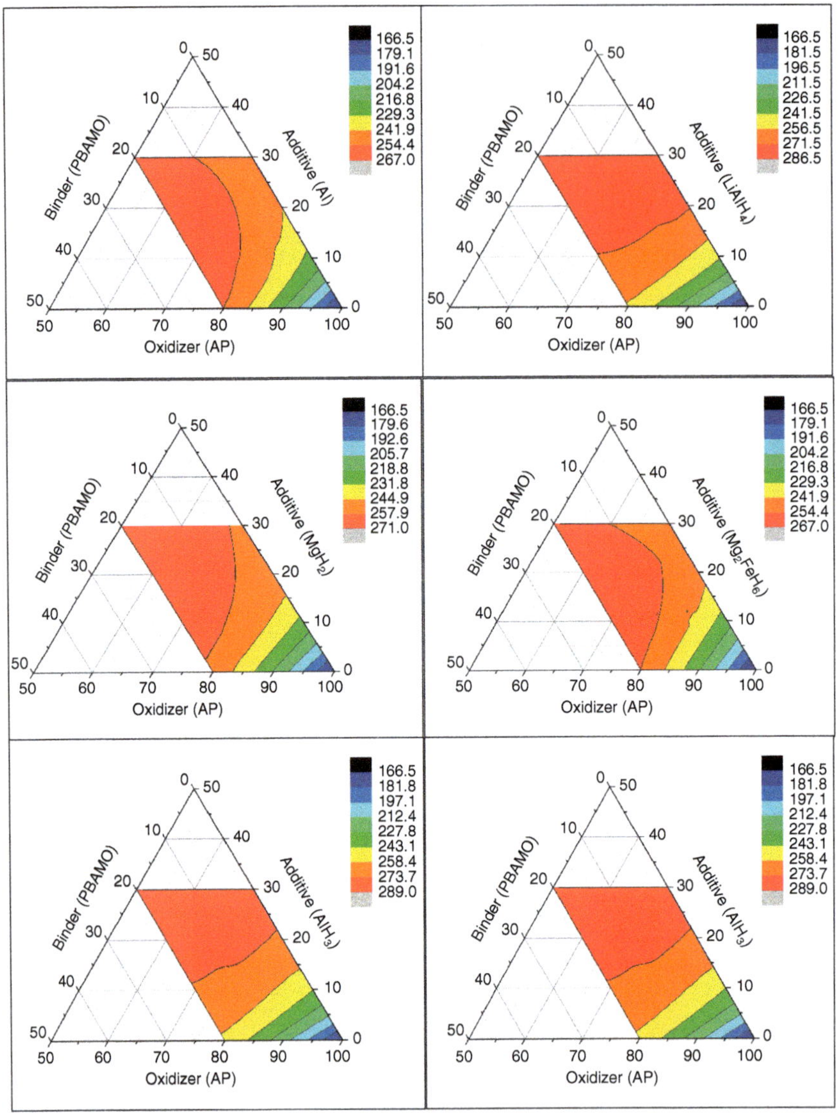

Figure 11.5 Ideal specific impulse evolution versus ingredients mass fractions of PBAMO based-propellant formulations.

with the mixture of PGN binder and aluminum, reaching a theoretical value of 2918.9 K. The improvement of combustion temperature of the studied optimal formulations, induced by the employment of active polymers, are in the following order [PGN] > [GAP] > [PBAMO] > [PAMMO]. This performance enhancement achieved by the used active binders may be due to the highly exothermic decomposition of the nitro and azido moieties, which release additional heat and contribute to increasing the combustion flame temperature. However, in terms of the average molecular weight of combustion products, it appears that the replacement of HTPB

Table 11.2 Optimal and maximal specific impulse of the investigated HTPB-based propellants.

	Constrained optimal composition (mass %)				Non constrained optimal composition (mass %)			
Additive	AP	Additive	HTPB	$I_{sp,opt}$ (s)	AP	Additive	HTPB	$I_{sp,max}$ (s)
Al	69	20	11	263.4	69	20	11	263.4
AlH_3	62	28	10	283.4	65	30	5	286.2
MgH_2	60	30	10	266.2	62	30	8	266.8
Mg_2FeH_6	69	20	11	263.7	69	20	11	263.7
$LiMgAlH_6$	60	30	10	275.9	84	16	0	280.0
$LiAlH_4$	60	30	10	281.7	64	30	6	283.4

Table 11.3 Optimal and maximal specific impulse of the investigated GAP-based propellants.

	Optimal composition (mass %)				Non constrained optimal composition (mass %)			
Additive	AP	Additive	GAP	$I_{sp,opt}$ (s)	AP	Additive	GAP	$I_{sp,max}$ (s)
Al	60	20	20	263.9	55	20	25	265.7
AlH_3	59	30	11	287.1	59	30	11	287.1
MgH_2	52	30	18	268.1	52	30	18	268.1
Mg_2FeH_6	62	18	20	264.0	57	18	25	264.9
$LiMgAlH_6$	52	30	18	276.7	52	30	18	276.7
$LiAlH_4$	55	30	15	284.3	55	30	15	284.3

Table 11.4 Optimal and maximal specific impulse of the investigated PGN-based propellants.

	Constrained optimal composition (mass %)				Non constrained optimal composition (mass %)			
Additive	AP	Additive	PNG	$I_{sp,opt}$ (s)	AP	Additive	PGN	$I_{sp,max}$ (s)
Al	58	22	20	256.8	39	22	39	262.4
AlH_3	54	30	16	285.9	54	30	16	285.9
MgH_2	50	30	20	265.2	43	30	27	266.4
Mg_2FeH_6	62	18	20	258.0	57	18	25	259.8
$LiMgAlH_6$	50	30	20	273.9	47	30	23	274.0
$LiAlH_4$	51	30	19	281.8	51	30	19	281.8

Table 11.5 Optimal and maximal specific impulse of the investigated PAMMO-based propellants.

	Constrained optimal composition (mass %)				Non constrained optimal composition (mass %)			
Additive	**AP**	**Additive**	**PAMMO**	$I_{sp,opt}$ **(s)**	**AP**	**Additive**	**PAMMO**	$I_{sp,max}$ **(s)**
Al	60	20	20	269.5	60	20	20	269.5
AlH_3	60	30	10	288.5	61	30	9	288.7
MgH_2	53	30	17	271.1	53	30	17	271.1
Mg_2FeH_6	62	18	20	268.7	60	18	22	269.1
$LiMgAlH_6$	53	30	17	279.9	53	30	17	279.9
$LiAlH_4$	58	30	12	286.5	58	30	12	286.5

Table 11.6 Optimal and maximal specific impulse of the investigated PBAMO-based propellants.

	Constrained optimal composition (mass %)				Non constrained optimal composition (mass %)			
Additive	**AP**	**Additive**	**PBAMO**	$I_{sp,opt}$ **(s)**	**AP**	**Additive**	**PBAMO**	$I_{sp,max}$ **(s)**
Al	50	30	20	257.6	54	20	26	269.2
AlH_3	59	30	11	288.5	59	30	11	288.5
MgH_2	50	30	20	270.5	49	30	21	270.5
Mg_2FeH_6	62	18	20	266.7	57	18	25	268.4
$LiMgAlH_6$	51	30	19	278.8	51	30	19	278.8
$LiAlH_4$	54	30	16	286.4	54	30	16	286.4

binder with energetic polymers led to an increase in the mean molar mass of the generated combustion species. This decrease of the propellant performance was compensated by a significant improvement in the adiabatic flame temperature, thus augmenting the resulting ratio of combustion temperature to molar mass, as highlighted in column 4 of Tables 11.7–11.11. According to the data given in this column, it appears that AlH_3 and $LiAlH_4$ based propellants exhibit the highest T_c/M_w ratio.

Regarding the CCPs at the nozzle throat section, as can be seen in Figure 11.6 and Tables 11.7–11.11, it was found that AlH_3-based propellants generated the highest levels of condensed phase particles, whereas composite formulations containing Mg_2FeH_6 produced a less amount of CCPs. This interesting property of magnesium iron hexahydride was slightly enhanced by the employment of energetic polymers rather than HTPB binder (CCPs content decreases from 11.5% to less than 9%). When

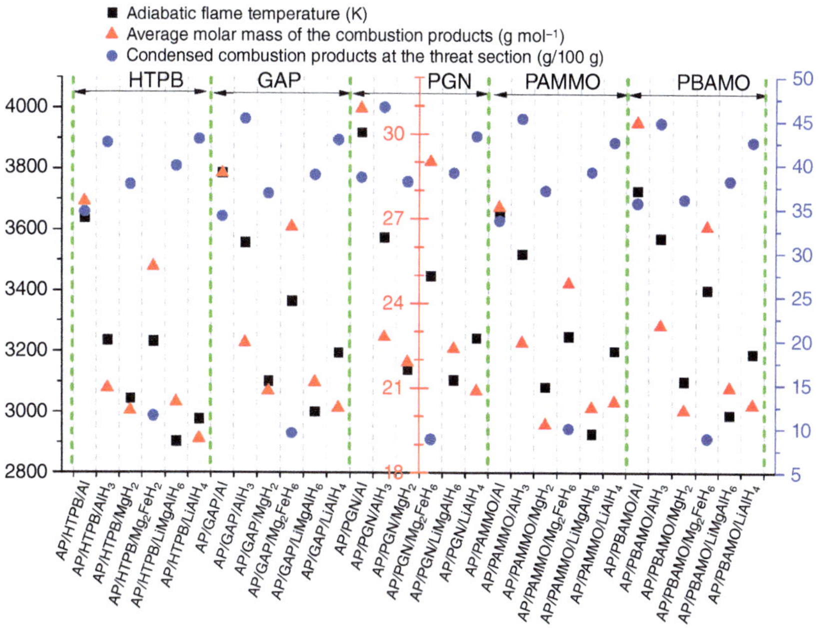

Figure 11.6 Ideal combustion characteristics of the investigated CSPs formulations.

Table 11.7 Average mass molar of combustion products and adiabatic flame temperature of the investigated HTPB-based propellants.

	Optimal composition (mass %)									
Additive	**AP**	**Add**	**HTPB**	T_c **(K)**	M_w **(g mol^{-1})**	T_c/M_w	**CCPs (%)**	ρ **(g cm^{-3})**	I_{sp} **(s)**	$I_{sp.v}$ **(s g cm^{-3})**
Al	69	20	11	3636.7	27.6	131.9	34.7	1.827	263.4	481
AlH_3	62	28	10	3235.3	21.0	153.7	42.6	1.624	283.4	460
MgH_2	60	30	10	3044.0	20.2	150.4	37.8	1.604	266.2	427
Mg_2FeH_6	69	20	11	3231.0	25.3	127.5	11.5	1.830	263.7	483
$LiMgAlH_6$	60	30	10	2902.0	20.5	141.5	39.9	/	275.9	/
$LiAlH_4$	60	30	10	2976.2	19.2	154.7	43.0	1.347	281.7	379

analyzing the combustion product species of the Mg_2FeH_6-based propellant at the throat area, it was found that almost all of the available iron in propellant composition was converted into gaseous species, including mainly iron chloride compounds. Moreover, a considerable mass fraction of magnesium was also found in form of gaseous molecules such as $Mg(OH)_2$ and $MgCl_2$. However, the large metal content of the other investigated propellants has been converted principally to very low volatility compounds such as their oxides. These latter are found in liquid or

Table 11.8 Average mass molar of combustion products and adiabatic flame temperature of the investigated GAP-based propellants.

Additive	Optimal composition (mass %)			T_c (K)	M_w (g mol^{-1})	T_c/M_w	CCPs (%)	ρ (g cm^{-3})	I_{sp} (s)	$I_{sp.v}$ (s g cm^{-3})
	AP	Add	GAP							
Al	60	20	20	3784.9	28.6	132.2	34.2	1.867	263.4	493
AlH_3	59	30	11	3556.8	22.6	157.1	45.3	1.695	283.4	487
MgH_2	52	30	18	3100.0	20.9	148.4	36.8	1.634	266.2	438
Mg_2FeH_6	62	18	20	3363.5	26.7	125.8	9.5	1.861	263.7	491
$LiMgAlH_6$	52	30	18	3000.7	21.2	141.6	38.9	/	275.9	/
$LiAlH_4$	55	30	15	3193.9	20.3	157.2	42.9	1.382	281.7	393

Table 11.9 Average mass molar of combustion products and adiabatic flame temperature of the investigated PGN-based propellants.

Additive	Optimal composition (mass %)			T_c (K)	M_w (g mol^{-1})	T_c/M_w	CCPs (%)	ρ (g cm^{-3})	I_{sp} (s)	$I_{sp.v}$ (s g cm^{-3})
	AP	Add	PGN							
Al	58	22	20	3918.9	30.9	127.0	38.6	1.924	263.4	505
AlH_3	54	30	16	3573.0	22.8	156.6	46.6	1.688	283.4	483
MgH_2	50	30	20	3138.5	21.9	143.3	38.1	1.655	266.2	441
Mg_2FeH_6	62	18	20	3446.9	29.0	118.9	8.8	1.907	263.7	495
$LiMgAlH_6$	50	30	20	3104.9	22.4	138.9	39.1	/	275.9	/
$LiAlH_4$	51	30	19	3243.1	20.9	155.2	43.3	1.386	281.7	391

Table 11.10 Average mass molar of combustion products and adiabatic flame temperature of the investigated PAMMO-based propellants.

Additive	Optimal composition (mass %)			T_c (K)	M_w (g mol^{-1})	T_c/M_w	CCPs (%)	ρ (g cm^{-3})	I_{sp} (s)	$I_{sp.v}$ (s g cm^{-3})
	AP	Add	PAMMO							
Al	60	20	20	3647.9	27.4	133.0	33.7	1.753	269.5	472
AlH_3	60	30	10	3519.3	22.6	155.4	45.3	1.654	288.5	477
MgH_2	53	30	17	3081.7	19.7	156.3	37.1	1.565	271.1	424
Mg_2FeH_6	62	18	20	3248.3	24.7	131.4	10.0	1.747	268.7	469
$LiMgAlH_6$	53	30	17	2927.1	20.3	143.9	39.2	/	279.9	/
$LiAlH_4$	58	30	12	3199.7	20.5	156.4	42.6	1.357	286.5	389

Table 11.11 Average mass molar of combustion products and adiabatic flame temperature of the investigated PBAMO-based propellants.

Additive	Optimal composition (mass %) AP	Add	PBAMO	T_c (K)	M_w (g mol^{-1})	T_c/M_w	CCPs (%)	ρ (g cm^{-3})	I_{sp} (s)	$I_{sp.v}$ (s g cm^{-3})
Al	50	30	20	3726.9	30.4	122.7	35.7	1.967	257.6	506.7
AlH_3	59	30	11	3570.8	23.2	154.0	44.8	1.716	288.5	495.1
MgH_2	50	30	20	3100.0	20.2	153.3	36.1	1.655	270.5	447.8
Mg_2FeH_6	62	18	20	3400.4	26.7	127.5	8.9	1.907	266.7	508.6
$LiMgAlH_6$	51	30	19	2990.0	21.0	142.4	38.2	/	278.8	/
$LiAlH_4$	54	30	16	3189.8	20.4	156.1	42.6	1.397	286.4	400.1

solid states under temperatures below their vaporization point. Volumetric specific impulse, which is one of the most important energetic properties of solid propellant formulations, was also calculated in this study and reported in Tables 11.7–11.11.

Considering the optimal aluminized HTPB-based composition as the baseline propellant formulation, the $I_{sp,v}$ improvement can be assessed as shown in Figure 11.7. As a first remark, it was noticed from this bar graph that all the energetic binder-based aluminized propellants displayed an improvement in the volumetric specific impulse except for PAMMO, with the greatest improvement obtained by the AP/PBAMO/Al mixture, which increases the $I_{sp,v}$ value by 5.3%. The negative contribution of PAMMO to the enhancement of the volumetric specific impulse is due to its relatively low density compared to the other active polymers. Regarding the positive contribution on $I_{sp,v}$ of the studied metal hydrides, it was found that only two hydrogenated additives provided some improvement in $I_{sp,v}$. Mg_2FeH_6 Exhibited a positive $\Delta I_{sp,v}$ for all the employed polymers, excluding the PAMMO binder, for which the highest value (5.7%) is obtained using PBAMO. This finding can be attributed to the energetic performance of Mg_2FeH_6, which is comparable to aluminum with a slightly higher density. The second interesting additive is aluminum hydride, which brings some benefit to the volumetric specific impulse when used with dense energetic polymers. Indeed, the low density of AlH_3 is compensated both by its high energy and by the relatively high density of the used energetic binders. Consequently, the volumetric specific impulse of the resulting propellants is improved.

As explained in Section 11.2.3, the ability of a given propellant formulation to reduce the hydrochloric acid emission can be evaluated by calculating the "HPHS" parameter given in Eq. (11.5). As shown in Tables 11.12–11.16 and Figure 11.8, the highest performance of this environmentally friendly criterion is achieved by the addition of complex rather than simple metal hydrides, showing very high values above 92% with all the investigated binders, which means that almost complete removal of hydrochloric acid from the exhausting combustion products

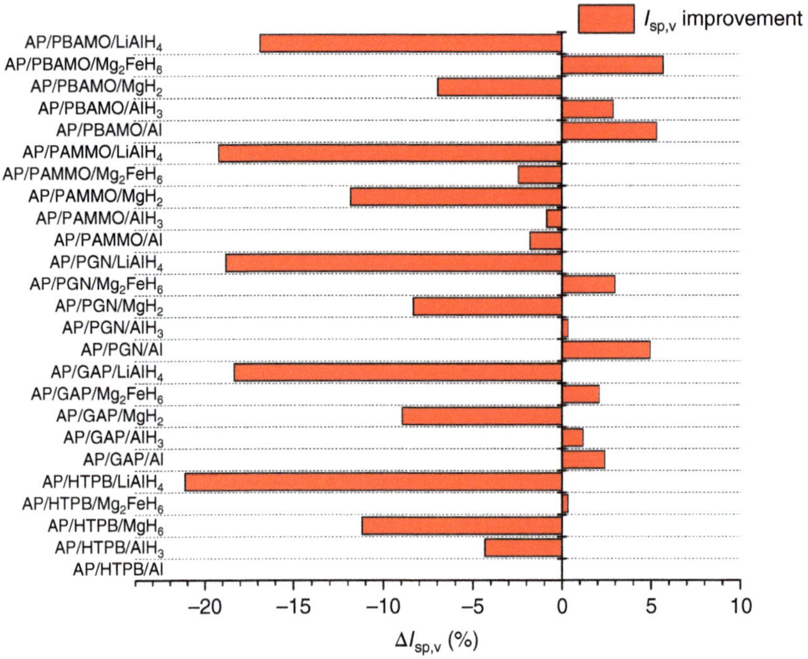

Figure 11.7 Volumetric specific impulse improvement of the investigated CSPs formulations.

Table 11.12 Hydrochloric acid elimination from the investigated HTPB-based propellants.

	Optimal composition (mass %)						
Additive	**AP**	**Additive**	**HTPB**	$I_{sp,opt}$ **(s)**	$I_{sp,max}$ **(s)**	**Cl → HCl (%)**	**HPHS(%)**
Al	69	20	11	263.4	263.4	98.5	1.5
AlH_3	62	28	10	283.4	286.2	98.6	1.4
MgH_2	60	30	10	266.2	266.8	88.7	11.3
Mg_2FeH_6	69	20	11	263.7	263.7	7.5	92.5
$LiMgAlH_6$	60	30	10	275.9	280.0	11.9	86.8
$LiAlH_4$	60	30	10	281.7	283.4	4.0	95.4

has occurred. When exploring the combustion products of composite propellants based on these three complex hydrides, it was found that almost all of their available chlorine content was converted to metal chloride salts such as LiCl, $FeCl_3$, and $MgCl_2$, thus decreasing the amount of chlorine ion that would be expected to form hydrochloric acid. Concerning the effect of the employed binders on the chlorine scavenging of AP-based aluminized propellants, it was found that PGN binder provided some improvement over the HTPB propellant, where the HPHS value

Table 11.13 Hydrochloric acid elimination from the investigated GAP-based propellants.

Additive	Optimal composition (mass %) AP	Additive	GAP	$I_{sp,opt}$ (s)	$I_{sp,max}$ (s)	Cl → HCl (%)	HPHS(%)
Al	60	20	20	263.9	265.7	95.9	4.1
AlH_3	59	30	11	287.1	287.1	98.7	1.3
MgH_2	52	30	18	268.1	268.1	84.6	15.4
Mg_2FeH_6	62	18	20	264.0	264.9	7.3	92.4
$LiMgAlH_6$	52	30	18	276.7	276.7	1.5	98.5
$LiAlH_4$	55	30	15	284.3	284.3	3.3	96.7

Table 11.14 Hydrochloric acid elimination from the investigated PGN-based propellants.

Additive	Optimal composition (mass %) AP	Additive	PGN	$I_{sp,opt}$ (s)	$I_{sp,max}$ (s)	Cl → HCl (%)	HPHS(%)
Al	58	22	20	256.8	262.4	87.6	12.1
AlH_3	54	30	16	285.9	285.9	98.8	1.2
MgH_2	50	30	20	265.2	266.4	86.8	13.1
Mg_2FeH_6	62	18	20	258.0	259.8	7.0	92.4
$LiMgAlH_6$	50	30	20	273.9	274.0	3.9	96.0
$LiAlH_4$	51	30	19	281.8	281.8	4.8	95.2

Table 11.15 Hydrochloric acid elimination from the investigated PAMMO-based propellants.

Additive	Optimal composition (mass %) AP	Additive	PAMMO	$I_{sp,opt}$ (s)	$I_{sp,max}$ (s)	Cl → HCl (%)	HPHS(%)
Al	60	20	20	269.5	269.5	98.5	1.5
AlH_3	60	30	10	288.5	288.7	98.7	1.3
MgH_2	53	30	17	271.1	271.1	85.4	14.6
Mg_2FeH_6	62	18	20	268.7	269.1	7.4	92.5
$LiMgAlH_6$	53	30	17	279.9	279.9	1.0	99.0
$LiAlH_4$	58	30	12	286.5	286.5	3.7	96.3

Table 11.16 Hydrochloric acid elimination from the investigated PBAMO-based propellants.

	Optimal composition (mass %)						
Additive	**AP**	**Additive**	**PBAMO**	$I_{sp,opt}$ **(s)**	$I_{sp,max}$ **(s)**	**Cl → HCl (%)**	**HPHS(%)**
Al	50	30	20	257.6	269.2	98.4	1.6
AlH_3	59	30	11	288.5	288.5	98.6	1.4
MgH_2	50	30	20	270.5	270.5	79.6	20.4
Mg_2FeH_6	62	18	20	266.7	268.4	7.3	92.1
$LiMgAlH_6$	51	30	19	278.8	278.8	0.9	99.1
$LiAlH_4$	54	30	16	286.4	286.4	2.6	97.4

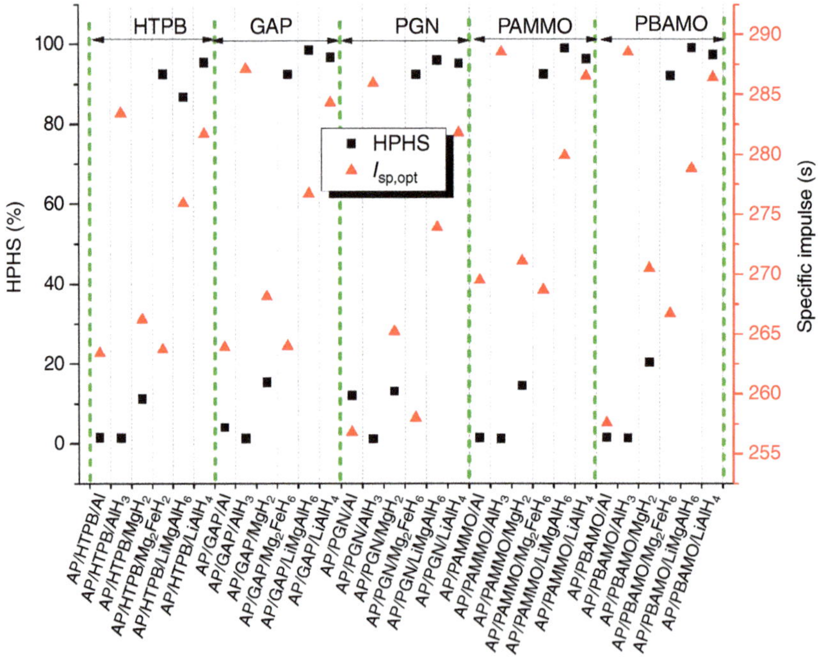

Figure 11.8 Energetic performance and chlorine scavenging effect of the investigated CSPs formulations.

increased from 1.5% to 12.1%. This improvement in hydrochloric acid elimination may be due to the low chlorine content required in optimal aluminized formulations based on energetic binder compared to that of HTPB. However, the employed active polymers did not provide any contribution to reducing HCl emission from aluminum hydride-based propellants (HPHS value was lowered to less than 1.5%).

This means that almost all chlorine content of AlH_3-based propellant was transformed into hydrochloric acid.

11.4 Conclusion

The promising characteristics of new energetic binders and the environmentally friendly properties of some metal hydrides have drawn attention to the possibility of using them in the next generation of composite solid propellants. This chapter provides interesting results concerning the theoretical performance of different propellant formulations based on various binders and enriched with a range of metal hydride additives. In terms of specific impulse, the three used azido-polymers (PAMMO, PBAMO, and GAP) have provided a positive contribution compared to the HTPB binder, and all metal hydride-based propellant formulations have shown enhancement over the aluminized mixtures for which the greatest performance was achieved by PAMMO. The adiabatic flame temperature of the studied propellant formulations increased when energetic binders are used, but decreased when metal hydrides are added rather than aluminum particles. In terms of volumetric specific impulse, it was found that the energetic binder-based aluminized formulations offered a positive contribution compared to the HTPB-based aluminized propellant. Compared to the reference propellant, it was revealed that only some metal hydride-based propellant formulations increased $I_{sp,v}$, including those containing energetic binders that were supplemented with AlH_3 or Mg_2FeH_6. The latter two additives generated the highest and lowest CCPs at the nozzle throat section, respectively. Furthermore, the addition of $LiAlH_4$ and Mg_2FeH_6 has resulted almost in the complete removal of hydrochloric acid from the exhausting combustion products, especially with the use of PGN binder. Moreover, the very poor chlorine removal characteristics of the HTPB-based aluminized propellant have been improved by the use of energetic binders. Based on these findings, it could be possible to rely on these energetic binder formulations enriched with hydrogenated additives for the transition to the next generation of solid propellants. For this purpose, further investigations dealing with ingredient compatibility, propellant stability, manufacturing process safety, and rocket operating reliability are needed to well select the more suitable formulations for practical applications.

References

1 Thakre, P. and Yang, V. (2010). *Solid Propellants. Encyclopedia of Aerospace Engineering*. Wiley.
2 Gromov, A.A. et al. (2020). Aluminized solid propellants loaded with metals and metal oxides: characterization, thermal behavior, and combustion. In: *Innovative Energetic Materials: Properties, Combustion Performance and Application*

(ed. W.Q. Pang, L.T. DeLuca, A.A. Gromov and A.S. Cumming), 157–182. Springer.
3 Korotkikh, A., Sorokin, I.V., Selikhova, E.A. et al. (2020). Ignition and combustion of composite solid propellants based on a double oxidizer and boron-based additives. *Russian Journal of Physical Chemistry B* 14 (4): 592–600.
4 Gajjar, P. and Malhotra, V. (2018). Advanced upper stage energetic propellants. In: *2018 IEEE Aerospace Conference*, 1–12. IEEE.
5 Yuan, J. et al. (2019). Aluminum agglomeration of AP/HTPB composite propellant. *Acta Astronautica* 156: 14–22.
6 Chalghoum, F. et al. (2020). Effect of complex metal hydrides on the elimination of hydrochloric acid exhaust products from high-performance composite solid propellants: a theoretical analysis. *Propellants, Explosives, Pyrotechnics* 45 (8): 1204–1215.
7 DeLuca, L.T. et al. (2007). Physical and ballistic characterization of AlH_3-based space propellants. *Aerospace Science and Technology* 11 (1): 18–25.
8 Lawrence, A.R. et al. (2019). Organically-capped, nanoscale alkali metal hydride and aluminum particles as solid propellant additives. *Journal of Propulsion and Power* 35 (4): 736–746.
9 Shark, S. et al. (2011). Theoretical performance analysis of metal hydride fuel additives for rocket propellant applications. In: *47th AIAA/ASME/SAE/ASEE Joint Propulsion Conference & Exhibit* (ed. J. Flamm), 5556. American Institute for Aeronautics and Astronautics (AIAA).
10 Maggi, F. et al. (2012). Theoretical analysis of hydrides in solid and hybrid rocket propulsion. *International Journal of Hydrogen Energy* 37 (2): 1760–1769.
11 Chen, H. et al. (2014). Perchlorate exposure and thyroid function in ammonium perchlorate workers in Yicheng, China. *International Journal of Environmental Research and Public Health* 11 (5): 4926–4938.
12 Parker, D.R.J.E.C. (2009). Perchlorate in the environment: the emerging emphasis on natural occurrence. *Environmental Chemistry* 6 (1): 10–27.
13 Ma, H. (2016). et al, Biological treatment of ammonium perchlorate-contaminated wastewater: a review. *Journal of Water Reuse and Desalination* 6 (1): 82–107.
14 Park, J.-W. (2006). et al, The thyroid endocrine disruptor perchlorate affects reproduction, growth, and survival of mosquitofish. *Ecotoxicology and Environmental Safety* 63 (3): 343–352.
15 Chen, Y. (2008). et al, Effects of maternal exposure to ammonium perchlorate on thyroid function and the expression of thyroid-responsive genes in Japanese quail embryos. *General and Comparative Endocrinology* 159 (2, 3): 196–207.
16 Pang, W. et al. (2020). *Innovative Energetic Materials: Properties, Combustion Performance and Application*. Springer.
17 Szklarek, S., Górecka, A., and Wojtal-Frankiewicz, A.J.S.o.T.T.E. (2022). The effects of road salt on freshwater ecosystems and solutions for mitigating chloride pollution – a review. *Science of The Total Environment* 805: 150289.

18 Lee, B.D. et al. (2017). A comparison study of performance and environmental impacts of chloride-based deicers and eco-label certified deicers in South Korea. *Cold Regions Science and Technology* 143: 43–51.

19 Bhowmik, D. et al. (2015). An energetic binder for the formulation of advanced solid rocket propellants. *Central European Journal of Energetic Materials* 12 (1): 145–158.

20 Betzler, F.M. et al. (2016). A new energetic binder: glycidyl nitramine polymer. *Central European Journal of Energetic Materials* 13 (2): 289–300.

21 Chen, H. et al. (2018). Exploring chemical, mechanical, and electrical functionalities of binders for advanced energy-storage devices. *Chemical Reviews* 118 (18): 8936–8982.

22 Davenas, A. (2012). *Solid Rocket Propulsion Technology*. Newnes, Pergamon Press.

23 Yuan, J. et al. (2020). Thermal decomposition and combustion characteristics of Al/AP/HTPB propellant. *Journal of Thermal Analysis and Calorimetry* 1–10.

24 Sutton, G.P. and Biblarz, O. (2016). *Rocket Propulsion Elements*. Wiley.

25 de Oliveira, J.I.S. et al. (2006). MIR/NIR/FIR characterization of poly-AMMO and poly-BAMO and their precursors as energetic binders to be used in solid propellants. *Propellants, Explosives, Pyrotechnics: An International Journal Dealing with Scientific and Technological Aspects of Energetic Materials* 31 (5): 395–400.

26 Li, H. et al. (2016). A comparison of triazole cross-linked polymers based on poly-AMMO and GAP: mechanical properties and curing kinetics. *Journal of Applied Polymer Science* 133 (17).

27 Agrawal, J.P. (2010). *High Energy Materials: Propellants, Explosives and Pyrotechnics*. Wiley.

28 Cheng, T. (2019). Review of novel energetic polymers and binders–high energy propellant ingredients for the new space race. *Designed Monomers and Polymers* 22 (1): 54–65.

29 Zhang, Z. et al. (2014). Thermal decomposition of energetic thermoplastic elastomers of poly(glycidyl nitrate). *Journal of Applied Polymer Science* 131 (21).

30 Doll, D.W. and Lund, G.K. (1992). Magnesium-neutralized propellant. *Journal of Propulsion and Power* 8 (6): 1185–1191.

31 Terry, B.C. et al. (2016). Removing hydrochloric acid exhaust products from high performance solid rocket propellant using aluminum-lithium alloy. *Journal of Hazardous Materials* 317: 259–266.

32 Badgujar, D. (2017). et al, New directions in the area of modern energetic polymers: an overview. 53 (4): 371–387.

33 Homburg, A., Meyer, R., and Köhler, J. (2008). *Explosivstoffe*. Wiley.

34 Oliveira, J.I.S.d. et al. (2007). Determination of CHN content in energetic binder by MIR analysis. *Combustion, Explosion, and Shock Waves* 17 (1): 46–50.

35 Diaz, E. et al. (2003). Heats of combustion and formation of new energetic thermoplastic elastomers based on GAP, PolyNIMMO and PolyGLYN. *Propellants, Explosives, Pyrotechnics: An International Journal Dealing with Scientific and Technological Aspects of Energetic Materials* 28 (3): 101–106.

36 McBride, B.J. (1996). *Computer Program for Calculation of Complex Chemical Equilibrium Compositions and Applications*, vol. 2. NASA Lewis Research Center.

37 Gordon, S. and McBride, B.J. (1994). *Computer Program for Calculation of Complex Chemical Equilibrium*, vol. 1311. NASA Reference Publication.

12

Combustion of Flake Aluminum with PTFE in Solid and Hybrid Rockets*

Gaurav Marothiya and P. A. Ramakrishna

Indian Institute of Technology Madras, Department of Aerospace Engineering, Chennai, 600036, India

12.1 Introduction

12.1.1 Solid Rockets

Solid propellant rockets consist of a casing, liner, solid propellant grain, nozzle, and igniter in their most basic form. Fuel and oxidizer are in the solid phase in a solid-propellant rocket and often require a binder (which is a fuel) to maintain these solid materials packed close to each other. The solid propellant burn rates are a crucial parameter that controls the mass flow rate and in turn, the thrust delivered by the solid rocket motor. The burn rates of a solid propellant are affected by various factors such as chamber pressure, initial temperature, and solid propellant composition. St. Robert's Law (also known as Vielle's Law) [1–4] provides a general description of the burn rate dependence on chamber pressure and is as depicted in Eq. (12.1).

$$\dot{r} = a \cdot P_c^n \tag{12.1}$$

$$a = a_0 e^{\sigma_P (T_1 - T_0)} \tag{12.2}$$

$$\sigma_P = \frac{\mathrm{d} \ln \dot{r}}{\mathrm{d}T} \approx \frac{2}{\dot{r}_2 + \dot{r}_1} \times \frac{\dot{r}_2 - \dot{r}_1}{T_2 - T_1} \tag{12.3}$$

where, $\dot{r}$ is regression rate normal to the solid propellant's burning surface, P_c is the chamber pressure inside the combustion chamber, a is the burn rate of solid propellant at unit pressure, which has the initial temperature sensitivity of the propellant embedded in it [2–5] (refer to Eq. (12.2)), and n is the burn rate pressure index. In Eq. (12.2), a and T_0 are the burn rate at unit pressure and temperature

* Parts of this article were taken and reworded from the below mentioned articles of the authors:
1. Gaurav, M. and Ramakrishna P.A. (2016). Effect of mechanical activation of high specific surface area aluminum with PTFE on composite solid propellant. *Combustion and Flame* 166 (1): 203–215, with permission of Elsevier.
2. Gaurav, M. and Ramakrishna P.A. (2018). Utilization of mechanically activated aluminum in hybrid rockets. *Journal of Propulsion and Power* 34 (5): 1206–1213, with permission of Indian Institute of Technology, Madras.

Nano and Micro-Scale Energetic Materials: Propellants and Explosives, First Edition.
Edited by Weiqiang Pang and Luigi T. DeLuca.

at standard conditions, respectively. T_1 is the initial temperature of the propellant at which burn rates are measured. Equation (12.3) can be used to evaluate the temperature sensitivity (σ_P) of the propellant. $\dot{r}_1$ and $\dot{r}_2$ in Eq. (12.3) are the burn rates at the temperature T_1 and T_2, respectively. A Strand Burner or a Crawford bomb maintained at various pressure can measure the values of burn rates at a particular initial temperature.

Homogeneous (double base) propellants and heterogeneous (composite) propellants are mainly the two types of solid rocket propellants. The ingredients (such as nitroglycerine and nitrocellulose) in double-base solid propellants are blended at the molecular level, whereas the ingredients in composite propellants are mixed at the macroscopic level. In the composite solid propellant, the ingredients (ammonium perchlorate [AP], ammonium nitrate [AN], aluminum [Al] etc.) are generally in powder form. They require a hydrocarbon-based binder (typically hydroxyl terminated poly-butadiene [HTPB], carboxyl terminated poly-butadiene [CTPB]), which also act as fuel in the composition. The combustion in composite solid propellants is partially diffusion-driven [3] because ingredients are mechanically mixed and partly premixed due to solid oxidizer combustion. This leads to a lower burn rate pressure index ($n = 0.3$–0.4) [2–4].

12.1.1.1 Need for High Burn Rates in Solid Rockets

A higher propellant loading in solid rocket stages of launch vehicles or missiles is advantageous since it reduces the structural coefficient of that stage. End burning grain is one approach to achieving a high propellant loading. The current propellant loading and thrust for various launch vehicles were taken from the literature [6–10] and tabulated in Table 12.1. Table 12.1 shows the computed burn rate required if an end-burning grain was employed in these stages of launch vehicles. It can be seen from Table 12.1 that the propellant loadings of the solid rocket stages are lower compared to an end-burning solid propellant grain (~96–98%). Table 12.1 also shows

Table 12.1 Propellant loading, thrust, and required burn rate for an end-burning grain used for solid rocket stage of various launch vehicles.

Name of solid rocket stage	L/D	Stage	Approximate current propellant loading	Thrust (kN)	Required burn rate for end-burning grain (mm s^{-1})
Space shuttle booster	12.64	Lower	75.6%	12 000	206.7
Ariane ES booster	10.36	Lower	78.2%	6 470	142.1
GSLV Mk III booster	8.05	Lower	76.1%	3 578	111.7
PSLV 1st stage	7.26	Lower	71.1%	4 860	120.6
PSLV 3rd stage	1.77	Upper	74%	244	20.5
Pegasus 2nd stage[a)]	2.4	Intermediate	75.6%	158	24.2
Pegasus 3rd stage[a)]	1.38	Upper	67.3%	32.7	9.74

a) As the combustion chamber length was unavailable for computing the propellant loading, it was extracted from the pictures available in Northrop-Grumman [7, 8].

Table 12.2 Variation of vacuum I_{sp} in the third stage of PSLV with the chamber pressure at nozzle exit area ratio equal to 52.

Chamber pressure (bar)	Vacuum I_{sp} (s)	Exit pressure (mbar)
60	311.37	112
40	3110.14	75
30	310.96	57
22	310.76	42

that the burn rate required for the lower stage of a launch vehicle is relatively high due to the high thrust requirement. Because of the high L/D ratios possible in these motors, the burning surface area available in lower-stage motors is relatively large. If they are to be replaced with an end-burning grain (small burning surface area), burn rates must be extremely high. However, if the emphasis is on the upper stages, both the L/D and the thrust required are reduced. As a result, the burn rates needed are less than 25 mm s^{-1}, which appears to be achievable.

Furthermore, if these solid propellants burn rates are accomplished at pressures lower than their current working pressures, it may contribute to a significant reduction in the structural coefficient (ζ_{str} in Eq. (12.4)). In Eqs. (12.4) and (12.5), $M_{Structural}$, $M_{Initial}$, and M_{Final} are the structural, initial and final masses of the stage, respectively, in Eq. (12.4). It could lead to a higher incremental velocity (ΔV) (refer Eq. (12.5)) of the stage, and it would be an added benefit of using high burn rate solid propellants as the end burning grain in the launch vehicles' upper stages.

$$\zeta_{str} = \frac{M_{Structural}}{M_{Initial}} \tag{12.4}$$

$$\Delta V = I_{sp} \ln \left(\frac{M_{Initial}}{M_{Final}} \right) \tag{12.5}$$

Another issue to consider is the performance (I_{sp}) as chamber pressures decrease. To investigate the effect of chamber pressure on the I_{sp}, a chemical equilibrium analysis for the current propellant composition of the PSLV 3rd stage at various chamber pressures at an area ratio of 52 was performed using NASA SP-273 [11]. The data required for the analysis (such as area ratio, thrust, the mass of stage, etc.) was obtained from SpaceFlight101.com [9]. Table 12.2 summarizes the results of the analysis. As it is an upper-stage motor that operates in near-vacuum conditions (above 200 km altitude, ambient pressure 5×10^{-9} bar), the exit pressures (as seen in Table 12.2) are much higher than the ambient pressure. As a result, high-pressure ratios across the nozzle are attainable even at low chamber pressures. The constraint here could be the diameter of the vehicle/nozzle exit. Therefore, individual vehicles would require careful consideration of these factors. Therefore, it is critical to obtain high burn rates for the upper stages of launch vehicles for higher propellant loadings according to the above considerations.

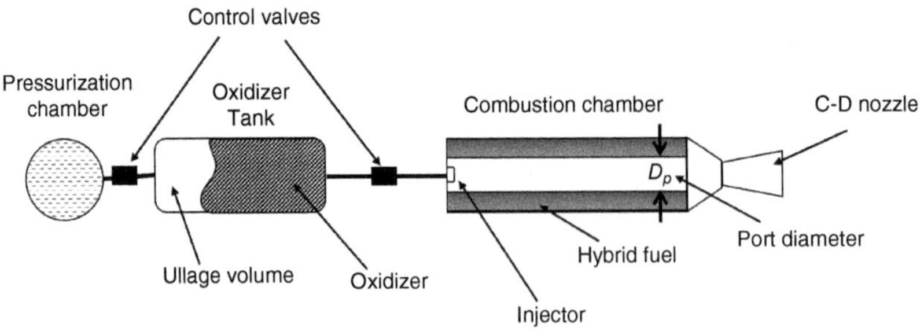

Figure 12.1 Schematic of the hybrid rocket. Source: Gaurav Marothiya and P.A. Ramakrishna.

12.1.2 Hybrid Rockets

A combination of liquid and solid rockets is known as a hybrid rocket. In a typical hybrid rocket, fuel is housed in the combustion chamber in the solid phase, while the oxidizer is stored separately in the liquid or gaseous phase [4], as shown in Figure 12.1. Since the fuel and oxidizer are compartmentalized, it is safer than solid rockets and has the ability to start–stop–restart. In comparison to liquid rockets, hybrid rockets require essentially one plumbing system for the oxidizer, reducing the system's complexity by half. Throttling is simpler in hybrid rockets than in liquid rockets because just the oxidizer flow needs to be controlled.

Furthermore, the combustion in hybrid rockets is primarily a diffusion-driven phenomenon because the oxidizer injected in the combustion chamber has to mix with solid fuel stored in the chamber. As a result, it is less prone to combustion instability due to minimal pressure dependency of the combustion. Hence it is safer than liquid rockets. A simple regression rate law for a typical hybrid rocket is provided by Eq. (12.6), where, $\dot{r}$ is the regression rate of fuel normal to the burning surface inside the combustion chamber. Oxidizer mass flux (G_{ox}) is the mass flow rate of oxidizer ($\dot{m}_{\text{ox}}$) divided by the port area (A_{p}) as shown in Eq. (12.6). In Eq. (12.6), $A_{\text{p}} = \pi D_{\text{p}}^2/4$ for single circular port and D_{p} is the port diameter of the fuel grain and also indicated in the Figure 12.1. Oxidizer mass flux index is represented as m in the Eq. (12.6).

$$\dot{r} = a \cdot G_{\text{ox}}^m = a \cdot \left(\frac{\dot{m}_{\text{ox}}}{A_{\text{p}}}\right)^m \tag{12.6}$$

Some drawbacks of hybrid rockets prevent them from being used in practical applications despite the many advantages over liquid and solid rockets. As mentioned earlier, the combustion in hybrid rockets is diffusion-driven, the low regression rates of fuel are one of the most significant drawbacks. To get a larger thrust, a high burning surface area would be necessary, which reduces the hybrid propellant's volumetric loading. The burn rate as given in Eq. (12.6) is seen to change as port area changes due to burning if the index m is not maintained around 0.5. This could result in changes to the oxidizer to fuel ratio (O/F). As a result, the specific impulse and thus the thrust could be constantly changing as the propellant burns. Further discussions on hybrid rockets will follow in Section 12.4.

The two sections above show section 1.1 and 1.2 that achieving high burn rates is essential in solid and hybrid rockets. As high burn rates can be achieved with the use of fine aluminum in the composition, however, the slow combustion of aluminum limits its use. Therefore, in this chapter, the primary focus would be on the increase of burn or regression rates in solid and hybrid rockets without compromising or perhaps increasing the performance of the rocket motors.

12.2 Aluminum Combustion in Composite Solid Propellant

As aluminum is utilized in pyrotechnics and explosives, it has been a very active topic of research worldwide. Aluminum has also been shown to boost the specific impulse of composite solid propellants due to the higher adiabatic flame temperature. During the combustion process in the rocket motor, aluminum oxidized to alumina (Al_2O_3). As the product gases pass through the nozzle with large area ratio, the alumina cools down and form agglomerates and are also known to decelerates the flow [1, 2, 4]. Therefore, these agglomerates are the primary disadvantage of using aluminum in composite solid propellants and cause two-phase flow losses and slag formation (when used in unison with a submerged nozzle [2–4, 12]).

12.2.1 Literature on Aluminum Combustion

The literature on aluminum combustion and the production of aluminum agglomerates is extensive. Beckstead [13, 14] provided a comprehensive analysis of aluminum combustion. He evaluated methods utilized by numerous researchers worldwide to ignite aluminum particles and investigate the influence of different oxidizing and nonoxidizing environments on aluminum combustion. Glassman et al. [15] proposed that diffusion of aluminum vapors and oxidizer controls aluminum combustion in the air. However, they say because the result of aluminum combustion is the formation of alumina (Al_2O_3), and condensed alumina dominates the combustion process and heat transmission to the fuel (aluminum), combustion of aluminum cannot be evaluated using a basic hydrocarbon combustion model.

Ermakov et al. [16] embedded a thermocouple into an aluminum particle and measured the ignition temperature as 2000–2100 K, concluding that the ignition occurred due to oxide shell failure rather than melting the shell; they observed a small tongue-shaped flame that originated at the crack and propagated throughout the particle surface. Mechanical cracking of the oxide shell, according to Lokenbakh et al. [17], can occur under varied heating and atmospheric conditions. They discovered that at high heating rates and moderately high temperatures (1300–1400 K), the oxide shell fractures and molten aluminum agglomerates in irregular forms, eventually forming a neck-like structure.

In addition, research has also been conducted by igniting aluminum particles in various oxidizing environments and studying the influence on the burning time of aluminum particles. Prentice [18] conducted studies in which the oxygen level was varied in an enclosed atmosphere, with the other gas being nitrogen or argon. They discovered that the concentration of oxygen significantly affected the burn time of aluminum particles and observed that the higher the concentration, the shorter the burn time. It is to be expected, as the concentration gradient is the primary driving force for the diffusion flame. Zennin et al. [19] burnt aluminum particles with diameters of 220 and 350 μm in air and a 100% CO_2 atmosphere. They also burned aluminum in O_2/N_2 and O_2/Ar environments, similar to the study performed by Prentice [18], but only at ~20% oxygen concentration. They concluded that CO_2 is just 20% as effective as O_2 as an oxidizer.

Many researchers worldwide have employed various types of aluminum to examine the combustion of aluminum and its effect on the burn rate of solid propellants [20–23]. They demonstrated that nano-aluminum significantly increases the burn rate of solid propellants. However, the purity of nano-aluminum was found to be very low due to the oxidation of nano-aluminum owing to the large specific surface area of nano-aluminum [22]. Moreover, the processing of the solid propellant would also be a paramount consideration with nano-aluminum since a large number of particles raise the solid propellant's viscosity. Verma and Ramakrishna [24] introduced Pyral, a flake-like particle with a specific surface area comparable to nano-aluminum. The Pyral used in their investigation was made in a ball mill with a stearic acid coating to prevent the formation of an oxide layer during ball milling. They discovered that the aluminum oxide level was minimal despite having a very high specific surface (~23 $m^2\ g^{-1}$). They also discovered that the burn rates of solid propellants made using Pyral were comparable to those of nano-aluminum with a similar specific surface area.

There has also been significant research on the combustion of activated aluminum powder and its effect on aluminum agglomeration. The influence of nickel-coated aluminum particles on the ignition behavior of aluminum has been examined by Shafirovich et al. [25] As the exothermic intermetallic reaction between aluminum and nickel takes place, they discovered that covering nickel over aluminum reduces the ignition delay of individual aluminum particles. Hahma et al. [26] have also conducted similar research. They coated the aluminum powder with tripotassium hexa-fluoroaluminate (K_3AlF_6), utilizing aqueous potassium hydrogen fluoride (KHF_2) as the fluoride source and the aluminum particles themselves as the aluminum source in the complex fluoride. The complex fluoride is created in situ by reacting aluminum with KHF_2. They burned the coated aluminum in several gas environments and found that the fluoride coating increased the burn rate most effectively in a carbon dioxide atmosphere. The burning rates of this fluoride-coated aluminum in a CO_2 environment even exceeded the burn rates in air for this powder. They have also observed that the fluoride coating also prevented the agglomeration of residue at all pressures and different gaseous environments. Osborne and Pantoya [27] conducted DSC and TG (Thermogravimetric) analyses on nanometer-sized PTFE (poly-tetra-fluoro-ethylene) and several types of aluminum powders. They discovered that the composite of PTFE/Al powder has a lower ignition temperature and higher exothermicity due to a pre-ignition (thermite) reaction (Eq. (12.7)) between aluminum and PTFE.

$$4n\text{Al} + 3(\text{C}_2\text{F}_4)_n \xrightarrow{\text{Thermite reaction}} 6n\text{C} + 4n\text{AlF}_3 + \text{Heat} \tag{12.7}$$

Sippel et al. [28, 29] combined PTFE and aluminum in a ball mill, and they have also discovered a similar pattern to that described by Osborne and Pantoya [27] with regards to DSC and TG analyses. They discovered that the strong reactivity and low ignition temperature of the PTFE–aluminum reaction limit the ball mill speed and powder yield. They used this PTFE/Al powder in composite solid propellants and discovered that it marginally improves the burn rates of the solid propellants over base composition. They conducted extensive research to demonstrate that using the above approach of depositing PTFE on aluminum reduces the size of the aluminum agglomeration particle.

In totality, from the above review of the literature on aluminum combustion following points can be deduced:

- Aluminum particle combustion is diffusion-driven, and the ignition of the particles is essentially influenced by the thickness of the aluminum oxide shell on the particle.
- Aluminum particle size reduction improves the solid propellant's burn rates. However, because of the large specific surface area available for oxidation in nano-sized aluminum, the nano-sized aluminum has a high concentration of aluminum oxide.
- Furthermore, because of the enormous number of nano-sized aluminum particles, the viscosity of the solid propellant rises during processing, preventing its usage in the industry.
- Activation of aluminum enhances the reactivity of aluminum and assists in the reduction of the agglomeration of aluminum particles.

12.3 Effect of Mechanical Activation in Composite Solid Propellants

As discussed earlier, the activation of aluminum with PTFE enhances the aluminum combustion and contributes to reducing the agglomeration. Also, the flake-like particles used by Verma and Ramakrishna [24], which has a comparable specific surface area as nano-aluminum with minimal aluminum oxide present in the powder, enhance the burn rates of the solid propellant significantly. It would be prudent to use this Pyral in tandem with PTFE and study its effects on the various performance and ballistic parameters of the composite solid propellants.

12.3.1 Experiments with Solid Propellants

Table 12.3 tabulates the source of all the ingredients used in this study for preparing the composite solid propellants. The purity of all the ingredients tabulated in

Table 12.3 Source of various ingredients used for preparing solid propellant.

Ingredient	Type	Source
HTPB ($(C_4H_6(OH)_2)_n$	Binder and Fuel	Anabond Ltd., Kanchipuram
AP (NH_4ClO_4)	Oxidizer	Tamilnadu Chlorates, Madurai
IPDI ($C_{12}H_{18}N_2O_2$)	Curing agent	Satish Dhawan Space Centre, Sriharikota
DOA ($C_{22}H_{42}O_4$)	Plasticizer	Satish Dhawan Space Centre, Sriharikota
PTFE powder $(C_2F_4)_n$ (<10 μm)	Catalyst for aluminum	Pragati Plastics Pvt. Ltd., New Delhi
Aluminum powder (Pyral)	Additive	Thangamani Trading Company, Chennai
IO (Fe_2O_3) (<5 μm)	Catalyst for AP	Sigma Aldrich, USA

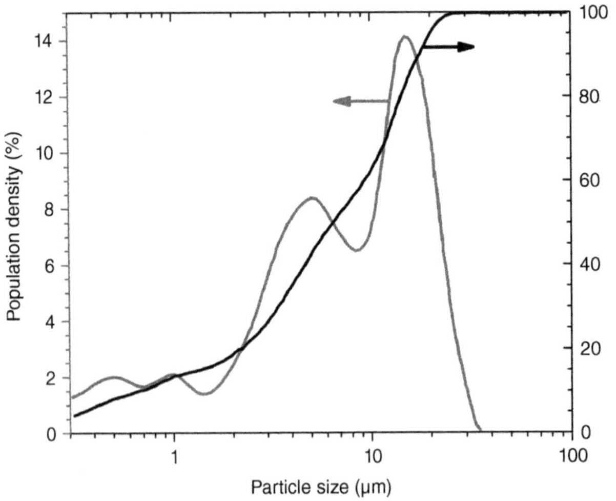

Figure 12.2 Particle size distribution of Pyral powder. Source: Gaurav Marothiya and P.A. Ramakrishna.

Table 12.3 was more than 99%. The aluminum powder chosen for mechanical activation was Pyral powder (which has a large specific surface area of ~22 $m^2 g^{-1}$), used by Verma and Ramakrishna [24]. Figure 12.2 depicts the particle size distribution of Pyral. The Pyral powder had a d_{50} of 8.23 μm, and if the flake-like particles of Pyral powder are assumed to represent a disc of 8.23 μm, the thickness of that disc is 33 nm.

12.3.1.1 Preparation of Mechanically Activated Pyral

Initially, a required amount of Pyral powder was mixed with the PTFE powder in the desired fraction of aluminum using a magnetic stirrer for almost 15 minutes to achieve uniformity. This uniformly mixed Al–PTFE powder was then transferred to the ball milling cups. For ball milling, stainless steel balls with a diameter of 4 mm were chosen, and for each 50 g of powder, 20 SS balls were used. Ball milling was used to bind the Pyral and PTFE particles mechanically. The rotational speed of the planetary ball mill (FRITSCH GmbH) was 300 rpm. Dry ice was used during this operation to maintain the temperature of the particles from rising above a certain threshold where the powder can self-ignite in the ball mill. The cold gases (CO_2) produced by the sublimation of dry ice were used for the purposes mentioned above. Ball milling was done in four-minute intervals, followed by a ten-minute rest period. This cycle of operations was repeated 15 times to obtain the mechanically activated Pyral powder at the end of the operation. No process control agent (PCA) was used during the ball milling.

12.3.1.2 Preparation of Propellants

All the ingredients listed in Table 12.3 were weighed using an electronic weighing scale with a least count of 10 mg. The solid ingredients were stored overnight in an oven at 333 K to remove moisture before being used to prepare the propellant. A 250 ml capacity beaker was used to mix the listed ingredients by hand for 15–20 minutes. The resulting propellant slurry was loaded to a sigma mixer which has a capacity of 200 g, and mixed thoroughly for an hour. The mixer was also linked to a vacuum pump, which was used to remove trapped gases from the propellant slurry while mixing. The propellant slurry was then transferred to a 50 mm diameter PTFE casting tube. This tube, along with the propellant, was inserted in a vacuum casting chamber to remove the trapped gases further. This casting chamber was designed with a hot water jacket on the circumference, which raised the temperature of the slurry to 328 K. Heating reduces the viscosity of the propellant slurry inside the chamber and further aided in the removal of trapped gases. The vacuum casting chamber maintained the vacuum of 50 Torr (~8000 Pa) for roughly one hour. After that, the degassed slurry was kept under a pressure of 2.5 bar for 10 days under ambient conditions (average temperature being 306 K). Several samples of dimensions 5 mm × 5 mm × 25 mm were cut from the solid propellant after curing it for conducting the burn rate evaluation experiments.

The solid loading of all propellants was fixed at 86%. To maintain the aluminum proportion consistent across all propellant formulations, PTFE was added at the expense of AP. The binder in all propellants was 10.7% HTPB monomer, 2.3% DOA (dioctyl adipate), and 1% IPDI (iso-phorone di-isocyanate), while AP was used in a coarse to fine ratio of 1 : 1. The particles sizes of course and fine AP used was 355 to 300 μm (sieve number 45 and 50 ISO standard) and 63–45 μm (sieve number 230 and 325 ISO Standard), respectively.

12.3.1.3 Experimental Setup

The burning rates of propellants were determined using a typical Crawford bomb in conjunction with a pressure transducer, as illustrated in Figure 12.3. The Crawford bomb has a free volume of 9.6×10^{-4} m^3, which is more than 2000 times larger than the propellant sample tested in the experiments. The pressure variation during the experiment was measured using a piezoelectric pressure transducer

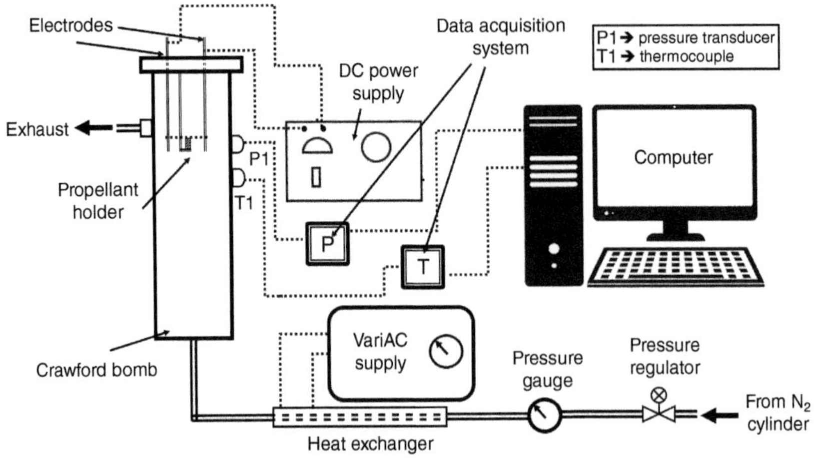

Figure 12.3 Schematic of Crawford bomb setup. Source: Gaurav Marothiya and P.A. Ramakrishna.

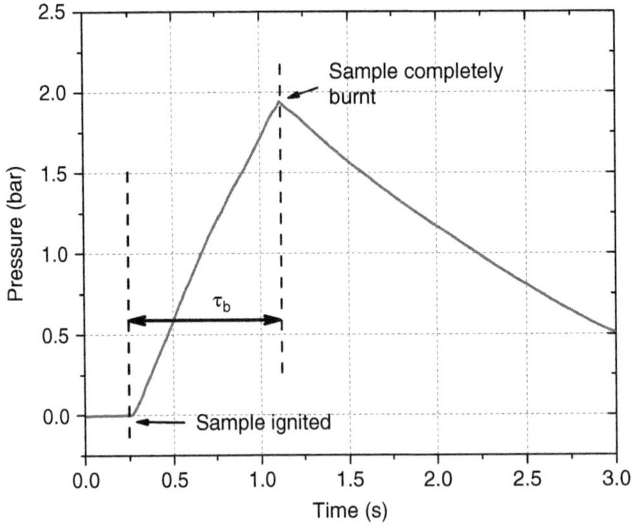

Figure 12.4 Pressure variation inside the Crawford bomb during the experiment. Source: Gaurav Marothiya and P.A. Ramakrishna.

(PCB-11B322). This pressure variation was plotted on a pressure–time graph, as shown in Figure 12.4. As the propellant begins to burn, the pressure inside the Crawford bomb rises due to rapid mass addition in the gas phase. After the propellant sample has been completely burned, the pressure measured using a piezoelectric pressure transducer inside the Crawford bomb decreases. This is attributed to the fact that the piezoelectric pressure transducer only measures the change in pressure inside the Crawford bomb.

Furthermore, once the propellant had been completely burned, the pressure in the Crawford bomb began to fall and eventually reached the constant value at which the experiments were being carried out. The pressure rises inside the Crawford bomb for the highest burn rates encountered is depicted in Figure 12.4 in this study. However, the pressure rise was around 1 atm gauge pressure for all other experiments. As the piezoelectric pressure transducer was used to measure the pressure rise inside the Crawford bomb, it will only detect the signal when any change in pressure is observed. Hence, at the end of sample burning, the pressure inside the Crawford bomb became zero, the electrical signal detected by the transducer also tended to zero. Therefore, the burn duration was calculated as the time between the rise and drop of the signal, as indicated in Figure 12.4 as τ_b. The resulting average burn rate is estimated since a known length of the propellant sample is used, and the associated τ_b is acquired as described above. Burn rates measured by this method was also cross verified by the other two methods (i) using a CCD camera (Ishitha and Ramakrishna [30]) and (ii) wire cutting method using a microsecond timer circuit (Verma and Ramakrishna [24]) of measuring the burn rates of the solid propellant in the lab.

To pressurize the Crawford bomb, commercially available nitrogen cylinders were used as the source. As depicted in Figure 12.3, a pressure regulator was used to keep the pressure within the bomb at a stable value. The temperature of the propellant sample could be raised to the desired temperature using a heat exchanger, as shown in Figure 12.3, to evaluate the temperature sensitivity of the propellants. The heat exchanger is just an insulated container with an electric heating coil inside. Using a VariAC (variable AC supply), the power supplied to the coil was changed. The required temperature inside the Crawford bomb (as well as in the sample) was obtained by injecting nitrogen into the heat exchanger at meager flow rates, and the temperature inside the propellant sample was measured using an imbedded thermocouple in the sample. Once the desired temperature in the propellant sample was achieved, the pressure inside the Crawford bomb was increased slowly to the desired pressure to avoid the thermodynamic heating of the propellant sample.

As illustrated in Figure 12.5, the propellant sample was placed on a propellant holder. Two electrodes are positioned in the propellant holder to aid in the ignition of the propellant sample. Silica grease, a known inhibitor has been applied to the propellant sample as shown in Figure 12.5 to ensure that the propellant burning surface area is $5 \times 5\,mm^2$ at all times. For ignition, a direct current of 22 V and 1 A was used. The Ni-chrome wire used for ignition has a diameter of 0.5 mm.

12.3.1.4 Experimental Procedure

The propellant sample was placed in the propellant holder (Figure 12.5), and the ignition connections were secured. To prevent the nonuniform burning, the non-burning surfaces of the sample were covered with silica grease as explained earlier in Section 12.3.1.3. The Crawford bomb was then filled with nitrogen gas from a nitrogen cylinder and pressure regulated to the required pressure. As previously explained, the pressure–time trace acquired from the pressure transducer output was utilized to calculate the combustion time to measure the burning

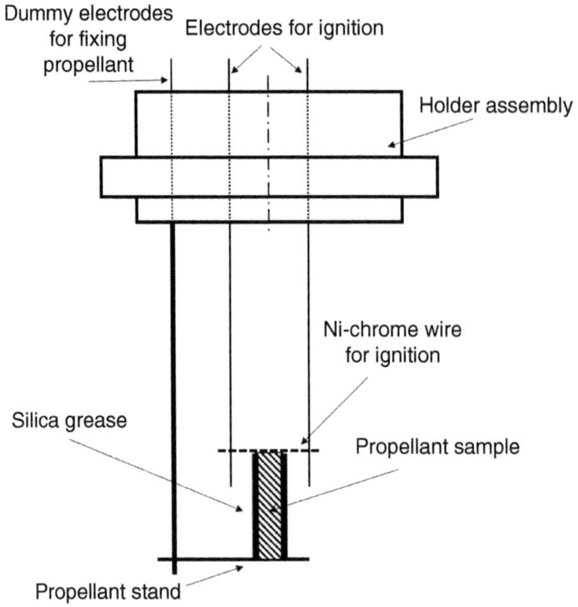

Figure 12.5 Schematic of the propellant holder for burn rate measurement. Source: Gaurav Marothiya and P.A. Ramakrishna.

rates. Experiments were conducted three times at each pressure to establish the repeatability of burning rate values. The reported data are the averages of these three measurements, and the highest dispersion in burn rate values was less than 1.8% of the average burn rate at that pressure.

12.3.2 Results and Discussions on Solid Rockets

12.3.2.1 Chemical Equilibrium Analysis

According to the literature review (Section 12.2.1), it would be beneficial to develop a method in which large specific surface area aluminum like Pyral (flake like aluminum powder) is utilized in conjunction with PTFE. This Pyral powder was mechanically activated according to the procedure outlined in Section 12.3.1.1 Prior to doing any physical assessments with the mechanically activated Pyral, a chemical equilibrium analysis with varied amounts of PTFE with aluminum was performed using NASA SP 273 software [11] to determine the optimum PTFE content for obtaining maximum specific impulse.

PTFE fraction was substituted at the expense of AP in the propellant. Figure 12.6 depicts vacuum I_{sp} with varied PTFE content in the propellant and the chamber pressure was assumed to be 70 bar. For this analysis, the nozzle area ratio was determined so that the flow is optimally expanded at sea level conditions. Figure 12.6 also depicts the variation of adiabatic flame temperature and molecular weight of product gases in the combustion chamber as the PTFE fraction in the solid propellant composition changes. As shown in Figure 12.6, a PTFE fraction of 3.18%

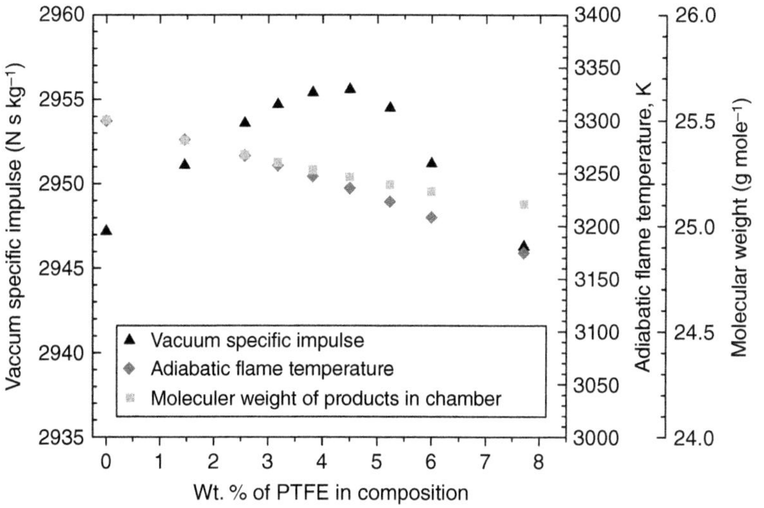

Figure 12.6 Various performance parameters versus PTFE fraction in solid propellant. Source: Gaurav Marothiya and P.A. Ramakrishna.

in the propellant composition resulted in the best I_{sp}, which corresponds to an Al/PTFE ratio of 85 : 15. The adiabatic flame temperature decreases as the PTFE content increases. This study restricts itself to the maximum PTFE content of 15% in aluminum. As previously stated, Sippel et al. [28] increased the PTFE proportion in the composition by decreasing the aluminum fraction. As a result, they noticed a significant decrease in the Al_2O_3 mass fraction in the products of thermochemical equilibrium calculations with an increase in the PTFE percentage in the propellant. As opposed to that, and as stated previously, PTFE is introduced here by decreasing the fraction of AP in the solid propellant to maintain the aluminum fraction constant while increasing the PTFE fraction. One can also include the PTFE fraction at the expense of the HTPB binder. However, the addition of PTFE at the expanse of HTPB in the actual compositions would increase the end-mix viscosity of the composition and make the processibility of the solid propellant very tedious. Hence in this study, the PTFE was added at the expense of AP in the composition.

12.3.2.2 DSC and TG Analysis of Mechanically Activated Pyral

DSC and TG analysis were performed in the nitrogen environment using a Netzsch STA449F3 Jupiter instrument at temperatures ranging from 30 to 900 °C. The experiment was conducted at a heating rate of 20 °C/min, and the results are presented in Figure 12.7. According to DSC and TG analysis, aluminum present in an activated aluminum powder containing 15% PTFE begins reacting with fluorine at significantly lower temperatures (~520 °C), which is an exothermic fluorination process, as indicated by Osborne and Pantoya [27]. The initial exotherm found in DSC analysis for nano-aluminum by Sippel et al. [28, 29] and Osborne and Pantoya [27] at 450 °C, was not observed for Pyral powder as seen in Figure 12.7. The explanation for this could be that the first exotherm for nano-aluminum and

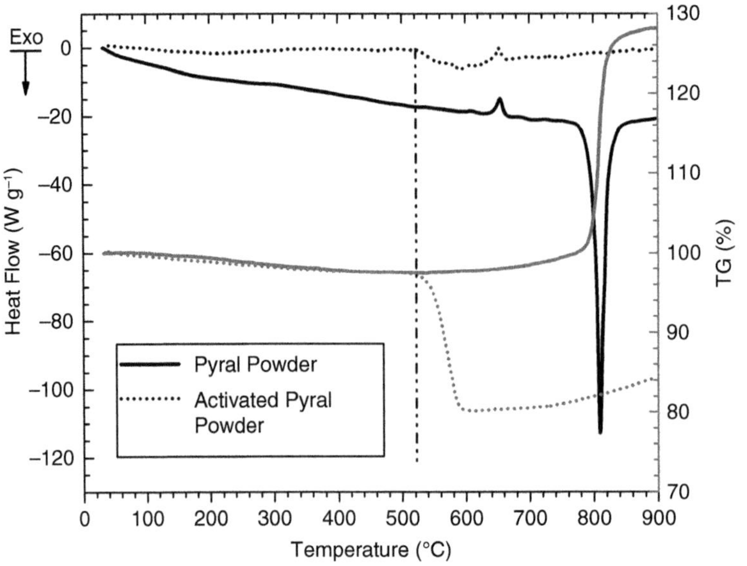

Figure 12.7 TGA and DSC analysis of Pyral powder activated with 15% PTFE in a nitrogen environment and at 20 °C per minute heating rates. Source: Gaurav Marothiya and P.A. Ramakrishna.

PTFE occurs at 450 °C due to fluorination of the oxide shell of nano-aluminum, as stated by Osborne and Pantoya [27]. They reported that the oxide percentage in the nano-aluminum employed in their investigation was 25%. These pre-ignition reactions aid in exposing the inner aluminum layer and facilitating the reaction between aluminum and PTFE, which is visible as a second exotherm at roughly 500 °C. In the case of Pyral, this pre-ignition exotherm peak is absent in DSC analysis due to the negligible presence of the oxide shell on the particle, as observed by Verma and Ramakrishna [24]. Sippel et al. [29] discovered a similar pattern in DSC and TG analysis for their ball-milled 70/30 Al/PTFE powder combination. The sharp exotherm seen for aluminum at 810 °C was caused by the reaction of aluminum with nitrogen and the creation of AlN, which can be seen as an increase in sample mass in TG analysis. Sippel et al. [28] reported similar results in other cases. The lack of a pronounced exotherm with Al–PTFE powder is most likely owing to the reaction of PTFE with aluminum, which begins at a considerably lower temperature. To prove it more evidently advanced technologies, such as thermal gravimetric mass spectrometry (TG-MS), TG-FTIR, or gas chromatography–mass spectrometry (GC–MS) are needed. However, due to unavailability of these instruments at the authors' end at the time of submitting this article, these tests could not be included in this chapter.

12.3.2.3 SEM Analysis of Mechanically Activated Pyral

Mechanical activation of Pyral was produced by the use of ball milling. Elemental analysis was performed using FEI Quanta 200 FEG EDAX to address whether PTFE is uniformly coupled with Pyral, and Figure 12.8 represents the results.

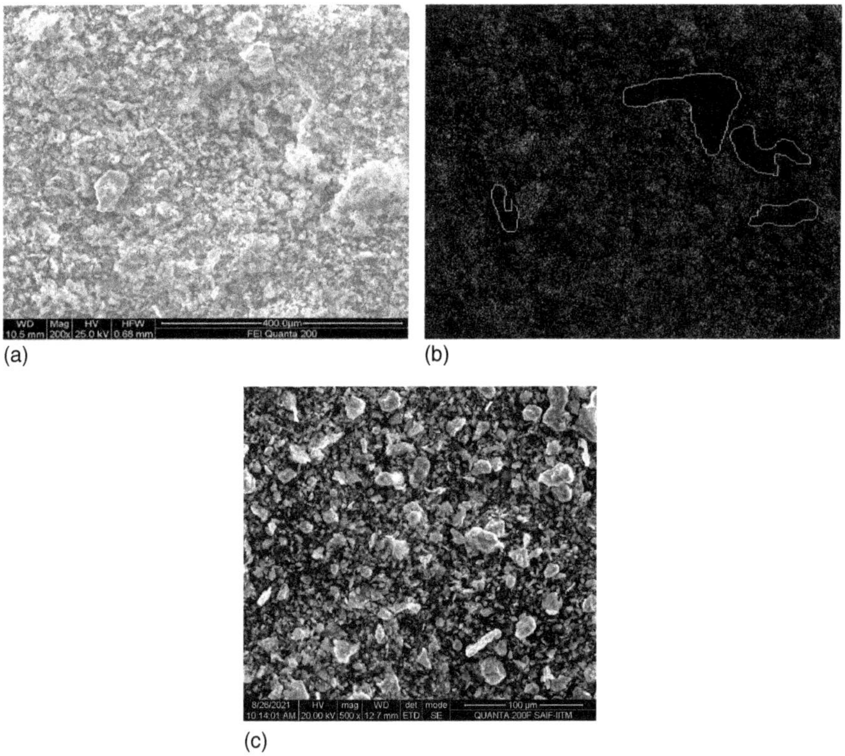

Figure 12.8 (a) SEM picture of the mechanically activated aluminum powder. (b) Corresponding elemental analysis of the activated aluminum particles picture (green is aluminum and red is fluorine [PTFE]). Source: reproduced with permission from Elsevier [31]. (c) SEM picture of Pyral powder before mechanical activation. Source: Gaurav and Ramakrishna.

As this analysis is elemental rather than compound-based, a unique element from each compound is identified, and a map is constructed, as shown in Figure 12.8b. The locations highlighted in Figure 12.8b are not in the same plane as the rest of the sample and so had low reflected counts for elemental analysis, and as a result, the region is dark. Figure 12.8a,b show that PTFE is uniformly bonded with aluminum powder.

12.3.2.4 Burn Rates and Temperature Sensitivity Analysis with Varying PTFE Fraction

Various solid propellant compositions (as indicated in Table 12.4) were developed to investigate the effect of aluminum powder's mechanical activation with PTFE. The results were obtained at pressures of 10, 30, 50, and 70 bar. To ensure that the results were repeatable, at least three samples were burnt at each of the above-indicated pressures. As previously stated, the highest variations in the value of burn rates measured at a single pressure were less than 1.8% from the mean for various compositions. Mix c1 was created as a base or control composition without any additives, using flaky aluminum powder with a high specific surface area ($\sim$22.5 $m^2\ g^{-1}$).

Table 12.4 The list of compositions prepared by varying PTFE fraction in composition.

Mix	Al (%)	PTFE % in propellant (% PTFE in Pyral)	AP (%)	Coarse to fine ratio of AP	Binder (%)	Measured density (theoretical density) ($kg\,m^{-3}$)	Remark
c1	18	0 (0)	68	1 : 1	14	1770 (1778)	Aluminum used was Pyral
c2	18	3.18 (15)	64.82	1 : 1	14	1775 (1782)	PTFE added at the time of mixing
c3	18	1.46 (7.5)	66.54	1 : 1	14	1770 (1779)	Activated Pyral powder with PTFE
c4	18	2.57 (12.5)	65.43	1 : 1	14	1772 (1781)	
c5	18	3.18 (15)	64.82	1 : 1	14	1771 (1782)	

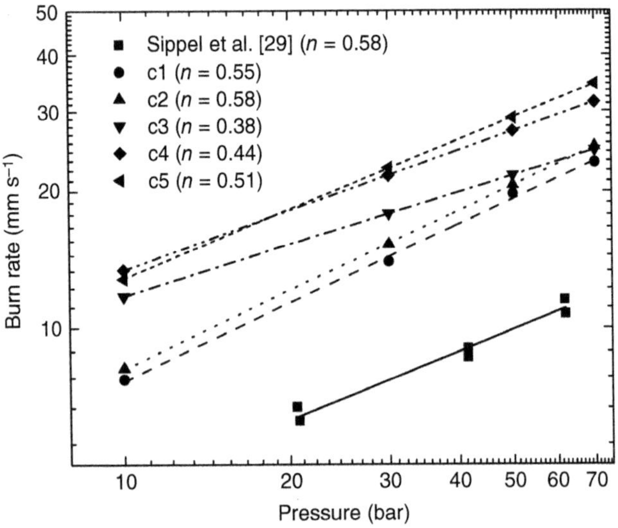

Figure 12.9 Burn rate versus pressure comparison of propellant with PTFE fraction variation at 30 °C. Source: Gaurav Marothiya and P.A. Ramakrishna.

This composition was identical to that prepared by Verma and Ramakrishna [24]. Figure 12.9 shows the burn rates versus pressure plot for various compositions tabulated in Table 12.4 at 30 °C. Mix c2 was developed by incorporating PTFE directly into the propellant while mixing the other ingredients, as shown in Table 12.4. As shown in Figure 12.9, the burn rates obtained with c2 are equal to those obtained with c1. According to the evidence provided in Figure 12.9, a simple addition of PTFE has no discernible effect on burn rates. This could explain that PTFE is not in close contact with aluminum due to a binder layer that separates PTFE from aluminum.

Compositions c3, c4, and c5 were made using the mechanically activated Pyral. However, the fraction of PTFE during mechanical activation of Pyral was varied, as shown in Table 12.4. Figure 12.9 also depicts the burn rates for compositions c3, c4, and c5. It was discovered that as the PTFE fraction increased, the burn rates increased considerably over the control composition c1. It could be explained as follows: when aluminum is not activated with PTFE, the aluminum particles get ejected from the burning surface during solid propellant combustion. As the ignition temperature (~1700–2200 K, Beckstead [13]) of aluminum is much larger compared to the temperature at the burning surface (~850 K as measured by Zanotti et al. [32] at 10 atm), it will melt and agglomerates till the time it reaches closer to the flame where this agglomerate gets ignited. However, in the case of mechanically activated aluminum, as the ignition temperature is ~793 K, the particles might ignite inside or very close to the burning surface. This increases the heat flux to the burning surface and increases the burning rates in the process.

In this study, the flake-like aluminum with a high specific surface area is mechanically activated with PTFE by ball milling. However, Sippel et al. [29] combined PTFE particles with spherical aluminum particles using ball milling in an Argon atmosphere. Due to the high specific surface area of Pyral, the contact boundary between the aluminum and PTFE would be higher, and hence the larger area available for the reaction between aluminum and PTFE. It could explain why the burn rates reported in this investigation are much higher than those reported by Sippel et al. [29], as illustrated in Figure 12.9.

The burn rate pressure index of mix c1, made using Pyral, which did not contain any PTFE, was higher than the pressure index of a similar composition prepared with micron-sized aluminum reported by Verma and Ramakrishna [24, 33]. Verma and Ramakrishna [24] explain the increased burn rate of c1 to heat release caused by aluminum combustion closer to the burning surface. This was the same line of reasoning provided by researchers [20–23, 34], who earlier reported similar results using nano-aluminum.

Figure 12.9 further shows that the burn rate pressure index initially reduces with activated aluminum powder (c3). However, the burn rate pressure index rises again as the PTFE fraction in the composition increases (composition c4 and c5). This is explained as follows: According to Osborne and Pantoya [27], Al–PTFE reactions are highly exothermic fluorination processes. The premixedness of the gaseous products emerging from the propellant surface is increased due to these processes. In propellant combustion, it is widely recognized that increasing premixing leads to a higher pressure index [1–4]. Therefore, when the PTFE fraction in composition increases, the overall pressure index also increases. Sippel et al. [28] discovered a similar pattern in the burn rate pressure index. They discovered that the addition of PTFE raises the burn rate pressure index from 0.36 for base propellant without PTFE to 0.58 for 30/70 Al/PTFE mix propellant. A detailed explanation of above is provided by Gaurav and Ramakrishna [31].

The temperature sensitivity of compositions c1, c3, c4, and c5 was also determined by evaluating burn rates at two different initial temperatures of 30 °C and 60 °C. As previously mentioned in Section 12.3.1.3, the desired temperature inside the

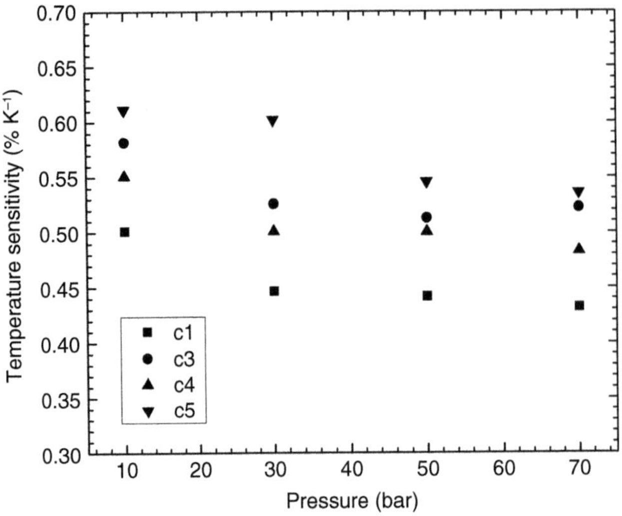

Figure 12.10 Variation of temperature sensitivity with pressure for the compositions listed in Table 12.4. Source: Gaurav Marothiya and P.A. Ramakrishna.

Crawford bomb was maintained by using a manually regulated variable AC voltage supply to the heating coil in the heat exchanger. Equation (12.3) was used to calculate temperature sensitivity. $\dot{r}_1$ and $\dot{r}_2$ were the burning rate of propellant at the initial temperatures of $T_1 = 30\,°\text{C}$ and $T_2 = 60\,°\text{C}$, respectively at a particular pressure. The temperature sensitivities for various composition listed in Table 12.4 plotted versus pressure are shown in Figure 12.10. The maximum uncertainty error in measuring the temperature sensitivity was ~3.8%. The temperature sensitivity evaluated for the composition c1 was higher than the one reported in the literature [2] (~0.32) for a similar aluminized composition with micron-sized aluminum. The reason for that could be the larger specific surface area of Pyral used here, which burn closer to the surface. Hence, it could increase the surface temperature and heat stored in the condensed phase, increasing the temperature sensitivity. Also, increased temperature sensitivities with increasing the fraction of PTFE were observed for composition c3–c5 (as depicted in Figure 12.10). This could be attributed to the fact that mechanical activation of aluminum reduces the ignition temperature of the aluminum (refer to Section 12.3.2.2) and could ignite much closer to the surface than Pyral powder. As a result, it could again increase the surface temperature and hence increase the temperature sensitivities. For a detailed discussion on this, the reader can refer to Gaurav and Ramakrishna [31].

12.3.2.5 Effect of Mechanical Activation of Pyral on Density, Viscosity, and Heat of Combustion of Propellant

Another advantage of using mechanically activated aluminum is that it reduces viscosity buildup in the propellant slurry. According to Sippel et al. [29], using nano-aluminum in a propellant is problematic since the viscosity of the propellant slurry is relatively high due to the small size of the aluminum which increases the

Table 12.5 Viscosity, density, and heat of combustion comparison.

Mix	Viscosity at 30 °C (cP)	Density obtained (kg m^{-3})	% of theoretical density	The heat of combustion (MJ kg^{-1})
c1	7.18×10^6	1770	99.5	−6.79, −6.83, −6.85[a)]
c5	5.02×10^5	1771	99.4	−7.35, −7.45, −7.48

a) Verma and Ramakrishna [24].

number of aluminum particles in the propellant. While Pyral, as reported by Verma and Ramakrishna [24], reduced propellant viscosity compared to nano-aluminum, it was still high. A Brookfield DV2T viscometer was used to measure the viscosity of the propellant slurry. The end of mix viscosity of compositions c1 and c5 were measured to compare the change in viscosity owing to activated aluminum powder, and the results are reported in Table 12.5.

Verma and Ramakrishna [33] analyzed the influence of density on solid propellant burn rates. They discovered that the actual amount of ingredients present differed from the expected amount of ingredients in the propellant, resulting in a considerable difference in realized density from theoretical density, especially when the mixer sizes were small. Sippel et al. [28] examined the cured density of propellants and discovered that the densities attained for their propellants were between 85% and 90% of theoretical densities. The density of propellant generated using mix c1 and c5 was determined using the water displacement method described by Verma and Ramakrishna [33] and is shown in Table 12.5 along with the end mix viscosity. The densities obtained for cured propellants with c1 and c5 deviate from the anticipated theoretical densities marginally. It assured that no blowholes were present in the propellant and that the fraction of ingredients in propellant was the same as the desired composition.

The heat of combustion for composition c5 was also measured using a bomb calorimeter, which contained a T-type copper–copper tungsten thermocouple with a 1 mm bead size to monitor the rise in water temperature. The heat of combustion for composition c5 was measured using three samples, and the results are shown in Table 12.5. As the composition of c1 is identical to that of Verma and Ramakrishna [24], the heat of combustion value is also listed in Table 12.6. When compared to a propellant that does not contain PTFE (mix c1), the heat of combustion with c5 is 9% higher. This indicates that the specific impulse could be higher with these propellants.

12.3.2.6 Effect of Mechanically Activation of Pyral on the Agglomeration of Aluminum

If one has to make use of these propellants in upper stages of launch vehicles, the other aspect other than I_{sp} that needs to be considered is the two-phase flow losses and slag accumulation. Due to aluminum agglomeration, there are primarily two types of losses in a solid rocket motor. The size of the agglomerates causes

Table 12.6 Comparison of agglomerate mass of the various components.

	Mix c1		Mix c5				
			Individual results				Cumulative results
Propellant sample weight (g)	9.2	10.9	10.5	10.8	11.1	12.5	44.9
Agglomerate mass (mg)	19	22	3	4	4	5	17
Percentage (%) of agglomeration mass	0.206	0.202	0.028	0.037	0.036	0.040	0.037

two-phase losses, while the mass of the agglomerates causes slag accumulation. The secondary flow losses in the nozzle will be more significant due to the large size of the agglomerates, and the slag deposition with the submerged nozzle will be higher, especially with the mass of agglomerates generated, which adds to the system's dead weight [35].

The fluorination reaction of aluminum reduces the agglomeration of aluminum particles in solid propellants as stated by Osborne and Pantoya [27], and Sippel et al. [28, 29]. As a result, experiments were carried out at 70 bar to collect the quenched agglomerates. A different type of propellant holder was devised to gather the agglomerates that can be contained in the same Crawford bomb (Figure 12.11). A small amount of propellant was placed in the holder and ignited from the bottom with the same Ni-chrome wire as previously described in Section 12.3.1.4. The propellant sample burned from bottom to top, and an ethanol-filled pan was placed beneath it to quench the fumes emitted by the burning propellant, as illustrated in Figure 12.11. The alumina agglomerates ejected from the burning of propellant was quenched and accumulated in the pan below. A hot air oven at 60 °C was used for drying these agglomerates. Jayaraman et al. [36] have also used a similar setup to acquire the quenched agglomerates from the combustion of aluminized solid propellants.

Compositions c1 and c5 were chosen for the agglomeration investigation, and experiments were performed on propellants weighing approximately 10 g. A Malvern Mastersizer 2000 device with water as the dispersion medium was used to measure the particle size distribution of agglomerates (shown in Figure 12.12). The particle size of agglomerates was found to be higher with activated aluminum. Sippel et al. [28, 29] investigated the particle size of the agglomerates after quenching them. They discovered that the average diameter of agglomerates with 70/30 Al/PTFE at 70 bar was one-third that of agglomerates with no PTFE in the propellant. Sippel et al. [29] used spherical particles for ball milling with PTFE, whereas flake-like aluminum was used here. The particle size of the agglomerates was found to be more prominent with flake-like aluminum than with spherical particles. As a result, the end particle size of the agglomerated residue using composition c5 may have a greater average particle size than those reported by Sippel et al. [29]. This could also be explained as follow; the PTFE fraction in the

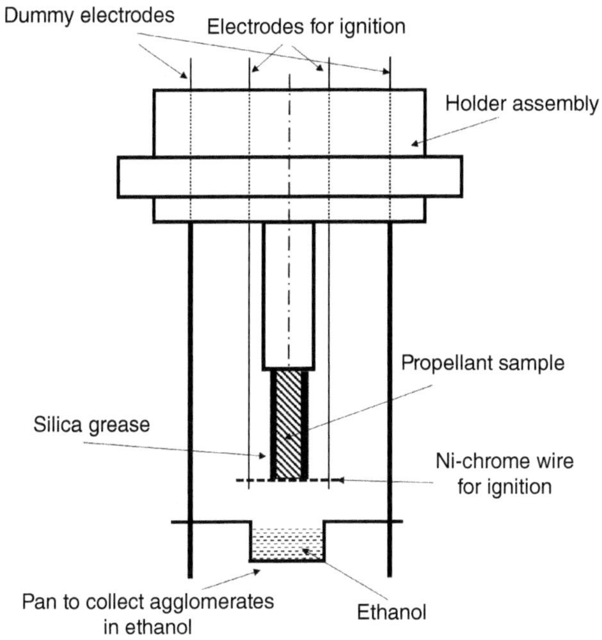

Figure 12.11 Schematic of the propellant holder for agglomerate collection. Source: Gaurav Marothiya and P.A. Ramakrishna.

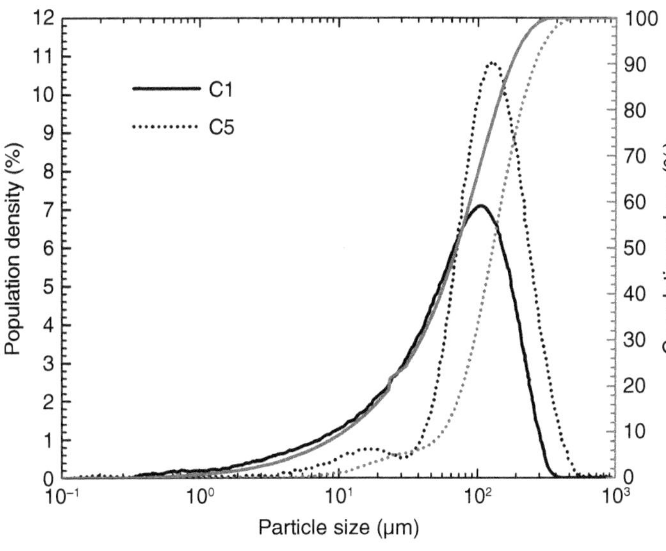

Figure 12.12 Particle size distribution of agglomerates after quenching. Source: Gaurav Marothiya and P.A. Ramakrishna.

propellant composition is not the stoichiometric with aluminum fraction, and it is in a fuel (aluminum) rich condition. Therefore, some agglomerates would still be present in the exhaust gas, which gets cooled down and generate agglomerates. It is also well documented in the literature that flake-like aluminum agglomerates more compared to spherical aluminum. Moreover, due to the presence of PTFE in tandem with the flake aluminum, a smaller composite of Al–PTFE might get converted to aluminum fluoride, and part of the larger composite might get agglomerated and remain in the exhaust, which gets collected in the ethanol bath. Furthermore, larger agglomerates' contribution could be higher in the particle size analysis, and D50 for agglomerates of composition c5 was also higher.

The mass of agglomerates collected per unit mass of propellant burned was also determined using a weighing scale with a least count of 1 mg, and the results are shown in Table 12.6. The agglomeration residue collected per unit solid propellant mass for composition c5 was approximately one-fifth of c1. The mass of agglomeration obtained for each sample was closer to the weighing balance's least count for mix c5. To reduce errors in the measurement, the cumulative mass of the agglomerates for mix c5 was also measured and summarized in Table 12.6. This reduced agglomeration mass could be attributed to a fraction of the agglomerates being aluminum fluoride and having a lower aluminum/aluminum oxide concentration. A more detailed discussion on the same can be found in Gaurav and Ramakrishna [31]. Thus, with the introduction of mechanically activated aluminum the slag accumulation especially when a submerged nozzle is used would be lower based on the above results. The two-phase flow losses could also be lower as the mass of the condensate matter in the exhaust is lower.

12.3.2.7 Redesigning the Upper Stages of Launch Vehicles

As mentioned earlier, one of the motivations for this study was to achieve a higher solid loading in solid rockets using an end-burning grain configuration. As discussed in Section 12.1.1.1, the end-burning grain configuration can only be employed in the upper stages of a launch vehicle since the burn rate required for the lower stage using end-burning configuration would be substantially large. Furthermore, when the expression for velocity increment (ΔV) (refer to Eq. (12.5)) is examined, the initial to final mass ratio for the lower stages is smaller than for the upper stages. As a result, even if propellant loading is enhanced in the lower stages, the effect on ΔV would be negligible. Therefore, it would be prudent to limit the application of the end-burning grain to the launch vehicle's upper stages. A detailed study has been provided by Gaurav and Ramakrishna [31].

Case Studies with Pegasus Upper Stages With propellant development as a backdrop, data for the existing Pegasus launch vehicle developed by Orbital ATK (now Northrop-Grumman), USA, was collected from the literature [7, 8, 10]. This information was used to compute the burn rate required for the propellant in the second and third stage motor of the Pegasus launch vehicle with an end-burning grain. The propellant in the second and third stages of the Pegasus vehicle was replaced with end-burning grains in a new design while keeping the sizes comparable to

Table 12.7 Comparison of current Pegasus design with the proposed new design.

Stage	Parameters	Existing design	Design-A
2nd	Propellant mass (metric tons)	3.9	4.48
	Structural coefficient (%)	9.1	8
3rd	Propellant mass (metric tons)	0.77	1.145
	Structural coefficient (%)	11.7	8.1
Overall	Payload in low earth orbit (LEO) (metric tons)	0.443	0.521
	Payload increment (kg)	0	78

the current configuration. *There would be minimal changes (like wing size and some other aerodynamic features) needed to the existing hardware of the current setup of the Pegasus launch vehicle apart from using a cartridge-loaded grain instead of a case-bounded one.* In this case study, the average thrust required was increased to maintain the grain burn time at the original value for the new design. The mass flow rate of the propellant was estimated by dividing the average thrust by the I_{sp} of the stage as provided by Northrop-Grumman [7, 8] and spaceflight101.com [10]. The I_{sp} was presumed to be the same as the existing value in this case, although it is a conservative approach given the improvement in I_{sp} seen earlier in Figure 12.6 with mechanically activated aluminum. After determining the propellant mass flow rate, the requisite burn rates were computed as mass flow rate divided by propellant density and burning surface area of the grain. Since the density of the propellant achievable on the large scale is closer to the theoretical density of composition, the propellant density considered for calculations was the theoretical density of mix c5 (1782 kg m^{-3}).

Since an end-burning grain is being used, the burning area was determined after the casing and liner thicknesses were taken into account (combined thickness 5 cm). The computations above yielded a required burn rate of 25.93 and 9.21 mm s^{-1} for the second and third stages with end-burning grains, respectively. The burn rate required for the second and third stage can be readily reached by using composition c5 at significantly lower pressure (40 bar for second stage and 5 bar for third stage [refer to Figure 12.9]) compared to the current configuration. Since these are upper-stage motors that must function in near-vacuum, the outlet pressures can be extremely low. This enables the use of a lower chamber pressure while keeping the current design's area ratio. Hence, the propellant mass loading increases while the structural coefficient decreases (from 9.1% to 8% for the second stage and from 11.7% to 8.1% for the third stage), when the dimensions and structural design of the second and third stages for design-A are kept identical to those of the current design, as shown in Table 12.7. Thus, it is conceivable to have an end-burning grain for the Pegasus launch vehicle's second and third-stage motor based on the requisite burn rates.

Incremental velocity, ΔV for each stage of the Pegasus launch vehicle was computed using Eq. (12.5). Equation (12.8) shows the incremental velocity of the stage, taking gravitational and aerodynamic losses for a stage into account [4]. Where, $\bar{g}$ is the average acceleration due to gravity after accounting for altitude change, t_b is the burn duration of the stage, and C_D and A are the vehicle's drag coefficient and projection area, respectively. The ρ and u represent the density of the atmosphere and the vehicle's velocity, respectively. In Eq. (12.8), ζ is the ratio of propellant mass to the stage's initial mass. These losses were not taken into account in this analysis. Because the burn time for design-A is the same as for the current design, the gravitational losses for design-A are the same as for the existing designs. Since design-A's hardware (external dimensions) is identical to that of the current design, the aerodynamic losses for design-A will be identical. As a result, while evaluating the ΔV increase, only Eq. (12.5) was considered. The overall ΔV of the vehicle increases as the structural coefficients of the second and third stages are reduced. This increase in ΔV is traded off for a greater payload, resulting in a 17.6% increase in payload over the existing design. This represents a considerable gain, with minimal alterations to the vehicle other than increased propellant loading in the third stage. There are possibilities for improvement of these numbers as the reduction in two-phase flow losses and slag losses due to the use of mechanically activated aluminum is not accounted for here. The other benefit not accounted for here is the fact that the stages are operated at lower pressure compared to the earlier case and this could lead to an improvement in the structural factor. With the present cost of launching a kilogram of satellite to SSO being ~118 000 dollars per kilogram [37], this would result in huge savings. A detailed discussion is also provided by Gaurav and Ramakrishna [38]

$$\Delta V = I_{sp} \ln\left(\frac{M_{\text{Initial}}}{M_{\text{Final}}}\right) - \bar{g}t_b - \frac{C_D A}{2M_{\text{Initial}}} \int_0^{t_b} \frac{\rho u^2}{(1 - \zeta t/t_b)} dt \tag{12.8}$$

12.4 Aluminum Combustion in Hybrid Rockets

As indicated in Section 12.1.2, hybrid rockets have several drawbacks that prohibit them from being widely used, with low regression rates being one of the most serious ones. To achieve higher thrust from a hybrid rocket engine, the requisite burning surface area must be considerable, which minimizes the volumetric loading of the hybrid rocket. Furthermore, as the fuel burns with an oxidizer inside the combustion chamber, the O/F ratio of the motor changes continuously, causing the specific impulse and thus the thrust to vary with burn time.

12.4.1 Literature Review

To overcome the disadvantages of hybrid rockets mentioned previously, a substantial amount of research has been conducted [39–44] on the use of aluminum, particularly ultrafine aluminum, to boost the regression rates of HTPB-based hybrid rockets. According to George et al. [39], the combination of 6% AP (98 μm) and 18% Al (28 μm) with HTPB-based binder increases the regression rate by more than

100% when compared to pure HTPB. They also discovered that the inclusion of aluminum increases the oxidizer mass flux index compared to pure HTPB fuel. Chiaverini et al. [41] observed that increasing the ultrafine aluminum fraction in HTPB grain from 4% to 20% enhanced the fuel regression rate by 15–45%. Risha et al. [42] observed a similar trend in the regression rate as Chiaverini et al. [41]. Risha et al. [42] used nano-sized aluminum particles (ALEX©) coated with Viton-A (a fluorocarbon-based elastomer) in an HTPB-based binder and observed a 120% improvement in fuel mass burning rate and linear regression rates over pure HTPB-based fuel at lower oxidizer mass flux (G_{ox}) (11 g cm^{-2} s^{-1}). It can be interpreted that, Viton-A dissociates to form fluorine-based species, which contribute to the rapid combustion of the nano-aluminum particles similar to PTFE. They also noticed an increase in combustion efficiency from 81% for pure HTPB to 90% when Viton-A coated ALEX© powder was used. Lips [43] employed a fluorine-based oxidizer with highly aluminized (60% and 80% aluminum in Poly-Urethane [PU]) fuel grain, which resulted in increased regression rate and improved combustion efficiency of the hybrid rocket motor.

In addition, research has advanced in the direction of using an alternate fuel (without additives), resulting in high regression rates compared to typical hybrid fuels like HTPB. Karabeyoglu et al. [45, 46] proposed using wax as a fuel in hybrid rockets and discovered a considerable increase in regression rates compared to conventional fuels. However, wax exhibits relatively poor mechanical properties. Kumar and Ramakrishna [47] have recently recommended using EVA (ethylene vinyl acetate) to overcome the drawback of wax as a fuel. They discovered that by combining wax with EVA in an 80 : 20 ratio, the mechanical characteristics of wax could be significantly improved, although a drop in regression rates was observed upon the use of it.

Based on the findings of Risha et al. [44], who employed a fluorine-based chemical to boost the reactivity of aluminum in hybrid rockets, it is conceived to employ mechanically activated Pyral powder, which as seen earlier was able to increase the burn rates of solid propellant to investigate the influence on the regression rate and combustion efficiency of the hybrid motor. Following the work of Kumar and Ramakrishna [47], combining wax–EVA with mechanically activated Pyral powder may aid in improving the regression rate and combustion efficiency while maintaining superior mechanical properties of the fuel grain.

12.4.2 Experiments with Mechanically Activated Pyral in Hybrid Rockets

12.4.2.1 Preparation of the Fuel Grain

All materials needed to make the hybrid fuel (wax, EVA, aluminum powder) were measured using a weighing scale with a minimum least count of 10 mg. The wax utilized in this study was a blend of 30% microcrystalline wax and 70% paraffin wax. Table 12.8 lists the sources of the additional ingredients involved in the hybrid rocket study apart from Pyral and PTFE used in the composite solid propellant study. The mechanical activation of aluminum, Pyral powder was carried out as described

Table 12.8 Source of various ingredients used for preparing hybrid fuel.

Fuel grain ingredient	Type	Source
Wax	Fuel	Mahatha Petroleum Private Limited, Chennai
EVA (ethylene vinyl acetate) (7350M)	Fuel	Taiwan
Aluminum (<25 μm)	Fuel	Thangamani Trading Company, Chennai

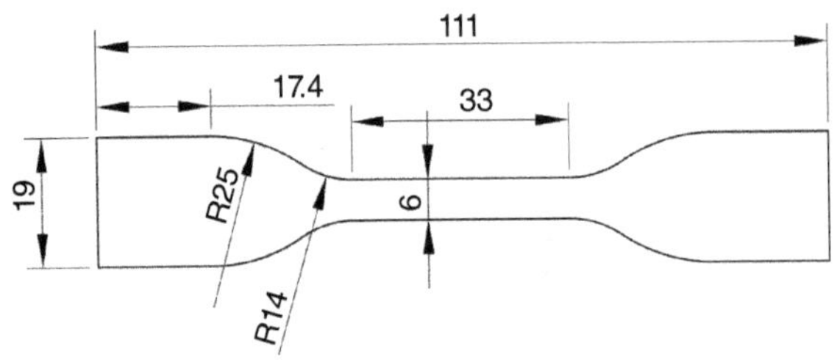

Figure 12.13 Schematic of a dog bone-shaped fuel specimen (ASTM D638 standard) for measuring the mechanical properties of fuel (all dimensions in mm). Source: Gaurav Marothiya and P.A. Ramakrishna.

in detail (refer to Section 12.3.1.1). The EVA used in this study is identical to Kumar and Ramakrishna [47] with an 18% vinyl acetate fraction.

EVA pellets were added to a specified amount of molten wax and heated until the EVA was uniformly emulsified into the wax. The wax–EVA binder contained a 4 : 1 wax–EVA ratio, the same as that adopted by Kumar and Ramakrishna [47]. After adding all of the other measured components (such as Pyral, mechanically activated Pyral or micron-sized aluminum), the mixture was constantly hand-stirred for 15 minutes. The slurry was then put into a polyvinyl chloride (PVC) pipe with the following dimensions: 49.5 mm outside diameter, 42 mm internal diameter, and a length of 150 mm. A process described by Kumar and Ramakrishna [48] was employed here to cure the wax–EVA (and other ingredients) mixture in the PVC pipe. The cured fuel cartridges were then trimmed to 134 mm lengths, with a 9 mm diameter hole drilled to serve as the fuel grain port.

12.4.2.2 The Process to Measure the Mechanical Properties

According to Kumar and Ramakrishna [47], wax-based propellants must be carefully evaluated because they have poor mechanical properties. In addition, samples were prepared for an experiment to evaluate the mechanical properties of the fuel grain. The procedure used to make the fuel slurry for this experiment was similar to the one described previously. The slurry was then placed on a silicon-oil-greased flat metal tray and into a sheet of a specified thickness as described by Kumar and Ramakrishna [48]. After that, samples of the requisite dimensions (as shown in Figure 12.13) were cut using a CNC (computer numerical control) machine. These dog bone-shaped (ASTM D638 standard) samples were tested in a ZWICK-ROELL 500N UTM machine for their tensile properties.

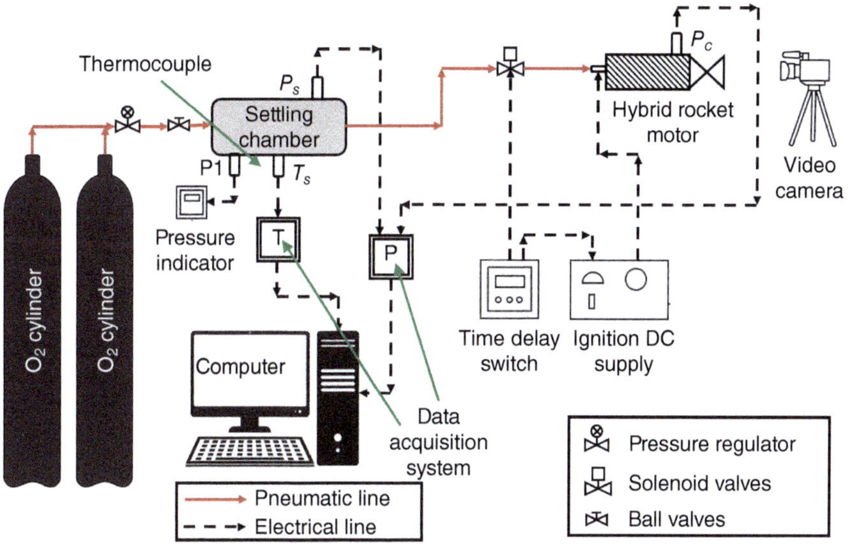

Figure 12.14 Schematic of the experimental setup used for hybrid rocket study. Source: Gaurav Marothiya and P.A. Ramakrishna.

12.4.2.3 Experimental Setup and Test Procedures

Figure 12.14 depicts the experimental setup used to conduct studies on hybrid rockets. It is a modified version of the setup developed by Kumar and Ramakrishna [49]. The mass flow rate of 30 g s^{-1} for the oxidizer was obtained using two commercially available oxygen cylinders. A solenoid valve controlled the flow of oxygen. As illustrated in Figure 12.14, the pressure transducer P1 was attached to a digital display. The pressure inside the settling chamber was measured using a piezoresistive pressure transducer (GE UNIX 5000) P2 and was used to determine the oxidizer mass flow rate. The pressure data was collected using the DAQ NI-9215 (sampling rate of 1000 Hz) from national instruments. The temperature inside the settling chamber was also measured using a t-type thermocouple (T1) and National Instruments' DAQ NI-9211.

The instantaneous mass flow rate ($\dot{m}_{ox,i}$) of the oxidizer was calculated using the pressure data ($P_{s,\,i}$) obtained from the pressure transducer (P_s) and the temperature T_s connected in the settling chamber (as depicted in Eq. (12.9)). $A_{t,\,ox}$ is the area of orifice used in the settling chamber (diameter 3.5 mm). The characteristic velocity for oxygen is denoted as C^*_{ox} and can be evaluated using Eq. (12.10). γ_{ox} and R_{ox} are specific heat ratio and gas constant for oxygen, respectively, in Eq. (12.10). $\overline{T}_s$ denotes the average temperature of the chamber as measured using a t-type thermocouple throughout the experiment. The estimated instantaneous mass flow rate of oxidizer was integrated over the complete period of combustion to obtain the total mass of oxidizer consumed throughout the burn time and the average mass flow rate of oxidizer ($\dot{m}_{ox,\,avg}$) was calculated by dividing the oxidizer mass by the entire burn time. Kumar and Ramakrishna [50] have explained this methodology in detail.

$$\dot{m}_{ox,i} = \frac{P_{s,i} A_{t,ox}}{C^*_{ox}} \tag{12.9}$$

$$C^*_{ox} = \frac{\sqrt{\gamma_{ox} R_{ox}}\sqrt{T_s}}{\gamma_{ox}\sqrt{\left(\frac{2}{\gamma_{ox}+1}\right)^{\frac{\gamma_{ox}+1}{\gamma_{ox}-1}}}} \tag{12.10}$$

Kumar and Ramakrishna [49, 50] demonstrated that the error in calculating the regression rate using the weight loss technique (interrupted burning) decreases as the burning time decreases for a lab-scale hybrid motor. Following that, a burn time of 0.5 seconds, identical to the one used by Kumar and Ramakrishna [47], was used for the regression rate measurement in the current study. A 0.5 seconds burn period was obtained using a sequential timer delay switch (SELEC PT-380) connected to the solenoid valves and the ignitor batteries. The timer delay switch connects the electrical circuit for a predetermined amount of time before disconnecting it. It regulates the timing of the ignition process and the opening and closing of the solenoid valve. The timer delay switch operated in the following manner: Ignition period is 0.7 seconds, and the solenoid valve controlling oxidizer flow is opened for 0.5 seconds. A tiny solid propellant (0.4 g) was used as an ignitor, and a Ni-chrome wire was passed through the solid propellant. A DC power supply was used to heat this Ni-chrome wire, igniting this solid propellant in the process. A video camera recorded the entire experiment. A piezoresistive pressure transducer P_c (GE UNIX 5000) was used to measure the combustion chamber's pressure to calculate the combustion efficiency, which will be discussed later in the Section 12.4.3.3.

As shown in Figure 12.15, the hybrid rocket motor employed in the experiments had a length of 134 mm and an inner diameter of 50 mm. These dimensions are same as the ones used in Kumar and Ramakrishna's [47, 49, 50]. The majority of motor components were constructed of mild steel. A graphite convergent nozzle with an 8 mm throat diameter was used. The nozzle throat diameter increased from 8 to 8.05 mm after 8–10 tests of 0.5 seconds each. As a result, the nozzle was changed every five experiments. As presented in Figure 12.15 and stated in Section 12.4.2.1, the fuel was cast in a PVC pipe with inner and outer diameters of 42 and 49.5 mm, respectively. The pressure transducer (P_c) was connected to the hybrid motor's nozzle end, and the pressure inside the combustion chamber was sensed through a hole (2 mm diameter) drilled in the nozzle, as shown in Figure 12.15. To establish the fuel regression behavior throughout a range of oxidizer mass fluxes, each fuel composition was burned at least three times to ascertain the regression rate or combustion efficiency.

$$d_{f,r} = \sqrt{\frac{4m_{f,r}}{\pi \rho_f L_g} + d_{i,r}^2} \tag{12.11}$$

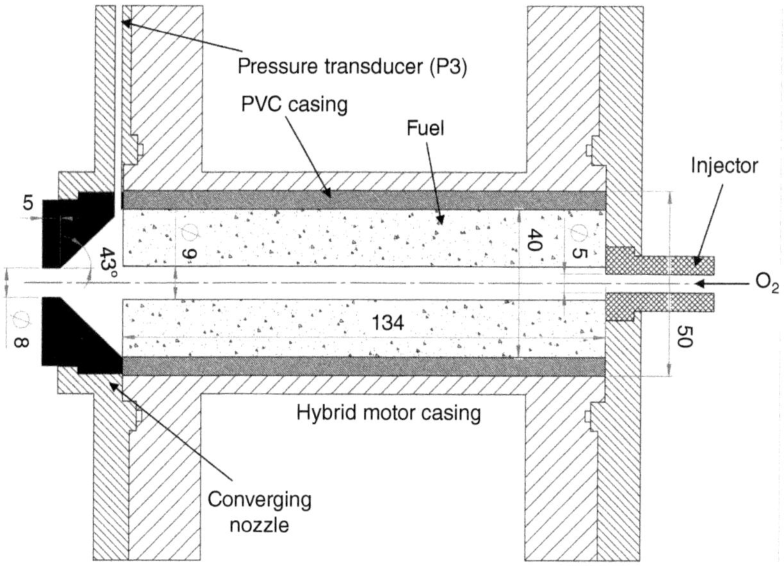

Figure 12.15 Schematic of the hybrid rocket motor. Source: Gaurav Marothiya and P.A. Ramakrishna.

$$\dot{r}_{h,\,r} = \frac{(d_{f,\,r} - d_{i,\,r})}{2\tau_{b,\,r}} \tag{12.12}$$

$$d_{i,\,r+1} = d_{f,\,r} \tag{12.13}$$

The weight loss technique was used to evaluate regression rates for the current investigation [46, 49]. The method utilizes the mass of the fuel, which is calculated by subtracting the final mass of fuel after a predetermined combustion time (0.5 seconds) from the initial mass of fuel before the combustion. The regression rate was determined using Eqs. (12.11) and (12.12). In Eq. (12.12), $m_{f,\,r}$, $d_{i,\,r}$, L_g, ρ_f, and $\tau_{b,\,r}$ are the total mass of fuel burned, the initial port diameter, the length of fuel grain, the density of fuel, and the burn duration, respectively. The burning was terminated every 0.5 seconds, and the mass fuel loss was recorded. The same motor was then utilized for the next 0.5 seconds interrupted burn test, which meant that the final port diameter of one test would be the initial port diameter of the next test, and so on (as given by Eq. (12.13)).

$$P_{c,avg} = \frac{\int_0^{\tau_b} P_{c,i}\,dt}{\tau_b} \tag{12.14}$$

$$C^*_{exp} = \frac{P_{c,avg} A_t}{(\dot{m}_{ox,avg} + \dot{m}_{f,avg})} \tag{12.15}$$

$$\eta_{comb} = \frac{C^*_{exp}}{C^*_{theo}} \tag{12.16}$$

The average chamber pressure ($P_{c,\,avg}$) was used to determine the combustion efficiency (η_{comb}) using Eqs. (12.14), (12.15), and (12.16), where $P_{c,\,i}$ is the instantaneous chamber pressure and τ_b is the period of burning. Equation (12.14) was used to calculate the experimental characteristic velocity (C^*_{exp}), where A_t is the nozzle throat area. The oxidizer and fuel average mass flow rates were $\dot{m}_{ox,avg}$ and $\dot{m}_{f,avg}$, respectively. As a result, the combustion efficiency (η_{comb}) was determined using Eq. (12.16), where theoretical characteristic velocity (C^*_{theo}) was computed using chemical equilibrium analysis (CEA) analysis by NASA-SP-273 [11].

12.4.3 Results and Discussions on Hybrid Rockets

As previously stated, the shortcomings of a hybrid rocket make it unsuitable for real-world applications. As a result, concerns like low regression rates, low combustion efficiency, and poor mechanical qualities must be addressed before using in the actual applications. It was also stated previously that additives such as nano-sized aluminum and fluorinated chemicals could enhance regression rates and combustion efficiency.

Initially, a CEA was performed on several aluminum/wax combinations using NASA SP-273 [11] to determine the best composition in terms of specific impulse (I_{sp}). The exit area to throat area ratio (A_e/A_t) of the supersonic nozzle employed in the CEA analysis was kept constant at 10.85. For all cases studied, the chamber pressure was 70 bar. As seen in Figure 12.16a, the O/F at which the best I_{sp} was attained was lower for the aluminized fuel when compared to a pure wax–EVA fuel. In hybrid rockets, a low O/F ratio is desirable since only the liquid oxidizer requires a pressurization system to be pumped into the combustion chamber. Hence, the mass of the oxidizer would reduce in tandem with the mass of pressurization systems, resulting in a reduction in the overall weight and size of the hybrid rocket system. Furthermore, an aluminized fuel has a higher density than oxygen (oxidizer), with a low O/F ratio, so the density specific impulse for aluminized fuels is higher, as illustrated in Figure 12.16b. A higher density specific impulse is desired since it increases the rocket motor's compactness and can be calculated using Eq. (12.17) for a hybrid rocket. According to Figure 12.16a, aluminized fuel has a higher density specific impulse at all O/F (below O/F = 3) than wax alone. Figure 12.16a,b show that the combination of wax and Pyral in a 60 : 40 ratio produces the highest I_{sp} of the three aluminized fuels (without PTFE). However, this has the greatest I_{sp} at larger O/F ratios (lower density specific impulse) when compared to a 50 : 50 mix of wax and Pyral, hence this mix with 60 : 40 ratio of wax and the Pyral was not investigated further. The combination with the ratio 40 : 60 for wax and Pyral has the best density specific impulse, as shown in Figure 12.16a,b. However, the high surface area of Pyral makes mixing difficult, and the viscosity buildup results in a nonuniform mixture. As a result, this particular ratio (60 : 40 Pyral and wax) was also not considered for further investigation.

$$\text{density specific impulse} = I_{sp} \cdot \frac{((\text{O/F} \cdot \rho_{ox}) + \rho_f)}{(1 + \text{O/F})} \tag{12.17}$$

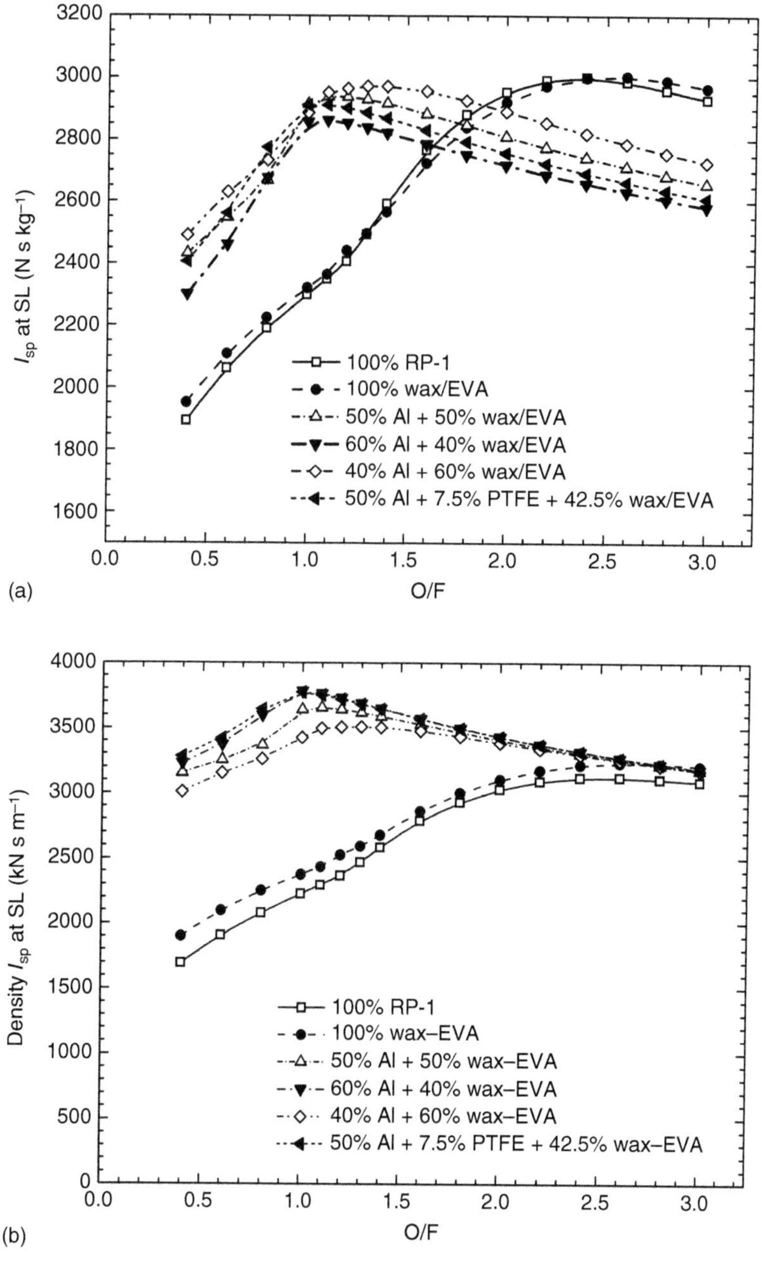

Figure 12.16 I_{sp} versus O/F (a) and density specific impulse versus O/F (b) at sea level (SL) for various combinations of aluminum and wax with oxygen as the oxidizer, at $P_c = 70$ bar and $A_e/A_t = 10.85$. Source: Gaurav Marothiya and P.A. Ramakrishna.

Table 12.9 List of various hybrid fuel compositions prepared.

Mix	Al (%)	PTFE % in composition (PTFE % in Pyral)	Wax + EVA (%)	Obtained density (theoretical density) ($kg\,m^{-3}$)	Regression rate law $\dot{r}_h = b \cdot G_{ox}{}^m$
H1[a)]	0	0 (0)	100	(908)	$0.061\,G_{ox}{}^{0.87}$
H2	50 (Pyral)	0 (0)	50	1352 (1359)	$1.148\,G_{ox}{}^{0.41}$
H3	50 (Pyral)	7.5 (15) (mechanically activated)	42.5	1448 (1453)	$0.809\,G_{ox}{}^{0.61}$
H4	50 (<25 μm Al)	0 (0)	50	1348 (1359)	$0.281\,G_{ox}{}^{0.55}$

a) Kumar and Ramakrishna [47].

As a result, the subsequent best combination of Pyral and wax in a 50 : 50 ratio was chosen for this investigation. Aside from that, a CEA analysis for I_{sp} and density-specific impulse was also performed by adding 7.5% PTFE at the expense of wax while maintaining the aluminum content constant in the composition. According to Figure 12.16a, the addition of PTFE changed the O/F ratio required for optimum I_{sp} to a lower O/F value when compared to the composition with a 50 : 50 wax ratio.

12.4.3.1 Effect of Activated Pyral on Regression Rate

The biggest disadvantage of hybrid rockets is the low regression rate. As a result, several fuel compositions, as shown in Table 12.9, were created to measure the regression rates. Composition H1 was taken from the literature [47] and was simply utilized for comparison. Composition H2 was made with 50% Pyral and 50% wax–EVA mix. Similar to Section 12.3, the Pyral powder was mechanically activated with PTFE. A Pyral:PTFE ratio of 85 : 15 was employed for mechanical activation, and composition H3 was prepared using 57.5% of this mechanically activated Pyral powder and 42.5% wax–EVA. These fuels were tested in the hybrid test facility shown in Figure 12.14 and described in Section 12.4.2. The uncertainty in measuring the regression rates, mass flux of oxidizer and oxidizer mass flux exponent were 2.68%, 2.56%, and 3.7%, respectively (for detailed analysis reader can refer to Gaurav and Ramakrishna [51]). Composition H4 was created to test the influence of different types of aluminum on the combustion efficiency of the hybrid motor, as described further down in the study.

Figure 12.17 indicates that the regression rate of composition H2 (50 : 50 Pyral and wax–EVA compositions) was much greater than that of composition H1 (100% wax–EVA). Almost 400% improvement in regression rate at higher oxidizer mass flux (~25 $g\,cm^{-2}\,s^{-1}$) reported in this investigation with the addition of Pyral was much more than that observed by Risha et al. [44]. However, it should be noted that the baseline fuel (composition H1) exhibited lower regression rates when compared to wax reported in the literature [45, 46] since it contained 20% EVA.

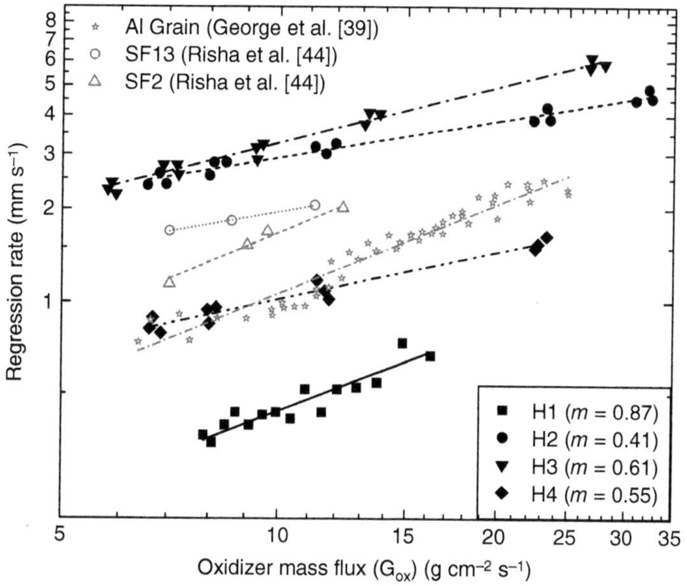

Figure 12.17 Regression rate versus G_{ox} for fuel composition tabulated in Table 12.2. Source: Gaurav Marothiya and P.A. Ramakrishna.

Similar results were reported by Kumar and Ramakrishna [47] and Maruyama et al. [52].

It can also be noted from Figure 12.17 that the regression rates with mechanically activated Pyral powder (composition H3) were approximately 35% higher than those with composition H2 (at $G_{ox} = 25\,g\,cm^{-2}\,s^{-1}$). The increased reactivity of activated Pyral aids in the rapid combustion of activated Pyral particles near the regressing surface of the fuel. This minimizes the time required for the activated Pyral particles to burn. As a result, the combustion of these particles will occur closer to the fuel surface than that of Pyral aluminum particles, similar to the case in solid propellants. It means that more heat is transmitted to the fuel surface than to the Pyral powder, increasing the regression rate seen for composition H3 over H2. The regression rates obtained in this investigation were also significantly higher than observed in the literature [39, 44], as indicated in Figure 12.17.

Table 12.9 also shows that the density of composition H3 was the largest, and thus the mass flow rate of fuel burned was the highest. The density of composition H3 was nearly 60% greater than that of pure wax (composition H1), so even if composition H3 and H1 had the same regression rates, the mass flow rate associated with composition H3 would be 60% more than that of H1. When compared to composition H1, this would increase the thrust generated by the motor with composition H3.

12.4.3.2 Effect on Mechanical Properties

A fuel grain must have acceptable mechanical characteristics to be used in any practical application utilizing hybrid rockets. Samples were prepared to determine the mechanical properties, as previously explained in Section 12.4.2.2. The mechanical

Table 12.10 Mechanical properties of the various compositions of hybrid rockets.

	Properties (deviation)		
	Maximum tensile strength (MPa)	**Percentage elongation (%)**	**Young's modulus (MPa)**
H1[a)]	3.7 (3%)	16.6 (5%)	63.6 (2%)
H2	4.3 (2.19%)	5.1 (12.3%)	175 (2.37%)
H3	2.67 (11.73%)	1.2 (11.93%)	211 (4.16%)

a) Kumar and Ramakrishna [47].

characteristics data of composition H1 determined by Kumar and Ramakrishna [47] were used as a benchmark. Samples of composition H2 and H3 were tensile tested in a ZWICK-ROELL 500N UTM machine at a strain rate of 5 mm/min. To ensure repeatability, three samples from each composition were examined to evaluate the mechanical properties. Table 12.10 summarizes the results obtained here and the results reported by Kumar and Ramakrishna [47]. The value in parenthesis in Table 12.10 is the highest dispersion obtained throughout these studies. Table 12.10 also shows that the percentage elongation of the fuel sample decreased from composition H1 to H3 due to an increase in solid particle loading. However, Young's modulus of the fuel samples improved from composition H1 to H3.

This decrease in elongation can be reasoned as follows; increased solid loading or particle size reduction of solids in the fuel grain increases the number of solid particles, increasing the number of contacts between solid particles and binder (wax–EVA). Because heterogeneous materials have inferior mechanical characteristics than homogeneous materials, a decline in homogeneity reduces the percentage elongation at break and the tensile strength. According to, the ideal mechanical parameters of the propellant grain for cartridge-loaded propellants desired are; ultimate tensile strength of 1–2 MPa, percent elongation of 6–25%, and Young's Modulus value of 30–40 MPa. While the mechanical properties of composition H2 are just about adequate for usage as a cartridge-loaded grain, composition H3's mechanical properties require improvement.

12.4.3.3 Effect on Combustion Efficiency

Another disadvantage of a hybrid rocket is its low combustion efficiency, as stated previously. One could argue that because the fuel had a large fraction of aluminum chosen and a small lab-scale motor was being used, the combustion efficiency would be lower than those reported by Kumar [53] in a similar lab-scale motor. As stated earlier and in literature [13–15], aluminum combustion is a slow process, and mechanical activation increases the reactivity of aluminum [25–29], allowing for ignition at a lower temperature than with aluminum alone. This may have some positive effects on the motor's combustion efficiency. An attempt was made to investigate the influence of different forms of aluminum (micron-sized, Pyral, and mechanically activated Pyral) on the combustion efficiency of the hybrid rocket motor. Compositions H2, H3, and H4 were used to investigate the combustion

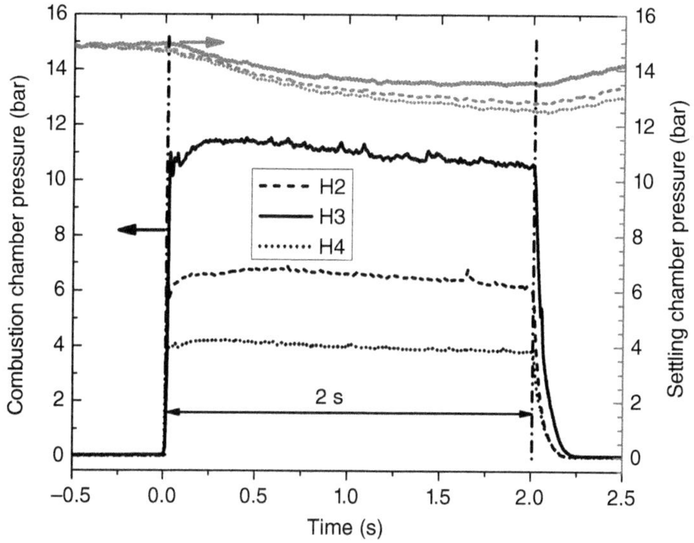

Figure 12.18 Pressure in the combustion chamber and settling chamber versus time curve for different fuel compositions. Source: Gaurav Marothiya and P.A. Ramakrishna.

efficiency in a lab-scale hybrid rocket motor. Composition H4 is similar to H2, except that micron-sized aluminum (particle size less than 25 μm) was used instead of Pyral powder.

The experimental setup described in conjunction with Figure 12.14 was used and all three fuel compositions (H2, H3, and H4) were examined. Experimentation time was extended to two seconds, and pressure–time curves were obtained by measuring pressure inside the combustion chamber and the settling chamber. Figure 12.18 illustrate the combustion chamber pressure on the left *y*-axis and settling chamber pressure on the right *y*-axis (gray color). It was discovered that composition H3 had the maximum chamber pressure. The combustion efficiency (η_{comb}), has been computed and presented in Table 12.11 together with the other characteristics of importance.

Table 12.11 shows that the mix made using micron-sized aluminum had the lowest combustion efficiency (composition H4). It was lower than the η_{comb} (61%) reported by Kumar [53] for wax–EVA fuel grain under identical conditions like the initial *L*/*D* of motor and oxidizer flow rate. The *L*/*D* ratio here represents the length of the grain divided by the initial port diameter. This could be because aluminum combustion is a slow process and the motor length is short (134 mm), the residence time in the motor is less than the combustion time of the aluminum particles. As a result, a considerable proportion of the aluminum particles would escape from the nozzle in an unburned or partially burnt state. The introduction of Pyral (composition H2) improved combustion efficiency. This was owing to the flake-like structure of the Pyral particles, which had a thickness of a few nanometers. As previously stated, the particle size of micron aluminum employed in this investigation was less than 25 μm, and the estimated average thickness of Pyral particles was 33 nm. From

Table 12.11 Various combustion parameters for different hybrid fuel compositions.

	H2			H3			H4		
Exp. no.	1	2	3	1	2	3	1	2	3
Total mass of fuel consumed (g)	35.2	39.7	38.2	63.3	58.3	59.1	23.1	16.6	17.1
Total mass of oxidizer consumed (g)	59	59	59	56	56	56	60	60	60
Average O/F ratio	1.68	1.49	1.52	0.89	0.96	1.02	2.59	3.61	3.5
$P_{c,\,avg}$ (bar)	7.5	8	7.8	11.9	11.4	11.5	5	4.6	4.7
$T_{c,\,theo}$ (K)	3801	3781	3790	3701	3695	3675	3466	3454	3550
τ_{res} (ms)	0.21	0.206	0.208	0.169	0.169	0.168	0.280	0.279	0.278
τ_{comb} (ms)[a]	0.237 (0.04 μm)			—			1.07 (22 μm)		
C^*_{theo} (m s^{-1})	1680	1703	1692	1713	1720	1721	1580	1508	1512
C^*_{exp} (m s^{-1})	800	822	806	1003	1006	1007	605	608	607
η_{comb} (%)[b]	47.6	48.2	47.8	58.6	58.5	58.3	38.3	40.6	40.1

a) Calculated based on Yetter et al. [54].
b) Combustion efficiency of the H1 was 61% [53].

Yetter et al. [54], the combustion time of the particles (particle size close to 22 μm and 40 nm) was acquired and listed in Table 12.11.

Equation (12.18) was used to compute the residence time (τ_{res}) for the particles in the rocket motor, where (L^*) is the characteristic length as provided in Eq. (12.19). The condition in which the particles have the least residence time in the combustion chamber at the onset of combustion was chosen to compare the three cases. In Eq. (12.19), the combustion chamber volume (V_c) was used as the initial port volume (smallest port diameter of 9 mm). It can also be noted from Table 12.11 that the residence time available for composition H4 is highest, while it is the shortest for composition H3. When these timings are compared, it is clear that the combustion period for Pyral is significantly less than that of micron-sized aluminum. Furthermore, the difference in combustion duration and residence time for micron-sized aluminum powder was greater than for Pyral particles. Consequently, when compared to Pyral particles, a more significant amount of unburnt micro-sized aluminum particles would be emitting from the nozzle in this case. This could explain the reported improvement in combustion efficiency for composition H2 when compared to H4.

$$\tau_{res} = \frac{L^*}{C_{exp}{}^*} \tag{12.18}$$

$$L^* = \frac{V_c}{A_t} \tag{12.19}$$

Furthermore, mechanically activated Pyral powder (composition H3) enhanced the combustion efficiency over composition H2. This could be because mechanically activated Pyral powder ignites at a lower temperature (520 °C) than Pyral powder (800 °C) (as depicted in Figure 12.7 and explained in Section 12.3.2.2). Mechanically activated Pyral particles would have a shorter combustion time than Pyral and micron-sized aluminum particles. As a result, the combustion efficiency improved even more with composition H3. The combustion efficiency obtained with composition H3 was similar to the combustion efficiency obtained with pure wax–EVA (61% by Kumar [53]) without compromising on the regression rates, and the combustion efficiency of H3 can further be enhanced to over 95% by using a bluff body and protrusion, as described by Kumar and Ramakrishna [47, 55, 56].

It can be argued that characteristic velocity (C^*) alters with O/F ratio, and therefore combustion efficiency should be stated at the highest C^* value. Figure 12.19 depicts the variation of C^* with the propellant's O/F ratio at pressures determined in the combustion chamber during investigations. The experimentally acquired C^* is also presented in Figure 12.19. Compositions H2 and H3 were operated extremely close to the value of O/F value at which the highest theoretical characteristic velocity (C^*_{theo}) is achieved, as shown in Figure 12.19. The difference between O/F for composition H2 and the optimum C^* is 1.2% for composition H2 and for composition H3 this O/F difference is 0.3%. However, this variation was significant (9%) for composition H4. This discrepancy in C^* can be decreased by either decreasing the oxidizer flow rate or increasing the fuel flow rate by increasing the motor length. However, changing either of them will increase the residence time of the aluminum particles in the hybrid motor. If the combustion efficiency of different types of aluminum is to be compared, the residence duration in the motor must be kept constant for all forms of aluminum. Considering the above, these tests can

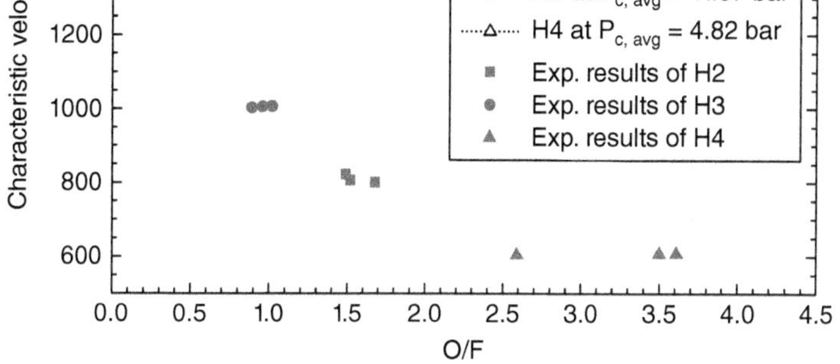

Figure 12.19 Theoretically and experimentally obtained C^* versus O/F for various compositions. Source: Gaurav Marothiya and P.A. Ramakrishna.

be utilized to demonstrate the favorable impacts of mechanical activation of Pyral powder on aluminum combustion.

12.4.3.4 Effect on the Exhaust Products

The exhaust gas analysis was the final feature that needed to be validated with aluminum combustion in hybrid rockets. In this section, the influence of mechanically activated Pyral powder, Pyral powder, and micron-sized aluminum on the exhaust products exiting the nozzle of hybrid rockets is investigated.

To accomplish this, a copper tape was wound around a metal rod and positioned in the direction of the rocket motor's exhaust plume to collect the exhaust products (which condensed on the copper tape). The distance between the nozzle outlet and rod stayed unchanged for all experiments at 50 cm, as indicated in Figure 12.20. The aluminum and alumina agglomerates in the exhaust gases were collected for three different mixes (Compositions H2, H3, and H4), and the mass of the agglomerates are listed in Table 12.12. The mass of particulate matter collected on the copper tape was measured with a 1 mg least count weighing balance. When compared to Pyral (H2) or micron-sized aluminum powder (H4), the particulate matter reduces in the case of mechanically activated Pyral powder (H3) (refer to Table 12.12). This reduction in the mass of agglomerates ejected out of hybrid rockets was due to again the increased reactivity of mechanically activated aluminum for aluminized, similar to the case of composite solid propellants in Section 12.3.2.6.

An SEM (scanning electron microscope) was used to capture images of the particles gathered on the copper tape. For this, an FEI Inspect-F SEM machine

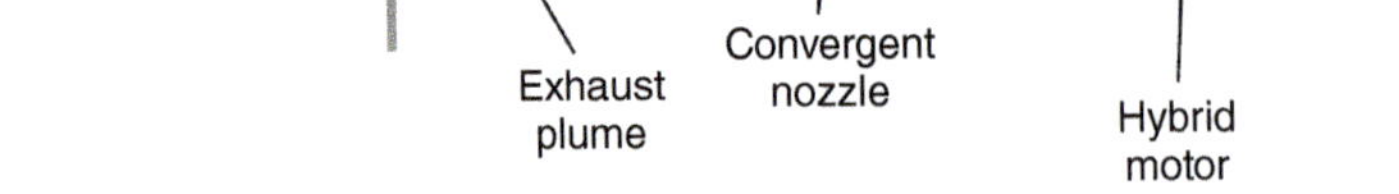

Figure 12.20 Schematic of the setup used for collecting the particulate matter from the exhaust of hybrid motor. Source: Gaurav Marothiya and P.A. Ramakrishna.

Table 12.12 Mass of particulate matter collected from the exhaust plume of the hybrid rocket motor.

	H2		**H3**		**H4**	
Exp. no.	1	2	1	2	1	2
Mass of particulate matter (mg)	16	17	10	10	37	38

Figure 12.21 SEM analysis (1500 times magnification) of particulate matter collected from the exhaust of composition (a) H2, (b) H3, and (c) H4. Source: Reproduced with permission from Indian Institute of Technology, Madras [51].

was used. Figure 12.21 depicts SEM images of exhaust products obtained during the combustion of Composition H2, H3, and H4. The particle sizes of the exhaust products condensed on the copper tape for Composition H3 (made with mechanically activated Pyral powder) were significantly smaller than those of composition H2 and H4 (which were prepared with nonactivated Pyral and micron-sized aluminum powder, respectively). By comparing Figure 12.21a,c, it was discovered that the particle size of exhaust products collected from the exhaust of propellants prepared with micron-sized aluminum (composition H4) was greater than that of the propellant prepared with nonactivated Pyral powder (composition H2). These results demonstrated a similar trend in particle sizes of exhaust products as in the mass of particulate matter for the three compositions.

One can also argue about slag accumulation in the hybrid motor, given that the hybrid fuel employed in this study contains 50% aluminum. When compared to the composition employed in the literature [39, 41–44], this could enhance slag accumulation. To understand this, it is pertinent to consider the total aluminum content of the aluminized hybrid propellant (fuel and oxidizer combined). At the ideal O/F ratio of 1.1, the overall proportion of aluminum content is roughly 24% (in terms of specific impulse). This aluminum percentage (24%) in the entire propellant composition is slightly higher than the one generally utilized in the solid propellant sector (roughly 18%). After scaling up, the problem of slag formation becomes critical. It is where the mechanically activated aluminum discussed in this study could come in handy because it decreases particulate matter in the exhaust.

12.5 Conclusions

The principal motivation behind this study was to minimize various disadvantages of aluminum combustion, such as low burning rates, agglomeration, two-phase losses in nozzle, and slag accumulation when the submerged nozzle was used in a rocket motor. The other objective of this study was to achieve high burn rates in the composite solid propellant for its use in the upper stages of the launch vehicles as an end burning grain, which would result in high propellant loading with end-burning grain and an improvement in the incremental velocity (ΔV) attained with the reduction in structural coefficient of the upper stage. In this direction, a high specific surface area aluminum named Pyral was mechanically activated using PTFE. Apart from enhancing the burn rates of composite solid propellants and evaluating various parameters, a design study has also been presented to increase the payload capacity of the Pegasus launch vehicle.

This mechanically activated aluminum powder was also used as an additive in the fuel for hybrid rockets. The effect of this powder was studied in eliminating a few of the main drawbacks (like low regression rates and low combustion efficiency) of the hybrid rocket motor and were compared with the results obtained using nonactivated aluminum powder from literature.

The below salient conclusions can be drawn from this investigation.

- The technique used for mechanical activation of Pyral powder was shown to have a pronounced improvement in the reactivity of the aluminum. It was due to the exothermic fluorination reaction occurring at a much lower temperature (~520 °C) than the ignition temperature (~800 °C) of nonactivated Pyral powder.
- This mechanically activated Pyral powder has shown great promise to be used in place of regular aluminum in a solid propellant to enhance composite solid propellant burn rate. It improves the burn rate substantially (~60%) over the propellant prepared with Pyral aluminum alone. The burn rates obtained at 70 bar with a 1 : 1 coarse to fine ratio of AP were 35.2 mm s^{-1}.
- When the mechanically activated Pyral powder was employed in the solid propellant, the alumina agglomeration mass in the solid propellant exhaust was significantly reduced. Using a composition prepared with mechanically activated Pyral powder in a rocket motor with a submerged nozzle is advantageous to reduce the slag formation in the motor.
- A design analysis was carried out to explore the feasibility of using the compositions mentioned earlier to enhance the payload capacity of launch vehicles. The increase in the payload was around 17.6% for the Pegasus (Orbital ATK, USA) launch vehicle (with no changes to existing hardware) when the compositions mentioned above were used as the end burning grain in the upper stages of these vehicles.
- Payload capacity was observed to increase when the upper stages of the launch vehicle were operated at lower chamber pressures than the current operating pressure using the composition prepared with mechanically activated Pyral powder.
- The 50% Pyral powder alone without any addition of PTFE in wax–EVA fuel grain itself enhances the regression rates significantly (~400%) over the fuel grain with pure wax–EVA as fuel grain in hybrid rockets.
- The mechanically activated Pyral powder was most effective in not only enhancing the regression rate of hybrid fuel (more than ~500% over wax and ~35% over composition prepared with nonactivated Pyral powder at higher oxidizer mass flux) but also the combustion efficiency over those observed with the fuel grain with only Pyral and wax–EVA.

- In the exhaust gas from the hybrid rocket, particulate matter mass and particle sizes were reduced when mechanically activated Pyral powder was used in the fuel grain.

Therefore, the mechanically activated aluminum enhances the ballistic performance of the solid and hybrid rockets and makes the system more compact by increasing the density-specific impulse of the rocket motor.

References

1 Kubota, N. (2015). *Propellants and Explosives: Thermochemical Aspects of Combustion*. Wiley.
2 Mukunda, H.S. (2017). *Understanding Aerospace Chemical Propulsion*. Bangalore, India: IK International Publishing House.
3 Williams, F.A., Barrère, M., and Huang, N.C. (1969). *Fundamental Aspects of Solid Propellant Rockets*. EWS Technivision Ltd.
4 Sutton, G.P. and Biblarz, O. (2016). *Rocket Propulsion Elements*. New York, USA: Wiley.
5 Hill, P.G. and Peterson, C.R. (1992). *Mechanics and Thermodynamics of Propulsion*. Addison-Wesley.
6 Nagappa, R. (2014). *Evolution of Solid Propellant Rockets in India*. Hyderabad, India: Defence Scientific Information and Documentation Center, DRDO, Metcalfe House.
7 Northrop-Grumman (2020). *Pegasus: Payload User's Guide*. Chandler, AZ: Northrop Grumman Corp.
8 Northrop-Grumman (2016). *Propulsion Product Catalog*. Chandler, AZ: Northrop Grumman Corp.
9 Spaceflight101. (2016). PSLV launch vehicle specification. http://spaceflight101.com/spacerockets/pslv (accessed October 2016).
10 Spaceflight101. (2016). Pegasus Xl launch vehicle. http://spaceflight101.com/spacerockets/pegasus-xl (accessed October 2016).
11 Gordon, S., and Mcbride B. J. (1976). Computer program for calculation of complex chemical equilibrium compositions, rocket performance, incident and reflected shocks, and Chapman–Jouguet detonations. *Tech. Rep. NASA-SP-273.*
12 Boraas, S. (1984). Modeling slag deposition in the space shuttle solid rocket motor. *Journal of Spacecraft and Rockets* 21 (1): 47–54. https://doi.org/10.2514/3.8606.
13 Beckstead, M.W. (2004). Internal aerodynamics in solid rocket propulsion: a summary of aluminum combustion. *Tech. Rep. EN-023*. NATO-RTO Educational Notes.
14 Beckstead, M.W. (2005). Correlating aluminum burning times. *Combustion Explosion and Shock Waves* 41 (5): 533–546. https://doi.org/10.1007/s10573-005-0067-2.
15 Glassman, I., Mellor A. M., Sullivan H. F., et al. (1970). A review of metal ignition and flame models. *AGARD Conference Proceedings no 52, Propulsion and Energetics Panel 34th Meeting 8th Colloquium held at the Aerospace Research*

Laboratories, Wright-Patterson Air Force Base, Dayton, Ohio USA, 13–17 October 1969, 19–41.

16 Ermakov, V.A., Razdobreev, A.A., Skorik, A.I. et al. (1982). Temperature of aluminum particles at the time of ignition and combustion. *Combustion Explosion and Shock Waves* 18 (2): 256–257. https://doi.org/10.1007/BF00789629.

17 Lokenbakh, A.K., Zaporina, N.A., Knipele, A.Z. et al. (1985). Effects of heating conditions on the agglomeration of aluminum powder in air. *Combustion Explosion and Shock Waves* 21 (1): 69–77. https://doi.org/10.1007/BF01471142.

18 Prentice, J.L. (1974). Combustion of laser-ignited aluminum droplets in wet and dry oxidizers. In: *AIAA 12th Aerospace Sciences Meeting* (30 January 1974), 74–146. Washington, DC: American Institute of Aeronautics and Astronautics.

19 Zennin, A.A., Glaskova, A.P., Leipunskyi, O.I. et al. (1969). Effects of metallic additives on the deflagration of condensed systems. *Symposium (International) on Combustion* 12 (1): 27–35. https://doi.org/10.1016/S0082-0784(69)80389-9.

20 Mench, M.M., Yeh, C.L., and Kuo, K.K. (1998). Propellant burning rate enhancement and thermal behavior of ultra-fine aluminium powders (Alex). In: *Energetic Materials: Production, Processing and Characterization* (ed. U. Teipel), 1–30. International annual conference of the Fraunhofer ICT.

21 DeLuca, L.T., Galfetti, L., Severini, F. et al. (2005). Burning of nano-aluminized composite rocket propellants. *Combustion Explosion and Shock Waves* 41 (6): 680–692. https://doi.org/10.1007/s10573-005-0080-5.

22 Dokhan, A., Price, E.W., Seitzman, J.M. et al. (2002). The effects of bimodal aluminum with ultrafine aluminum on the burning rates of solid propellants. *Proceedings of the Combustion Institute* 29 (2): 2939–2946. https://doi.org/10.1016/S1540-7489(02)80359-5.

23 Jayaraman, K., Anand, K.V., Chakravarthy, S.R. et al. (2009). Effect of aano-aluminium in plateau-burning and catalyzed composite solid propellant combustion. *Combustion and Flame* 156 (8): 1662–1673. https://doi.org/10.1016/j.combustflame.2009.03.014.

24 Verma, S. and Ramakrishna, P.A. (2013). Effect of specific surface area of aluminum on composite solid propellant burning. *Journal of Propulsion and Power* 29 (5): 1200–1206. https://doi.org/10.2514/1.B34772.

25 Shafirovich, E., Bocanegra, P.E., Chauveau, C. et al. (2005). Ignition of single nickel-coated aluminum particles. *Proceedings of the Combustion Institute* 30 (2): 2055–2062. https://doi.org/10.1016/j.proci.2004.08.107.

26 Hahma, A., Gany, A., and Palovuori, K. (2006). Combustion of activated aluminum. *Combustion and Flame* 145 (3): 464–480. https://doi.org/10.1016/j.combustflame.2006.01.003.

27 Osborne, D.T. and Pantoya, M.L. (2007). Effect of Al particle size on the thermal degradation of Al/teflon mixtures. *Combustion Science and Technology* 179 (8): 1467–1480. https://doi.org/10.1080/00102200601182333.

28 Sippel, T.R., Son, S.F., and Groven, L.J. (2013). Altering reactivity of aluminum with selective inclusion of polytetrafluoroethylene through mechanical activation. *Propellants, Explosives, Pyrotechnics* 38 (2): 286–295. https://doi.org/10.1002/prep.201200102.

29 Sippel, T.R., Son, S.F., and Groven, L.J. (2014). Aluminum agglomeration reduction in a composite propellant using tailored Al/PTFE particles. *Combustion and Flame* 161 (1): 311–321. https://doi.org/10.1016/j.combustflame.2013.08.009.

30 Ishitha, K. and Ramakrishna, P.A. (2015). Reducing agglomeration of ammonium perchlorate using activated charcoal. *Propellants, Explosives, Pyrotechnics* 40 (6): 838–847.

31 Gaurav, M. and Ramakrishna, P.A. (2016). Effect of mechanical activation of high specific surface area aluminium with PTFE on composite solid propellant. *Combustion and Flame* 166 (1): 203–215. https://doi.org/10.1016/j.combustflame.2016.01.019.

32 Zanotti, C., Volpi, A., Bianchessi, M. et al. (1992). Measuring thermodynamic properties of burning propellants. In: *Progress in Astronautics and Aeronautics* (ed. L.T. DeLuca, E.W. Price and M. Summerfield). USA: American Institute of Aeronautics and Astronautics Inc.

33 Verma, S. and Ramakrishna, P.A. (2014). Dependence of density and burning rate of composite solid propellant on mixer size. *Acta Astronautica* 93 (1): 130–137. https://doi.org/10.1016/j.actaastro.2013.07.016.

34 Galfetti, L., DeLuca, L.T., Severini, F. et al. (2007). Pre and post-burning analysis of nano-aluminized solid rocket propellants. *Aerospace Science and Technology* 11 (1): 26–32. https://doi.org/10.1016/j.ast.2006.08.005.

35 Chaturvedi, A.K., Kumar, S., and Chakraborty, D. (2015). Slag prediction in submerged rocket nozzle through two-phase CFD simulations. *Defence Science Journal* 65 (2): 99. https://doi.org/10.14429/DSJ.65.7147.

36 Jayaraman, K., Chakravarthy, S.R., and Sarathi, R. (2011). Quench collection of nano-aluminium agglomerates from combustion of sandwiches and propellants. *Proceedings of the Combustion Institute* 33 (2): 1941–1947. https://doi.org/10.1016/j.proci.2010.06.047.

37 Spaceflight. (2016). Schedule and pricing. http://www.spaceflight.com/schedule-pricing (accessed October 2016).

38 Gaurav, M. and Ramakrishna, P.A. (2017). Enhancement of aluminum reactivity to achieve high burn rate for an end burning rocket motor. *Propellants, Explosives, Pyrotechnics* 42 (7): 816–825. https://doi.org/10.1002/prep.201600304.

39 George, P., Krishnan, S., Varkey, P.M. et al. (2001). Fuel regression rate in hydroxyl-terminated-polybutadiene/gaseous-oxygen hybrid rocket motors. *Journal of Propulsion and Power* 17 (1): 35–42. https://doi.org/10.2514/2.5704.

40 Chiaverini, M.J. (2007). Review of solid-fuel regression rate behavior in classical and nonclassical hybrid rocket motors. In: *Fundamentals of Hybrid Rocket Combustion and Propulsion* (ed. M.J. Chiaverini and K.K. Kuo), 37–125. American Institute of Aeronautics and Astronautics Inc.

41 Chiaverini, M.J., Serin, N., Johnson, D.K. et al. (2000). Regression rate behavior of hybrid rocket solid fuels. *Journal of Propulsion and Power* 16 (1): 125–132. https://doi.org/10.2514/2.5541.

42 Risha, G., Ulas, A., Boyer, E., et al. (2001). Combustion of HTPB-based solid fuels containing nano-sized energetic powder in a hybrid rocket motor. *37th*

Joint Propulsion Conference and Exhibit (08 July 2001). Salt Lake City, UT, USA: American Institute of Aeronautics and Astronautics.

43 Lips, H.R. (1977). Experimental investigation on hybrid rocket engines using highly aluminized fuels. *Journal of Spacecraft and Rockets* 14 (9): 539–545. https://doi.org/10.2514/6.2003-4593.

44 Risha, G., Boyer, E., Wehrman, R., et al. (2003). Nano-sized aluminum and boron-based solid fuel characterization in a hybrid rocket engine. *39th AIAA/ASME/SAE/ASEE Joint Propulsion Conference and Exhibit* (20 July 2003). Huntsville, Alabama, USA: American Institute of Aeronautics and Astronautics.

45 Karabeyoglu, M.A., Cantwell, B.J., and Altman, D. (2002). Combustion of liquefying hybrid propellants: part 1, general theory. *Journal of Propulsion and Power* 18 (3): 610–620. https://doi.org/10.2514/2.5975.

46 Karabeyoglu, M.A., Cantwell, B.J., and Altman, D. (2001). Development and testing of paraffin-based hybrid rocket fuels. *37th Joint Propulsion Conference and Exhibit, Joint Propulsion Conferences* (08 July 2001). Salt Lake City, UT, USA: American Institute of Aeronautics and Astronautics Inc.

47 Kumar, R. and Ramakrishna, P.A. (2016). Studies on eva-based wax fuel for launch vehicle applications. *Propellants, Explosives, Pyrotechnics* 41 (2): 295–303. https://doi.org/10.1002/prep.201500172.

48 Kumar, R. and Ramakrishna P.A. (2015). Improving the mechanical properties of paraffin wax. I. P. India, IDF-1215, filed 16 September 2015 and issued 10 October 2020.

49 Kumar, R. and Ramakrishna, P.A. (2013). Issues related to the measurement of regression rate of fast-burning hybrid fuels. *Journal of Propulsion and Power* 29 (5): 1114–1121. https://doi.org/10.2514/1.B34757.

50 Kumar, R. and Ramakrishna, P.A. (2014). Measurement of regression rate in hybrid rocket using combustion chamber pressure. *Acta Astronautica* 103 (1): 226–234. https://doi.org/10.1016/j.actaastro.2014.06.044.

51 Gaurav, M. and Ramakrishna, P.A. (2018). Utilization of mechanically activated aluminum in hybrid rockets. *Journal of Propulsion and Power* 34 (5): 1206–1213. https://doi.org/10.2514/1.B36846.

52 Maruyama, S., Ishiguro, T., Shinohara, K. et al. (2011). Study on the mechanical characteristics of paraffin-based fuel. In: *47th AIAA/ASME/SAE/ASEE Joint Propulsion Conference & Exhibit* (31 July 2011), 306–307. San Diego, CA: American Institute of Aeronautics and Astronautics.

53 Kumar, R. (2014). Regression rate studies using wax as a hybrid fuel. PhD thesis. Indian Institute of Technology, Madras, India.

54 Yetter, R.A., Risha, G.A., and Son, S.F. (2009). Metal particle combustion and nanotechnology. *Proceedings of the Combustion Institute* 32 (2): 1819–1838. https://doi.org/10.1016/j.proci.2008.08.013.

55 Kumar, R. and Ramakrishna, P.A. (2014). Enhancement of hybrid fuel regression rate using a bluff body. *Journal of Propulsion and Power* 30 (4): 909–916. https://doi.org/10.2514/1.B34975.

56 Kumar, R. and Ramakrishna, P.A. (2014). Effect of protrusion on the enhancement of regression rate. *Aerospace Science and Technology* 39 (1): 169–178. https://doi.org/10.1016/j.ast.2014.09.001.

13

Effect of Nanometal Additives on The Ignition of Al-Based Energetic Materials

Alexander G. Korotkikh[1,2], *Ivan V. Sorokin*[1,3], *Vladimir E. Zarko*[3], and *Vladimir A. Arkhipov*[2]

[1] *Tomsk Polytechnic University, School of Power Engineering, 30, Lenin Ave., Tomsk 634050, Russia*
[2] *Tomsk State University, Research Institute of Applied Mathematics and Mechanics, 36, Lenin Ave., Tomsk 634050, Russia*
[3] *Voevodsky Institute of Chemical Kinetics and Combustion Siberian Branch of the Russian Academy of Sciences, Department of Physics and Chemistry of High Energy Systems, 3, Institutskaya St., Novosibirsk 630090, Russia*

13.1 Introduction

Modern energetic materials (EMs) used in rocket engines contain crystals of oxidizers, nitramines (ammonium perchlorate [AP] and ammonium nitrate AN, RDX, HMX), combustible binders (CBs) based on butadiene or tetrazole rubbers, components of energy-intensive fuels in the form micro-sized powders of aluminum, magnesium, beryllium, boron, etc., the mass content of which ranges from 5 to 20 wt% [1–8]. Metallic components are added to rocket solid propellants to increase the temperature and total heat of combustion, which provides an increase in the specific impulse of the engine compared to metal-free propellants.

Aluminum is widely used as a metal fuel in solid rocket propellants, due to its relatively low cost and sufficiently high reactivity, with a specific heat of combustion of ~31 MJ kg^{-1} [9]. As a rule, aluminum particles of the powders have a protective cover (oxide layer), which reduces the rate of their oxidation and can lead to incomplete combustion of the metal in fast processes. A possible solution to this problem can be various coatings of the surface of aluminum particles, which make it possible to reduce the effect of the oxide layer of Al_2O_3 (organic films, thin layers in the form of various metals or their oxides) [10, 11]. The disadvantage of this approach is the negative effect of covers on the heat of combustion of aluminum particles. The use of metals with a higher chemical activity and/or better characteristics of ignition and combustion (for example, copper, nickel, cerium, boron, etc.) is a worthy alternative to conventional solutions, since such substances are not only promoters of ignition and more complete combustion particles of aluminum, but also directly combustible components of EM [7, 8, 12, 13].

Boron is highly desirable as an energy-intensive fuel for solid propellants. The specific heat of combustion of boron is 58.1 MJ kg^{-1} [9], and significantly exceeds the

Nano and Micro-Scale Energetic Materials: Propellants and Explosives, First Edition.
Edited by Weiqiang Pang and Luigi T. DeLuca.

specific heat of aluminum combustion. However, when boron is heated, the formed molten oxide layer on the surface of boron particles prevents its oxidation and slows down the ignition process, reducing the rate of oxidant diffusion and chemical reactions. In addition, the oxidation of boron particles requires twice as much gaseous oxidizer as aluminum and, during its combustion; the formation of larger particles of agglomerates, boron carbides is possible, which affects its combustion efficiency [8, 9, 14, 15].

In this regard, practical interest are energy-intensive fuels based on aluminum (Al–Me, Al–B) in the form of alloys or mechanical mixtures of different dispersity (from micro to nanosized particles), which can be used in various component compositions of EM. Analysis of experimental data showed [8, 9, 16, 17] that boron-based fuels, in particular aluminum borides, have a high density and specific heat of combustion (at the level of boron). Particles of aluminum borides AlB_x and mechanical Al/B mixtures are easier to ignite and burn with greater completeness than boron particles [16–21]. In this case, the heat of combustion of boron-based metal fuels is higher than the heat of combustion of aluminum.

At present, research is being carried out in the field of combustion of boron-containing, bimetal individual particles and powder systems in various gaseous media and conditions, and the problems of their application in the composition of energy compositions based on various oxidizers and CB are being solved. Boron is known to have a longer ignition delay than most other metal fuels. An effective method of stimulating the boron ignition is the use of active metals and their oxides, as well as compounds with a low ignition temperature, high-energy density, and high oxygen balance (for example, nitramines). The most frequently used catalysts for the combustion of boron particles are oxides of copper, iron, bismuth, and zirconium [22–25]. Previous studies have shown that the oxide layer on the surface of boron particles is destroyed by contact with carbon dioxide, ammonia NH_3 or water vapor, which are formed during the EM combustion.

A number of works are devoted to the study of the combustion characteristics of EM compositions, in which micron-sized aluminum powders are partially or completely replaced by nanopowder (NP) of aluminum and another metals (for example, iron, magnesium). This method allows to multiply the burning rate of solid propellant, ensure its stable ignition and increase the completeness of combustion of the metal. In addition, in recent years, special attention has been paid to metal borides and mechanical mixtures based on aluminum, titanium, magnesium, and boron. The research results showed that the use of mixtures of metals with boron reduces the accumulation of a liquid oxide layer B_2O_3 on the surface of particles and stimulates their combustion [14, 16, 23, 26, 27]. Despite the possibility of obtaining promising results, there is a limited number of publications in this area, which present experimental data on the temporal characteristics of the ignition and combustion of single boron-based particles, and the compositions of condensed combustion products.

The characteristics of oxidation of metallic fuels, ignition, and combustion of EMs containing metal powders are the determining factors for ensuring stable and complete combustion of propellant, analyzing modes, and ensuring optimal operation of various propulsion systems and gas generators. Ignition serves as the initial stage of operation of the gas generating device, and in accordance with the technical regulations and the purpose of this device, requirements arise to ensure the appropriate ignition characteristics. In particular, requirements for minimizing the mass of the igniting composition are established for rocket systems. For other applications, it is important to ensure a minimum or, conversely, an increased delay in the ignition of the charge. Problems of this kind are solved by a detailed study of the ignition mechanism under a specific thermal effect on varied component composition of EM.

This chapter presents the results of an experimental study of the oxidation characteristics for mechanical mixtures based on aluminum/metal and aluminum/boron NPs, as well as ignition of the EM samples, containing an oxidizer, CB, and various energy-intensive additives, under conditions of intense heat supply from an external radiation source. This chapter also includes the EM ignition kinetics derived from the experimental data.

13.2 Thermal Behavior of Metal NPs and EM Compositions

Oxidation of metal particles is important from the point of view of completeness and rate of its combustion when used in the EM composition. As objects of research, we selected nanopowders of aluminum, iron, titanium, and copper, obtained by the method of electric explosion of conductors (LLC "Advanced Powder Technologies," Russia), and amorphous boron powder. These powders have a high specific surface area of the particles (from 7 to 16 $m^2\,g^{-1}$) and have high chemical activity in contact with atmospheric oxygen. Therefore, these powder systems can be considered as promising combustible components in solid propellants and hybrid fuels.

Using TESCAN MIRA 3 LMU scanning electron microscope, the shape and structure of particles of the used metal and amorphous boron powders have been studied. The micrographs of powders are shown in Figure 13.1. Micrographs of powders were used in determining the average particle diameter d and calculating the specific surface area S_{sp} of the powders (Table 13.1). The table also contains data on the mass content of the active element C_{Me} in the metal powder. The used metal powders have the average particle diameter of about 70–100 nm, and boron powder is 210–240 nm. The particles form clusters in the form of grape bunches (agglomerates), the size of which can exceed several microns. To compare the dispersed characteristics and oxidation parameters of the metal and boron NPs, we used a microsized aluminum powder μAl, the characteristics of which are well studied.

In the thermal kinetic study of the EM decomposition characteristics, we used solid compositions, which were prepared by batch mixing of components (oxidizer powder, metal powder, and combustible binder) followed by continuous pressing and polymerization in a drying oven. The resulting cylindrical EM samples 10 mm in diameter and 5 mm in height contained 64.6 wt% bimodal AP (sieve fractions finer than 50 and 165–315 μm in a weight ratio of 40/60), 19.7 wt%. SKDM-80 butadiene rubber and 15.7 wt% metal powder. Mechanical mixtures of NPs of Alex with 2 wt%

Figure 13.1 Micrographs of the initial metal and amorphous boron powders. (a) Alex, (b) Fe, (c) B, (d) Ti, (e) Cu, and (f) μAl.

Table 13.1 Average diameter, specific surface area of particles and active element content for metal and boron powders.

Me powder	*d* (nm)	S_{sp} ($m^2 g^{-1}$)	C_{Me} (wt%)
Alex	90–110	15.5	90.0
B	210–240	10.0	99.5
Fe	90–110	7.7	92.0
Ti	60–80	13.8	99.8
Cu	50–70	12.0	85.0
μAl	7400	0.3	98.5

Notice: *d* – average diameter; S_{sp} – specific surface area; C_{Me} – active element content (weight percent).

boron, iron, titanium, or copper were used as a bimetal fuel. Boron, iron, titanium, or copper powder was introduced into the mixture by partially replacing Alex. To prepare a mechanical mixture of powders, we used unsifted powders (their appearance is shown in Figure 13.1). Before introducing the mixture of powders into the EM, we carefully mixed them for 15 minutes.

Using the simultaneous thermal analyzer Netzsch STA 449 F3 Jupiter, a series of TG–DSC-measurements was carried out to determine the characteristics of the oxidation for the metal and amorphous boron NPs, as well as the decomposition for the EM samples. In the thermal analysis, we used ceramic crucibles, into which a fuel sample was placed or a weighed portion of metal and boron powders weighing ~5–6 mg was poured. The samples were heated in a furnace from 30 to 1300 °C with a constant heating rate of 10 °C min^{-1}. in argon and air. The obtained results of the thermal analysis of metal and boron powders in the form of TG lines (characterizing the change in mass) and DSC lines (characterizing the change in the specific heat flux) during heating and oxidation, depending on the temperature, are shown in Figure 13.2.

Also, the values of the characteristic temperatures of the onset T_{on}, intensive T_{int}, and end T_{end} of the oxidation process, as well as the specific heat released Q_s as a result of complete oxidation of the samples, were determined. The main parameters of oxidation for metal and boron powders are presented in Table 13.2.

The oxidation process of aluminum powders is carried out in two stages, which differ in temperature ranges and specific heat released. Alex NP begins to oxidize in air at a temperature of ~570 °C due to the diffusion of the oxidizer through the porous oxide layer. The maximum specific heat flux at the first stage is characteristic at a temperature of 600 °C. In this case, the maximum rate of change in the sample mass is ~1.6 μg s^{-1} in the temperature range of 590–610 °C. Endothermic melting of aluminum occurs at a temperature of ~660 °C with a subsequent increase in the mass of the sample. When the core of aluminum particles melts on the surface of the oxide layer, thermal stresses arise due to an increase in excess pressure or a decrease in the density of the melt, as a result of which cracks form on the surface of the

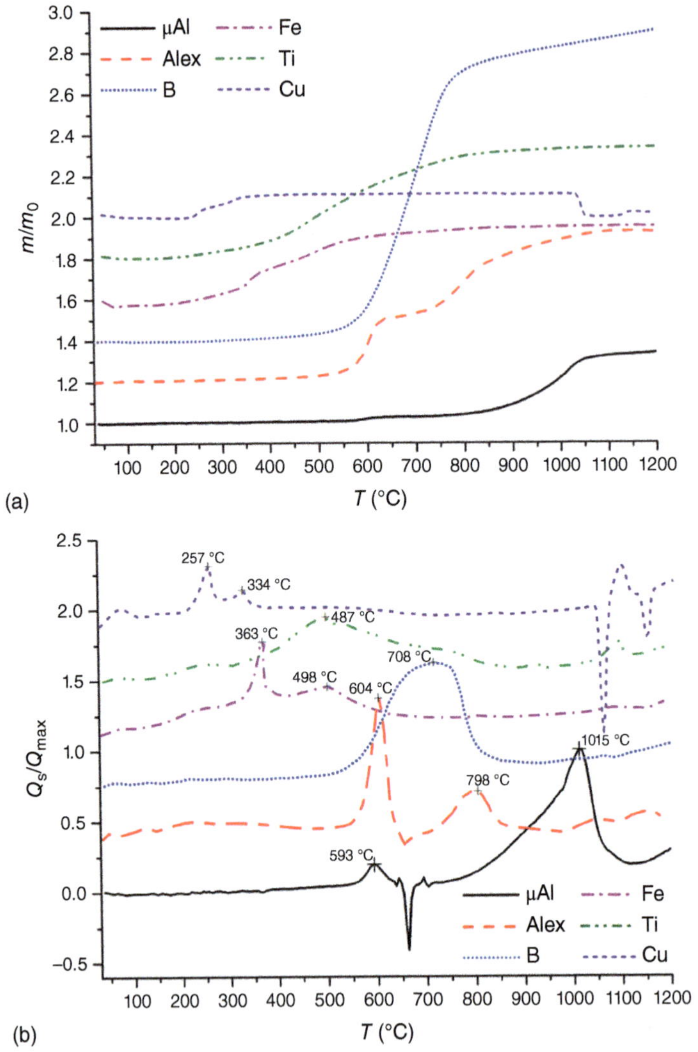

Figure 13.2 TG (a) and DSC (b) lines of metal and boron powders in air.

oxide layer and the release of active metal, which reacts with atmospheric oxygen. At a temperature of ~720 °C, the rate of the Alex NP oxidation reaction slows down due to the accumulation of the oxide layer on the particle surface and the phase transition of Al_2O_3 oxide. At the second stage, the oxidation rate slows down, while a less intense increase in the sample mass with heat release is observed at a temperature of the second peak of ~800 °C. Upon subsequent heating of the Alex NP to a temperature of 1200 °C, the oxidation of aluminum significantly slows down.

For micro Al, the first stage of oxidation begins at a temperature of 540–620 °C with an insignificant increase in the sample mass due to the diffusion of the oxidizer through the amorphous oxide layer and its interaction with the active metal. During

Table 13.2 Temperatures of the onset, intensive and end of oxidation, the specific heat effect of the reaction for the metal and boron powders.

	Temperature (°C)			
Me powder	T_{on}	T_{int}	T_{end}	Q_s ($J\,g^{-1}$)
Alex, stage 1	570	604	628	4 070
stage 2	724	798	842	
B	559	708	798	21 550
Fe	158	363	583	1 550
Ti	99	487	828	8 330
Cu	237	257	354	660
μAl	800	1015	–	3 700

heating and melting of aluminum, the formation of cracks on the surface of the oxide layer of large particles, the reaction rate increases due to the contact of the active metal with the oxidizing agent. The second stage of oxidation is carried out at higher temperatures of ~800–1060 °C in comparison with the Alex NP. In this temperature range, an intense increase in the sample mass is noted, which is accompanied by an intense heat release at a temperature of 1015 °C.

The oxidation process of amorphous boron proceeds in one stage and begins at a temperature of ~560 °C. The maximum specific heat flux is observed at a temperature of ~710 °C. The maximum rate of change in the sample weight is ~9 $\mu g\,s^{-1}$ in the temperature range of 650–750 °C. In the temperature range of 560–800 °C, the sample weight increases by 2.3 times relative to the initial weight of the sample. When boron is heated above 800 °C, the oxidation process and the growth of mass slows down. A decrease in the oxidation rate is associated with a slowdown in the diffusion rate due to a significant increase in the thickness of the B_2O_3 oxide layer on the particle surface, as well as the formation of a liquid oxide shell and the beginning of its evaporation. As noted by the authors in [27], the rate of decrease in the thickness of the B_2O_3 layer due to oxide evaporation strongly depends on the temperature.

The oxidation process of titanium and iron NPs begins at a relatively low temperature (~100 and 160 °C, respectively) due to the decomposition of the passivating coating (hexane) of metal particles, which is applied to the surface of the particles during the production of the powder, and the contact interaction of the active metal with oxygen air. The maximum specific heat flux during the oxidation of Ti and Fe NPs is characteristic at temperatures of ~490 and 360 °C, respectively.

For the EM samples, containing 13.7 wt% Alex NP and 2 wt% additive of metal or boron, the characteristic temperatures of decomposition were determined when the samples were heated in argon at a constant rate (Table 13.3). The TG analysis data show that the decomposition process with a slight change in the sample weight (up to ~2%) begins at a temperature of ~100–150 °C due to the onset of decomposition of combustible binder (Figure 13.3). In the temperature range of 150–215 °C, a

Table 13.3 Temperatures of the onset, intensive and end of decomposition for EM containing Alex/Me NP.

EM sample with Me	Temperature (°C)		
	T_{on}	T_{int}	T_{end}
15.7%Alex	200	318	354
Alex/2%B	180	320	355
Alex/2%Ti	190	327	368
Alex/2%Fe	170	295	350
Alex/2%Cu	160	290	340

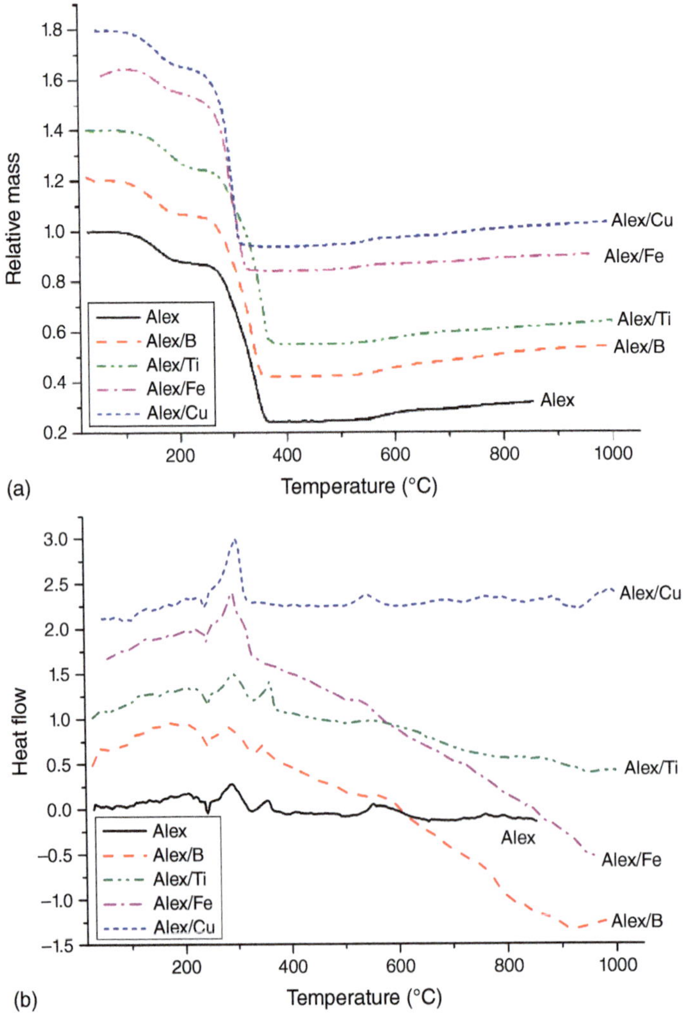

Figure 13.3 TG (a) and DSC (b) lines of EM containing Alex/Me NP.

decrease in the EM mass by ~10% of the initial mass is observed, which corresponds to the temperature of the onset of AP decomposition. The decomposition rate of ammonium perchlorate slows down when heated to 255 °C, because at a temperature of 240 °C, a polymorphic transformation of the AP crystal lattice (from rhombic to cubic) occurs. This transition is accompanied by heat absorption and is displayed as an endothermic peak on thermograms at a temperature of ~245 °C. The decomposition of the bulk of AP and combustible binder for the EM composition proceeds in the temperature range of 280–400 °C with a decrease in the mass of the samples by 66–88 wt%. The maximum rate of change in the sample mass is ~11 $\mu g\,s^{-1}$ in the temperature range of 300–315 °C. In this case, the main stage of oxidation of metal or boron powders is observed at temperatures above 600 °C, which is accompanied by a slight increase in the sample weight in the TG curves.

Note the steep slope of the TG lines in the range of 280–330 °C and a decrease in the characteristic temperature of intense decomposition by ~30 °C for EM compositions containing NP of Alex/2% Fe and Alex/2% Cu, as compared to Alex-based EM. With the introduction of iron or copper particles into the composition of EM with Alex, the process of propellant decomposition is intensified upon heating due to the formation of a metal oxide on the surface of the particles and its contact with AP crystals. This catalytic effect is significantly manifested in metal powder systems with a high specific surface area and reactivity during heating of an energetic material [7, 8, 25].

Thus, in the thermal kinetic study of the oxidation characteristics for metal nanopowder systems and decomposition of EM, the dependences of the oxidation of metal, boron NPs and decomposition of solid-propellant components on the heating temperature were obtained, which makes it possible to predict the behavior of EM during its heating, pyrolysis, and ignition, as well as to evaluate the completeness of oxidation of metallic fuel when certain conditions are reached.

13.3 Ignition Characteristics of EM

Ignition of EMs containing a mixture of metal and boron NPs is a multistage process, including inert heating of the propellant layer, decomposition, and interaction of the oxidizer and binder with metal particles, accompanied by intense heat release and the appearance of a flame. In determining the ignition characteristics of EM, experimental methods are mainly used, which allow obtaining the necessary data for solving mathematical models and problems of ignition and combustion of metalized solid propellants, as well as predicting the timing characteristics of EM ignition in a wide range of heat fluxes.

In the experimental study of ignition characteristics, we used cylindrical EM samples 10 mm in diameter and 5 mm in height, which contained bidispersed AP, butadiene rubber, and 15.7 wt% bimetal powder. As a bimetal powder, we used mechanical mixtures of Alex with 2 wt% boron, iron, titanium, and copper NPs. We also used mechanical mixtures of Alex with 5, 7, and 13 wt% boron NPs, which in terms of mass content correspond to the phase compositions of aluminum borides.

The ignition of the EM sample containing bimetal NP was carried out on an experimental stand consisting of a continuous CO_2 laser with a maximum power of 200 W

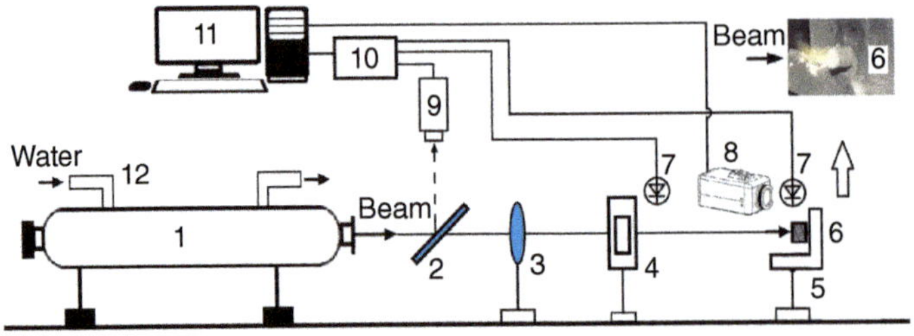

Figure 13.4 The schematic diagram of experimental setup based on CO_2 laser: 1 – CO_2 laser; 2 – beam-splitting mirror; 3 – lens; 4 – shutter; 5 – sample holder; 6 – sample; 7 – photodiodes; 8 – video camera; 9 – thermoelectric sensor of radiation power; 10 – analog to digital converter (ADC); 11 – PC; 12 – cooling system.

with a radiation wavelength of 10.6 μm and a system for measuring ignition parameters (Figure 13.4) [25]. The diameter of the incident laser beam on the end surface of the sample was approximately equal to the diameter of EM. The experiments were carried out in air at room temperature and atmospheric pressure. The ignition delay time t_{ign} of the EM samples was determined from the difference in the times of changes in the signals of the photodiodes, one of which registered the opening of the electromagnetic shutter (through which the laser beam passed), and the second, the appearance of a flash (flame) over the end surface of the EM sample. The temperature on the surface of the reaction layer of the propellant during its heating with subsequent ignition was recorded using a Jade J530 SB thermal imaging camera in the wavelength range of 2.5–2.7 μm.

A series of experiments have been performed to measure the ignition characteristics of the EM samples in air with a change in the heat flux density q from 60 to 220 W cm^{-2}. The experimental dependences of the EM ignition delay time t_{ign} on the heat flux density q incident on the propellant sample are obtained. The measurement results in the form of experimental points and lines of approximation dependences (as $t_{ign} = A\,q^{-B}$ function, Table 13.4) are shown in Figure 13.5. To assess the

Table 13.4 Fitted constants and determination coefficient.

EM sample	$A{\cdot}(10^4)$	B	R^2
μAl-EM	5.35	1.42 ± 0.04	0.99
Alex-EM	14.33	1.71 ± 0.05	0.99
Alex/2%B-EM	8.42	1.59 ± 0.09	0.97
Alex/5%B-EM	9.54	1.67 ± 0.19	0.95
Alex/7%B-EM	3.41	1.39 ± 0.21	0.92
Alex/13%B-EM	4.61	1.44 ± 0.07	0.97
Alex/2%Ti-EM	7.75	1.56 ± 0.10	0.98
Alex/2%Fe-EM	8.60	1.63 ± 0.30	0.85
Alex/2%Cu-EM	12.77	1.71 ± 0.04	0.99

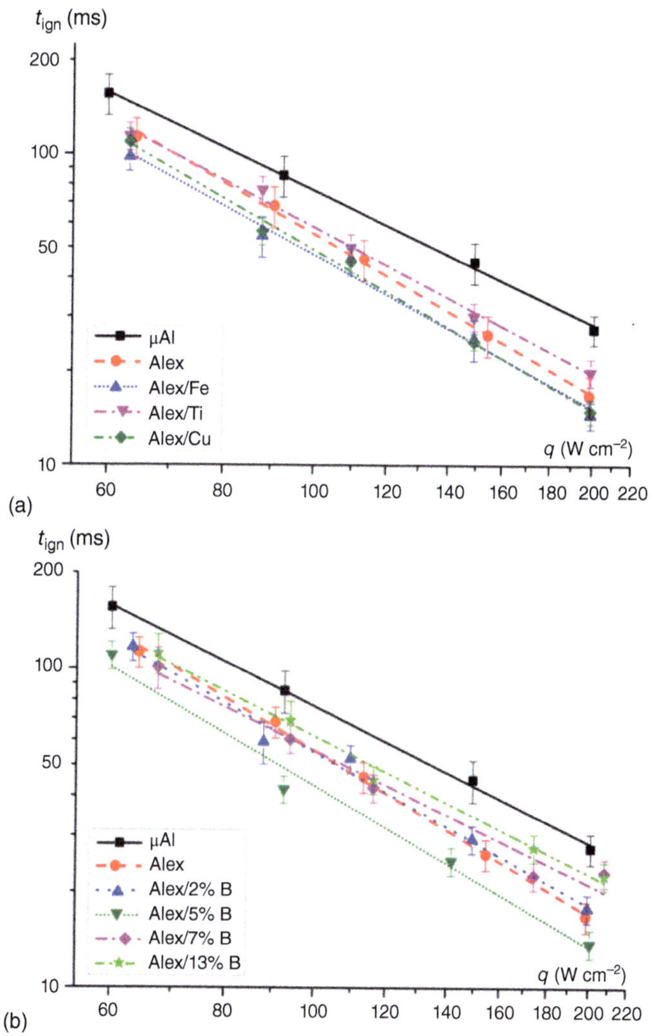

Figure 13.5 Ignition delay time on the heat flux density for the EM samples with Alex/Me (a) and Alex/B (b) powders.

efficiency of using bimetal fuel in the EM composition, we used the K_{ign} coefficient (Table 13.5), which is equal to the ratio of the ignition delay time of the EM sample containing µAl to the ignition delay time EM, containing bimetal NP.

The analysis of the obtained results showed that the complete replacement of µAl powder with the Alex NP in the EM composition leads to a 1.2–1.7 times decrease in the ignition delay time and ignition energy density $E_{ign} = t_{ign} \cdot q$ in the entire range of heat flux density $q = 60–220\,\mathrm{W\,cm^{-2}}$, which confirms the previously obtained data [2, 7] on the effect of the size and the specific surface area of aluminum particles for the period of heating and formation of the EM reaction layer with radiant heat supply.

Table 13.5 Coefficient K_{ign} for the EM samples with Alex/Me and Alex/B NPs.

Ratio t_{ign} for EM samples	K_{ign} at q	
	60 W cm^{-2}	210 W cm^{-2}
t_{ign} (μAl-EM)/t_{ign}(Alex-EM)	1.2	1.7
t_{ign} (μAl-EM)/t_{ign}(Alex/2% B-EM)	1.3	1.6
t_{ign} (μAl-EM)/t_{ign}(Alex/2% Fe-EM)	1.4	1.9
t_{ign} (μAl-EM)/t_{ign}(Alex/2% Ti-EM)	1.2	1.5
t_{ign} (μAl-EM)/t_{ign}(Alex/2% Cu-EM)	1.3	2.0
t_{ign} (μAl-EM)/t_{ign}(Alex/5% B-EM)	1.5	2.1
t_{ign} (μAl-EM)/t_{ign}(8.7% Alex/7% B-EM)	1.4	1.4
t_{ign} (μAl-EM)/t_{ign}(2.7% Alex/13% B-EM)	1.2	1.3

Partial replacement of Alex by 2 wt% iron, titanium, or copper NPs leads to a decrease in the ignition delay time of the EM sample and in the source ignition energy density by 1.4–1.9, 1.2–1.5, and 1.3–2.0, respectively, relative to the t_{ign} of EM with μAl. Furthermore, at the use of Alex/2% Fe and Alex/2% Cu NPs the ignition delay time t_{ign} of EM reduces by 10–16% compared to the t_{ign} of the EM sample with Alex. The presence of an oxide layer and a developed specific surface area of nanoparticles of iron and copper leads to an increase in the rate of AP decomposition during the period of heating and the formation of a heated surface layer, contributes to an increase in the rate of outflow of oxidizer decomposition products from the surface of the reaction layer and the rate of gas-phase chemical reactions with decomposition products of combustible binder.

Partial replacement of Alex by 2, 7, and 13 wt% boron NP leads to a decrease in t_{ign} by a factor of 1.3–1.6, 1.4, and 1.2–1.3 times, respectively, in comparison with the t_{ign} for the μAl-based EM. It should be noted that when using Alex/2% B in EM composition, the ignition delay time is the same in comparison with the Alex-based EM sample and within the measurement error (less than 10%). The maximum reduction in the ignition delay time by 16–20% was obtained for the EM sample, containing Alex/5% B, relative to the EM sample with Alex. When the sample surface is heated to 450 °C, boron particles are covered with a molten oxide layer, which reacts with aluminum particles ($B_2O_3 + 2Al \rightarrow 2B + Al_2O_3$) to release heat. Perhaps, there is the maximum energy release at Alex/B = 10/5 ratio, which increases surface temperature and reduces the heating time and formation of the reaction layer. Thus, a change in the mass content of boron in Alex/B NPs makes it possible to significantly modify the ignition characteristics of the EM sample. Also of interest are systems based on Ti/B, Ni/B, and Fe/B NPs, which can be considered as modifiers of the EM ignition, the ignition characteristics of which are presented in [26].

Visualization of the ignition process made it possible to establish the main stages of EM ignition. For example, consider the main ignition stages for several EM samples, containing μAl, Alex, and Alex/B, when heated at a constant q = 68 W cm^{-2}.

When the electromagnetic shutter is opened, laser radiation falls on the end surface of the EM sample. In "hot spots" with a maximum heat flux, a rapid local heating of the sample is carried out with the formation of a reaction layer of propellant and decomposition its components. At the moment of release of gaseous decomposition products of components near the surface of the EM sample with μAl (Figure 13.6a), a cloud with a diameter of ~8 mm (t = 202 ms) forms. The release of single luminous aluminum particles begins at t = 226 ms, which contribute to the initiation of a flash near the sample surface (t = 229 ms). A visible flame is formed within 3 ms (t = 231 ms). During the formation of the visible flame zone (τ = 234–275 ms) in the form of a "mushroom" with ~60 mm long, there is an emergence of clusters of aluminum particles from the EM surface (t = 275 ms), which is accompanied by a

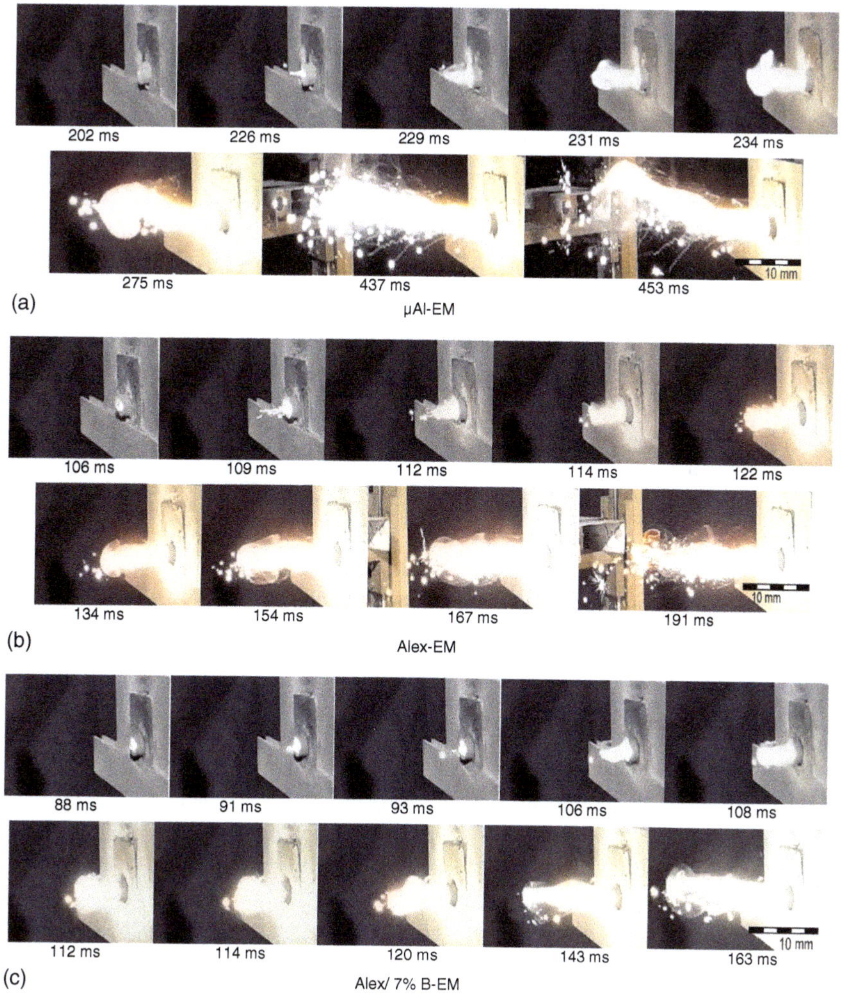

Figure 13.6 High-speed video frames of the EM ignition. (a) μAl-EM, (b) Alex-EM, and (c) Alex/7%B-EM.

sharp increase in temperature near the surface. By the time of 437 ms, the length of the flame zone reaches ~160 mm. The period of reaching the stationary combustion mode of the EM sample is several hundred milliseconds.

Ignition of the EM sample, containing Alex NP, (Figure 13.6b) is carried out more intensively than the μAl-based EM, due to the high reactivity of Al nanoparticles with a lower ignition temperature upon contact with an oxidizer and the rapid formation of a reaction layer of sample in "hot spots", accompanied by an intense outflow of decomposition products of components and accumulations of nanoparticles from the surface of the EM sample. The period of formation of glow in the places of "hot spots" and the release of aluminum nanoparticles from the sample surface is ~6 ms (t = 106–112 ms), and the formation of a visible flame is ~28 ms (t = 134 ms). The combustion of single aluminum particles occurs at the boundary of their surface due to the diffusion of the oxidizer, and the combustion temperature of the particles is determined by the melting temperature of the oxide layer, which significantly depends on the particle size. It is known, that with a decrease in the diameter of Al particles from 100 μm to 100 nm, the combustion temperature on the surface decreases by 1350 °C and amounts to ~1000 °C in an oxygen-containing environment [28].

Ignition of the EM sample, containing Alex/7%B NP, (Figure 13.6c) has the same ignition mechanism with the Alex-based EM. When the sample is heated in the region of maximum heat flux, a flash is formed due to the interaction of the decomposition products of AP and a combustible binder (t = 93–106 ms). In this case, luminous Al and B particles emerge from the sample surface (t = 108–120 ms), which intensifies the heat supply to the surface layer of the sample due to exothermic reactions of oxidation of particles near the propellant surface and the glow of the formed flame zone. The period of formation of a visible flame on the EM sample surface is ~24 ms (t = 112 ms).

13.4 Kinetic Parameters of Ignition

According to the solid-phase model of the ignition of condensed systems, the ignition moment is determined by the condition when the rate of heat supply from chemical reactions in the solid phase should exceed the rate of heat removal into the depth of the sample. When an EM sample is ignited, the substance is converted with the release of heat, which is described by the Arrhenius dependence on temperature including the activation energy and the ignition temperature. The ignition temperature of the solid propellant on the sample surface is determined from the condition of equality of the rates of heat arrival from chemical reactions in the reaction volume and the flow rate deep into the cold sample.

The surface layer of the EM sample is heated to the temperature T_{ign0} as an inert substance when solid propellant is ignited by a constant radiation heat flux. When the temperature increases to T_{ign}, the reaction layer is formed on the EM surface, in which heat is released due to exothermic chemical reactions during the decomposition of the oxidizer and fuel. It leads to a sharp increase in the temperature near the EM sample surface and the appearance of a flame. The moment of EM ignition is determined by the following condition: the heat input rate as a result of chemical reactions in the solid phase Q_1 will exceed the rate of heat transfer to the depth of the EM sample Q_2:

$$Q_1 \geq Q_2. \tag{13.1}$$

There is a temperature distribution from T_{ign0} to the ambient temperature T_0 in the heated layer of the EM sample at the moment t_{ign} when the flame appears above the propellant surface. The predominant heat release due to chemical reactions in the solid phase occurs at temperatures close to T_{ign}, i.e. in a narrow reaction layer with a thickness of x_1. When the EM sample is ignited, the substance is converted with the heat release, which is described by the Arrhenius temperature dependence:

$$Q_1 = Q \cdot \rho \cdot x_1 \cdot z \cdot \exp\left(-\frac{E}{RT_{ign}}\right) \tag{13.2}$$

where Q is the specific (per unit weight) heat of the chemical reactions, J kg^{-1}; ρ is the density of EM, kg m^{-3}; $x_1 = \frac{2\lambda RT_{ign}^2}{Eq}$ is the thickness of the zone of chemical reactions, m; z is the pre-exponential factor, s^{-1}; E is the effective activation energy of the ignition, J mol^{-1}; R is the universal gas constant, J (mol K)$^{-1}$; λ is the thermal conductivity of solid propellant, W (m K)$^{-1}$; q is the radiative heat flux density, W cm^{-2}.

The rate of conductive heat transfer from the chemical reaction zone deep into the EM sample is determined by the heat flux density of the laser radiation (convective heat loss to the environment is excluded):

$$Q_2 = -\lambda \frac{\partial T}{\partial x} \tag{13.3}$$

The condition of the EM sample ignition takes the following form:

$$Q \cdot \frac{2\lambda \rho RT_{ign}^2}{Eq} \cdot z \cdot \exp\left(-\frac{E}{RT_{ign}}\right) = -\lambda \frac{\partial T}{\partial x} \tag{13.4}$$

The thermal conductivity equation in a one-dimensional formulation for the solid-phase ignition model in the heated zone of the sample is given by:

$$\frac{\partial T}{\partial t} = a\frac{\partial^2 T}{\partial x^2} + \frac{Q}{c} z \exp\left(-\frac{E}{RT}\right) \tag{13.5}$$

where a is thermal diffusivity coefficient, m^2 s^{-1}; x is coordinate, m; c is specific thermal capacity, J (kg K)$^{-1}$.

Initial and boundary conditions for Eq. (13.5):

$$\begin{aligned}
&T(x, 0) = T_0 = \text{const}, \qquad 0 \leq x < \infty \\
&-\lambda \frac{\partial T}{\partial x} = q = \text{const} > 0, \qquad x = 0, t > 0 \\
&\frac{\partial T}{\partial x} \to 0, \quad x \to \infty, \qquad t > 0
\end{aligned} \tag{13.6}$$

The analytical solution of boundary value problems (13.5) and (13.6) with respect to the ignition delay time t_{ign} of the EM sample is written as [29]:

$$t_{ign} = 0.35\left(1 - \frac{T_0}{T_{ign}}\right)^2 \frac{Ec}{RQz} \exp\left(\frac{E}{RT_{ign}}\right), \tag{13.7}$$

or

$$\ln\left(\frac{t_{ign}}{(1 - T_0/T_{ign})^2}\right) = \ln\left(\frac{0.35 \cdot Ec}{RQz}\right) + \frac{E}{RT_{ign}}. \tag{13.8}$$

The ignition temperature on the EM sample surface is derived from the equality of two heat fluxes: the heat released from chemical reactions in the reaction layer and the heat transfer into the depth of the cold sample. It can be written as [29]:

$$T_{ign} = T_0 + 1.2q\sqrt{\frac{t_{ign}}{\lambda \cdot c \cdot \rho}} \tag{13.9}$$

The activation energy E is defined as the tangent of slope angle tgα of the $\ln\left(\frac{t_{ign}}{(1-T_0/T_{ign})^2}\right)$ versus $\frac{1}{T_{ign}}$ line. Intersection of curve with the axis $\frac{1}{T_{ign}}$ determines the temperature value of T_{ign0}. The multiplication of the thermal effect on the pre-exponent $Q{\cdot}z$ is determined using the equality $\ln\left[0.35\frac{E \cdot c}{R \cdot Q \cdot z}\right] = -\frac{E}{R \cdot T_{ign0}}$.

Using the obtained experimental dependences of the EM ignition delay time on the heat flux density $t_{ign}(q)$, the kinetic parameters of solid-propellant ignition were calculated – the activation energy, the multiplication of the reaction heat and the pre-exponent, as well as the ignition temperatures according to the method presented in [30]. In calculating the constants of formal kinetics, the following thermophysical parameters for the EM samples were used: density $\rho = 1.87\,g\,cm^{-3}$, specific heat $c = 1.24\,kJ\ (kg\,K)^{-1}$, thermal conductivity $\lambda = 0.66\ W\ (m\,K)^{-1}$. Table 13.6 shows the calculated kinetic parameters and ignition temperatures for the studied EM compositions.

Table 13.6 Calculated kinetic parameters and temperatures of ignition of the EM samples with Al and Alex/Me powders.

EM sample	E (kJ mol^{-1})	$Q \times z$ (W g^{-1})	T_{ign0} (K)	T_{ign} (K) at $q = 60$–$200\,W\,cm^{-2}$
μAl-EM	54.1	8.69×10^8	518	527–631
Alex-EM	119	1.82×10^{16}	498	503–546
Alex/2%B-EM	84.0	4.42×10^{12}	486	496–555
Alex/2%Fe-EM	106	1.86×10^{15}	480	486–531
Alex/2%Ti-EM	79.0	9.41×10^{11}	493	501–565
Alex/2%Cu-EM	125	1.80×10^{17}	488	492–531
Alex/5%B-EM	108	4.24×10^{15}	474	482–525
Alex/7%B-EM	56.8	5.17×10^9	477	490–575
Alex/13%B-EM	61.4	1.12×10^{10}	492	502–585

Relatively low values of the activation energy for EM ignition were obtained for the EM samples with μAl, Alex/7%B, and Alex/13%B powders (E = 54–62 kJ mol^{-1}) by using the accepted calculation procedure and the method of processing the results. Note that the values of the constants of formal kinetics and temperatures found by this method do not correspond to the true values of the reaction kinetics, but serve as the matching coefficients in the calculations of the ignition characteristics taking into account the given solid-phase model.

13.5 Conclusion

1. The thermal kinetic study of the oxidation characteristics of metal powder systems in air and the decomposition of metalized-EM containing AP and butadiene rubber has been carried out. It is shown that the use of bimetal fuel based on Alex/2%Fe and Alex/2%Cu NPs provides a decrease in the characteristic temperature of intense decomposition and an increase in the maximum decomposition rate of EM components.
2. The use of Alex/2%Fe and Alex/2%Cu NPs in the EM sample reduces the temperatures of the onset and intense decomposition by 30–40 °C due to a possible catalytic effect, which decreases the onset temperature of the high-temperature decomposition of AP. The addition of boron NP to the Alex-based EM does not change the decomposition temperatures for the sample.
3. The influence of bimetal fuel Al/B, Al/Fe, Al/Ti, and Al/Cu was experimentally studied and the timing characteristics of the ignition of EM on the heat flux density were determined. On the basis of the obtained experimental data, the constants of formal kinetics are calculated – the activation energy and the multiplication of the heat effect of the reaction and the pre-exponent. It was shown, that a change in the component composition of a bimetal fuel makes it possible to significantly modify the ignition characteristics of the EM sample – the ignition delay time, the ignition temperature, and the ignition energy density.
4. The use of Alex/2%Fe and Alex/2%Cu NPs in the EM sample decreases the ignition delay time and ignition energy density by 10–16% in the range of the heat flux density of 60–220 W cm^{-2} compared to the t_{ign} of the EM sample with Alex. The presence of an oxide layer and a developed specific surface area of nanoparticles of iron and copper leads to an increase in the decomposition rate of AP during the period of heating and the formation of a heated surface layer, contributes to an increase in the outflow rate of oxidizer decomposition products from the surface of the reaction layer and the rate of gas-phase chemical reactions with decomposition products of combustible binder. The use of Alex/5%B NP in EM reduces the ignition delay time and ignition energy density by 16–20% in comparison with Alex-based EM, which makes it possible to predict the prospects of using boron nanoparticles in fuels and propellants.
5. With the use of high-speed visualization and thermal imaging, the physical reaction and the development of flame processes on the surface of the EM sample with a variable composition of the bimetal fuel under radiant heating have been

studied. The influence of the component composition of the bimetal fuel on the time period of the formation of the heated layer and the outflow of gasification products with metal particles from the surface of the reaction layer, the speed of propagation of the flame zone, and the ignition delay of the EM samples are established.

Acknowledgments

The reported study was supported by RFBR according to the research project No. 20-03-00588.

References

1 Ahmad, S.R. and Cartwright, M. (2015). *Laser Ignition of Energetic Materials*. Wiley.

2 DeLuca, L.T. (2018). Overview of Al-based nanoenergetic ingredients for solid rocket propulsion. *Defence Technology* 14: 357–365. https://doi.org/10.1016/j.dt.2018.06.005.

3 Elbasuney, S. (2017). Combustion characteristics of extruded double base propellant based on ammonium perchlorate/aluminum binary mixture. *Fuel* 208: 296–304. https://doi.org/10.1016/j.fuel.2017.07.020.

4 Sergienko, A.V., Popenko, E.M., Slyusarsky, K.V. et al. (2019). Burning characteristics of the HMX/CL-20/AP/polyvinyltetrazole binder/Al solid propellants loaded with nanometals. *Propellants, Explosives, Pyrotechnics* 44: 217–223. https://doi.org/10.1002/prep.201800204.

5 Gong, L., Li, J., Li, Y., and Yang, R. (2020). Combustion properties of composite propellants based on two kinds of polyether binders and different oxidizers. *Propellants, Explosives, Pyrotechnics* 45: 1634–1644. https://doi.org/10.1002/prep.202000041.

6 Li, Y., Xie, W., Wang, H. et al. (2020). Investigation on the thermal behavior of ammonium dinitramide with different copper-based catalysts. *Propellants, Explosives, Pyrotechnics* 45: 1607–1613. https://doi.org/10.1002/prep.202000065.

7 Korotkikh, A.G., Glotov, O.G., Arkhipov, V.A. et al. (2017). Effect of iron and boron ultrafine powders on combustion of aluminized solid propellants. *Combustion and Flame* 178: 195–204. https://doi.org/10.1016/j.combustflame.2017.01.004.

8 Huseynov, S.L. and Fedorov, S.G. (2015). *Nanopowders of Aluminum, Boron, Aluminum and Silicon Borides in High-Energy Materials*. Moscow: Torus Press.

9 Yanovskiy, L.S. (2009). *Energy-Intensive Fuel for Aircraft and Rocket Engines*. Moscow: Fizmatlit.

10 Valluri, S.K., Schoenitz, M., and Dreizin, E. (2019). Fluorine-containing oxidizers for metal fuels in energetic formulations. *Defence Technology* 15 (1): 1–22. https://doi.org/10.1016/j.dt.2018.06.001.

11 Yagodnikov, D.A., Andreev, E.A., Vorob'ev, V.S., and Glotov, O.G. (2006). Ignition, combustion, and agglomeration of encapsulated aluminum particles in a composite solid propellant. I. Theoretical study of the ignition and combustion of aluminum with fluorine-containing coatings. *Combustion, Explosion, and Shock Waves* 42 (5): 534–542. https://doi.org/10.1007/s10573-006-0085-8.

12 Belal, H., Han, C.W., Gunduz, I.E. et al. (2018). Ignition and combustion behavior of mechanically activated Al-Mg particles in composite solid propellants. *Combustion and Flame* 194: 410–418. https://doi.org/10.1016/j.combustflame.2018.04.010.

13 Dossi, S. and Maggi, F. (2019). Ignition of mechanically activated aluminum powders doped with metal oxides. *Propellants, Explosives, Pyrotechnics* 44: 1312–1318. https://doi.org/10.1002/prep.201900036.

14 Ao, W., Wang, Y., Li, H. et al. (2014). Effect of initial oxide layer on ignition and combustion of boron powder. *Propellants, Explosives, Pyrotechnics* 39: 185–191. https://doi.org/10.1002/prep.201300079.

15 Ulas, A., Kuo, K.K., and Gotzmer, C. (2001). Ignition and combustion of boron particles in fluorine-containing environments. *Combustion and Flame* 127 (1-2): 1935–1957. https://doi.org/10.1016/S0010-2180(01)00299-1.

16 Liang, D., Xiao, R., Liu, J., and Wang, Y. (2019). Ignition and heterogeneous combustion of aluminum boride and boron-aluminum blend. *Aerospace Science and Technology* 84: 1081–1091. https://doi.org/10.1016/j.ast.2018.11.046.

17 Korotkikh, A.G., Arkhipov, V.A., Slyusarsky, K.V., and Sorokin, I.V. (2018). Study of ignition of high-energy materials with boron and aluminum and titanium diborides. *Combustion, Explosion, and Shock Waves* 54: 350–356. https://doi.org/10.1134/S0010508218030127.

18 Bulanin, F.K., Sidorov, A.E., Kiro, S.A. et al. (2020). Ignition of metal boride particle-air mixtures. *Combustion, Explosion, and Shock Waves* 56: 57–62. https://doi.org/10.1134/S0010508220010074.

19 Korotkikh, A.G. and Sorokin, I.V. (2020). Study of the chemical activity of metal powders based on aluminum, boron and titanium. *AIP Conference Proceedings* 2212: 020029. https://doi.org/10.1063/5.0000838.

20 Sun, Y., Ren, H., Jiao, Q. et al. (2020). Oxidation, ignition and combustion behaviors of differently prepared boron-magnesium composites. *Combustion and Flame* 221: 11–19. https://doi.org/10.1016/j.combustflame.2020.07.022.

21 Sun, Y., Ren, H., Jiao, Q. et al. (2020). Ignition and combustion behavior of sintered-B/MgB_2 combined with KNO_3. *Propellants, Explosives, Pyrotechnics* 45: 997–1004. https://doi.org/10.1002/prep.201900367.

22 Xi, J., Liu, J., Wang, Y. et al. (2014). Metal oxides as catalysts for boron oxidation. *Journal of Propulsion and Power* 30: 47–53. https://doi.org/10.2514/1.B35037.

23 Liu, J.-Z., Xi, J.-F., Yang, W.-J. et al. (2014). Effect of magnesium on the burning characteristics of boron particles. *Acta Astronautica* 96: 89–96. https://doi.org/10.1016/j.actaastro.2013.11.039.

24 Liu, L.-J., He, G.-Q., and Wang, Y.-H. (2012). Thermal reaction characteristics of the boron used in the fuel-rich propellant. *Journal of Thermal Analysis and Calorimetry* 114: 1057–1068. https://doi.org/10.1007/s10973-013-3119-y.

25 Korotkikh, A.G., Sorokin, I.V., Selikhova, E.A., and Arkhipov, V.A. (2020). Effect of B, Fe, Ti, Cu nanopowders on the laser ignition of Al-based high-energy materials. *Combustion and Flame* 222: 103–110. https://doi.org/10.1016/j.combustflame.2020.08.045.

26 Korotkikh, A.G. and Sorokin, I.V. (2021). Effect of Me/B-powder on the ignition of high-energy materials. *Propellants, Explosives, Pyrotechnics* 46: 1709–1716. https://doi.org/10.1002/prep.202100180.

27 Pivkina, A.N., Muravyev, N.V., Monogarov, K.A. et al. (2018). Comparative analysis of boron powders obtained by various methods. I. Microstructure and oxidation parameters during heating. *Combustion, Explosion, and Shock Waves* 54 (4): 450–460. https://doi.org/10.1134/S0010508218040093.

28 Sundaram, D. Sr., Purib, P., and Yang, V. (2016). A general theory of ignition and combustion of nano- and micron-sized aluminum particles. *Combustion and Flame* 169: 94–109. https://doi.org/10.1016/j.combustflame.2016.04.005.

29 Zarko, V.E. and Knyazeva, A.G. (2020). Determination of kinetic parameters of exothermic condensed phase reaction using the energetic material ignition delay data. *Combustion and Flame* 221: 453–461. https://doi.org/10.1016/j.combustflame.2020.08.022.

30 Arkhipov, V.A. and Korotkikh, A.G. (2012). The influence of aluminum powder dispersity on composite solid propellants ignitability by laser radiation. *Combustion and Flame* 159 (1): 409–415. https://doi.org/10.1016/j.combustflame.2011.06.020.